MARINE ECOLOGY

A Comprehensive, Integrated Treatise on Life in Oceans
and Coastal Waters

Volume I   ENVIRONMENTAL FACTORS

Volume II   PHYSIOLOGICAL MECHANISMS

Volume III   CULTIVATION

Volume IV   DYNAMICS

Volume V   OCEAN MANAGEMENT

MARINE ECOLOGY

A Comprehensive, Integrated Treatise on Life in Oceans
and Coastal Waters

Editor
OTTO KINNE

Biologische Anstalt Helgoland
Hamburg, Federal Republic of Germany

VOLUME V

Ocean Management

Part 4: Pollution and Protection of the Seas: Pesticides, Domestic Wastes,
and Thermal Deformations

A Wiley–Interscience Publication

1984
JOHN WILEY & SONS
Chichester · New York · Brisbane · Toronto · Singapore

Copyright © 1984 by John Wiley & Sons, Ltd.

All rights reserved.

Library of Congress Cataloging in Publication Data:

(Revised for vol. V, pts. 3 & 4)
Kinne, Otto.
    Marine ecology.
    Vol. V published: Chichester, New York, Wiley.
    Includes bibliographies.
    CONTENTS: v. I. Environmental factors. 3 v.—v. II.
Physiological mechanisms. 2. v—[etc.]—v. V. Ocean
management. 4. v
    1. Marine ecology—Collected works. I. Title.
QH541.5.S3K5   574.5'2636   79-121779

ISBN 0 471 902179

British Library Cataloguing in Publication Data:

Marine ecology.
    Vol. V: Ocean management
    Pt. 4
    1. Marine ecology
    I. Kinne, Otto
    574.5'2636   QH 541.5.S3

    ISBN 0 471 90217 9

Typeset by Preface Ltd, Salisbury, Wiltshire
Printed by The Pitman Press, Bath, Avon.

FOREWORD
to
VOLUME V: OCEAN MANAGEMENT

'Ocean Management', the last volume of *Marine Ecology*, describes and evaluates all the essential information available on structures and functions of interorganismic coexistence; on organic resources of the seas; on pollution of marine habitats and on the protection of life in oceans and coastal waters. The volume consists of four parts:

Part 1

Zonations and Organismic Assemblages

Chapter 1: Introduction to Part 1—Zonations and Organismic Assemblages
Chapter 2: Zonations
Chapter 3: General Features of Organismic Assemblages in the Pelagial and Benthal
Chapter 4: Structure and Dynamics of Assemblages in the Pelagial
Chapter 5: Structure and Dynamics of Assemblages in the Benthal
Chapter 6: Major Pelagic Assemblages
Chapter 7: Specific Pelagic Assemblages
Chapter 8: Major Benthic Assemblages
Chapter 9: Specific Benthic Assemblages

Part 2

Ecosystems and Organic Resources

Chapter 1: Introduction to Part 2—Ecosystems and Organic Resources
Chapter 2: Open-Ocean Ecosystems
Chapter 3: Coastal Ecosystems
        3.1: Brackish Waters, Estuaries and Lagoons
        3.2: Flow Patterns of Energy and Matter

Chapter 4: World Resources of Marine Plants
       4.1: Benthic Plants
       4.2: Planktonic Plants
Chapter 5: World Resources of Fisheries and their Management

Part 3

Pollution and Protection of the Seas:
Radioactive Materials, Heavy Metals and Oil

Chapter 1: Introduction to Part 3—Radioactive Materials, Heavy
       Metals, and Oil
Chapter 2: Contamination due to Radioactive Materials
Chapter 3: Heavy Metals and their Compounds
Chapter 4: Oil Pollution and its Management

Part 4

Pollution and Protection of the Seas:
Pesticides, Domestic Wastes and Thermal Deformations

Chapter 5: Introduction to Part 4—Pesticides, Domestic Wastes, and
       Thermal Deformation
Chapter 6: Pesticides and Technical Organic Chemicals
Chapter 7: Domestic Wastes
Chapter 8: Thermal Deformations

The culmination of *Marine Ecology*, Volume V has its roots in and draws much of its basic substance from, the preceding volumes of the treatise:

Volume I ('Environmental Factors') is concerned with the most important environmental factors operating in oceans and coastal waters and their effects on microorganisms, plants and animals.

Volume II ('Physiological Mechanisms') reviews the information available on the mechanisms involved in the synthesis and transversion of organic material; in thermoregulation, ion- and osmoregulation; in evolution and population genetics; and in organismic orientation in space and time.

Volume III ('Cultivation') comprehensively assesses the art of maintaining, rearing, breeding and experimenting with marine organisms under environmental and nutritive conditions which are, to a considerable degree, controlled.

Volume IV ('Dynamics') summarizes and critically evaluates the knowledge available on the production, transformation and decomposition of organic matter in the marine environment, as well as on food webs and population dynamics.

Of necessity somewhat heterogeneous in concept and coverage, 'Ocean Management' introduces the readers to fields of applied marine ecology, i.e. to man's use of oceans and coastal waters for his own ends. In order to provide a solid basis for a sound assessment of the sea's man-supporting qualities, Parts 1 and 2 deal comprehensively with the basic multi-specific units encountered and with the resources which they constitute. After summarizing our current knowledge on the large variety of organismic groupings in the form of zonations and assemblages, and after considering the structures and functions of major marine ecosystems, the significance of these units and their components as resources utilizable for food or for raw materials are reviewed in depth. Parts 3 and 4, finally, evaluate man's potentially destructive impact. They focus on the different facets of pollution and critically assess measures—currently applied and considered practicable in the future—for protecting life in oceans and coastal waters from detrimental human influences.

Selected as early as 1965, the title 'Ocean Management' still seems too ambitious, if not somewhat misleading—even though, as anticipated, a host of new data and many new and important insights into the machinery of ecological systems have been brought to light in the meantime. Environmental management requires the concerted, judicious, responsible application of science and technology for the protection and control of those properties of ecosystems, species, resources, or areas that are regarded as absolute requirements for the continued support of civilized human societies[*]. Our knowledge has remained insufficient for an objective, exact definition of such properties. Hence, maintenance of a high degree of organismic and environmental diversity, and maximum possible conservation of natural conditions is deemed essential to avoid or to reduce irreversible long-term damage.

The means and aims of ocean management receive detailed attention in Part 2. Our capacity for true management is still restricted to narrowly-defined areas and to specific organisms, i.e. to a few heavily exploited or otherwise especially endangered species. Marine ecosystem management remains problematic; it is in need of much more basic ecological knowledge than is at present available.

The recent trend of using English as the international scientific language—in itself of great significance for international communication and cooperation—has often been disadvantageous to scientists with insufficient command of that language. It has frequently diminished the representativeness and distorted the emphasis of the total information actually available. While organizing and editing *Marine Ecology*, I have therefore attempted to include scientists from countries which made important contributions to the field of marine ecology, but whose scientists largely use non-English languages. In this way I wanted to underline the international status and significance of marine ecological research and the need to draw from different sources in order to provide the best possible representation of the state-of-the-art. Of course, I had to pay for this: an enormous amount of time and effort had to be invested in translation and manuscript improvement.

While I am writing this last foreword of the treatise, *Marine Ecology* is nearing completion. It is my sincere wish to thank again all those who have supported me during the many years of planning, carrying out and finalizing this *magnum opus*. From 1965 to 1981, the work on *Marine Ecology* has taken up most of my evenings, weekends and holidays. Who can blame me for feeling relieved?

As was the case with previous volumes of this treatise, I have received much help and advice while working on Volume V. With profound gratitude I acknowledge the close and fruitful cooperation with all contributors; the support, patience and confidence of the publishers; and, last but not least, the technical assistance of Monica Blake, Angela Giraldi, Alice Langley, Julia Maxim, Seetha Murthy, Sherry Stansbury and Helga Witt.

O.K.

[*]KINNE, O. 1980: *14th European Marine Biology Symposium 'Protection of Life in the Sea'*: Summary of symposium papers and conclusions. *Helgoländer Meeresunters.*, **33**, 732–761.

CONTENTS
OF
VOLUME V, PART 4

Chapter 5 **Introduction to Part 4—Pollution and Protection of the Seas:**
Pesticides, Domestic Wastes and Thermal Deformations *O. Kinne* 1619
  (1) General aspects 1619
  (2) Comments on Chapters 6 to 8 1620

Chapter 6 **Pesticides and Technical Organic Chemicals** . *W. Ernst* 1627
  (1) Introduction 1627
  (2) Analysis of organic pollutants 1628
    (a) Sampling 1629
      Water 1629
      Sediments 1633
      Biota 1633
    (b) Extraction 1633
      Water 1633
      Sediments 1633
      Biota 1634
    (c) Clean-up 1634
    (d) Separation, identification and quantification . . . 1634
      Limits of detection 1635
      Unidentified and natural compounds 1638
    (e) Conclusions 1639
  (3) Bioconcentration and biomagnification 1639
    (a) Bioconcentration 1640
    (b) Biomagnification 1642
    (c) Sediments and particulate matter 1645
    (d) Conclusions 1650
  (4) Degradation 1651
    (a) Pathways of degradation 1651
    (b) Transformation in animals 1652
      Metabolites of PCBs and DDT 1652
      Other compounds 1655
    (c) Transformation in micro-organisms 1656
    (d) Conclusions 1659
  (5) Occurrence and distribution of organic chemicals . . . 1659
    (a) Sea water and surface slicks 1660
      Water 1660
      Surface slicks 1663
      Organisms 1664

Sediments 1665
(b) Conclusions 1665
(6) Effects on living systems 1675
(a) Acute toxicity 1675
Factors affecting toxicity 1675
(b) Long-term toxicity 1676
Mixed phytoplankton—compared with single species . . 1686
Geographic differences 1686
Combined effects 1690
Miscellaneous effects 1690
Effects on man 1692
(c) Conclusions 1692
(7) Future prospects 1692
Literature cited 1695

Chapter 7 **Domestic Wastes** . . . . . . *D. J. Reish* 1711
(1) Introduction 1711
(a) Sewage treatment 1712
(2) Identification and quantification 1715
(a) Chemical analyses 1715
(b) Geological analysis 1717
(c) Biological surveys and monitoring 1719
Water column 1721
Benthos 1722
Macrofauna 1723
Meiofauna 1724
Microfauna 1724
Macrophyta 1724
Micro-organisms 1725
Fish 1726
Data analysis 1727
(3) Fate of domestic waters discharged into the marine environment . . 1730
(a) Dispersion 1730
(b) Water column 1733
(c) Interaction with sediments 1736
(d) Benthic communities 1740
Estuaries 1740
Intertidal 1742
Subtidal 1743
Indicator-organism concept 1746
(4) Effects of domestic wastes on marine organisms . . . 1747
(a) Liquid wastes 1747
(b) Diseases of marine organisms 1749
(c) Movement of toxicants through a marine food chain . . 1754
(5) Management of municipal waste disposal 1756
Literature cited 1759

Chapter 8 **Thermal Deformations** . *P. R. O. Barnett and B. L. S. Hardy* 1769
  (1) Introduction 1769
    (a) Cold effluents 1771
    (b) Heated effluents 1771
        Power station effluents 1773
        Natural gas liquefaction 1779
        Desalination plant effluents 1779
    (c) Pertinent literature 1781
    (d) Scope of the chapter 1782
  (2) Instrumentation and methods of monitoring 1782
    (a) Thermometers, thermistors and thermographs . . . 1782
    (b) Infrared techniques 1784
    (c) Other methods 1786
    (d) Modelling 1787
  (3) *In situ* distribution 1788
    (a) Fate of buoyant thermal effluents 1788
    (b) Negatively buoyant effluents 1795
        Heated effluents 1795
        Desalination effluents 1796
        Cold effluents 1796
  (4) Biological effects 1796
    (a) Entrainment effects 1797
        Bacteria 1798
        Phytoplankton 1799
        Zooplankton 1802
        Fish 1806
    (b) Discharge area effects 1807
        Phytoplankton 1808
        Zooplankton 1809
        Fish 1809
        Phytobenthos 1816
        Zoobenthos 1820
    (c) Synergistic and non-thermal effects 1831
        Synergistic effects of temperature and salinity . . 1831
        Desalination effluents 1831
        Metal ion and synergistic effects 1833
        Chlorine and synergistic effects 1834
        Current flow and impingement effects . . . 1836
        Pressure effects 1837
    (d) Ecosystem effects 1837
  (5) Management of waste-heat impacts 1840
    (a) Changing emphasis from environmental to energy conservation . 1841
    (b) Improved industrial methods and energy use . . . 1842
        Power station efficiencies 1842
        Combined heat and power generation . . . 1843
        Energy use 1847
        Desalination techniques 1847

(c) Reducing impingement and entrainment 1848
(d) Power station condenser temperatures 1849
(e) Reducing heated effluent temperatures before discharge . . 1849
    Dilution of cooling water 1849
    Cooling in special ponds, lakes, reservoirs or canals . . . 1850
    Wet cooling towers 1852
    Dry cooling towers 1852
(f) Reducing the impact of thermal discharges 1854
    Types of discharge 1854
    Siting considerations 1858
(g) Criteria for mitigating the effects of temperature changes on
    organisms 1860
    Temperature elevations 1860
    Cold shock following heated discharges 1871
    Cold effluents 1872
(6) Management of waste-heat uses 1872
(a) Thermal aquaculture 1873
    Use of heated effluent in aquaculture 1874
    Problems in the use of heated effluents in aquaculture . . 1885
    Holding methods in thermal aquaculture 1898
(b) Sport fishing 1902
(c) Agricultural and horticultural uses 1904
    Soil warming 1904
    Greenhouse heating 1906
    Animal shelters 1909
(d) Other constructive uses of thermal effluents . . . 1910
    Heated effluents 1910
    Cold effluents 1911
(7) Conclusions and the future 1911
(a) Conclusions 1911
(b) The future 1914
Literature cited 1926

Author index 1965

Taxonomic Index 1985

Subject Index 1993

CONTRIBUTORS

to

VOLUME V, PART 4

BARNETT, P. L. O. *Scottish Marine Biological Association, Dunstaffnage Marine Research Laboratory, P.O. Box 3, Oban, Argyll PA34 4AD, Scotland.*

ERNST, W. *Institut für Meeresforschung, Am Handelshafen 12, D-2850 Bremerhaven 1, Federal Republic of Germany.*

HARDY, B. L. S. *Scottish Marine Biological Association, Dunstaffnage Marine Research Laboratory. Editor, contributors and publishers regret to announce that Dr Hardy died on 23 January 1984.*

KINNE, O. *Biologische Anstalt Helgoland (Zentrale), Notkestraße 31, D-2000 Hamburg 52, Federal Republic of Germany.*

REISH, D. J. *Department of Biology, California State University, Long Beach, California 90840, USA.*

OCEAN MANAGEMENT

Marine Ecology Vol. V, Part 4
Edited by Otto Kinne
© 1984 John Wiley & Sons Ltd

5. INTRODUCTION TO PART 4— POLLUTION AND PROTECTION OF THE SEAS: PESTICIDES, DOMESTIC WASTES AND THERMAL DEFORMATIONS

O. KINNE

(1) General Aspects

This book—the 13th and last one in the treatise *Marine Ecology*—continues and completes the overview of the present status of our knowledge on the pollution and protection of marine and brackish waters and their biota. While Part 3 focused on general considerations and assessments of the state-of-the-art (Chapter 1), radioactive materials (Chapter 2), heavy metals (Chapter 3) and oil (Chapter 4), Part 4—after a brief introduction (Chapter 5)—reviews the effects of pesticides and technical organic chemicals (Chapter 6), domestic wastes (Chapter 7) and thermal deformations (Chapter 8).

Pesticides and technical organic chemicals comprise the most dangerous groups of pollutants. They represent artificial, man-made materials which are largely or entirely foreign to nature. Marine ecosystems have no or only limited capacities for metabolizing and degrading such compounds and their derivatives. Hence pesticides and technical organic chemicals released into marine and fresh waters tend to accumulate and to cause long-term effects.

The most important means for ocean management is here—as with other toxic pollutants—to reduce further accumulation and to avoid critical damage to life in the seas by a total banning of the most detrimental toxic substances from disposal into marine waters. A number of organizational and managerial steps have already been taken by major industrial nations towards this goal. International conventions have identified particularly toxic materials and listed them on so-called 'black lists'. Less acutely dangerous substances require permission for disposal ('grey-list' substances). For details, consult Chapters 1 and 6.

Domestic wastes add large amounts of organic materials to marine waters, especially near outfalls of major coastal population centres. The ecological consequences of such additions range from beneficial (fertilization) through harmless or neutral (rapid self-cleaning of receiving water bodies) to detrimental (eutrophication, odour nuisance, etc.). A major danger from large amounts of domestic wastes disposed of in coastal waters stems from pollutants (e.g. heavy metals, pesticides) which constitute inherent components of domestic wastes. Negative, i.e. detrimental, effects of domestic-waste disposal near sewer outlets have been documented. Usually they turned out to be reversible at the population and ecosystem levels upon reduction or termination of the pollution impact.

Thermal deformations, i.e. significant man-made temperature changes in natural habitats, may exert damage by adding heat to estuarine or coastal waters, especially in the tropics and subtropics. Here, the biota often live already near their upper thermal limits. The major source of thermal deformations near the coasts—power stations—inflict damage due to physical entrainment, pollution and supranormal temperatures. While damage due to heat disposal into rivers, estuaries and coastal waters has remained locally restricted and reversible, man's general additions to the earth's heat load and his interference with global heat dissipation processes requires control and management in order to avoid negative long-term consequences.

(2) Comments on Chapters 6 to 8

Chapter 6: Pesticides and Technical Organic Chemicals

Modern man employs large amounts and numbers of pesticides and technical organic chemicals for manufacturing a wide variety of materials used in industry and food production (agriculture, aquaculture). A total of about 60 000 different organic chemicals are at present being used, and several hundred new products enter the market every year. Many, if not most, of these substances will eventually find their way into coastal waters and oceans. Added to the complex natural chemistry of waters, sediments and organisms, these system-foreign compounds form a potential source of ecological impairment.

Natural ecosystems have only limited capacities for degrading or mineralizing pesticides and technical organic chemicals. Hence, many of these compounds tend to accumulate in water, sediment and biota, causing a variety of—as yet insufficiently known—modifications that may ultimately result in detrimental ecological effects. Manufacture of defined target chemicals produces a large variety of organic by-products. Often insufficiently identified, many of these are toxic, cannot be industrially recycled and hence pose additional problems of disposal and pollution. The fungicide pentachlorophenol, for example, contains as impurities tetrachlorophenol, other lower chlorinated phenols, chlorinated diphenyl ethers, dibenzofurans, dihydroxybiphenyls and phenoxyphenols which are precursors of highly toxic dioxins (p. 1628).

The first ecological concern about the dangerous potential of pesticides and technical organic chemicals as world-wide pollutants dates back to the finding of traces of DDT in marine organisms collected far away from the coasts. The capacity of marine ecosystems for metabolizing, degrading and mineralizing these man-made compounds has been overestimated. We are now witnessing the accumulation of these substances and their derivatives at all levels of marine, fresh water and terrestrial ecosystems.

The large and still increasing number of industrial chemicals and their derivatives, the formation of new compounds following pollutant release and the endless possibilities of *in situ* compound combinations render impossible and impracticable the determination of substance-specific ecological consequences except for the most dangerous compounds. Since sensitive analytical identification methods are available thus far only for a comparatively small number of substances, attention has focused on selected groups of compounds. These include persistent chemicals such as polychlorinated biphenyls, and other chlorinated hydrocarbons, chlorinated phenols and chlorinated pesticides as well as organophosphates and carbamates.

The ecological *in situ* consequences of these groups of pollutants must be recorded and evaluated in long-term, extensive monitoring programmes. The large annual output of chemicals is exemplified by the following figures: vinyl chloride, $10 \cdot 5 \times 10^6$ tons; trichloroethylene, 10^6 tons; PCBs, 10^4–10^5 tons; chlorinated phenols, $1 \cdot 5 \times 10^5$ tons; synthetic pesticides, more than 10^5 tons. In natural sea water, concentrations of artificial organic chemicals usually are of the order of ng l^{-1}; in sediments and biota they vary from μg kg^{-1} to mg kg^{-1}. Especially in sea water, such low levels create problems of procedural contamination during sampling and analysis, and require sophisticated methods of separation, identification and quantification.

Chapter 6 covers the analysis of organic pollutants, bioconcentration and biomagnification, degradation, *in situ* occurrence, and distribution and pollution effects on living systems.

While modern methods of analysis facilitate determinations down to levels of ng l^{-1} in water, and to ng g^{-1} in sediments and organisms, most of the numerous investigations carried out thus far are restricted to very few compounds (PCBs, DDT and other chlorinated pesticides). More refined techniques—notably glass capillary column gas chromatography and mass spectrometry—have recently enlarged our analytical capacity and will shed more light on the complex chemical situation in the seas.

Bioconcentration—pollutant accumulation from ambient water to supranormal levels in organisms—is counteracted by mechanisms of elimination, i.e. active excretion (faeces, urine) and other metabolic activities, and by passive release via body surfaces. Biomagnification—pollutant accumulation to supranormal levels via the food chain—may, in addition, be counteracted by selective mortality, behaviour, degradation and a variety of dissipation mechanisms, including migration and passive distribution. The data on, and significance of, biomagnification are still a matter of scientific dispute. While accumulation, due to both bioconcentration and biomagnification, tends to outweigh elimination, the resulting dynamic equilibrium level depends on the properties and ambient concentrations of the pollutants concerned, the counteractive potentials of the organisms involved and environmental factors, such as temperature, salinity and water movement. Accumulation, storage, remobilization and related processes also take place in non-living components of ecosystems—particularly floating matter and sediments. Again, the dynamics of these processes require more intensive investigation.

Degradation, i.e. progressive transformation by biotic and abiotic processes of a complex substance to increasingly less complex materials, may ultimately lead to complete mineralization, viz. the formation of inorganic end products. However, as has already been pointed out, the capacity of natural ecosystems for degrading pesticides and technical organic chemicals is usually quite limited. Furthermore, in the course of degradation, intermediate products may be formed which inflict more negative effects on health, resources, amenities or ecosystems than their parent compounds. While a number of degradation pathways have been brought to light, in many cases the analytical techniques at hand allow neither proper identification of the derivatives formed, nor sufficiently accurate assessments of degradation rates and capacities. There is much need for more studies under *in situ* conditions.

The occurrence and distribution of organic chemicals in the seas depends on input localities, times and rates, as well as on physico-chemical and biological *in situ* mechanisms of degradation and dissipation. In general, pollutant concentrations decrease from the coast to the open sea, and values recorded in sea water parallel those determined in sediments and organisms. The trapping function of sediments may cause long-lasting effects in heavily contaminated areas, especially in estuaries and mud flats. Insufficiencies in pollutant analysis are reflected by the continuing detection of 'new' pollutants in the marine environment.

Responses of living systems reveal that rates of reproduction and early stages of ontogenetic development (embryo, larva, early juvenile) provide particularly sensitive criteria for assessing negative pollutant effects. Frequently used criteria are mortality, growth, metabolic rate (including photosynthesis and enzyme activities), behaviour and reproduction. As has been pointed out repeatedly, mortality data—while providing a quick, general assessment of pollutant toxicity—yield information insufficient for a detailed evaluation of ecological consequences under *in situ* conditions (Chapter 1).

From extensive literature studies, the reviewer draws the following conclusions: (i) The ecological impact of pesticides and technical organic chemicals cannot be judged on the basis of toxicological data alone. There is need for more information on the rates of bioconcentration and degradation and on pollutant effects regarding reproductive potentials and competitive dynamics. (ii) The high complexity of ecosystems and analytical limitations make the assessment of ecologically meaningful responses to pollution extremely difficult. (iii) There are indications for increasing intoxication of estuarine and coastal areas. (iv) While present *in situ* levels of pesticides and technical organic chemicals do not seem to pose an acute threat to most members of the marine biota, our knowledge on long-term sublethal effects—especially at the population and ecosystems levels—is still very limited. (v) Since organic chemicals may exert deleterious effects on life in the seas and on man, careful continuous assessments of their *in situ* fates and effects are absolutely necessary. (vi) Newly produced chemicals must be thoroughly examined for possibly undesirable side-effects before general application.

Chapter 7: Domestic Wastes

Over the last 100 yr the amounts of domestic wastes discharged into the marine environment have increased dramatically, especially near rapidly growing, large centres of urbanization. The first warnings and documentations that rivers and estuaries may suffer from the consequences were published at the beginning of this century. However, whether domestic wastes may critically damage marine ecosystems has remained a controversial issue.

Since huge marine water masses are very poor in nutrients, some authors have argued that sewage disposal may actually be beneficial: it augments the nutrient supply, accelerates primary and secondary production rates and, ultimately, increases the annual fishery yields. On the other hand, domestic wastes are usually released in large amounts in restricted areas. Hence, the receiving ecosystems may not be able to metabolize the wastes quickly enough. The ensuing waste accumulation has been shown to inflict damage at the individual and population levels and to distort ecosystem dynamics.

In addition, domestic wastes usually contain a variety of toxic compounds.They serve as a vehicle for other types of pollutants, especially heavy metals and detergents. Large amounts of domestic wastes are released into coastal areas already subject to severe deformation due to a variety of man's activities. Thus they tend to augment the total resulting human impact.

Documented ecological damage to the marine biota at or near the sites of large domestic waste discharges includes the following: deterioration of benthic communities; changes in species composition, abundance and distribution; increased incidences of disease (e.g. in fishes); and impairments of organismic functions, such as fecundity, reproduction and stress endurance. Heavy domestic waste disposal augments the level of ambient organic substances, reduces the concentrations of dissolved oxygen and thus causes eutrophication.

Thus far, these and related consequences of domestic-waste pollution have remained restricted locally and turned out to be reversible quickly and effectively after terminating or diminishing the pollution impact, in some cases after removing major toxic components or reducing their total load. While coastal marine ecosystems may be capable of utilizing large amounts of sewage, the continuing increase in the growth of metropolitan centres near the coasts will enhance the problems already encountered. This perspective underlines the need for long-term research programmes aimed at a more complete assessment of stress due to domestic wastes, continuous monitoring of pollution types and loads and of the carrying capacities of the receiving ecosystems as a function of time.

On the basis of the information at hand, the most important conclusions are: (i) there is need for eliminating or reducing the levels of toxic substances, preferably at the source but also in the sewage-treatment process; (ii) the site for a sewage outfall must be carefully determined on the basis of ecological knowledge; (iii) large sewage outlets should be placed in deep waters away from the coast; and (iv) maximum sewage dispersion is of great importance for reducing detrimental effects; several small, isolated outlets will cause less harm than large amounts of sewage released at one point.

Of considerable promise are present plans in the United States for discharging large amounts of sewage sludge through pipes at water depths of 300–400 m. Should it turn out that such sludge disposals do not exert harmful effects on the local marine biotas, a solution to domestic-waste pollution would be in sight: municipal wastes would undergo secondary treatment with at-source control for toxic substances, eliminating or reducing these to acceptable levels; the liquid wastes would then be piped into shallow coastal waters and the sludge into deeper, further-away waters (p. 1758).

For monitoring the potential impact of domestic-waste release, measurements have included pH, suspended solids, BOD, nutrients, chlorinated hydrocarbons, trace organics and metal constituents. Among the important assay organisms the polychaete *Capitella capitata* has played a key role (p. 1747). The indicator-organism concept is discussed on pp. 1746–1747.

As an important measure for managing domestic-waste disposal, the US Congress has enacted the 'Clean Water Act'. Requiring mandatory secondary treatment of all sewage discharged, this act represents the most ambitious effort to date for managing municipal waste effluents into marine waters. Since the primary aim of the act is to protect the environment, it is of basic significance for ecologists to determine whether or not secondary sewage treatment does indeed serve this end. Secondary treatment may, in fact, augment eutrophication by making primary nutrients available in a more effective chemical form (p. 1758).

Chapter 8: Thermal Deformations

The term 'thermal deformations' refers to man-made temperature changes in natural habitats. The major sources that cause significant artificial heating or cooling of marine waters are industrial users. Most of these utilize water for cooling and hence release waste heat. The major waste-heat producers are electricity-generating power stations. Very few industrial plants remove heat from natural waters and hence release waters cooler than the ambient natural water.

In addition to adding or removing heat from natural waters, man affects environmental temperatures in a variety of other ways, for example, by releasing large amounts of heat into the atmosphere and by large-scale changes in terrestrial ecosystems, such as deforestation or land reclamation. The increasing waste-heat production by man is causing concern. How much heat can be added to our waters and atmosphere without negative effects on climate and biota?

Based on the reviews on temperature effects on marine and brackish water organisms presented in Volume I (see also Volumes II–IV), Chapter 8 focuses on ecological consequences, management and use of waste-heat production by power stations located near rivers or estuaries or at the coast. The chapter reviews comprehensively and critically (i) sources of thermal deformations, (ii) dissipation of thermal effluents, (iii) biological effects of entrainment and of thermally deformed water discharges, (iv) management of the ecological impact of thermally deformed waters and (v) the use of waste heat, especially in thermal aquaculture, sport fishing and agriculture.

As is the case with other facets of environmental pollution, the distinction between the effects of natural temperature fluctuations and of man-made thermal deformations is often difficult, and requires extensive studies over long periods of time. Nevertheless, such distinction is the basic prerequisite for *in situ* impact assessment and for the development of environmental management procedures. Long-term field studies must record continuously a variety of abiotic and biotic parameters as a function of time. They must be combined with experimental research in natural waters and in the laboratory, paying particular attention to the population and ecosystem levels. The compensatory mechanisms available to individuals, populations and ecosystems require more attention than they have received thus far. These mechanisms (regulation, adaptation, selection) may effectively counteract environmental stress, thus masking biological consequences until immediately before critical damage.

The degree of biological damage due to thermal deformations depends on the duration and degree of stress exposure. The duration is a function of the quantity of thermally deformed water, intensity of water mixing and water movement, and the amount of heat loss to adjacent waters and the atmosphere. The degree of stress depends on the thermal impact and concomitant effects of other stresses, such as salinity (i.e. in desalination plants) or chemical pollutants (e.g. chloride). Of great importance is the thermal state of the receiving water relative to the ecological potential of its biota. While the same amount of heat addition may be beneficial in the coldest winter months, it may cause a catastrophe in the warmest summer months. Buoyant thermal discharges tend to form surface plumes. Three phases of disposal are recognized: discharge as a jet (near-field phase), spreading at the water surface (mid-field phase) and final mixing with ambient water (far-field phase). In unusual cases, the effluent may be negatively buoyant, i.e. sink below the surface due to higher density (lower temperature than ambient and/or higher salinity).

The effects of elevated temperatures on marine and brackish organisms have received detailed attention (Volume I and the present chapter). Power stations may affect aquatic life in two principal ways: (i) transport by water movement into the intake culverts, i.e. entrainment in the station's cooling-water system during which high, deformed temperatures plus other stresses (including physical injuries) are effective; and (ii) exposure to deformed temperatures and discharge-water pollutants near the point of water release. Generalizations regarding biological consequences are difficult because thermal deformations can cause greatly different effects ranging from heat death, through a large variety of sublethal responses (Volume I) to beneficial effects in cases where otherwise cold damage due to extreme winter conditions may have prevailed. Power-station effects have been shown to vary as a function of seasonal, geographical, locational, hydrographic, climatic, topographic and biological conditions. Stations with open-circuit cooling affect organisms thermally, physically and chemically and thus tend to exert the greatest possible impact. Organisms are entrained, impinged on intake screens, exposed to chemical (e.g. biocide) and physical stress (pumping, turbulence) in addition to thermal impact in the condenser tubes, before being released in a plume of warm water which cools and changes rapidly as it mixes with receiving ambient waters. Sometimes the most serious effects are not due to heat but to chemical stress and mechanical damage during entrainment. Synergistic effects tend to complicate the comprehension of the ultimate biological and ecological consequences. In general, negative effects due to heated-water release are maximal in subtropical and tropical regions where marine and estuarine organisms tend to live close to their upper lethal temperature limits.

Management of detrimental effects of heated-water release involves: (i) technical improvements in the efficiency of power generation; (ii) combination of heat production with heat utilization, especially in aquaculture (pp. 1873–1902), horticulture and agriculture; (iii) careful siting of power stations on the basis of ecological knowledge; (iv) use of closed-circuit cooling systems employing cooling ponds or lakes, wet or dry cooling towers or combinations of these; and (v) development of alternative energy sources which do not add to the global heat load. In the long run we have no choice but to develop efficient mechanisms for controlling and managing man's global energy and heat budgets.

While man's general effects on regional and global climates are presently barely detectable, they will increase in the future. Thus accumulation of carbon dioxide gas in the atmosphere is trapping growing amounts of the long-wave radiation dissipating from the earth's surface into outer space ('greenhouse' effect). Doubling the present atmospheric carbon dioxide concentration could lead to a global rise in surface temperatures of $3 \cdot 6 \, C°$ (p. 1915). Other potentially detrimental effects may result from large emissions of nitrogen oxides from nuclear explosions which break down ozone in the stratosphere by photochemical reactions (p. 1916). Some grandiose but still speculative schemes that would effect the world climate are mentioned on p. 1914.

The information at hand indicates that thermal deformations have caused definitely detrimental effects on estuarine and coastal waters and their biota. However, these effects have remained locally restricted and are fully reversible at the population and ecosystem levels. More serious and less reparable long-term damage may result from global temperature increases via waste-heat production in general combined with augmented levels of atmospheric carbon dioxide.

Marine Ecology Vol. V, Part 4
Edited by Otto Kinne
© 1984 John Wiley & Sons Ltd

6.   PESTICIDES AND TECHNICAL ORGANIC CHEMICALS

W. ERNST

(1) Introduction

Pollution of the sea by organic chemicals was recognized 15–20 yr ago when highly accumulating materials such as DDT were found in marine organisms, far away from the location of intended application. Subsequently, man has become aware of the fact that organic chemicals can be transported over long distances by water movement, wind and precipitation, finally accumulating in oceans and coastal waters as an ultimate sink. The discharged materials do not undergo degradation or even mineralization as rapidly as previously expected. Unfortunately, the occurrence of long-lived 'persistent' chemical species in the sea has received insufficient attention and their effects implemented by the special features of marine environments have long been considerably underestimated, although they have been discussed in previous publications (RUIVO, 1972; JOHNSTON, 1976; KINNE and BULNHEIM, 1980; GERLACH, 1981).

More recently, aspects of organic marine pollution, including the occurrence, fate and biological effects, have been studied by numerous investigators. However, the majority of the studies were restricted to certain types of compounds, 'classical' organic pollutants such as DDT, their primary degradation products, some chlorinated pesticides and—since their discovery as marine pollutants by JENSEN (1966, 1972)—the polychlorinated biphenyls (PCBs). With growing attention and concern about the increasing amounts of organic wastes discharged via different routes, and with the help of more sensitive and confirmatory analytical tools, the array of organic compounds identified as marine pollutants is steadily increasing.

Organic compounds are subjected to specific usage patterns resulting in their continuous or increasing production or in their abolishment and replacement. Consequently, the number and concentration of actual pollutants may vary considerably. Many of the 60 000 organic chemicals at present in use (MAUGH, 1978) may finally reach estuaries and oceans. It seems unlikely that a detailed critical assessment of the specific risks can ever be accomplished for such a large number of compounds, since their complex interactions with components of the marine ecosystem are poorly understood and appropriate test methods for the evaluation of their ecotoxicological behaviour are not sufficiently developed. Obviously, the more persistent chemicals deserve special attention and their effects must be studied over long periods of time. High priority should be given to identifying and effect-testing of such compounds. The majority of organic substances may finally be subjected to more or less rapid degradation of their parent structures, sometimes even to complete mineralization. Nevertheless, while present in their active forms they might be harmful, especially to the biota in estuarine and coastal waters.

Many chemicals are manufactured in large quantities every year (KORTE, 1980). Examples are vinyl chloride 10.5×10^6 tons, trichloroethylene 10^6 tons (PEARSON and McCONNELL, 1975), PCBs 10^4–10^5 tons (PEAKALL, 1975), chlorinated phenols 1.5×10^5 tons (NILSSON and co-authors, 1978) and synthetic pesticides in the 10^5 tons range. Chemicals produced in such large amounts are technical products, and hence contain varying and sometimes high quantities of often numerous known and unknown impurities. In addition to the well defined and identified compounds, the impurities may endanger life when released. Pentachlorophenol, for example, which is used as a fungicide, contains as impurities tetrachlorophenol, other lower chlorinated phenols, chlorinated diphenyl ethers, dibenzofurans, dihydroxybiphenyls and phenoxyphenols—the last named being precursors of highly toxic dioxins (NILSSON and co-authors, 1978). Similar patterns of by-products may exist with other compounds.

Along with the chemicals produced for certain ends, organic wastes are generated which can often not be recycled and, therefore, pose problems with regard to their disposal. If discharged into the sea in large amounts (NATIONAL RESEARCH COUNCIL, 1975), such wastes may endanger marine ecosystems, at least until sufficient degradation has occurred. Other processes, such as incineration at sea of organochlorines, usually lead to a number of new products, although admittedly at low rates.

Although many chemicals may potentially act as marine pollutants, only those mentioned here are of actual significance in the sea. They may be arranged in the following groups: chlorinated pesticides, organophosphates, carbamates, chlorinated aliphatic hydrocarbons, chlorinated aromatic hydrocarbons, chlorinated phenols, polychlorinated biphenyls and phthalic esters.

On the basis of the information at present available, this chapter attempts to assess the significance of organic chemicals as potential sources of pollution in oceans and coastal waters. The pertinent literature has been selected from this point of view. Contributions likely to provide a reliable fundament for prediction were considered preferentially. Mainly papers of marine concern were utilized; however, information on limnic environments has been included where considered necessary.

(2) Analysis of Organic Pollutants

It is the task of analytical chemistry to provide reliable data on the pollution load, in its broadest sense, of oceans and coastal waters. However, in view of the immense number of potential pollutants two questions arise: (i) which compounds can be determined on the basis of currently available analytical capabilities?; and (ii) which compounds need to be analysed in order to assess the ecological situation?

While increasing refinement of instrumental analytical chemistry (and neglect of costs) should make it possible to detect and identify numerous organic pollutants at or below the ng l^{-1} level in sea water—provided the structure of the chemical species to be determined is known—risk assessment requires only those compounds to be identified and quantified which exert toxic or undesirable effects on the marine biota or man. A compilation of priority pollutants, comprising some classes of organic compounds (Table 6-1), gives some indication for particularly dangerous water-related chemicals (EPA, 1979).

On the other hand, many pollutants are modified in the marine environment, e.g. by radiation, hydrolysis or metabolization, evaporation and sorption to sediment and biota. In these processes the original structure of the pollutant is more or less altered and the resulting product may display new physical, chemical and toxic properties. Such history or 'fate' may modify a well known pollutant into one or more 'new' or unknown compounds which usually escape detection in instrumental analysis. It is, therefore, a great challenge for the environmental analyst to elucidate pathways of pollutants in different marine compartments in order to quantify realistic levels to which ecosystems are exposed, and to work out reliable risk evaluations. The levels of organic contaminants in sea water usually are in the range of ng l^{-1} and those in sediments and biota in the μg kg^{-1} and sometimes even the mg kg^{-1} range. Such low levels, especially in sea water, make sampling and analytical procedures extremely vulnerable to interfering external contamination and, therefore, require special precautions.

The low *in situ* levels further command procedures such as the 'clean up' of extracted pollutants from co-extracted natural substances, e.g. lipids, wax esters and pigments, that interfere with instrumental analytical steps. The simultaneous occurrence of an array of contaminants in the samples usually requires a pre-separation into certain classes of compounds in order to avoid errors in interpretations. In most cases, gas chromatography (GC) is the final separation step; the identity of pollutants may be checked either by GC on columns of different polarity, partly after derivatization or by mass spectrometry. A general flow scheme for the analytical procedure is given in Fig. 6-1.

(a) Sampling

Water

Sample volumes may vary between 1 and 1000 l according to sample location and the nature of the substances to be analysed. Analysis of trace amounts of organics in open ocean waters usually requires large volumes.

Large-volume water samplers, as described by BODMAN and co-authors (1961), can operate throughout the water column. A modification of the BODMAN sampler was described by DELAPPE and co-authors (1980) for collecting sea water volumes of 90 l, allowing extreme low-level detection of organohalogens. Background levels using this type of sampler were 6 to 60 pg l^{-1}, e.g. of PCBs in an equivalent volume of 40 l.

Other methods of large-volume sea water analysis involve either continuous liquid–liquid extraction (AHNOFF and JOSEFSSON, 1974) or water pumping through adsorption columns containing adsorbents for organic compounds. Adsorbents such as macroreticular resins, e.g. Amberlite XAD-2 (RILEY and TAYLOR, 1969; HARVEY, 1972; OSTERROHT, 1974) or polyurethane foam (UTHE and co-authors, 1972, 1974; MUSTY and NICKLESS, 1974, 1975; DELAPPE and co-authors, 1980) retain organic non-polar compounds satisfactorily. A flow velocity of up to 5 bed volumes min^{-1} allows a small-sized column to extract several hundred litres of water within a few hours.

Devices for shipboard operation have been described by DELAPPE and co-authors (1980) for adsorptive collection on polyurethane foam and by EHRHARDT (1976) for Amberlite XAD-2 resins. The depth for collecting water by this method is limited because of the operation of pumps from shipboard; EHRHARDT (1976), for example,

Table 6-1

Water-related priority pollutants (Based on the ENVIRONMENTAL PROTECTION AGENCY, 1979)

1. Pesticides
Acrolein
Aldrin
Chlordane
DDD
DDE
DDT
Dieldrin
Endosulfan and endosulfan sulphate
Endrin
Heptachlor
Heptachlor epoxide
Hexachlorocyclohexane (α, β, δ isomers)
γ-Hexachlorocyclohexane (lindane)
Isophorone
TCDD
Toxaphene

2. PCBs and Related Compounds
Polychlorinated biphenyls
2-Chloronaphthalene

4. Halogenated Ethers
Bis(chloromethyl) ether
Bis(2-chloroethyl) ether
Bis(2-chloroisopropyl) ether
2-Chloroethyl vinyl ether
4-Chlorophenyl phenyl ether
4-Bromophenyl phenyl ether
Bis(2-chloroethoxy)methane

5. Monocyclic Aromatics
Benzene
Chlorobenzene
1,2-Dichlorobenzene
1,3-Dichlorobenzene
1,4-Dichlorobenzene
1,2,4-Trichlorobenzene
Hexachlorobenzene
Ethylbenzene
Nitrobenzene
Toluene
2,4-Dinitrotoluene

Anthracene
Fluoranthene
Phenanthrene
Benzo[a]anthracene
Benzo[b]fluoranthene
Benzo[k]fluoranthene
Chrysene
Pyrene
Benzo[ghi]perylene
Benzo[a]pyrene
Dibenzo[a]anthracene
Indeno[1,2,3-cd]pyrene

8. Nitrosamines and Miscellaneous Compounds
Dimethylnitrosamine
Diphenylnitrosamine
Di-n-propylnitrosamine
Benzidine
3,3'-Dichlorobenzidine
1,2-Diphenylhydrazine (hydrazobenzene)
Acrylonitrile

3. Halogenated Aliphatic Hydrocarbons

Chloromethane (methyl chloride)
Dichloromethane (methylene chloride)
Trichloromethane (chloroform)
Tetrachloromethane (carbon tetrachloride)
Chloroethane (ethyl chloride)
1,1-Dichloroethane (ethylidine chloride)
1,2-Dichloroethane (ethylene dichloride)
1,1,1-Trichloroethane (methylchloroform)
1,1,2-Trichloroethane
1,1,2,2-Tetrachloroethane
Hexachloroethane
Chloroethene (vinyl chloride)
1,1-Dichloroethene (vinylidene chloride)
Trichloroethene
Tetrachloroethene (perchloroethylene)
1,2-Dichloropropane
Hexachlorobutadiene
Hexachlorocyclopentadiene
Bromoethane (methyl bromide)
Bromodichloromethane
Dibromochloromethane
Tribromomethane (bromoform)
Dichlorodifluoromethane
Trichlorofluoromethane

2,6-Dinitrotoluene
Phenol
2-Chlorophenol
2,4-Dichlorophenol
2,4,6-Trichlorophenol
Pentachlorophenol
2-Nitrophenol
4-Nitrophenol
2,4-Dinitrophenol
2,4-Dimethyl phenol
p-Chloro-m-cresol
4,6-Dinitro-o-cresol

6. Phthalate Esters

Phthalate Esters: dimethyl, diethyl, di-n-butyl, di-n-octyl, bis(2-ethylhexyl) and butyl benzyl

7. Polycyclic Aromatic Hydrocarbons

Acenaphthene
Acenaphthylene
Fluorene
Naphthalene

9. Priority Pollutants of Pulp and Paper Industries*

Abietic acid
Dehydroabietic acid
Isopimaric acid
Pimaric acid
Oleic acid
Linoleic acid
Linolenic acid
9,10-Epoxystearic acid
9,10-Dichlorostearic acid
Monochlorodehydroabietic acid
Dichlorodehydroabietic acid
3,4,5-Trichloroguaiacol
Tetrachloroguaiacol

* After CLAVES (1979).

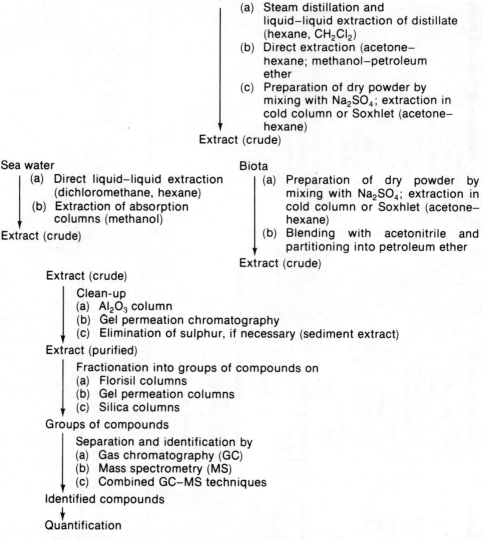

Sediment
  (a) Steam distillation and
      liquid–liquid extraction of distillate
      (hexane, CH_2Cl_2)
  (b) Direct extraction (acetone–
      hexane; methanol–petroleum
      ether
  (c) Preparation of dry powder by
      mixing with Na_2SO_4; extraction in
      cold column or Soxhlet (acetone–
      hexane)
Extract (crude)

Sea water
  (a) Direct liquid–liquid extraction
      (dichloromethane, hexane)
  (b) Extraction of absorption
      columns (methanol)
Extract (crude)

Biota
  (a) Preparation of dry powder by
      mixing with Na_2SO_4; extraction in
      cold column or Soxhlet (acetone–
      hexane)
  (b) Blending with acetonitrile and
      partitioning into petroleum ether
Extract (crude)

Extract (crude)
  Clean-up
  (a) Al_2O_3 column
  (b) Gel permeation chromatography
  (c) Elimination of sulphur, if necessary (sediment extract)
Extract (purified)
  Fractionation into groups of compounds on
  (a) Florisil columns
  (b) Gel permeation columns
  (c) Silica columns
Groups of compounds
  Separation and identification by
  (a) Gas chromatography (GC)
  (b) Mass spectrometry (MS)
  (c) Combined GC–MS techniques
Identified compounds

Quantification

Fig. 6-1: Flow scheme for analysis of water, sediment and biota. (Original.)

used a depth of 5 m below ship bottom for sampling. For depth down to 200 m a new apparatus has been designed (EHRHARDT, 1980) using the XAD-2 sorption column and a battery-driven pumping system in a pressure-equilibrated compartment. An XAD-2 adsorption system in an automatic buoy with an internal power supply system has also been found useful (EHRHARDT, 1978).

   In summary, sampling of large volumes of sea water by adsorptive extraction facilitates shipboard operations because of small apparatus size, lower costs and great flexibility in the volume of water to be sampled. The method is less vulnerable to contamination than direct solvent extraction of water. On the other hand, the sampling depth is limited and the array of substances to be extracted is confined to the adsorbants available. For the analysis of coastal sea water usually smaller water volumes are required.

Glass bottles with volumes from 1 to 10 l are used for sub-surface water; solvent extractions should be carried out immediately after sampling in the same bottle (GRASSHOFF, 1976; STADLER and SCHOMAKER, 1976, 1977a; ERNST and WEBER, 1978). This method is simple, cheap and reliable and may be used preferably for rapid screening of coastal and estuarine waters.

For 'fate studies' it might be necessary to investigate surface slicks and surface waters; both concentrate a number of pollutants. Sea-water surface samples may be obtained as described by GARRETT (1965), PARKER and co-authors (1968), HATCHER and PARKER (1974), LARSSON and co-authors (1974), MIGET and co-authors (1974), STADLER and SCHOMAKER (1977b) and HAMILTON and CLIFTON (1979).

Sediments

Sediments are collected by means of van Veen grab sampler, Shipek bottom sampler or Reineck box corer. These samplers collect surface sediments up to a depth of about 20 cm. A study on the distribution of DDT compounds in bottom sediments in a strongly polluted area off Los Angeles (USA) revealed that these compounds are mainly bound to the top 10 cm (MACGREGOR, 1976). However, some compounds occur deeper and DDMU was still present at 36 cm. Other compounds may exhibit a different distribution; exact evaluation, therefore, requires proper depth selection. The sediment samples collected are deep-frozen on shipboard until analysis according to the analytical flow scheme shown in Fig. 6-1.

Biota

Animals are sampled by grab, dredging or trawling. Preparations of animals are preferentially carried out on shipboard, and tissues are deep-frozen until analysis. General precautions, also for plankton sampling, have been discussed by GRICE and co-authors (1972). For adequate comparison of pollutant levels in animals from different locations (evaluations of trends), the animals, e.g. fishes, should be selected at appropriate size classes and in sufficient numbers to allow statistical data treatment.

(b) Extraction

Water

Extraction of the collected water samples can be carried out on shipboard or—after preservation of samples—in land-based laboratories by liquid–liquid extraction. Organic solvents such as chloroform, hexane, dichloromethane and pentane are used. For the extraction of pollutants adsorbed to XAD-2 resins, aqueous methanol has been proposed (EHRHARDT, 1976). The extracts are then processed according to the general scheme shown in Fig. 6-1.

Sediments

Extraction involving various solvent systems can be made with wet and sodium sulphate-dried sediments (MESTRES and co-authors, 1974; EDER, 1976; HARVEY and GIAM, 1976; MACGREGOR, 1976). Sulphur extracted along with the pollutants may interfere with some peaks in gas chromatography; its elimination can be achieved by reaction with copper (TEICHMAN and co-authors, 1978) or mercury (GOERLITZ and LAW, 1971, 1974) in purified extracts.

Biota

Extraction of whole organisms or dissected tissues is accomplished by blending the wet or sodium sulphate-dried material with solvents, such as acetonitrile, and subsequent partitioning into hexane. Another effective method, usable as routine, is to mortar-mill the deep-frozen tissue with four times its weight of sodium sulphate to give a dry powder, which is then eluted in a column with n-hexane–acetone (ERNST and co-authors, 1974).

(c) Clean-up

While pollutant levels in the biota are higher than those in water and sediments, the organismic matrix—consisting of fatty material, pigments, wax esters, etc.—requires thorough clean-up procedures. Especially the elimination of lipids is of major significance in the analysis of marine species. In essence, four groups of clean-up procedures can be distinguished: (i) column chromatography; (ii) solvent partition; (iii) decomposition procedures; (iv) high-performance liquid chromatography. Chromatography on aluminium oxide columns (HOLDEN and MARSDEN, 1969) and gel permeation chromatography (STALLING and co-authors, 1972, 1973) are superior to solvent partition techniques. Although decomposition procedures, such as microscale alkali treatment, offers some advantages for confirmation of alkali-stable compounds, some components are lost by this treatment (HUTZINGER and co-authors, 1974). This is true also for sulphuric acid treatment (JENSEN and co-authors, 1972a), which is simple and rapid and especially suitable for the very stable PCBs and some other chlorinated pesticides (JANSSON and co-authors, 1979). Fractionating into compound classes can be achieved by column chromatography on Florisil (MILLS, 1959; REYNOLDS, 1969; SÖRENSEN, 1973; ERNST and co-authors, 1974) or silica gel (HOLDEN and MARSDEN, 1969; HUTZINGER and co-authors, 1974; DAWSON, 1976).

(d) Separation, Identification and Quantification

The final step in the determination and confirmation of pollutants is gas chromatography (GC) using an electron-capture detector (ECD) and gas chromatography combined with mass spectrometry (GC–MS). The ECD facilitates the analyses of minute amounts of organohalogen compounds in marine matrices. The use of GC techniques has become common and is well documented. There are hardly any special features in the GC analysis of marine samples, except for the sampling and clean-up methods reported. An extremely large number of GC column packings are available. Some have been used successfully in numerous analyses; GC techniques have been refined by the development of DC-200, OV-101, OV-17, QF-1 and SE-30 on Chromosorb W. With the use of the glass capillary column technique a number of 'new' compounds became known. In contrast to the limited feasibilities of confirmatory procedures in GC, combined GC–MS offers unequalled capabilities in the confirmatory analysis of pollutants, and in the elucidation of unknown compounds. Selective ion-monitoring techniques make the mass spectrometer an excellent 'compound-selective' detector for GC, monitoring several ions continuously. Along with the consolidation of GC–MS techniques there is a trend to reduce maintenance costs and to computerize data acquisition and interpretation for automatic GC–MS analysis on the basis of library search systems which include GC retention data. Regulatory programmes in environmental pollution control utilize these modern techniques. Thus the use of GC–MS is now mandatory for the analysis of many substances dubbed 'priority pollutants' (Table 6-1) by the US Environmental Protection Agency (BROOKS and MIDDLEDITCH, 1979). The applications of GC–MS in environmental pollution control—dealing with pesticides, halogenated hydrocarbons and halogenated aromatic industrial pollutants, including technical details—have been reviewed extensively by MCFADEN (1973), SAFE and HUTZINGER (1973), BROOKS and MIDDLEDITCH (1979) and SAFE (1979).

Special problems are involved in the analysis of PCBs—the generic term for compound mixtures derived from biphenyl by various degrees of chlorination. More than 100 out of 209 possible compounds (Table 6-2) are formed; their mixtures are classified as different types according to chlorine content. In gas chromatographic analysis every PCB type exhibits an individual peak pattern, and quantification of PCBs in environmental samples is usually made by pattern recognition using packed columns. However, 'environmental' PCB patterns may differ markedly from those of the original individual PCB types owing to mixing of the different types in unknown proportions and to partial degradation. Attempts have been made to reduce the analytical difficulties involved by perchlorination of the PCB mixtures, resulting in the formation of only one compound, decachlorobiphenyl. However, the disadvantage of this method is the loss of information with respect to the type of PCBs and, therefore, their possible toxicological significance; other chlorinated products simultaneously formed, such as chlorinated naphthalenes, have to be differentiated.

At present, high-resolution glass capillary column chromatography provides the most suitable method for separation into single compounds. It is superior to other methods, especially in studies on the fate and persistence of these compounds in oceans and coastal waters.

Limits of Detection

Limiting factors for detection and quantification are instrumental sensitivity and quality of clean-up procedures. Electron-capture detectors are sensitive in the pg range, and GC–MS techniques require substance quantities in the low ng or sub-ng range if selective detection methods are employed; PCBs, as a mixture of compounds, can be measured by GC at a lower limit of about 250 pg per injection. On the basis of sample size, matrix and instrumental sensitivity, the lowest detectable concentrations for the majority of organic compounds may be ng l^{-1} for water samples and ng g^{-1} for sediments and biota. In view of the toxicity of the compounds in question, there is at present no need for further refinement of detection techniques, but rather for stability of instrumental performance and improved precautionary measures to avoid contamination at the laboratory level. The general outline of procedural steps for pollution analysis concerns only non-polar, non-volatile compounds; other substance classes require special treatment. For analytical methods for various classes of substances excellent literature

Table 6-2

Numbering and structure of PCB compounds (After BALLSCHMITER and ZELL, 1980a; reproduced by permission of Springer-Verlag)

No.	Structure	No.	Structure	No.	Structure	No.	Structure	No.	Structure
Monochlorobiphenyls		**Tetrachlorobiphenyls**		**Pentachlorobiphenyls**		**Hexachlorobiphenyls**		**Heptachlorobiphenyls**	
1	2	40	2,2',3,3'	82	2,2',3,3',4	128	2,2',3,3',4,4'	170	2,2',3,3',4,4',5
2	3	41	2,2',3,4	83	2,2',3,3',5	129	2,2',3,3',4,5	171	2,2',3,3',4,4',6
3	4	42	2,2',3,4'	84	2,2',3,3',6	130	2,2',3,3',4,5'	172	2,2',3,3',4,5,5'
Dichlorobiphenyls		43	2,2',3,5	85	2,2',3,4,4'	131	2,2',3,3',4,6	173	2,2',3,3',4,5,6
4	2,2'	44	2,2',3,5'	86	2,2',3,4,5	132	2,2',3,3',4,6'	174	2,2',3,3',4,5,6'
5	2,3	45	2,2',3,6	87	2,2',3,4,5'	133	2,2',3,3',5,5'	175	2,2',3,3',4,5',6
6	2,3'	46	2,2',3,6'	88	2,2',3,4,6	134	2,2',3,3',5,6	176	2,2',3,3',4,6,6'
7	2,4	47	2,2',4,4'	89	2,2',3,4,6'	135	2,2',3,3',5,6'	177	2,2',3,3',4',5,6
8	2,4'	48	2,2',4,5	90	2,2',3,4',5	136	2,2',3,3',6,6'	178	2,2',3,3',5,5',6
9	2,5	49	2,2',4,5'	91	2,2',3,4',6	137	2,2',3,4,4',5	179	2,2',3,3',5,6,6'
10	2,6	50	2,2',4,6	92	2,2',3,5,5'	138	2,2',3,4,4',5'	180	2,2',3,4,4',5,5'
11	3,3'	51	2,2',4,6'	93	2,2',3,5,6	139	2,2',3,4,4',6	181	2,2',3,4,4',5,6
12	3,4	52	2,2',5,5'	94	2,2',3,5,6'	140	2,2',3,4,4',6'	182	2,2',3,4,4',5,6'
13	3,4'	53	2,2',5,6'	95	2,2',3,5',6	141	2,2',3,4,5,5'	183	2,2',3,4,4',5',6
14	3,5	54	2,2',6,6'	96	2,2',3,6,6'	142	2,2',3,4,5,6	184	2,2',3,4,4',6,6'
15	4,4'	55	2,3,3',4	97	2,2',3',4,5	143	2,2',3,4,5,6'	185	2,2',3,4,5,5',6
		56	2,3,3',4'	98	2,2',3',4,6	144	2,2',3,4,5',6	186	2,2',3,4,5,6,6'
		57	2,3,3',5	99	2,2',4,4',5	145	2,2',3,4,6,6'	187	2,2',3,4',5,5',6

Trichlorobiphenyls

No.	Positions
16	2,2',3
17	2,2',4
18	2,2',5
19	2,2',6
20	2,3,3'
21	2,3,4
22	2,3,4'
23	2,3,5
24	2,3,6
25	2,3',4
26	2,3',5
27	2,3',6
28	2,4,4'
29	2,4,5
30	2,4,6
31	2,4',5
32	2,4',6
33	2',3,4
34	2',3,5
35	3,3',4
36	3,3',5
37	3,4,4'
38	3,4',5
39	3,4,5

No.	Positions
58	2,3,3',5'
59	2,3,3',6
60	2,3,4,4'
61	2,3,4,5
62	2,3,4,6
63	2,3,4',5
64	2,3,4',6
65	2,3,5,6
66	2,3',4,4'
67	2,3',4,5
68	2,3',4,5'
69	2,3',4,6
70	2,3',4',5
71	2,3',4',6
72	2,3',5,5'
73	2,3',5',6
74	2,4,4',5
75	2,4,4',6
76	2',3,4,5
77	3,3',4,4'
78	3,3',4,5
79	3,3',4,5'
80	3,3',5,5'
81	3,4,4',5

No.	Positions
100	2,2',4,4',6
101	2,2',4,5,5'
102	2,2',4,5,6'
103	2,2',4,5',6
104	2,2',4,6,6'
105	2,3,3',4,4'
106	2,3,3',4,5
107	2,3,3',4',5
108	2,3,3',4,5'
109	2,3,3',4,6
110	2,3,3',4',6
111	2,3,3',5,5'
112	2,3,3',5,6
113	2,3,3',5',6
114	2,3,4,4',5
115	2,3,4,4',6
116	2,3,4,5,6
117	2,3,4',5,6
118	2,3',4,4',5
119	2,3',4,4',6
120	2,3',4,5,5'
121	2,3',4,5',6
122	2',3,3',4,5
123	2',3,4,4',5
124	2',3,4,5,5'
125	2',3,4,5,6'
126	3,3',4,4',5
127	3,3',4,5,5'

No.	Positions
146	2,2',3,4',5,5'
147	2,2',3,4',5,6
148	2,2',3,4',5,6'
149	2,2',3,4',5',6
150	2,2',3,4',6,6'
151	2,2',3,5,5',6
152	2,2',3,5,6,6'
153	2,2',4,4',5,5'
154	2,2',4,4',5,6'
155	2,2',4,4',6,6'
156	2,3,3',4,4',5
157	2,3,3',4,4',5'
158	2,3,3',4,4',6
159	2,3,3',4,5,5'
160	2,3,3',4,5,6
161	2,3,3',4,5',6
162	2,3,3',4',5,5'
163	2,3,3',4',5,6
164	2,3,3',4',5',6
165	2,3,3',5,5',6
166	2,3,4,4',5,6
167	2,3',4,4',5,5'
168	2,3',4,4',5',6
169	3,3',4,4',5,5'

No.	Positions
188	2,2',3,4',5,6,6'
189	2,3,3',4,4',5,5'
190	2,3,3',4,4',5,6
191	2,3,3',4,4',5',6
192	2,3,3',4,5,5',6
193	2,3,3',4',5,5',6

Octachlorobiphenyls

No.	Positions
194	2,2',3,3',4,4',5,5'
195	2,2',3,3',4,4',5,6
196	2,2',3,3',4,4',5,6'
197	2,2',3,3',4,4',6,6'
198	2,2',3,3',4,5,5',6
199	2,2',3,3',4,5,6,6'
200	2,2',3,3',4,5',6,6'
201	2,2',3,3',4',5,5',6
202	2,2',3,3',5,5',6,6'
203	2,2',3,4,4',5,5',6
204	2,2',3,4,4',5,6,6'
205	2,3,3',4,4',5,5',6

Nonachlorobiphenyls

No.	Positions
206	2,2',3,3',4,4',5,5',6
207	2,2',3,3',4,4',5,6,6'
208	2,2',3,3',4,5,5',6,6'

Decachlorobiphenyl

No.	Positions
209	2,2',3,3',4,4',5,5',6,6'

Table 6-3

Selected methods for analysis of pollutants in marine samples (Original)

Type of pollutant	Sample[*]	Gas chromatography	Combined gas chromatography–mass spectrometry	Clean-up procedures	General methods
PCB	W	1, 3, 5	1	5	3, 4, 19, 24, 26
	S	1	1		
	B	1, 20, 24	1	9, 20	
Halogenated pesticides	W	1, 3, 5	1	5	3, 6, 10, 19, 26
	S	1	1		
	B	1, 9, 20, 25	1, 6, 7, 8, 11	9, 20	
Phthalate esters	W	1		1	17, 18
	S	1		1	
	B	1, 15, 16		1, 15	
Halogenated aliphatic hydrocarbons (low molecular weight)	W	1, 2, 14			19
	S	2, 14			
	B	2, 13, 14			
Halogenated phenols	W	21			19
	S	22			
	B	21			
Halogenated aromatics	B	12			19
Miscellaneous (pesticides, technical products)					23, 25

[*] W, water; S, sediment; B, biota.
[†] 1, GOLDBERG (1976); 2, MURRAY and RILEY (1973a); 3, RISEBROUGH and co-authors (1976); 4, HUTZINGER and co-authors (1974); 5, DAWSON (1976); 6, DOUGHERTY (1980); 7, DEMETER and co-authors (1978); 8, DELEON and co-authors (1978); 9, ERNST and co-authors (1974); 10, DAMICO (1972); 11, SCHAEFER (1974); 12, BAUMANN OFSTAD and co-authors (1978); 13, DICKSON and RILEY (1976); 14, PEARSON and McCONNELL (1975); 15, GIAM and co-authors (1975); 16, TAKESHITA and co-authors (1977); 17, SCHWARTZ and co-authors (1979); 18, BAKER (1978); 19, VAN HALL (1979); 20, JENSEN and co-authors (1972a); 21, RUDLING (1970); 22, EDER and WEBER (1980); 23, FISHBEIN (1975); 24, BALLSCHMITER and ZELL (1980a); 25, ZELL and BALLSCHMITER (1980); 26, NEU and co-authors (1978).

reviews exist; a number of contributions dealing with procedural work and having special references to relevant marine pollutants are compiled in Table 6-3, thus being selective rather than comprehensive.

Unidentified and Natural Compounds

As is well known from the analysis of marine samples, gas chromatography and mass spectrometry indicate the presence of substances which do not match any of the authentic substances used for comparison and which do not yield quantities high enough for elucidation of their chemical structures. The assumption that compounds of unknown structure and quantity exist in the sea has been emphasized by LUNDE and co-authors (1975), LUNDE and STEINNES (1975) and LUNDE and co-authors (1976). They found that the levels of non-polar organically bound halogen in marine samples were greater than the total organic halogen attributable to known organochlorine compounds. These results provide some evidence that an improvement in procedural steps may lead to the detection of 'new' substances.

In recent years efforts have been made in a comparatively new field of research, the isolation and identification of halogenated compounds from marine sources (SIUDA and DeBERNARDIS, 1973; FAULKNER and ANDERSEN, 1974; RINEHART and co-authors, 1975; SCHEUER, 1978). Hundreds of low-molecular-weight compounds could be identified in marine organisms; owing to their structure they are likely to be extracted along with halogen containing pollutants and care has to be taken in view of possible interferences in gas chromatographic peak identification.

(e) Conclusions

Modern analytical techniques allow the determination of organic pollutants down to ng l^{-1} levels for water and ng g^{-1} levels for sediments and biota. However, the vast majority of an almost incalculable number of investigations have dealt with a very narrow sector out of the numerous compounds that may be potential pollutants, namely the PCB group, DDT and some chlorinated pesticides. Only recently could a greater variety of substantial patterns be achieved by more refined analytical techniques, such as glass capillary column gas chromatography and mass spectrometry. Careful application of these techniques will help to provide a more comprehensive insight into the actual situation regarding pollutant loads. From a more practical viewpoint, progress in the analytical chemistry of environmental chemicals should at least keep pace with the toxicologist's requirements. New classes of naturally occurring organic halogenated compounds in marine organisms have been discovered in recent years. These findings will have to be considered in pollutant analysis in order to avoid confusion in the interpretation of analytical results.

(3) Bioconcentration and Biomagnification

Enrichment of chemicals in organisms has been termed accumulation, bioaccumulation, bioconcentration, biomagnification, etc., in the literature. For clarity, it was considered necessary to define these terms. In this chapter we distinguish between direct uptake of compounds from water and uptake via food chains.

Organic substances dissolved in sea water can penetrate marine organisms. 'Bioconcentration' is involved where the concentration of the test substance in a specific tissue increases beyond that in the ambient water. The term 'bioconcentration' will also be used when substances are adsorbed rather than absorbed by organisms; this is the case when algae, for example, associate with pollutants. The term 'biomagnification' includes pollutant accumulation in or on prey, in addition to bioconcentration, subsequently leading to increased concentrations at higher trophic levels. The processes of bioconcentration and biomagnification of organic compounds depend not only on the physicochemical properties of the substance itself and its concentration in water, but—in a fundamental way—also on the physiological properties of the organisms involved, e.g.

biochemical composition, metabolic activity and capabilities of pollutant elimination. Because organic pollutants exhibit varying degrees of lipid solubility, storage sites for lipophilic compounds, such as adipose tissues, are critical and govern the total pollutant burden of an organism.

(a) Bioconcentration

Bioconcentration is naturally counteracted in animals by excretory processes via faeces, urine and losses from the body surface or metabolic activity, all four being summarized as elimination. Where appreciable rates of metabolism prevail, the eliminated products may consist of the parent compound as well as of metabolites formed in the organism. Bioconcentration and elimination can be characterized by the following equations:

$$\frac{dC_A}{dt} = k_1 C_W - k_2 C_A \tag{1}$$

where C_A = concentration of pollutant in the animal (ng g^{-1}); C_W = concentration of pollutant in the water (ng g^{-1}); k_1 = rate constant for uptake (h^{-1}); and k_2 = rate constant for elimination (disappearance) (h^{-1}).

$$C_{A_t} = \frac{k_1}{k_2} \cdot C_W (1 - e^{-k_2 t}) \tag{2}$$

where C_{A_t} = concentration of pollutant in animals at time t. For sufficient long exposure times ($t \rightarrow \infty$), Eqn. 2 will become

$$C_{A_{st}} = \frac{k_1}{k_2} \cdot C_W = BCF \cdot C_W \tag{3}$$

where $C_{A_{st}}$ = maximum attainable concentration of pollutant in animal = steady-state concentration (ng g^{-1}); and BCF = bioconcentration factor.

As is apparent from eqn 3, the quotient of k_1 and k_2 is numerically identical with the bioconcentration factor and, therefore, is a measure for describing quantitatively the bioconcentration potential of compounds. Combining eqns. 2 and 3, the uptake curve can be described by

$$C_{A_t} = C_{A_{st}} (1 - e^{-k_2 t}) \tag{4}$$

From eqn 4 it can be recognized that, when animals are exposed to an organic chemical dissolved at a constant concentration in water, it will take more than 3 half-lives to attain about 90% of the theoretical steady-state level in that animal. On the other hand, the half-lifetime is becoming an important measure in bioconcentration studies.

The process of elimination can be expressed by

$$-\frac{dC_A}{dt} = k_2 C_A \tag{5}$$

which after integration becomes

$$C_{A_t} = C_{A_0} e^{-k_2 t} \tag{6}$$

where C_{A_0} = concentration of pollutant in animals at the end of the exposure period.

$$\ln C_{A_t} = \ln C_{A_0} - k_2 t \qquad (7)$$

Logarithmic transformation to eqn. 7 is useful for the determination of the elimination rate constant k_2 from experimental data. For $C_{A_t} = C_{A_0}/2$, the half-life can be determined according to $t_{1/2} = \ln 2/k_2$, where $t_{1/2}$ = half-life time of elimination (disappearance) (h^{-1}).

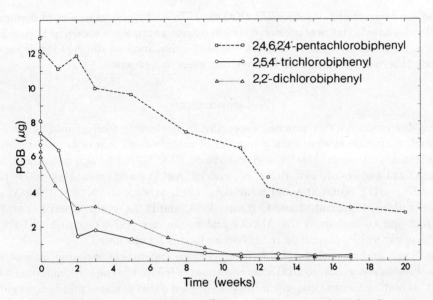

Fig. 6-2: *Nereis virens*. Elimination of different PCB compounds. (After GOERKE and ERNST, 1977; reproduced by permission of Pergamon Press.)

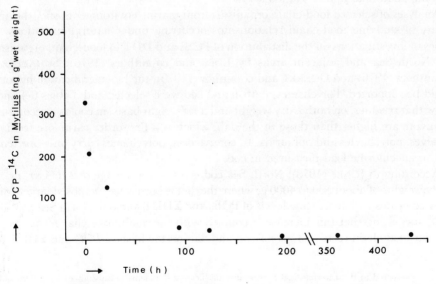

Fig. 6-3: *Mytilus edulis*. Depuration of [^{14}C]pentachlorophenol. (After ERNST, unpublished; original.)

It should be stressed at this point that the kinetic relationships are first order, which may not fit all experimental data but will be helpful in bioconcentration testing, at least at the screening level. The determination of depuration half-lives is a practical, although indirect, measure of bioconcentration. Plotting the logarithm of an initial 50% loss $(T_{1/2})$ of chlorobiphenyls and chlorinated pesticides versus log BCF, the regression equation is

$$\log \text{BCF} = 1 \cdot 327 \log T_{1/2} + 1 \cdot 186 (r = 0 \cdot 988) \, \sqrt{(T_{1/2}[h])}$$

in the common mussel *Mytilus edulis* (ERNST, 1979*). Typical examples of depuration of chlorobiphenyls and pentachlorophenol in marine animals are shown in Figs. 6-2 and 6-3. For slowly eliminating substances, 'half-lives' may increase during long depuration periods (Fig. 6-2) and initial levels may also exert an influence.

(b) Biomagnification

The discussion on this process, especially in freshwater food chains, is contrary, although it appears evident from a number of contributions that organochlorine compounds such as dieldrin (MACEK and co-authors, 1970; GRZENDA and co-authors, 1971; PETROCELLI and co-authors, 1975a, b), endrin (ARGYLE and co-authors, 1973; JACKSON, 1976), DDT (GRZENDA and co-authors, 1970; MACEK and KORN, 1970; MACEK and co-authors, 1970; ERNST and GOERKE, 1974) and PCBs (NESTEL and BUDD, 1975; GRUGER and co-authors, 1976; MAYER and co-authors, 1977; GOERKE and ERNST, 1977) appear to bioaccumulate readily when food is administered.

JARVINEN and co-authors (1977) conducted an experiment exposing fathead minnows, *Pimephales promelas*, to DDT concentrations of $0 \cdot 5$ and $2 \cdot 0$ μg l^{-1}; simulating a food chain, at both concentrations, the minnows were fed a diet of clams that had previously been exposed to a DDT concentration of $2 \cdot 0$ μg l^{-1} until the steady state. After an exposure period of 266 d the percentage of DDT caused by the food source was $62 \cdot 1$% for minnows exposed to $0 \cdot 5$ μg l^{-1} and $27 \cdot 6$% for those exposed to $2 \cdot 0$ μg l^{-1} of DDT in water proving additional uptake via food.

Analyses of selected food-chain organisms from marine environments offer the opportunity of studying food-chain relationships occurring under natural conditions. In a series of investigations on the distribution of PCBs and DDT in food-chain organisms in the North Sea and adjacent areas by EDER and co-authors (1976), SCHAEFER and co-authors (1976) and GOERKE and co-authors (1979), the biomagnification hypothesis could be supported. Polychaetes, flatfish and cod were selected and Tables 6-4 and 6-5 show that residues, on both a dry weight and a fat weight basis, in cod as a second-order carnivore are higher than those in the dab, which is a first-order carnivore feeding on bivalves, polychaetes and ophiurids. In comparison, polychaetes carry only one tenth of the organochlorine load measured in cod.

According to JONES (1978), North Sea cod grow fast at an age of 2 to 4 yr, i.e. from body weights of about 500 to 4000 g, where the diet is becoming composed largely of fish. This coincides with increasing levels of PCBs and ΣDDT found in North Sea cod (Table 6-4), suggesting that this increase is coupled with the qualitative changes in the food. Similar food-chain relationships for cod and dab on the basis of PCB and ΣDDT levels

---

* Paper presented at the International Symposium on Testing of Chemical Substances for Ecotoxicological Evaluation, Munich, May 1979.

Table 6-4

Gadus morhua. Levels of PCBs and ΣDDT in liver of North Sea cod in relation to body weights. Values are in mg kg^{-1} on fat weight basis (After SCHAEFER and co-authors, 1976; reproduced by permission of Verlag Paul Parey, Hamberg and Berlin)

Body weight (g)	PCBs	ΣDDT
86	9·3	0·96
243	8·9	1·3
471	10	3·8
917	21	4·2
2628	22	5·1

Table 6-5

Levels of PCBs and ΣDDT in selected food-chain members of the North Sea. Values are in mg kg^{-1} on dry and lipid weight basis (After SCHAEFER and co-authors, 1976; reproduced by permission of Verlag Paul Parey, Hamburg and Berlin)

Food-chain members	PCBs		ΣDDT	
	Dry	Lipid	Dry	Lipid
Polychaeta				
Polynoidae	0·10	2·2	0·02	0·42
Polyphysia crassa	0·085	3·8	0·0035	0·16
Aphrodite aculeata	0·22	13	0·0078	0·44
Pisces				
Limanda limanda	0·86	7·7	0·17	1·5
Gadus morhua	1·6	24	0·30	4·4

in their livers may be concluded from an ICES monitoring programme in the North Sea (1977a). Likewise, analyses of polychaetes and flatfish from the Skagerrak (Table 6-6) show again a clear predator–prey relationship (EDER and co-authors, 1976). Similar results were obtained when investigating invertebrates and flatfish collected in the Weser estuary, although in this case the species are not linked to a common food chain (GOERKE and co-authors, 1979).

Bioconcentration of highly recalcitrant molecules may result in elevated pollutant levels in animals, even when present at minute concentrations in water. The bioconcentration factors (*BCF*) for a number of compounds, as determined in laboratory experiments, are listed in Table 6-7a-d. The *BCF* covers a wide range, from about 10 to 10^5. From the higher values it can be concluded that, if the exposure time is long enough to reach maximum values (steady state), the levels attained may be sufficiently high to produce toxic effects or to make the sea food unpalatable for man. For these reasons prior assessment of the bioconcentration potential of new substances or those already in use is desirable.

Bioconcentration potentials may be derived from laboratory experiments, preferably under standardized test conditions. Essentially there are two modes of exposure: the static system assay and the continuous flow assay. Static systems are easier to maintain

Table 6-6

Levels of PCBs and ΣDDT (mg kg^{-1} fresh and lipid weight basis) in polychaetes and fishes from Skagerrak (After EDER and co-authors, 1976; not copyrighted)

Organisms	PCBs		ΣDDT	
	Wet	Lipid	Wet	Lipid
Polychaeta				
Neoleanira tetragona	0·017–0·020	1·4–1·6	0·004–0·005	0·35–0·45
Pisces				
Glyptocephalus cynoglossus				
(Pleuronectidae)	0·026–0·039	10–37	0·008	3·0–6·5

and are cheaper. However, difficulties occur with this method if highly volatile compounds that will readily disappear from the exposure tank have to be tested. The volatility of some compounds is shown in Fig. 6-4. Static exposure may be used when short exposure periods are adequate and when expensive radiolabelled compounds have to be used as test substances in the case when metabolites are assumed to be formed or if other analytical procedures are not feasible. Both static and flow procedures, should provide approximate steady-state *BCF*s.

In order to reduce exposure times, BRANSON and co-authors (1975) proposed an accelerated test for *Salmo gairdnerii*, using computer-aided determination of the rate constants k_1 and k_2 (eqn 1) after an exposure period of 5 d and an elimination period of 28 d. For the test substance tetrachlorobiphenyl good agreement was observed between the results of accelerated test and a continuous exposure test run over a period of 42 d. *BCF*s as derived from laboratory experiments may involve some uncertainties with respect to *in situ* extrapolation. In experiments with caged mussels, *Mytilus edulis* (Fig. 6-5), ERNST and co-authors (1982) were able to demonstrate that the *BCF* of α-HCH, γ-HCH and pentachlorophenol, obtained under natural estuarine conditions, agreed well with laboratory data. During exposure the concentration of the substances in the ambient water was monitored to calculate *BCF* values. *Lanice conchilega* polychaetes exposed under the same conditions behaved similarly with regard to γ-HCH and pentachlorophenol, but not to α-HCH, for which the natural bioconcentration was found to be 4-fold higher.

In recent years, physico-chemical data—such as water solubilities and partition coefficients—have been increasingly applied in the prediction of bioconcentration phenomena. It was found that bioconcentration is inversely related to the water solubilities of organic compounds (LU and METCALF, 1975) in fresh water species. A similar relationship has been reported by ERNST (1977) for the common mussel *Mytilus edulis*. As far as solubility data were available, these results are compiled, together with findings of other authors, in Fig. 6-6, although bioconcentration data were obtained with different species and under different conditions, mostly varying temperatures; however, these are not very likely to exert pronounced effects on bioconcentration.

Partition coefficients, P, obtained in the n-octanol–water system, are increasingly used to predict the bioconcentration potential according to the following equations:

$$\log BCF = 0.542 \log P + 0.124 \qquad \text{NEELY and co-authors (1974) (fish)}$$
$$\log BCF = 0.85 \log P - 0.7 \qquad \text{VEITH and co-authors (1979) (fish)}$$
$$\log BCF = 0.74 \log P - 0.53 \qquad \text{ERNST (1979a)}^* \text{ (mussel)}$$

As has been demonstrated by LEO and co-authors (1971) and LEO (1975), partition coefficients can be calculated from molecular fragment constants. Relationships between partition coefficients and water solubility have been reported by CHIOU and co-authors (1977). Although partition coefficients hold considerable promise for predictions, they have limited value for compounds undergoing substantial metabolism, which is a limiting factor (TULP and HUTZINGER, 1978). The relationship between $\log P$ and $\log BCF$ for different bivalves is shown in Fig. 6-7.

(c) Sediments and Particulate Matter

The relationships between dissolved organics in particular matter in the sea and sediments may be treated under the term 'association', since, very often, one cannot distinguish between adsorbed and absorbed organic compounds.

Both association of pollutants with particulate matter, and sediments and algae, are highly significant from an ecological point of view: algae are readily available to grazing animals and, therefore, facilitate transfer of pollutants into the lower trophic levels. Similarly, sediment-associated organic chemicals are a source of pollutants for benthic animals, and evidence for their remobilization by shrimp, crabs and polychaetes has been reported by several authors, e.g. NIMMO and co-authors (1971), COURTNEY and LANGSTON (1978), FOWLER and co-authors (1978) and ELDER and co-authors (1979).

Sediments can concentrate organic chemicals to levels several orders of magnitude higher than in the surrounding water. Additional loads of pollutants in sediments may be achieved from particulate matter settling from the water column. ELDER and FOWLER (1977) found PCB concentrations amounting to 17 mg kg^{-1} dry weight in faecal pellets from natural populations of the euphausiid *Meganyctiphanes norvegica,* collected in the western Mediterranean Sea. They calculate a delivery rate of PCBs to sediments of 1·4 to 4·1 μg m^{-2} yr^{-1} by *M. norvegica* which, however, represents only 1 to 5% of the total zooplankton biomass of that area. In comparison, pollutant transport by settling particulate matter was highly variable in the Kiel Bight, PCB input rates being in the range 24 to 112 μg m^{-2} yr^{-1} (OSTERROHT and SMETACEK, 1980).

The consequence of severe sediment pollution could be demonstrated in a costal area off Palos Verdes, USA. After the discharge of large quantities of DDT wastes over a period of 20 yr until 1970 into the coastal waters off Los Angeles (MACGREGOR, 1974), the bottom of the shallow-water area off Palos Verdes was shown to be heavily contaminated by DDT compounds; total DDT levels in excess of 150 mg kg^{-1} dry sediment (MACGREGOR, 1976) were measured. As a consequence, bottom-dwelling fish caught in 1970 in this area contained up to 1030 mg kg^{-1} of DDT in their livers and about 2600 mg kg^{-1} in their fat; the levels declined greatly in specimens taken further to the north, south and offshore from Los Angeles. A biomonitoring system, installed near a submarine outfall off Palos Verdes, revealed DDT-contaminated bottom sediments to be the likely source of DDT in the water column. Mussels from comparatively uncontaminated areas

---

* See footnote on p. 1642.

Table 6-7

Bioconcentration factors (*BCF*) of organic chemicals in different species: based on laboratory experiments. Unless stated otherwise, BCF values refer to whole animals (Compiled from the sources indicated)

Substance	Species	Concentration (μg l^{-1})	Exposure period (d)	BCF	Source
	Clams				
DDT	Mercenaria mercenaria	0·1	5	1260	Butler (1971)
DDT	Mya arenaria	0·1	5	8800	Butler (1971)
	Fishes				
DDT	Lagodon rhomboides	0·1	3	5700	Hansen and Wilson (1970)
DDT		0·1	7	8500	Hansen and Wilson (1970)
DDT		0·1	14	38 000	Hansen and Wilson (1970)
DDT	Lagodon rhomboides	1·0	3	2800	Hansen and Wilson (1970)
DDT		1·0	7	6900	Hansen and Wilson (1970)
DDT		1·0	14	10 600	Hansen and Wilson (1970)
DDT	Micropogon undulatus	0·1	3	3300	Hansen and Wilson (1970)
DDT		0·1	7	8000	Hansen and Wilson (1970)
DDT		0·1	14	11 000	Hansen and Wilson (1970)
DDT	Micropogon undulatus	1·0	3	2300	Hansen and Wilson (1970)
DDT		1·0	7	6900	Hansen and Wilson (1970)
DDT		1·0	14	12 000	Hansen and Wilson (1970)
	Mussel				
DDD	Mytilus edulis	0·05	7	9120	Ernst (1977)
Dieldrin	Mytilus edulis	0·17	7	1570	Ernst (1977)
	Oyster				
Dieldrin	Crassostrea virginica	0·5	7	2880	Mason and Rowe (1976)
Dieldrin		9·0	7	2070	Mason and Rowe (1976)
	Clams				
Dieldrin	Mercenaria mercenaria	0·5	5	760	Butler (1971)
Dieldrin	Mya arenaria	0·5	5	1740	Butler (1971)
Aldrin	Mercenaria mercenaria	0·5	5	380	Butler (1971)
Aldrin	Mya arenaria	0·5	5	4600	Butler (1971)
	Mussel				
Endrin	Mytilus edulis	0·17	7	1920	Ernst (1977)

Compound	Species	Concentration	n	BCF	Reference
Endrin	Oyster				
	Crassostrea virginica	0·1	7	2640	MASON and ROWE (1976)
		50·0	7	2780	MASON and ROWE (1976)
Endrin	Clams				
	Mercenaria mercenaria	0·5	5	480	BUTLER (1971)
Endrin	Mya arenaria	0·5	5	1240	BUTLER (1971)
α-Endosulfan	Mussel				
	Mytilus edulis	0·14	7	600	ERNST (1977)
Hepachlor	Clams				
	Mercenaria mercenaria	0·5	5	2600	BUTLER (1971)
Heptachlor	Mya arenaria	0·5	5	220	BUTLER (1971)
Heptachlorepoxide	Mussel				
	Mytilus edulis	0·22	7	1700	ERNST (1977)
Chlordane	Oyster				
	Crassostrea virginica	6·2	4	3200–8300	PARRISH and co-authors (1976)
Chlordane	Fishes				
	Cyprinodon variegatus	24·5	4	12600–18700	PARRISH and co-authors (1976)
Chlordane	Lagodon rhomboides	6·4	4	3000–7500	PARRISH and co-authors (1976)
Mirex	Shrimp				
	Penaeus duorarum	0·09	28	1780–3000	TAGATZ and co-authors (1975)
Mirex	Fish				
	Cyprinodon variegatus	0·09	28	10330–14440	TAGATZ and co-authors (1975)
Kepone	Cyprinodon variegatus	0·8	28	4500	HANSEN and co-authors (1977)
Methoxychlor	Clams				
	Mercenaria mercenaria	1·0	5	470	BUTLER (1971)
Methoxychlor	Mya arenaria	1·0	5	3000	BUTLER (1971)
γ-HCH	Mussel				
	Mytilus edulis	0·9–1·0[1]	8	139	ERNST (1979a)
γ-HCH	Polycheates				
	Lanice conchilega	0·58[1]	8	1240	ERNST (1979a)
γ-HCH	Nereis virens	1·0	17–21	440–480	GOERKE and ERNST (1980)
γ-HCH	Shrimp				
	Penaeus duorarum	0·13–0·62	4	84	SCHIMMEL and co-authors (1977a)
γ-HCH	Fish				
	Cyprinodon variegatus	41·9–108·7	4	490	SCHIMMEL and co-authors (1977a)
α-HCH	Mussel				
	Mytilus edulis	0·89	7	106	ERNST (1977)
α-HCH	Polychaete				
	Lanice conchilega	0·18	8	2750	ERNST (1979a)

[1]Steady-state concentrations.

Table 6-7—continued

Substance	Species	Concentration ($\mu g\ l^{-1}$)	Exposure period (d)	BCF	Source
PCP	Mussel				
	Mytilus edulis	0·3[1]	8	390	ERNST (1979a)
	Polychaetes				
	Lanice conchilega	0·09[1]	8	3830	ERNST (1979a)
PCP	Oyster				
	Crassostrea virginica	2·5–25·0	28	41–78	SCHIMMEL and co-authors (1978)
PCBs					
Aroclor 1254	Crassostrea virginica	5·0	168	85 000	LOWE and co-authors (1972)
2,3-Dichlorobiphenyl	Crassostrea virginica	0·0055–0·06	65	1200	VREELAND (1974)
3,4,2'-Trichlorobiphenyl	Crassostrea virginica	0·0055–0·06	65	6200	VREELAND (1974)
2,5,2',5'-Tetrachlorobiphenyl	Crassostrea virginica	0·0055–0·06	65	7400	VREELAND (1974)
2,3,2',5'-Tetrachlorobiphenyl	Crassostrea virginica	0·0055–0·06	65	11 000	VREELAND (1974)
2,3,4,2',5'-Pentachlorobiphenyl	Crassostrea virginica	0·0055–0·06	65	27 000	VREELAND (1974)
2,4,5,2',4',5'-Hexachlorobiphenyl	Crassostrea virginica	0·0055–0·06	65	48 000	VREELAND (1974)
Aroclor 1254	Fish				
	Lagodon rhomboides	5	35	21 800	HANSEN and co-authors (1971)
Aroclor 1254	Fish				
	Leiostomus xanthurus	1·0	56	27 000	HANSEN and co-authors (1971)
Perchloroethylene	Limanda limanda (muscle)	30–300	3–35	5–9	PEARSON and McCONNELL (1975)
	Limanda limanda (liver)	30–300	3–35	200–400	PEARSON and McCONNELL (1975)
Hexachlorobutadiene	Mussel				
	Mytilus edulis	1·3–1·6	38–50	900–2000	PEARSON and McCONNELL (1975)
	Fishes				
Hexachlorobutadiene	Limanda limanda (muscle)	1·6	27–39	700	PEARSON and McCONNELL (1975)
Hexachlorobutadiene	Limanda limanda (liver)	1·6	27–39	10 000	PEARSON and McCONNELL (1975)
Hexachlorobutadiene	Pleuronectes platessa (muscle)	1·7	21–106	500	PEARSON and McCONNELL (1975)
Hexachlorobutadiene	Pleuronectes platessa (liver)	1·7	21–106	7000	PEARSON and McCONNELL (1975)
Fenitrothion	Mussel				
	Mytilus edulis	0·18–13	14	78–130	McLEESE and co-authors (1979)
	Clam				

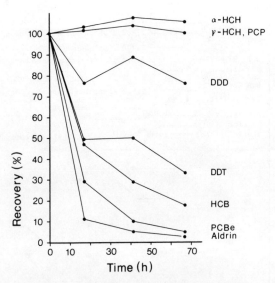

Fig. 6-4: Volatility of different pesticides from sea water aquaria under experimental conditions: 4 l sea water; 32–33‰ S; 10 °C; aeration, 40 ml min^{-1}; initial concentration, 0·5 μg l^{-1}. PCBe = pentachlorobenzene. (Based on ERNST, 1977.)

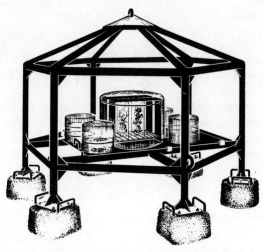

Fig. 6-5: Cage for exposing experimental animals (mussels, polychaetes) to *in situ* conditions in an estuary. (After ERNST and co-authors; 1982).

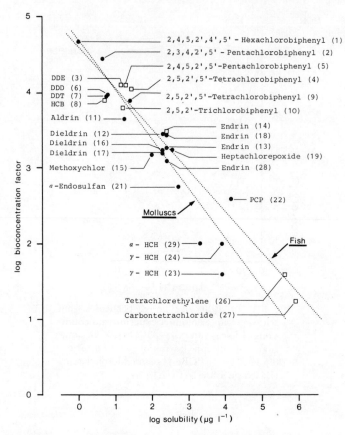

Fig. 6-6: Correlation between water solubility and bioconcentra-
tion factors of organic chemicals in fishes and bivalves. (After
ERNST, 1980; reproduced by permission of Biologische Anstalt
Helgoland.)

were introduced into the monitoring system at various depths for 13 wk. After that
period DDT levels were directly related to bottom proximity (YOUNG and co-authors,
1976).

Algae may concentrate DDT and PCB in the range 10^4 to 10^5-fold—about the
ambient concentration in water (COX, 1972; RICE and SIKKA, 1973; HARDING and
PHILLIPS, 1978). Interestingly, DDT uptake is obviously a passive process: dead algal
cells remove the compound at similar rates from the medium (RICE and SIKKA, 1973).

(d) Conclusions

Accumulation processes become apparent in sediments, particulate matter and biota.
The latter process can be related to physical properties of the pollutants, such as water
solubility and water–octanol partition coefficients. These data permit an estimate of
bioconcentration potentials. A wide range of bioconcentration factors, from 10 to 10^5,
have been found in a great variety of organisms. Under steady-state conditions maxi-

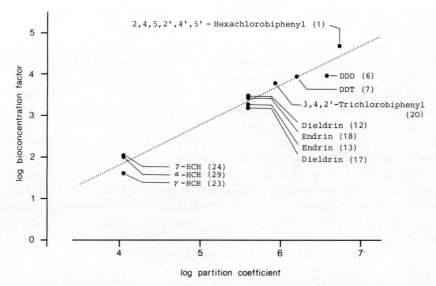

Fig. 6-7: Correlation between partition coefficients and bioconcentration factors of organic chemicals in bivalves. (After ERNST, 1980; reproduced by permission of Biologische Anstalt Helgoland.)

mum bioconcentration factors can be used for predicting pollutant levels in both biota and water. Data on biomagnification along food chains are still a matter of dispute. For highly persistent chemicals with elevated bioconcentration potentials, biomagnification should be considered ecologically significant. Elimination has been found to be an important alternative measure for determining the bioconcentration potential. However, degradation has its share in the total resulting pollutant disappearance from the biota.

(4) Degradation

(a) Pathways of Degradation

Once released into the marine environment a chemical substance is subject to a variety of degradation processes which may change its parent structure and even lead to total mineralization. In terms of environmental protection, the latter is, of course, most desirable. In the course of structural modification transformation products may be formed that exert higher degrees of toxicity or have greater bioconcentration potentials than the parent compounds. Moreover, such degradation products can escape detection if they are either unknown or do not fit into conventional analytical procedures, thus obscuring the total real pollutant loads.

Degradation in the sea may occur by radiation, chemical reactions and via a variety of biochemical pathways. Chemical transformations—mainly hydrolytic reactions—are confined to a limited number of chemicals, e.g. to some types of insecticidal organophosphates, such as parathion and malathion (WOLFE and co-authors, 1975; WEBER, 1976). Photochemical degradation is limited to substances in the uppermost layers of the water column. The literature available does not indicate that substantial amounts of pollutants are subject to this type of degradation in sea water, although photodegradation should be occurring, especially in surface layers. Photochemical processes are considered to be ecologically significant, since they may enhance toxicity. For example, aldrin and dieldrin are transformed into their corresponding photoisomers, which are several times more toxic to fish than the parent compounds (KHAN and co-authors, 1973).

Biological degradation has, thus far, received most attention, but even here the knowledge at hand is insufficient in view of the great ecological importance of these processes. Biological degradation has been studied experimentally in different animal species and in micro-organisms. Radiolabelled substances are preferably used in these studies, because this is often the only way to follow their fate at the low concentrations that have to be applied in the case of highly toxic compounds. In general, the degradation potentials vary in different locations, depending on species composition and physical and chemical environmental conditions. The terms 'degradation' or 'persistence' of a given substance should, therefore, be used only in connection with defined locations and should be expressed as a function of time. The latter is an important factor in the safety assessment of chemicals, since many are readily degradable.

(b) Transformation in Animals

BRODIE and MAICKEL (1962) concluded that fishes lack drug-metabolizing activities and are unable to conjugate phenols, even though a number of fish possessed measurable quantities of microsomal glucuronyl transferase. In contrast, other authors were able to demonstrate glucuronide formation in a number of fishes (see review by ADAMSON, 1967). Later, additional metabolic pathways, similar to those in mammals, were explored, including hydroxylation, acetylation, methylation and conjugation with sulphuric acid, glycine and taurine. Detailed studies on the capabilities of oxidative metabolism in marine animals, using tissue homogenates and subcellular fractions, such as mitochondria and microsomes, confirmed the presence of mixed-function oxygenase (MFO) activities in these species as well as the occurrence of significant levels of microsomal cytochrone P-450, a haemoprotein, which plays a central role in the metabolism of xenobiotics (ELMAMLOUK and co-authors, 1974; POHL and co-authors, 1974; WILLIS and ADDISON, 1974; ELMAMLOUK and GESSNER, 1976; BEND and co-authors, 1977, 1979; AHOKAS, 1979; LEE, 1981). The majority of degradation pathways have been elaborated in laboratory studies, after pollutant administration via water food or injection. As shown in Figs. 6-8 to 6-15, few data are available for marine species, especially invertebrates; data for metabolism in fishes originate mostly from experiments with freshwater species and were included in the scheme for comparison. Only results involving particular identified metabolites were considered.

Metabolites of PCBs and DDT

Degradation products of DDT and PCB compounds, such as phenolic metabolites of DDE, as well as the methyl sulphones of DDE and PCBs, have been discovered in animals such as seals (SUNDSTRÖM and co-authors, 1975; JENSEN and JANSSON, 1976; Figs. 6-8 and 6-9). Metabolic degradation frequently resulted in the formation of polar metabolites; however, these could not be identified in most cases. Table 6-8 reveals that even 2,4,6,2',4'-pentachlorobiphenyl undergoes moderate degradation when fed to the

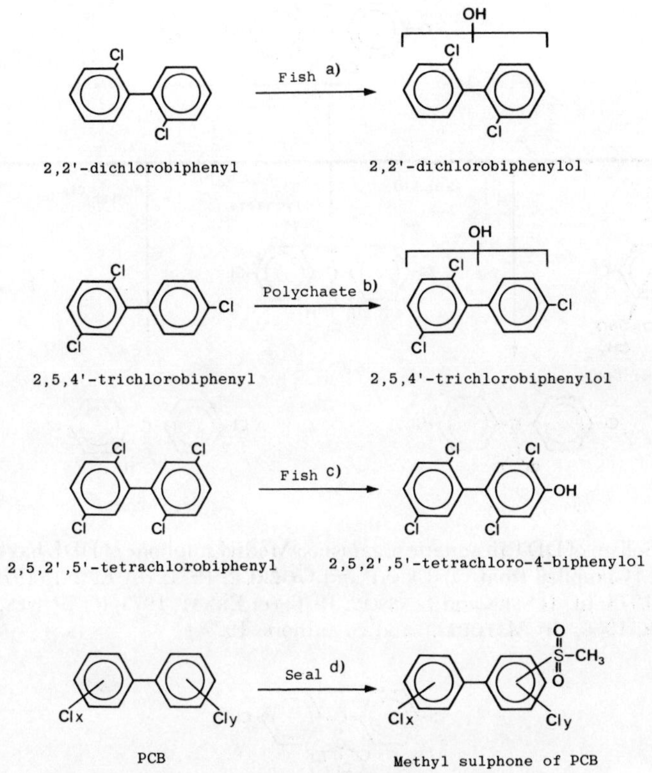

Fig. 6-8: Metabolites of various PCB compounds. Methyl sulphone of PCB has been detected in wild animals. [Compiled
from (a) HERBST and co-authors, 1976; (b) ERNST and co-
authors, 1977; (c) MELANCON and LECH, 1976; (d) JENSEN and
JANSSON, 1976.]

polychaete *Nereis virens*, whereas di- and trichlorobiphenyl were extensively metabolized
(ERNST and co-authors, 1977) and hydroxylated compounds are the most probable
metabolites (Fig. 6-8). BEND and co-authors (1973) injected the lobster *Homarus
americanus* with 2,4,5,2′,5′-pentachlorobiphenyl and found polar metabolites in green
gland and faeces, but could not elucidate their chemical structure. Hydroxylation of
dichlorobiphenyl was observed in fish and a defined conjugated hydroxy metabolite of
tetrachlorobiphenyl was isolated by MELANCON and LECH (1976) from the bile of
rainbow trout after exposure to the chemical (Fig. 6-8). Degradation of DDT has been
studied in a variety of animals including fishes, polychaetes and molluscs (Fig. 6-9).
The metabolites identified basically retained the parent 4,4′-dichlorobiphenyl methane
structure with modifications of the aliphatic moiety. The very low biodegradability
of the DDT structure is imparted by 4,4′-chloro atoms; substitution of chlorine results
in more readily degradable molecules as measured by reduced bioconcentration or
ecological magnification (Table 6-9) (METCALF, 1977).

Relationships between structure and biodegradability have also been proposed for
PCBs; for rapid transformation the presence of two vicinal hydrogen atoms is required in

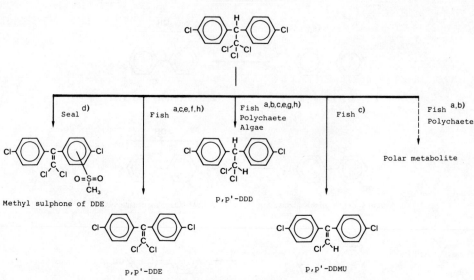

Fig. 6-9: Metabolism of DDT in aquatic organisms. Methyl sulphone of DDE has been detected in wild animals. [Compiled from (a) ERNST and GOERKE, 1974; (b) ERNST, 1970; (c) ADDISON and WILLIS, 1978; (d) JENSEN and JANSSON, 1976; (e) ERNST, 1973; (f) BOWES, 1972; (g) KEIL and PRIESTER, 1969; (h) MITCHELL and co-authors, 1977.]

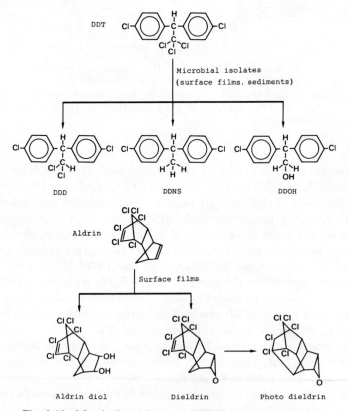

Fig. 6-10: Metabolic pathways of DDT and cyclodienes under oceanic conditions. (Reprinted with permission from PATIL and co-authors, 1972. Copyright 1972 American Chemical Society.)

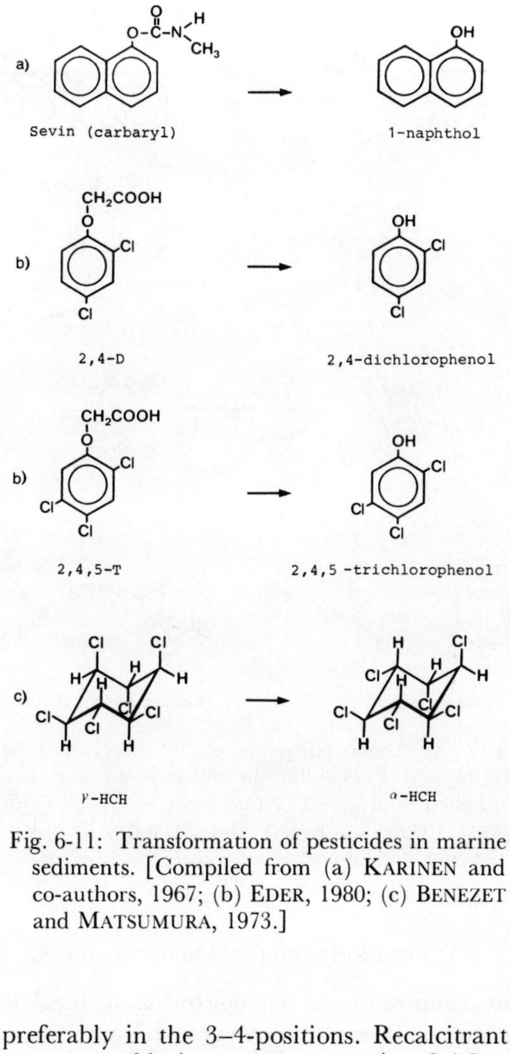

Fig. 6-11: Transformation of pesticides in marine
sediments. [Compiled from (a) KARINEN and
co-authors, 1967; (b) EDER, 1980; (c) BENEZET
and MATSUMURA, 1973.]

the PCB molecule, preferably in the 3–4-positions. Recalcitrant PCB molecules are characterized by six or more chlorine atoms occupying 2,4,5-, 2,3,4,5- and 2,3,5,6-positions (JENSEN and SUNDSTRÖM, 1974; SCHULTE and ACKER, 1974). Compounds in accordance with the structures supposed to be highly persistent have been found in the liver of marine fishes, comprising penta-, hexa- and octachlorobiphenyls (BALLSCHMI-TER and co-authors, 1978).

Other Compounds

A number of compounds—Sevin, naphthol, malathion parathion, phthalates and chlorophenols—are readily metabolized by hydrolysis and conjugation reactions and thus eliminated (Figs. 6-11 to 6-15). The herbicides 2,4-D and 2,4,5,-T are metabolized in the dogfish shark *Squalus acanthias,* being more than 90% transformed into correspond-ing taurine conjugates, following intravenous injection of the compound (GUARINO and co-authors, 1977).

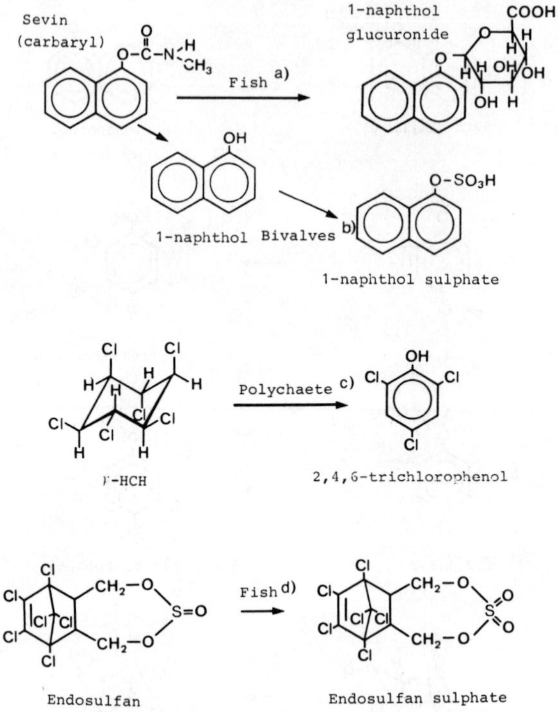

Fig. 6-12: Metabolic pathways of the insecticides Sevin, γ-HCH and Endosulfan in different aquatic species. [Compiled from (a) STATHAM and co-authors, 1975; (b) ERNST, 1979b; (c) ERNST and GEORKE, in press; (d) SCHIMMEL and co-authors, 1977b.]

(c) Transformation in Micro-organisms

Micro-organisms are assumed to play a major role in the breakdown of environmental chemicals and are preferentially used in biodegradability testing (EPA, 1975). SAYLER and co-authors (1978) found PCB-degrading bacteria associated with PCB-containing sediments and in sea water. Compared with marine samples, the PCB-degrading bacteria and PCBs exhibited higher concentrations in estuarine waters and sediments of Chesapeake Bay (USA). Two of the most dominant forms—*Pseudomonas* and *Vibrio* populations—could be identified as potential PCB degraders. SAYLER and co-authors (1978) concluded that a PCB-degrading potential exists in the natural microbial flora of marine habitats, although the results did not demonstrate naturally occurring PCB degradation by pure cultures of marine or estuarine bacteria. Similar observations were made by JUENGST and ALEXANDER (1976) who investigated the degradation of [14C]-labelled DDT by marine bacteria. They found 47 of 110 isolated bacteria to be capable of transforming [14C]DDT to the extent of more than 5% into water-soluble metabolites in a yeast extract–seawater medium, but none metabolized more than 10%. In contrast, model ecosystems containing surface sea water, sand and a variety of organic compounds were unable to transform [14C]DDT into water-soluble products.

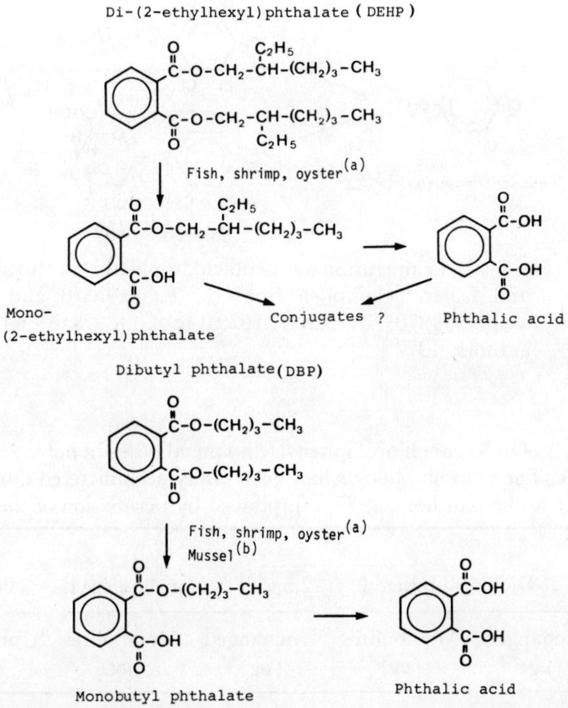

Di-(2-ethylhexyl)phthalate (DEHP)

Fig. 6-13: Degradation of phthalate esters in fishes and bivalves. [Compiled from (a) WOFFORD and co-authors, 1981; (b) ERNST, 1982.]

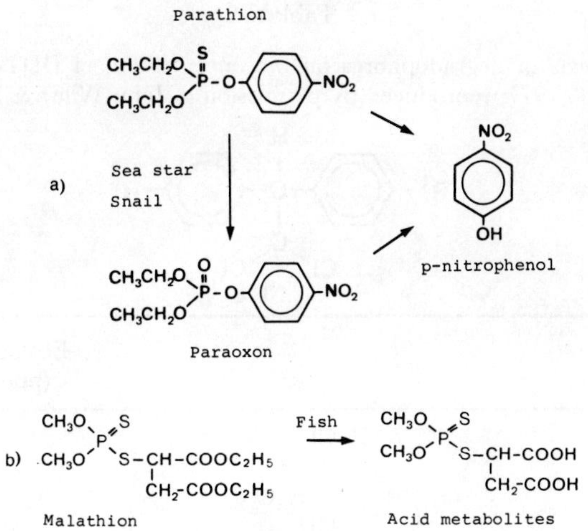

Fig. 6-14: Metabolization of insecticidal organo-phosphates by invertebrates and fishes. [Compiled from (a) GARNAS and CROSBY, 1979; (b) COOK and co-authors, 1976.]

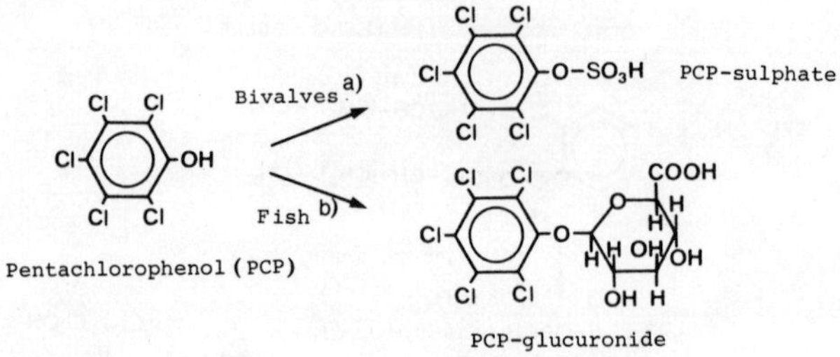

Fig. 6-15: Conjugation of pentachlorophenol by bivalves and fishes. [Compiled from (a) KOBAYASHI and co-authors, 1970; (b) ERNST, 1979a; (c) GLICKMAN and co-authors, 1977.]

Table 6-8

Neireis virens. Levels of different chlorobiphenyls and metabolites in polychaetes during elimination of the compounds. The chlorobiphenyls had been orally administered during 3 preceding weeks (After ERNST and co-authors, 1977; reproduced by permission of the Pergamon Press)

Weeks of elimination	2,2'-Dichlorobiphenyl		2,5,4'-Trichlorobiphenyl		2,4,6,2',4'-Pentachlorobiphenyl	
	Unchanged $(\mu g)^{**}$	Metabolites $(\mu g)^{**}$	Unchanged $(\mu g)^{**}$	Metabolites $(\mu g)^{**}$	Unchanged $(\mu g)^{**}$	Metabolites $(\mu g)^{**}$
1	2·07	2·3	0·36	6·05	8·79	2·33
2	1·30	1·74	0·06	1·39	9·52	2·31
3	1·20	1·92	0·07	1·75	7·56	2·39
9	0·36	0·40	0·02	0·43	6·73	0·67*

* Week 8.   ** µg < equivalent per animal per week.

Table 6-9

Gambusia affinis. Effects of degradophores on bioconcentration of DDT-type analogues (After METCALF, 1977; reproduced by permission of John Wiley & Sons, Inc.)

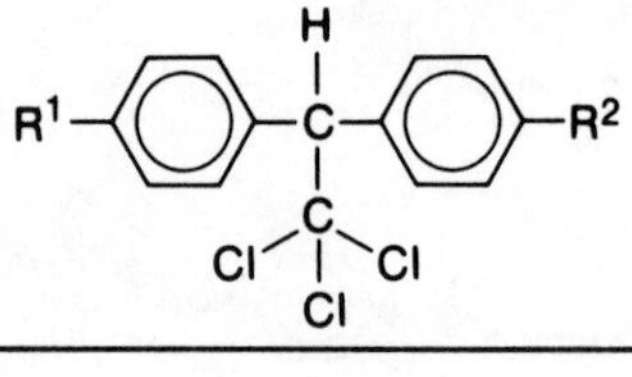

R^1	R^2	Ecological magnification (ppm fish/ppm H_2O)
Cl	Cl	84 500
CH_3O	CH_3O	1 545
C_2H_5O	C_2H_5O	1 536
CH_3	CH_3	140
CH_3S	CH_3S	5·5
CH_3O	CH_3S	310
CH_3	C_2H_5O	400
Cl	CH_3	1 400

Similar results were obtained by PATIL and co-authors (1972) with chlorinated insecticides such as DDT, dieldrin, aldrin and endrin in sea water. In sea water, and even in polluted estuarine waters, the insecticides are not metabolized when incubated for 30 d. Sandy sediments taken from the shelf, silts and muds of estuaries and bays exhibit no appreciable degradation activity and only a lagoon sediment showed a low level of degradation (Fig. 6-10). CAREY and HARVEY (1978) incubated black march mud with raw sea water and $[^{14}C]2,5,2',5'$-tetrachlorobiphenyl for 45 d, but were unable to detect radioactive products other than the parent compound. When using a sandy-beach sediment under aerobic conditions they found the first signs of metabolic activity after 10 d; however, only 1% of the extracted activity after 25 d appeared to be a polar metabolite, considered to be a lactone acid by the authors. Incubation of the tetrachlorobiphenyl with sea water under aerobic conditions yielded 2 to 4% metabolism within 3 d, but this level did not increase within several weeks of incubation. Likewise, 2,5,2'-trichlorobiphenyl was metabolized in sea water to the extent of 1·4%.

It can be concluded from these experiments that degradation of the pollutants concerned may occur in marine bacteria in laboratory experiments, but at very low rates. In view of the uncertainties created by extrapolation of laboratory experiments to nature, the question remains unanswered of whether or not highly persistent compounds, such as PCBs and DDT, can be metabolized by natural marine bacteria populations. This fact is of considerable significance when using results of degradability tests conducted under favourable incubation conditions and for predicting the degradability of these compounds under oceanic conditions.

Degradation of more polar compounds, enhanced by micro-organisms, has been observed with Sevin, an insecticidal carbamate used for predator control in oyster beds. This compound is readily hydrolysed to 1-naphthol, which is further transformed into various products of unknown structure and CO_2 (KARINEN and co-authors, 1967; LAMBERTON and CLAEYS, 1970). The herbicides 2,4-D and 2,4,5-T are similarly hydrolyzed to the corresponding chlorophenols in estuarine sediments (EDER, 1980). Another type of transformation has been found with γ-HCH in a marine sediment where an isomerization led to α-HCH (BENEZET and MATSUMURA, 1973) (Fig. 6-11).

(d) Conclusions

Various pathways of degradation of xenobiotic compounds have been explored. We now have some, but not sufficient, insight into the relevant metabolic features of marine organisms. There are gaps in our knowledge of degradation routes and the degree to which degradation takes place under particular environmental conditions. In many cases the analytical techniques available do not allow complete identification of the metabolites formed and a proper estimate of degradation potentials. This, however, is a prerequisite for a reliable interpretation of bioconcentration studies and environmental analyses. In order to provide reliable predictions of the fate of organic pollutants we need more investigations concerned with *in situ* conditions or at least conditions comparable to the situation prevailing in the natural marine environment.

(5) Occurrence and Distribution of Organic Chemicals

Continuous measurements of actual concentrations of pollutants in water, sediment and biota are an essential prerequisite for determining the occurrence and distribution of organic chemicals, as well as for an assessment of the risks involved for man and the ecosysystem and for an elucidation of pollutant pathways and sinks in nature. On a global scale, organic chemicals released into the marine environment are distributed according to prevailing physical, chemical and biological factors.

(a) Sea water and Surface Slicks

For the water in oceans and adjacent areas there are comparatively few data on the occurrence and distribution of organic pollutants. Because of the low levels (in the ng l^{-1} range), sampling techniques and subsequent analytical procedures are extremely vulnerable to accidental contamination of the samples. In view of quantitative differences in pollution levels in different compartments, it is desirable to distinguish between water and surface slicks. Estuarine and nearshore waters carry higher pollutant loads than open-ocean waters owing to riverborne material, dumping activities and direct inlets. Surface slicks, often present in the form of a coherent film, are formed by convection, diffusion and bubbling during calm weather. They tend to accumulate organic pollutants at comparatively high concentrations. The concentration process is due to the chemical constituents of slicks, consisting of saturated and unsaturated fatty acids, such as palmitic and myristic acids and long-chain polyunsaturated acids, esterified C_{12}–C_{16} alcohols and high concentrations of phosphorus compounds (GARRET, 1967; WILLIAMS, 1967; SCHULTZ and QUINN, 1972). Surface films sampled off the Swedish west coast contained triglycerides, free fatty acids and wax esters as dominating constituents (LARSSON and co-authors, 1974).

Water

Concentrations of PCBs and chlorinated pesticides are compiled in Table 6-10 for different regions, representing coastal areas and open-ocean waters. The sampling depth was 1 to 2 m—in some instances up to 1500 m (SCURA and MCCLURE, 1975) or 4000 m (ELDER and VILLENEUVE, 1977). Few data are available on the occurrence of pentachlorophenols and phthalate esters (Tables 6-11 and 6-12).

In 1970 it was found by JENSEN and co-authors (1972b) that EDC tar components are widely distributed over the North Atlantic Ocean, in both water and organisms. There are strong indications that the chlorinated compounds originated from the dumping of industrial by-products of vinyl chloride production. Tar contains a wide spectrum of hydrocarbons and 80 different substances could be identified in this product; 15 of the 80 were aromatic compounds, most of them chlorinated; the others were chlorinated aliphatic low-molecular-weight compounds (JENSEN and co-authors, 1975). The compounds are highly volatile but mid-water and bottom-water analyses revealed that these compounds are not confined to the surface water, although the gas chromatographic peak pattern exhibited some changes with depth. There are few data on specific compounds of this group of chemicals; a more detailed study in Liverpool Bay was conducted by PEARSON and MCCONNELL (1975). In open-ocean waters concentrations of volatile hydrocarbons are in the ng l^{-1}-range (MURRAY and RILEY, 1973b; LOVELOCK, 1975), whereas the levels in estuarine waters may be much higher, owing to local emissions (Table 6-13). Because these halocarbons are highly volatile and thus have a short lifetime in water, they are local contaminants in estuarine and coastal waters. Some halogenated hydrocarbons are obviously naturally produced in oceans, e.g. CH_3I,

Table 6-10

Concentrations of polychlorinated biphenyls and chlorinated pesticides in sea water (Compiled from the sources indicated)

Compound	Location, year[a]	Concentration[b] (ng l^{-1})	Source
Polychlorinated biphenyls (PCBs)	North Sea, 1974 (German Bight)	2·0–4·7	STADLER and ZIEBARTH (1975)
	North Sea, 1975 (German Bight)	0·71–3·6 (2·1)	STADLER (1977)
	Firth of Taye, 1973	5–20	HOLDEN (1973b)
	English Channel, 1974	0·2–0·3	DAWSON and RILEY (1977)
	Irish Sea, 1974	0·2–1·5	DAWSON and RILEY (1977)
	Oslo Fjord, 1973	(1·1; 1·6)	LUNDE and co-authors (1975)
	North Atlantic Ocean, 1973	(1·3; 0·8)	HARVEY and co-authors (1974b)
	North Atlantic Ocean, n.r.	0·02–0·2 (0·16)	GIAM and co-authors (1978b)
	Sargasso Sea, 1973	<0·9–3·6 (1·1)	BIDLEMAN and OLNEY (1974)
	Baltic Sea, 1975	0·3–3·0	OSTERROHT (1977)
	Western Baltic Sea, 1975	3·9[c] (1·1)	STADLER (1977)
	Mediterranean Sea, 1975	<0·2–8·6 (2·0)	ELDER and VILLENEUVE (1977)
	NW Mediterranean coast, 1975	1·5–38 (13)	ELDER (1976)
	Central Mediterranean coast, 1976/77 (Tiber estuary)	9–1000[d] (135–297)	PUCCETTI and LEONI (1980)
	NE Pacific Ocean, 1974 (Southern California coast)	2·3–35·6 (9·2)	SCURA and McCLURE (1975)
	Mississippi delta, n.r.	1·7–3·3 (2·45)	GIAM and co-authors (1978b)
pp'-DDT	North Sea, 1974 (German Bight)	0·1–0·7	STADLER and ZIEBARTH (1975)
	North Sea, 1975 (German Bight)	0·11–0·63 (0·33)	STADLER (1977)
	English Channel, 1974	0·01–0·03	DAWSON and RILEY (1977)
	Irish Sea, 1974	0·05–0·25	DAWSON and RILEY (1977)
	Sargasso Sea, 1973	0·15–0·5	BIDLEMAN and OLNEY (1974)
	Baltic Sea, 1975	0·1–0·2	OSTERROHT (1977)
	Western Baltic Sea, 1975	0·37[d] (0·15)	STADLER (1977)
	NE Pacific Ocean, 1974 (Southern California coast)	<0·1–0·7	SCURA and McCLURE (1975)
	Mississippi delta, n.r.	0·01–2·9 (1·7)	GIAM and co-authors (1978b)
	Hawaii, 1971	9·0[e]	BEVENUE and co-authors (1972)
pp'-DDE	North Sea, 1975 (German Bight)	0·2[c] (0·11)	STADLER (1977)
	Baltic Sea, 1975	0·5–5·2	OSTERROHT (1977)
	NE Pacific Ocean, 1974 (Southern California coast)	<0·1–1·2	SCURA and McCLURE (1975)
	Hawaii, 1971	0·3[e]	BEVENUE and co-authors (1972)
pp'-DDD	North Sea, 1975 (German Bight)	0·57[c]	STADLER (1977)
	Western Baltic Sea, 1975	0·24[c] (0·12)	STADLER (1977)
	Hawaii, 1971	3·6[e]	BEVENUE and co-authors (1972)
Dieldrin	North Sea, 1974 (German Bight)	0·1–0·6	STADLER and ZIEBARTH (1975)
	North Sea, 1975 (German Bight)	(0·18)	STADLER (1977)
	Western Baltic Sea, 1975	0·35[c] (0·13)	STADLER (1977)
	Hawaii, 1971	1·0[e]	BEVENUE and co-authors (1972)
γ-Hexachloro-cyclohexane (γ-HCH)	North Sea, 1975 (German Bight)	1·9–16·5 (5·6)	STADLER (1977)
	Western Baltic Sea, 1975	3·4–7·0 (5·0)	STADLER (1977)
	Hawaii, 1971	0·9[e]	BEVENUE and co-authors (1972)

[a] n.r., Year not reported; [b] average values in parentheses;
[c] maximum value; [d] data expressed as decachlorobiphenyl; [e] single value.

Table 6-11

Concentrations of chlorophenols in coastal waters (Compiled from the sources indicated)

Compound	Location, year	Concentration (ng l^{-1})	Source
2,6-Dichlorophenol	Weser estuary, 1977	0·3–1·8	WEBER and ERNST (1978)
2,4,6-Trichlorophenol	Weser estuary, 1977	0·4–10·4	WEBER and ERNST (1978)
2,3,4,5-Tetrachlorophenol	Weser estuary, 1977	0·5–2·9	WEBER and ERNST (1978)
2,3,4,6- and/or 2,3,5,6-	Weser estuary, 1977	1·2–86·4	WEBER and ERNST (1978)
Tetrachlorophenol	German Bight, 1977	0·08–0·14	WEBER and ERNST (1978)
Pentachlorophenol	Weser estuary, 1977	49–496	ERNST and WEBER (1978)
Pentachlorophenol	German Bight, 1977	<2–8	ERNST and WEBER (1978)
Pentachlorophenol	San Luis Pass West Galveston Bay, 1980	4·3–11	MURRAY and co-authors (1981)

CH$_3$Cl, CH$_3$Br, CCl$_4$ (LOVELOCK and co-authors, 1973; LOVELOCK, 1975). Methyl iodide has been found in all ocean waters examined and marine algae are considered to be a primary source; its annual production has been estimated at 40 megatons (LOVELOCK and co-authors, 1973). Concentrations of methyl iodide in waters over beds of the kelp *Laminaria digitata* were found to be 1000 times higher than in open oceans (LOVELOCK, 1975). It was concluded that also oceanic CH$_3$Cl and CCl$_4$ are not very likely to be derived from direct industrial emissions, and annual production rates were estimated at 1·9 megatons for CCl$_4$ and 28 megatons for CH$_3$Cl. Small amounts of these compounds were found to be produced by the marine plant *Asparagopsis armate* (MCCONNELL and FENICAL, 1977). A great number of halogen compounds seem to be formed biosynthetically (SIUDA and DEBERNARDIS, 1973; WHITE and HAGER, 1975); the occurrence of haloforms in edible sea weeds has been reported by MOORE (1978).

Table 6-12

Concentrations of phthalic acid esters in sea water (Compiled from the sources indicated)

Compound	Location, year[a]	Concentration (ng l^{-1})	Source
Di-(2-ethylhexyl) phthalate (DEHP)	North Atlantic Ocean, n.r.	0·1–6·3	GIAM and co-authors (1978b)
	Gulf of Mexico, 1973/74 (open gulf)	6–39	GIAM and co-authors (1976)
	coastal waters)	165–500	GIAM and co-authors (1976)
Diethyl phthalate (DEP)	Baltic Sea, 1978	1·0–2·1	EHRHARDT and DERENBACH (1980)
Dibutyl phthalate (DBP)	Kiel Bight, 1978	59–203	EHRHARDT and DERENBACH (1980)
	Gulf of Mexico, 1973/74	87	GIAM and co-authors (1976)

[a] n.r., Year not reported.

Table 6-13

Concentrations of aliphatic chlorinated hydrocarbons in sea water (Compiled from the sources indicated)

Compound	Location, year	Concentration (ng l^{-1})[a]	Source
Chloroform	N.E. Atlantic Ocean, 1972	4–13	MURRAY and RILEY (1973b)
	Liverpool Bay, 1972/73	1000[b]	PEARSON and McCONNELL (1975)
	Back River Estuary, 1977 (Maryland, USA)	2124[c]	HELZ and HSU (1978)
Trichloroethylene	N.E. Atlantic Ocean, 1972	5–11	MURRAY and RILEY (1973b)
	Liverpool Bay, 1972/73	300; 3600[b]	PEARSON and McCONNELL (1975)
Tetrachloroethylene	N.E. Atlantic Ocean, 1972	0·2–0·8	MURRAY and RILEY (1973b)
	Back River estuary, 1977 (Maryland, USA)	820[c]	HELZ and HSU (1978)
Perchloroethylene	Liverpool Bay, 1972/73	120; 2600[b]	PEARSON and McCONNELL (1975)
1.1.1-Trichloroethane + carbon tetrachloride	Liverpool Bay, 1972/73	250; 3300[b]	PEARSON and McCONNELL (1975)
Carbon tetrachloride	N.E. Atlantic Ocean, 1972	0·12–0·26	MURRAY and RILEY (1973b)
Hexachlorobutadiene	Liverpool Bay, 1972/73	4; 30[b]	PEARSON and McCONNELL (1975)
α-Hexachlorocyclohexane (α-HCH)	North Sea, 1975 (German Bight)	2·0–20 (6·5)	STADLER (1977)
	Western Baltic Sea, 1975	4·7–8·7 (6·8)	STADLER (1977)

[a] As the concentration unit, ng l^{-1} was used without correction when μg kg^{-1} sea water were reported.
[b] Maximum value.
[c] Calculated from HELZ and HSU (1978) at 2·92 g kg^{-1} salinity.

Surface slicks

In slicks of the Sargasso Sea, 3·8 to 19·3 ng l^{-1} of PCBs (Aroclor 1260) and 0·2 to 0·7 ng l^{-1} of p,p'-DDT were found by BIDLEMAN and OLNEY (1974). Similar concentrations occurred in slicks of the California Current: 4·9 ng l^{-1} of PCBs, 0·3 ng l^{-1} of p,p'-DDT, 0·1 ng l^{-1} of DDE, and in the North Central Pacific Gyre: 3·3 and 3·5 ng l^{-1} of PCBs but less than 0·02 and 0·01 ng l^{-1} of p,p'-DDT and p,p'-DDE, respectively (WILLIAMS and ROBERTSON, 1975). Concentrations of organochlorine compounds were higher in slicks sampled in near-shore regions than those of the open sea: 450 to 4200 ng l^{-1} of PCBs in Narragansett Bay (DUCE and co-authors, 1972), 11 to 90 ng l^{-1} of PCBs, 2·1 to 15 ng l^{-1} of p,p'-DDT, 0·1 to 1·8 ng l^{-1} of p,p'-DDE in California waters (WILLIAMS and ROBERTSON, 1975). Slicks of the western Baltic Sea contained 30 to 500 ng l^{-1} of PCBs, 1·2 to 4 ng l^{-1} of p,p'-DDT and 0·3 to 0·5 ng l^{-1} of dieldrin (STADLER and ZIEBARTH, 1976). In the German Bight, STADLER (1977) determined the following mean values in the uppermost 1 mm layer: 36 ng l^{-1} of PCBs, 1·5 ng l^{-1} of p,p'-DDT,

0·56 ng l^{-1} of p,p'-DDE, 1·24 ng l^{-1} of p,p'-DDD, 0·70 ng l^{-1} of dieldrin, 14 ng l^{-1} of α-HCH and 16 ng l^{-1} of γ-HCH. Hexachlorobenzene concentrations ranged up to 10 ng l^{-1} in Mediterranean coastal waters (PUCCETTI and LEONI, 1980).

Organisms

A survey of the vast literature on the regional and species-specific distribution of organic pollutants reveals that the majority of them are chlorinated compounds; phthalate esters at present appear to be the only group of substances without chlorine under the heading of this chapter. In Tables 6-14 to 6-18 a selection of regional and substantial patterns of pollutant distributions is presented which can be ascribed to polychlorinated biphenyls and DDT compounds (Table 6-14), cyclodienes (Tables 6-15 and 6-16), hexachloro-cyclohexanes (Table 6-15), chlorinated benzenes (Table 6-17), chlorinated aliphatic hydrocarbons (Table 6-18) and miscellaneous compounds, including phthalate esters (Table 6-16). The ranking of regional levels gives an indication of the relative severity of pollution; coastal areas of heavily industrialized countries with high population densities are most affected. The Baltic Sea, North Sea and Mediterranean Sea are typical examples and contrast sharply with the Atlantic Ocean.

Examples of high local pollution loads in organisms are shown in Tables 6-17 and 6-19. Sprat caught in a polluted Norwegian fjord exhibited decreasing levels of chlorinated benzenes and styrenes with increasing distance from the source of pollution (Table 6-17). The discharge of large quantities of DDT compounds into the marine environment off Palos Verdes Peninsula in southern California (USA) has caused long-lasting pollution in a defined area. This was established by monitoring fishes (Table 6-19). Three species with different feeding habits were analysed from 1970 to 1977: Dover sole, *Microstomus pacificus*, feeding on sedimentary invertebrates; black perch, *Embiotoca jacksoni*, feeding on small organisms attached to rocks and kelp bass, *Paralabrax clathratus*, which feeds on fish; and invertebrates in the water column (SMOKLER and co-authors, 1979). Table 6-19 shows that residues of kelp bass and black perch significantly decreased over the study interval, while those of Dover sole did not. From these findings it can be concluded that the high DDT load (Fig. 6-16) of the sediments is the source of the persistent contamination.

The ubiquitous distribution of PCBs and DDT compounds and their comparatively high levels in fatty tissues, such as liver and seal blubber, support experimental data on lipophilic behaviour and persistence. More recently, some more persistent compounds could be detected, such as those of the chlordane group and the toxaphene group. These chemicals were found in fish (MIYAZAKI and co-authors, 1980) and seals (JANSSON and co-authors, 1979) and even in remote areas they could be detected in fish liver (BALLSCHMITER and ZELL, 1980b); in general they appear to be similarly distributed as PCBs and DDT compounds. As with other types of organic chemicals, surprisingly high levels of a herbicide (CNP, Table 6-16) were recorded by means of GC–MS techniques. Only low levels of pentachlorophenol, phthalate esters and aliphatic chlorinated hydrocarbons turned out to be present in marine biota, although the latter two classes of compounds are produced at a rate of 10^6 tons yr^{-1}. Pentachlorophenol is readily metabolized by conjugation reactions and phthalate esters undergo degradation to phthalic acid and other products in aquatic animals. Degradation of aliphatic organohalogens are not reported to occur in marine organisms. On the basis of their concentration in water (Table 6-13) and their levels in different organisms (PEARSON and MCCONNELL, 1975), a very low bioconcentration factor (BCF) can be expected for these types of compounds. These findings have been strongly supported by DICKSON and RILEY (1976), who reported a BCF of 2 to 25 for marine organisms. Methyl iodide was most strongly accumulated (240 times) in fish brain.

Significant relationships exist between size/weight and lipid and residue levels in marine and freshwater fishes for PCBs and DDT (ANDERSON and EVERHART, 1966; ANDERSON and FENDERSON, 1970; REINERT, 1970; BACHE and co-authors, 1972; STENERSEN and KVALVAG, 1972; YOUNGS and co-authors, 1972; ERNST and co-authors, 1976; SCHAEFER and co-authors, 1976; STOUT, 1980). Such a relationship has been reported by LUCKAS and co-authors (1978) to exist between age/size and residue levels also in the liver of cod from the Baltic Sea; and relationships between pollutant levels in wet cod liver, C (mg kg^{-1}), and the size of the fish, L (cm), have been reported by SCHNEIDER and OSTERROHT (1977) to be as follows:

$$\log C_{\text{PCB}} = -2 \cdot 52 + 2 \cdot 02 \log L \pm 0 \cdot 23$$
$$\log C_{\Sigma\text{DDT}} = -3 \cdot 81 + 2 \cdot 64 \log L \pm 0 \cdot 29$$

In some cases residues in edible marine food were so high that the sale of these products has been forbidden (DYBERN and JENSEN, 1978), e.g. cod livers from the Kattegat area with high PCB levels.

Sediments

Pollutants dissolved in sea water can be accumulated in bottom sediments by sorption processes describable by the Freundlich isotherm equation. In general, the sorption potential increases with an increase in hydrophobic character of the chemical; for neutral organic compounds the organic matter or carbon content of soils is regarded to be the principal adsorbent (KENAGA, 1975). However, organic compounds may also be sorbed independent of the carbon content by various mineral fractions of bottom sediments, such as clays. The significance of sorbed pollutants has been stressed in connection with bottom-living animals; sorbed chemicals can act as a reservoir and can be released again to the water column, thus maintaining pollutant levels over long periods of time. With DDT and PCBs, an accumulation factor of 10^3 to 10^4 is very likely to occur in coastal sediments. The occurrence of various compounds in sediments of different locations is shown in Table 6-20. Areas of disposal of industrial waste or sewage sludge (Figs. 6-16 and 6-17) can accumulate persistent chemicals, such as DDT and PCB compounds, to high levels; these can be maintained over long periods of time and are reflected in the bottom fauna such as fishes (MACGREGOR, 1974; SMOKLER and co-authors, 1979) and molluscs (HALCROW and co-authors, 1974).

(b) Conclusions

The occurrence and distribution of a number of chlorinated organic pollutants on a global scale emphasizes the effectiveness of distributional mechanisms. High pollutant persistence is a key prerequisite for such behaviour. High levels of pollutant residues—due to local, coastal contaminations of the water—decrease towards the open sea, either as a consequence of degradation or of their escape from analytical recording.

Table 6-14

Levels of PCBs and ΣDDT (ng g^{-1} wet weight) in marine organisms from different locations (Compiled from the sources indicated)

Organism	ΣDDT	Polychlorinated biphenyls (PCBs)	Locality	Year of sampling	Source[b]
Plankton	21–101 (51)	80–2200 (500)	Firth of Clyde, near-shore	1971/72	1
	2–50 (9·5)	10–920 (99)	Firth of Clyde, off-shore	1971/72	1
		90–3050 (1723)	Gulf of St. Lawrence	1972	2
		40–700 (178)	Baltic Sea, Turku Archipelago	1974	4
Zooplankton		7–450	North Atlantic	1970/71	3
Kelp					
Sargassum	20–120 (60)	10–20	North Atlantic, open sea	1970	3
Molluscs					
Mytilus edulis		<100–600	Scotland, east coast (estuary)	1970	5
Mytilus edulis		10–310	North Sea	1976	6
Mytilus edulis	24	117	Baltic Sea	1974/75	7
Mytilus edulis		110–1920 (565)	Mediterranean, coast of France	1972	8
Mytilus galloprovincialis	27–43	93–233 (172)	Mediterranean, Ligurian Sea	1977/78	9
Mytilus galloprovincialis	30–35	61–100 (78)	Mediterranean, Sicily	1976/77	10
Cerastoderma edule	<1[a]	11	North Sea, Weser estuary	1976	11
Mya arenaria	5·9[a]	44	North Sea, Weser estuary	1976	11
Loligo forbesi	12–59 (27)	83–180 (129)	English Channel	1971	18
Crustaceans					
Pandalus borealis	<1	8–20	North Atlantic, West Greenland	1975	12
Pandalus borealis (abdominal muscle)	1–3	12–24	North Sea, Norwegian Trough	1972	13
Pandalus borealis	7	18	North Atlantic, Iceland	1971	14
Nephrops norvegicus	1·2–4·5	4·4–55 (16)	Mediterranean, Sicily	1976/77	10
Nephrops norvegicus	1·7–10	21–157 (73)	Mediterranean, Ligurian Sea	1977/78	9
Squilla empusa	9	10	Gulf of Mexico	1975	15
Pandalus jordani	1–5	11–69 (25)	Pacific, Oregon coast	1971/72	16
Euphausia sp.	33	3	Antarctic	1975	17

Fishes

Species (tissue)			Location	Year	Ref.
Gadus morhua (muscle)	6–11 (8·5)	33–66 (51)	North Sea	1972	19
(liver)	320–1800 (922)	310–8100 (3444)	North Sea	1975	19
(muscle)	<3–13	10–82	North Sea		6
(liver)	220–1900 (303–942)	1200–32 000 (1650–10 600)	North Sea		6
(muscle)	8·5	43–57	North Sea, Dogger Bank	1975	20
(muscle)	32–120	33–96	Baltic Sea	1969/71	21
(liver)	4400–13 000	1400–4900	Baltic Sea	1969/71	21
(muscle)	21–62	43–99	Baltic Sea	1974/75	7
(liver)	220–27 000 (2080–18 665)	1400–12 000 (4600–10 100)	Baltic Sea	1975/75	7
(muscle)		20	North Atlantic, Nova Scotia	1970	22
(muscle)		550	North Atlantic, Bay of Fundy		23
(muscle)	11	38	North Atlantic, Georges Bank	1971	14
(liver)	2700	22 000	North Atlantic, Georges Bank	1971	14
(muscle)	170	2	North Atlantic, Iceland	1971	14
(liver)		730	North Atlantic, Iceland	1971	14
(muscle)	<3–13	19–71	North Atlantic, Iceland	1975	12
Clupea harengus (muscle)	80–1460 (313–410)	170–3000 (457–890)	North Sea	1972	19
(liver)	10–55 (25)	29–110 (67)	North Sea	1972	19
(muscle)	44–78 (60)	130–330 (185)	Baltic Sea	1969/71	21
(muscle)	110–7700 (854–1735)	44–3400 (440–688)	Baltic Sea	1974/75	7
(muscle)	180–1400 (43–676)	210–940 (450–550)	North Atlantic		12
(muscle)	<5–29	10–80	North Sea	1975	6
(muscle)	6–170	40–130	North Sea	1976	6
(liver)	11–960 (27–361)	12–300	North Sea	1976	16
(whole)	25	146	Pacific, Oregon coast	1971	
Thunnus thynnus (muscle)	6·6–51 (23)	9–44 (22)	Mediterranean, Sicily	1976/77	10
(muscle)	40; 221	95; 407	Mediterranean		20
(liver)	1091; 4172	1526; 8598	Mediterranean		20

Table 6-14—continued

Organism	ΣDDT	Polychlorinated biphenyls (PCBs)	Locality	Year of sampling	Source[b]
Thunnus alalunga (muscle)	28	40	North Atlantic	1970/71	20
Coryphaena equiselis (liver)	95	1100	North Atlantic		14
Coryphaena hippurus (muscle)	3	10	North Atlantic		
Mammals					
Hydrurga leptonyx (blubber)	81	43	Antarctic	1975	17
Phoca hispida (muscle)	5–27	4–42	Arctic, Canada coast	1972	24
(liver)	10–163 (46)	3–113	Arctic, Canada coast		
(blubber)	536–2855 (1198)	50–1930	Arctic, Canada coast		
Pusa hispida (blubber)	170–2450 (610–1010)	1000–6000	Arctic		25
Phoca vitulina (liver)	60–250 (140)	60–2700 (963)	North Sea	1974–1976	26
(blubber)	2200–27 200 (4600–10 300)	27 300–564 000 (71 000–167 000)			
Phoca hispida (blubber)		600–1300 (900)	Greenland, west coast		27
Erignatus barbatus (blubber)		600–3000 (1800)	Greenland, west coast		27
Cystophora cristata (blubber)		300–4900 (2700)	Greenland, west coast		27

* Average values in parenthesis.

[a] DDD and DDE only.

[b] 1, WILLIAMS and HOLDEN (1973); 2, WARE and ADDISON (1973); 3, HARVEY and co-authors (1972); 4, LINKO and co-authors (1979); 5, HOLDEN and TOPPING (1971/72); 6, ICES (1977a); 7, ICES (1977b); 8, DELAPPE and co-authors (1973); 9, CONTARDI and co-authors (1979); 10, AMICO (1979); 11, GOERKE and co-authors (1979); 12, ICES (1977c); 13, EDER and co-authors (1976); 14, HARVEY and co-authors (1974a); 15, GIAM and co-authors (1978a); 16, CLAEYS and co-authors (1975); 17, RISEBROUGH and co-authors (1976); 18, ERNST and co-authors (1976); 19, SCHAEFER and co-authors (1976); 20, ALZIEU (1976); 21, JENSEN and co-authors (1972a); 22, ZITKO (1971); 23, ZITKO and co-authors (1972); 24, BOWES and JONKEL (1975); 25, ADDISON and SMITH (1974); 26, DRESCHER and co-authors (1977); 27, CLAUSEN and co-authors (1974); 28, YAMAGASHI and co-authors (1978); 29, ERNST and WEBER (1978); 30, ERNST and co-authors (1980); 31, MIYAZAKI and co-authors (1982); 32, JANSSON and co-authors (1979); 33, ZITKO and co-authors (1974); 34, BALLSCHMITER and ZELL (1980b); 35, MURRAY and co-authors (1981); 36, WICKSTROM and co-authors (1981); 37, ANDERSSON and BLOMKVIST (1981).

Table 6-15

Levels of dieldrin, γ-HCH and α-HCH (ng g⁻¹ wet weight) in marine organisms of different locations (Compiled from the sources indicated)

Species	Dieldrin	γ-HCH	α-HCH	Locality	Year of sampling	Source[a]
Molluscs						
Cerastoderma edule	1·5	2·4	0·9	North Sea, Weser estuary	1976	11
Mya arenaria	4·5	2·8	3·1	North Sea, Weser estuary	1976	11
Mytilus galloprovincialis	0·2–1·3	52		Mediterranean, Sicily	1976/77	10
Mytilus edulis				Baltic Sea	1974/75	7
Mytilus edulis	1–13	1–12	1–5	North Sea	1976	6
Crustaceans						
Crangon crangon	0·5	1·8	2·3	North Sea, Weser estuary	1976	11
Nephrops norvegicus	0·1–0·4			Mediterranean, Sicily	1976/77	10
Pandalus borealis	1–2	≤1	<1–2	West Greenland	1975	12
Fishes						
Gadus morhua (muscle)	1–3	<1	<1	North Sea	1976	6
(liver)	40–680	6–50 (20)	18–46 (29)			
(muscle)	<1	<1		Baltic Sea	1974/75	7
(liver)	77–420					
(muscle)	<1–4	<1–10		North Atlantic, Iceland	1975	12
(liver)	8–62	5–75				
Clupea harengus (muscle)	5–7	<1	<1	North Sea	1976	6
(liver)	8–120	≤1	<1–2	Baltic Sea	1974/75	7
(muscle)	6–37	2–10		Atlantic	1975	12
(muscle)	14–15	4				
Mammals						
Erignatus barbatus (blubber)		7–640 (53)		Greenland, west coast	—	27
Phoca hispida (blubber)		2–25 (6)		Greenland, west coast	—	27
Phoca vitulina (blubber)	40–900 (140–540)	40–980 (270–360)		North Sea	1974–1976	26
(liver)	10–24 (17)	5–6				

For references, see footnote to Table 6-14.

Table 6-16

Concentrations of various compounds in marine animals (Compiled from the sources indicated)

Substance	Concentration (ng g^{-1})	Organism	Location, year[a]	Source[b]
Hexachlorobenzene (HCB)	1–3	Herring *Clupea harengus* (muscle)	North Sea, 1972	19
	5–7	Herring *Clupea harengus* (liver)	North Sea, 1972	19
	0·5–1	Cod *Gadus morhua* (muscle)	North Sea, 1972	19
	3–130	Cod *Gadus morhua* (liver)	North Sea, 1972	19
	8–18	Herring *Clupea harengus* (muscle)	Baltic Sea, 1974/75	7
	0·49	Flounder (species unknown) (whole)	San Luis Pass, West Galveston Bay, 1980	35
	0·65	Longnose killifish *Fundulus similis* (whole)	San Luis Pass, West Galveston Bay, 1980	35
	0·88	Brown shrimp *Penaeus aztecus* (whole)	San Luis Pass, West Galveston Bay, 1980	35
	7·5	Fish *Dissostichus el+ginoides* (liver)	South Georgia shelf, Antarctica (n.r.)	34
Pentachlorophenol (PCP)	72–766	Polychaete *Lanice conchilega* (whole)	Weser estuary, 1976/77	29
	3–7	Actinian *Sargatia troglodytes* (whole)	Weser estuary, 1976/77	29
	5·9	Common mussel *Mytilus edulis*	Weser estuary, 1977	30
	7·5	Brown shrimp *Penaeus aztecus* (whole)	San Luis Pass, West Galveston Bay, 1980	35
	2·6	Flounder (species unknown) (whole)	San Luis Pass, West Galveston Bay, 1980	35
	5·3	Longnose killifish *Fundulus similis* (whole)	San Luis Pass, West Galveston Bay, 1980	35

Chlordane	20–50	Cod *Gadus morhua* (liver)	Baltic Sea (n.r.)	36
Oxychlordane	3	Goby-fish *Acanthogobius flavimanus* (whole)	Tokyo Bay, 1978	31
Chlordane	6–9	Goby fish *Acanthogobius flavimanus* (whole)	Tokyo Bay, 1978	31
Nonachlor	8–18	Goby-fish *Acanthogobius flavimanus* (whole)	Tokyo Bay, 1978	31
Sum of oxychlordane, chlordane and nonachlor	600	Herring *Clupea harengus*	Baltic Sea, 1978	32
	10 000	Grey seal *Halichoerus gryphus* (found dead) (blubber)	Baltic Sea, 1974/77	32
Toxaphene (chlorinated terpenes)	11 000	Grey seal *Halichoerus gryphus* (found dead) (blubber)	Baltic Sea, 1974/77	32
	13 000	Herring *Clupea harengus*	Baltic Sea, 1978	32
	68	Fish *Dissostichus eleginoides* (liver)	South Georgia shelf, Antarctica (n.r.)	34
1,3,5-Trichloro-2-(4-nitrophenoxy)benzene (CNP)	1040	Common mussel *Mytilus edulis*	Tokyo Bay, 1977	28
Polybrominated biphenyl ethers (PBBE)	900–1200	Eel *Anguilla anguilla* (muscle)	Klosterfjorden Bay, Sweden, 1979/80	37
Di-(2-ethylhexyl) phthalate (DEHP)	135	Starfish *Luidia clathrata* (whole)	Gulf of Mexico, Mississippi delta (n.r.)	15
	3	Blue crab *Callinectes sapidus rathpun* (muscle)	Gulf of Mexico, Mississippi delta (n.r.)	15
	9	Sand trout *Cynoscion nothus* (muscle)	Gulf of Mexico, Mississippi delta (n.r.)	15
	8	Shrimp *Squilla empusa* (whole)	Gulf of Mexico (open Gulf) (n.r.)	15
	<1–8	Catfish *Galeichthys felis* (muscle)	Gulf of Mexico (open Gulf) (n.r.)	15

[a] n.r., Year not reported.
[b] For references see footnote to Table 6-14.

Table 6-17

Chlorinated hydrocarbons (μg g^{-1} on fat weight basis) in sprat *Clupea sprattus* and cod *Gadus morhua* from contaminated areas of the Norwegian coast, 1974–1975 (Based on LUNDE and BAUMANN OFSTAD, 1976; BAUMANN OFSTAD and co-authors, 1978)

Chlorinated hydrocarbon	*Clupea sprattus* (oil)	*Gadus morhua*	
		Muscle	Liver
Hexachlorobenzene	0·04–16·0	8·3–141	79–208
Pentachlorobenzene	0·01–3·7	1·9–5·5	5·8–24
Tetrachlorobenzene	<0·01–0·4	0·2–0·4	0·4–1·4
Trichlorobenzene	<0·01–0·5	0·1–0·9	1·1–4·0
Heptachlorostyrene	<0·02–4·0	6·1–92	44–121
Octachlorostyrene	<0·1–11·2	10–361	223–675
PCBs	0·7–3·8	9–70	40–66

Table 6-18

Chlorinated aliphatic hydrocarbons (ng g^{-1} wet weight) in marine organisms of British estuaries (After PEARSON and McCONNELL, 1975; reproduced by permission of the Royal Society)

Substance	Algae	Invertebrates	Fishes		Mammals[a]	
			Muscle	Liver	Blubber	Liver
Trichlorethylene	16–23	0·05–16	0·8–11	2–56	2·5–7·2	3–6·2
Tetrachlorethylene	13–23	0·05–15	<0·1–11	1–41	0·6–19	0–3·2
Chloroform	n.r.[b]	0·02–180	5–50	6–18	7·6–22	0–12
Hexachlorobutadiene	0·6–8·9	0·06–7	0·03–2·6	0·2–2	0·4–3·6	0–0·8

[a] Farne Islands.
[b] n.r., not reported.

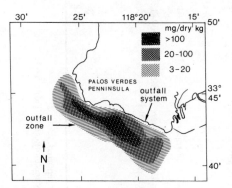

Fig. 6-16: Total DDT concentration in marine surface sediments off Palos Verdes Peninsula, California, USA; 1973. (After SMOKLER and co-authors, 1979; reproduced by permission of Pergamon Press.)

Table 6-19

Concentration of total DDT (mg kg⁻¹ wet weight) in muscle of fishes, collected in the discharge zone (see Fig. 6-16) (After SMOKLER and co-authors, 1979; reproduced by permission of Pergamon Press)

Dover sole

	1970 Fall	1971 Spring	1971 Fall	1972 Spring	1972 Fall	1973 Spring	1973 Fall	1974 Spring	1974 Fall	1975 Spring	1975 Fall
Median		4·7	5·6	17	8·1	6·7	25	13	12	14	
Range		1–75	4–15	3–31	3–50	5–13	8–39	0·2–45	0·3–31	0·1–98	
% > 5 mg kg⁻¹		50	80	86	80	67	100	85	82	83	
No. of samples	0	18	5	7	5	3	6	20	17	23	0

Black perch

	1970 Fall	1971 Spring	1971 Fall	1972 Spring	1972 Fall	1973 Spring	1973 Fall	1974 Spring	1974 Fall	1975 Spring	1975 Fall	1976 Spring	1976 Fall	1977 Spring
Median	8·0	18	9·1	9·3	25	24	4·5	6·8	5·7	6·3	2·9	4·0	3·3	3·2
Range	4–22	6–29	4–49	0·6–65	2–77	2–87	0·9–39	2–20	2–28	0·8–11	1–13	1–10	2–7	1–9
% > 5 mg kg⁻¹	75	100	71	73	85	92	46	59	63	55	44	23	18	20
No. of samples	4	2	7	44	7	24	24	22	19	20	9	13	11	25

Kelp bass

	1970 Fall	1971 Spring	1971 Fall	1972 Spring	1972 Fall	1973 Spring	1973 Fall	1974 Spring	1974 Fall	1975 Spring	1975 Fall	1976 Spring	1976 Fall	1977 Spring
Median	3·3	2·6	1·8	4·6	12	8·3	2·6	3·2	2·0	3·8	7·7	0·26	0·84	3·4
Range	3–8	0·8–47	0·8–5	0·8–13	5–47	3–22	2–8	0·2–8	0·2–15	0·2–19	6–9		0·8–0·9	0·7–67
% > 5 mg kg⁻¹	33	33	0	40	100	73	33	7	33	45	100	0	0	40
No. of samples	3	3	7	15	5	11	9	15	9	11	2	1	2	10

Table 6-20

Organic pollutants in marine sediments of different origin (Compiled from the sources indicated)

Pollutant	Concentration (ng g^{-1}, dry weight)	Location, year	Source
PCBs	100–2000	Japan, Osaka Bay, 1974	HIRAIZUMI and co-authors (1975)
PCBs	110	Mediterranean, off Monaco	FOWLER and co-authors (1978)
PCBs (Aroclor 1016 and 1242)	3–2035	Raritan Bay—lower New York Bay, 1977	STAINKEN and ROLLWAGEN (1979)
PCBs (Aroclor 1254)	10–2890	UK, Firth-of-Clyde	HALCROW and co-authors (1974)
PCBs (Aroclor 1260)	71–2322	Severn estuary, 1978	COOKE and co-authors (1979)
Biphenyl	8·8–144		
PCBs	1·3–775	Greece, Saronikos Gulf	DEXTER and PAVLOU (1973)
ΣDDT	7·1–1893		
PCBs	14–28	Central North Sea, Norwegian Trough, 1972/73	EDER (1976)
ΣDDT	0·16–0·36		
PCBs	13·4–16·7	South Adriatic coast	VILILIC and co-authors (1979)
ΣDDT	1·3		
PCBs (Aroclor 1254)	30–1000	N.W. Mediterranean Continental shelf between Sete and Fos/Mer, 1972	MESTRES and co-authors (1975)
DDT $(o,p'- + p,p'-)$	20–500		
DDE	20–30		
Dieldrin	80–380		
Heptachlor	4–500		
PCBs (Clophen A 60)	134–212	Western Baltic Sea. Eckernförde Bight, 1978	MÜLLER and co-authors (1980)
ΣDDT	28–46		
γ-HCH	0·33–0·96		
Di-(2-ethylhexyl) phthalate	75–159		
PCBs (Aroclor 1260)	0·25–0·78	San Luis Pass, West Galveston Bay, 1980	MURRAY and co-authors (1981)
Hexachlorobenzene	0·05–1·5		
Di-(2-ethylhexyl) phthalate	13–170		
Diethyl phthalate	<2–9		
Dibutyl phthalate	<0·15–0·93		
Pentachlorophenol	0·18–0·26		
Dichlorophenols	0·06–2·09[a]	Weser estuary, 1977/78	EDER and WEBER (1980)
Trichlorophenols	0·02–3·0[a]		
Tetrachlorophenols	0·02–4·7[a]		
Pentachlorophenol	0·1–20·4[a]		

[a] Wet weight.

Typically, the concentrations of pollutants in water are reflected by organisms and sediments. Pollution of sediments by waste disposal may result in long-lasting contamination of the afflicted areas. The continuing detection of 'new' pollutants, accomplished by more sophisticated analytical methods, shows clearly that our present inventory of pollutants is incomplete. There is need to enlarge our analytical capabilities.

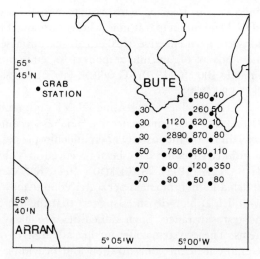

Fig. 6-17: PCB distribution pattern (ng g^{-1} dry weight) in sediments of a dumping area for sewage sludge. (After HALCROW and co-authors, 1974; reproduced by permission of Pergamon Press.)

(6) Effects on Living Systems

(a) Acute Toxicity

Pesticides conceived specifically for fighting specific pests may be detrimental to other forms of life, and aquatic life is especially sensitive to most pesticides and to various other organic chemicals. It has, therefore, become common practice to test newly produced organic chemicals for their potential toxicity to fishes as test organisms.

The first step in toxicity testing is the estimation of acute toxicity, i.e. the estimation of the LC_{50}, the concentration of a pollutant in water that is lethal to 50% of the exposed test organisms within 24, 48 or 96 h. In these tests only the toxicant concentration in water is measured, contrary to toxicity testing in mammals, for example, where a known dose is administered. However, toxic effects are related to the concentration of the pollutant associated with specified tissues and in the case of substances with a high bioconcentration potential, LC_{50} values tend to decrease with increasing exposure time. A minimum period of 20 d, covering continuous exposure for 10 d and a subsequent period of 10 d for observation, has, therefore, been proposed by EISLER (1970a) for toxicity tests with the fish *Fundulus heteroclitus*.

Factors Affecting Toxicity

The effect of temperature on the toxicity of numerous chemicals to fishes is well documented (HOLDEN, 1973a). In general, toxicity increases with increasing temperature, although in some cases the opposite might be the case. The 24-h LC_{50} for endrin, e.g. in rainbow trout, is 15, 5·3 and 2·8 μg l^{-1} at 1·6, 7·2 and 12·7° C, respectively (MACEK, 1969, in HOLDEN, 1973a). Examining an estuarine fish, EISLER (1970a) reported a sharp increase in toxicity with rising temperature when testing the organophosphates DDVP and methylparathion, whereas organochlorines such as DDT, endrin and heptachlor were most effective at 20° to 25 °C. For assessing thermal effects on toxicity levels, the study of organismic responses to temperature is essential. For details consult Volume I: BRETT (1970), GARSIDE (1970), GESSNER (1970), KINNE (1970) and OPPENHEIMER (1970).

While increasing salinities produced increasing DDVP and methylparathion toxicity in estuarine fishes (EISLER, 1970a), no general trend has become apparent yet with regard to the effects of salinity on the toxicity of organochlorine pesticides. Responses of marine organisms to salinity variation have been reviewed in Volume I: GESSNER and SCHRAMM (1971), HOLLIDAY (1971), MACLEOD (1971), KINNE (1971).

The size of the test fish and their biomass per water volume may have an influence on LC_{50} values. Tests with fish at high densities suggest that the fish could remove significant amounts of the test substance. Such difficulties are likely to be avoidable in continuous-flow systems. These systems offer further advantages over static systems in producing a more contant toxicant concentration and providing flushing of excreted material during the test. Because of physiological variations between the test species used and variations in the experimental lay-out of bioassays in different laboratories, test results are not always comparable (HOLDEN, 1973a).

As is apparent from Table 6-21, the LC_{50} values recorded are far above the pollutant concentrations likely to occur in the marine environment, except in cases of accidental spills or dumping in restricted areas. Sublethal long-term effects at low ambient pollutant concentrations are ecologically more meaningful. Nevertheless, while LC_{50} data have only limited value in predicting the detrimental potential of environmental chemicals, they are useful in establishing a first, rough relative index of pollutant toxicity.

(b) Long-term Toxicity

In toxicity studies under conditions resembling those prevailing in the marine environment—long exposure times at low substance concentrations—it has become apparent that toxic effects might occur at concentrations of a toxicant which are well below the LC_{50} value. A time-dependent mortality occurs for dieldrin in fish, e.g. Fig. 6-18 displays significant mortalities after 32 wk at subacute concentrations, suggesting that a critical concentration of the toxicant in the tissues has accumulated after prolonged exposure times. Similarly, time-dependent toxic effects have been reported by other authors, e.g. for 1,1,2-trichloroethane, dieldrin, pentachlorophenol and 3,4-dichloroaniline in worms, molluscs, crustaceans and fish, partly including their developmental states (ADEMA and VINK, 1981), and for Aroclor 1016 and chlorinated naphthalenes (Halowax 1000 and Halowax 1099) on the larval development of the mud crab (LAUGHLIN and co-authors, 1977).

These observations, among others, demonstrate that chronic toxicity testing deserve special attention in predicting long-term effects of pollutants. The goal of chronic toxicity testing is to estimate the highest toxicant concentrations tolerable without apparent effects over long exposure periods, using the most sensitive criteria available at critical life stages, such as mortality, growth, hatchability, egg fragility, spawning and structure deformations. In order to establish 'safe concentrations' for aquatic life, 'safety factors' of 0·1 to 0·01 had been used in connection with acute toxicity data. However, an

Table 6-21

Acute toxicity levels of organic compounds in marine organisms (Compiled from the sources indicated)

Substance	Species	LC$_{50}$ (mg l^{-1})			Conditions[a]	Source[b]
		24 h	48 h	96 h		
Aroclor 1016	Brown shrimp *Penaeus aztecus*			0·0105c	F	1
	Grass shrimp *Palaemonetes pugio*			0·125c	F	1
	American oyster *Crassostrea virginica*			0·0102c	F	1
	Pinfish *Lagodon rhomboides*			0·10d	F	1
Aroclor 1221	Killifish *Fundulus heteroclitus*	25		0·032e	F	2
Aroclor 1242	Brown shrimp *Crangon crangon*		1·0		S	3
	Armed bullhead *Agonus cataphractus*		>10		S	3
Aroclor 1248	Brown shrimp *Crangon crangon*		0·3–1·0		S	3
Aroclor 1254	Brown shrimp *Crangon crangon*		3–10		S	3
	Grass shrimp (juvenile) *Palaemonetes pugio*			0·0061–0·0078	S	19
Clophen A 30	Brown shrimp *Crangon crangon*		0·3–1		S	3
	Cockle *Cardium edule*		3		S	3
Clophen A 40	Brown shrimp *Crangon crangon*		1–3·3		S	3
Clophen A 50	Brown shrimp *Crangon crangon*		3·3–10		S	3
DDT	Harpacticoid *Nitocra spinipes*			0·03	S	4
	Brown shrimp *Crangon crangon*		0·0033–0·01		S	3
	Bleak *Alburnus alburnus*			0·08	S	4
	Plaice *Pleuronectes platessa*		0·3–1		S	3
	Striped killifish *Fundulus majalis*	0·003	0·002		S	5
	Longnose killifish *Fundulus similis*	0·0055	0·0055	0·001	F	6
	Spot (juvenile) *Leiostomus xanthurus*	0·005	0·002		F	6
	White mullet (juvenile) *Mugil curema*	0·0008	0·0004		F	6
	Striped mullet *Mugil cephalus*	0·004–0·007	0·0009–0·006	0·0009–0·003	S	5
	American eel *Anguilla rostrata*	0·007	0·006	0·004	S	5
	Mummichog *Fundulus heteroclitus*	0·011	0·005	0·005	S	5
	Atlantic silverside *Menidia menidia*	0·004	0·0004	0·0004	S	5
	Bluehead *Thalassoma bifasciatum*	0·017	0·004	0·007	S	5

Table 6-21—continued

Substance	Species	LC$_{50}$ (mg l^{-1})			Conditions[a]	Source[b]
		24 h	48 h	96 h		
Methoxychlor	Atlantic silverside *Menidia menidia*	0·044	0·044	0·033	S	5
	Bluehead *Thalassoma bifasciatum*	0·014	0·013	0·013	S	5
	Striped killifish *Fundulus majalis*	0·038	0·034	0·03	S	5
	Spot (juvenile) *Leiostomus xanthurus*	0·03	0·03		F	6
	White mullet (juvenile) *Mugil curema*	0·055	0·055		F	6
	Striped mullet *Mugil cephalus*	0·063	0·063	0·063	S	5
	Americal eel *Anguila rostrata*	0·025	0·025	0·012	S	5
	Mummichog *Fundulus heteroclitus*	0·037–0·085	0·035–0·062	0·035–0·057	S	5
Kelthane	Grass shrimp *Crangon franciscorum*	1·29^f	0·6^f		S	7
γ-HCH	Brown shrimp *Crangon crangon*		0·001–0·0033		S	3
	Pink shrimp *Penaeus duorarum*			0·00017	F	8
	Grass shrimp *Palaemonetes pugio*			0·0044	F	8
	Atlantic silverside *Menidia menidia*	0·023	0·020	0·009	S	5
	Bluehead *Thalassoma bifasciatum*	0·014	0·014	0·014	S	5
	Striped killifish *Fundulus majalis*	0·028	0·028	0·028	S	5
	Spot (juvenile) *Leiostomus xanthurus*	0·03	0·03		F	6
	White mullet *Mugil curema*	0·03	0·03		F	6
	Longnose killifish *Fundulus similis*	0·3	0·24		F	6
	Striped mullet *Mugil cephalus*	0·075	0·071	0·066	S	5
	Americal eel *Anguilla rostrata*	0·07	0·07	0·056	S	5
	Mummichog *Fundulus heteroclitus*	0·066	0·06	0·06	S	5
	Sheepshead minnow *Cyprinodon variegatus*			0·104	F	8
Dieldrin	Brown shrimp *Crangon crangon*		0·01–0·033	0·00093	S	3
	Pink shrimp *Penaeus duorarum*			0·0114	F	9
	Grass shrimp *Palaemonetes pugio*			0·0125	F	9
	American oyster *Crassostrea virginica*		0·01–0·033		F	9
	Shore crab *Carcinus maenas*		3·3		S	3
	Armed bullhead *Agonus cataphractus*				S	3
	Atlantic silverside *Menidia menidia*	0·01	0·005	0·005	S	5
	Bluehead *Thalassoma bifasciatum*	0·007	0·006	0·006	S	5

Pesticide	Species					
	Striped killifish *Fundulus majalis*	0·009	0·007	0·004	S	5
	Striped mullet *Mugil cephalus*	0·025	0·025	0·023	S	5
	American eel *Anguilla rostrata*	0·008	0·004	0·0009	S	5
	Mummichog *Fundulus heteroclitus*	0·020	0·009	0·005	S	5
	Spot (juvenile) *Leiostomus xanthurus*	0·0055	0·0055		F	6
	White mullet *Mugil curema*	0·0078	0·0071		F	6
	Plaice *Pleuronectes platessa*		0·0044	0·004	S + F	20
	Sheepshead minnow *Cyprinodon variegatus*			0·0236	F	9
Aldrin	Atlantic silverside *Menidia menidia*	0·045	0·02	0·013	S	5
	Bluehead *Thalassoma bifasciatum*	0·015	0·015	0·012	S	5
	Striped killifish *Fundulus majalis*	0·058	0·026	0·017	S	5
	Striped mullet *Mugil cephalus*	0·126	0·1	0·1	S	5
	American eel *Anguilla rostrata*	0·018	0·005	0·005	S	5
	Mummichog *Fundulus heteroclitus*	0·022	0·016	0·008	S	5
	Spot (juvenile) *Leiostomus xanthurus*	0·0082	0·0055		F	6
	White mullet *Mugil curema*	0·0031	0·0028		F	6
Endrin	Striped mullet *Mugil cephalus*	0·0007	0·0003	0·0003	S	5
	American eel *Anguilla rostrata*	0·0011	0·0006	0·0006	S	5
	Mummichog *Fundulus heteroclitus*	0·0018–0·0056	0·0007–0·0042	0·0006–0·0015	S	5
	Striped killifish *Fundulus majalis*	0·0018	0·0007	0·0003	F	6
	Longnose killifish *Fundulus similis*	0·0003	0·0003		S	5
	Atlantic silverside *Menidia menidia*	0·0005	0·00008	0·00005	S	5
	Bluehead *Thalassoma bifasciatum*	0·0006	0·0005	0·0001	F	6
	Spot (juvenile) *Leiostomus xanthurus*	0·0044	0·0006		F	6
	White mullet *Mugil curema*	0·0026	0·0026		F	6
Heptachlor	Atlantic silverside *Menidia menidia*	0·055	0·025	0·003	F	6
	Bluehead *Thalassoma bifasciatum*	0·0048	0·003	0·0008	S	5
	Striped killifish *Fundulus majalis*	0·024	0·006	0·032	S	5
	Striped mullet *Mugil cephalus*	0·017	0·08	0·194	S	5
	American eel *Anguilla rostrata*	0·05	0·043	0·01	S	5
	Mummichog *Fundulus heteroclitus*	0·224	0·208	0·05	S	5
Endosulfan	Pink shrimp *Penaeus duorarum*	0·071	0·049	0·00004	F	10
	Brown shrimp *Crangon crangon*	0·083	0·067		S	3
	Grass shrimp *Palaemonetes pugio*		0·01	0·0013	F	10
	Armed bullhead *Agonus cataphractus*		0·033–0·1		S	3

Table 6-21—continued

Substance	Species	LC$_{50}$ (mg l^{-1})			Con-ditions[a]	Source[b]
		24 h	48 h	96 h		
	Spot (juvenile) *Leiostomus xanthurus*	0·0009	0·0006		F	6
	Pinfish *Lagodon rhomboides*			0·00009	F	10
	White mullet *Mugil curema*	0·005		0·0003	F	10
	Striped mullet *Mugil cephalus*		0·0006		F	6
Chlordane	White mullet *Mugil curema*	0·043	0·0055	0·00038	F	10
	Grass shrimp *Palaemonetes pugio*			0·0048	F	6
	Sheepshead minnow *Cyprinodon variegatus*			0·0245	F	11
	Pinfish *Lagodon rhomboides*			0·0064	F	11
	Eastern oyster *Crassostrea virgnica*			0·0062 (EC$_{50}$)	F	11
	Pink shrimp *Penaeus duorarum*			0·0004	F	11
Malathion	Brown shrimp *Crangon crangon*	0·082	0·33–1·0		S	3
	American eel *Anguilla rostrata*		0·082	0·082	S	5
	Mummichog *Fundulus heteroclitus*	0·13–0·81	0·08–0·44	0·08–0·4	S	5
	Atlantic silverside *Menidia menidia*	0·315	0·315	0·125	S	5
	Bluehead *Thalassoma bifasciatum*	0·033	0·027	0·027	S	5
	Striped killifish *Fundulus majalis*	0·28	0·25	0·25	S	5
	Spot (juvenile) *Leiostomus xanthurus*	0·55	0·55		F	6
	White mullet *Mugil curema*	0·95	0·57		F	6
	Striped mullet *Mugil cephalus*	>0·96	0·55	0·55	S	5
Methidathion	Lobster *Homarus americanus*		0·014		S	12
Parathion	Brown shrimp *Crangon crangon*		0·0033–0·01		S	3
	Cockle *Cardium edule*		3·3–10		S	3
	Plaice *Pleuronectes platessa*		0·03–0·1		S	3
Methyl parathion	Atlantic silverside *Menidia menidia*	24·8	21·9	5·7	S	5
	Bluehead *Thalassoma bifasciatum*	98·0	88·0	12·3	S	5
	Striped killifish *Fundulus majalis*	29·0	19·4	13·8	S	5
	Striped mullet *Mugil cephalus*	39·0	26·3	5·2	S	5
	American eel *Anguilla rostrata*	27·6	22·4	16·9	S	5
	Mummichog *Fundulus heteroclitus*	>85·2	85·2	58·0	S	5

Compound	Species			Type	No.
Ethyl parathion	Brown shrimp *Crangon crangon*	0·0033–0·01		S	3
	Cockle *Cardium edule*	3·3–10		S	3
Phosphamidon	Lobster *Homarus americanus*	0·107			12
Dimethoate	Pink shrimp *Pandalus montagni*	>0·033		S	3
	Brown shrimp *Crangon crangon*	0·0003–0·001		S	3
	Cockle *Cardium edule*	>3·3		S	3
	Shore crab *Carcinus maenas*	>3·3		S	3
Atrazin	Brown shrimp *Crangon crangon*	10–33		S	3
	Cockle *Cardium edule*	>100		S	3
	Shore crab *Carcinus maenas*	>100		S	3
Simazine	Brown shrimp *Crangon crangon*	>100		S	3
	Cockle *Cardium edule*	100		S	3
Toxaphene	Spot (juvenile) *Leiostomus xanthurus*	0·0022	0·001	F	6
	White mullet *Mugil curema*	0·0055	0·0055	F	6
Mirex	Spot (juvenile) *Leiostomus xanthurus*	>2	>2	F	6
Kepone	Grass shrimp *Palaemonetes pugio*		0·121	F	13
	Blue crab *Callinectes sapidus*		>0·21	F	13
Fentin acetate	Brown shrimp *Crangon crangon*	>33		S	3
Diquat	Brown shrimp *Crangon crangon*	>10		S	3
	Cockle *Cardium edule*	>10		S	3
Paraquat	Brown shrimp *Crangon crangon*	>10		S	3
	Cockle *Cardium edule*	>10		S	3
Pentachlorophenol	Grass shrimp *Palaemonetes pugio*		>0·515		14
	Brown shrimp *Penaeus aztecus*		>0·195		14
	Brown shrimp *Crangon crangon*		1·79		18
	Brown shrimp *Crangon crangon* Larvae		0·11	S + F	18
	Adult		10	S	20
	Decapod *Palaemon elegans* Larvae		0·08	S	18
	Adult		10·39	S	18
	Harpacticoid *Nitrocra spinipes*		0·27	S + F	4
	Common mussel *Mytilus edulis*		18	S + F	20
	Worm *Ophryotrocha diadema* Larvae	1–10	0·62	S + F	20
	Adult	1–40	1·20		20
	Killifish *Fundulus similis*		>0·306	S + F	14
	Plaice *Pleuronectes platessa* Larvae	0·09	0·06	S + F	20
	Adult	0·25	0·15	S + F	20
	Goby fish *Gobius minutus*	0·45			20

Table 6-21—*continued*

Substance	Species	LC$_{50}$ (mg l^{-1})				Con-ditions[a]	Source[b]
		24 h	48 h	96 h			
Dichlobenil	Pinfish *Lagodon rhomboides*			0·053		S	14
	Brown shrimp *Crangon crangon*		3·3–10			S	3
	Cockle *Cardium edule*		>100			S	3
	Shore crab *Carcinus maenas*		10			S	3
Carbon tetrachloride	Dab *Limanda limanda*			*ca.* 50		F	15
Trichloroethylene	Dab *Limanda limanda*			16		F	15
Perchloroethylene	Dab *Limanda limanda*			5		F	15
1,2-Dichloroethane	Dab *Limanda limanda*			115		F	15
Trichloroethane	Dab *Limanda limanda*			33		F	15
1,1,2-Trichloroethane	Common mussel *Mytilus edulis*			80		S + F	20
	Worm *Ophryotrocha diadema*			190		S + F	20
	Plaice *Pleuronectes platessa*		45			S + F	20
Hexachlorobutadiene	Dab *Limanda limanda*			0·45		F	15
1,2-Dichlorobenzene	Shrimp *Palaemonetes pugio* or *Penaeus setiferus*	14·3	10·3	9·4		S	16
1,4-Dichlorobenzene	Shrimp *Palaemonetes pugio* or *Penaeus setiferus*		129·2	69·0		S	16
Polychlorinated paraffins 40–71% Cl	Bleak *Alburnus alburnus*			>5000–10 000		S	4
Phenol	Cockle *Cardium edule*		>500			S	3
	Flounder *Platichthys flesus*		33–100			S	3
	Pink shrimp *Pandalus montagui*		17·5			S	3
	Shrimp *Crangon septemspinosa*			1·8		S	21
	Brown shrimp *Crangon crangon*		23·5			S	3
p-sec-Butylphenol				1·3		S	21
o-sec-Butylphenol				2·4		S	21
o-tert.-Butylphenol				5·2		S	21
m-tert.-Butylphenol				1·7		S	21
p-tert.-Butylphenol				0·9		S	21
p-Hexylphenol				0·6		S	21
p-Heptylphenol				1·1		S	21
p-tert.-Octylphenol				0·3		S	21
p-Nonylphenol				0·15		S	21
p-Dodecylphenol						S	21

Cresol	Brown shrimp *Crangon crangon*	10–100		S	3
	Cockle *Cardium edule*	>100		S	3
	Armed bullhead *Agonus cataphractus*	10–33		S	3
	Plaice *Pleuronectes platessa*	10–33		S	3
3,4-Dichloroaniline	Common mussel *Mytilus edulis*		9·5	S + F	20
	Brown shrimp *Crangon crangon*		2·3	S + F	20
	Plaice *Pleuronectes platessa*	6·5	4·6	S + F	20
Dimethyl phthalate	Bleak *Alburnus alburnus*	100–115		S	4
	Harpacticoid *Nitocra spinipes*		62	S	4
Dinonyl phthalate	Harpacticoid *Nitocra spinipes*	300		S	4
Diisobutyl phthalate	Harpacticoid *Nitocra spinipes*		3·0	S	4
Di-(2-ethylhexyl) phthalate	Harpacticoid *Nitocra spinipes*		>300		4
Butyl benzyl phthalate	Mysed shrimp *Mysidopsis bahia*		0·9		17
Butyl benyl phthalate	Sheepshead minnow *Cyprinodon variegatus*		3·0		17
Butylated monochlorodiphenyl ether	Harpacticoid *Nitocra spinipes*		0·17	S	4

[a] S, Static exposure; F, Flow-through test.

[b] 1, HANSEN and co-authors (1974); 2, KINTER and co-authors (1972); 3, PORTMAN and WILSON (1971); 4, LINDEN and co-authors (1979); 5, EISLER (1970b); 6, HOLDEN (1973a); 7, KHORRAM and KNIGHT (1977); 8, SCHIMMEL and co-authors (1974); 9, PARRISH and co-authors (1977a); (1977b); 10, SCHIMMEL and co-authors (1978); 11, PARRISH and co-authors (1976); 12, McLEESE and METCALF (1979); 13, SCHIMMEL and WILSON (1977); 14, SCHIMMEL and co-authors (1978); 15, PEARSON and McCONNELL (1975); 16, CURTIS and co-authors (1979); 17, GLEDHILL and co-authors (1980); 18, VAN DIJK and co-authors (1977); 19, ROESIJADI and co-authors (1976); 20, ADEMA and VINK (1981); 21, McLEESE and co-authors (1981).

[c] EC_{50} for shell growth.

[d] No mortality at this concentration.

[e] Significant mortality in 42d.

[f] Calculated from authors' logarithmic values.

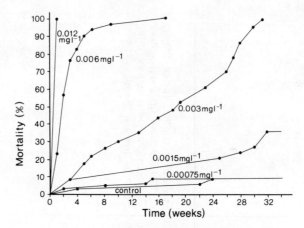

Fig. 6-18: *Poecilia latipinna*. Time-dependent mortality of sailfin molly kept in flowing sea water containing dieldrin. At the lowest concentrations (0·00075 and 0·0015 mg l^{-1}) more than 50% of the fish survive. (After LANE and LIVINGSTON, 1970; reproduced by permission of the American Fisheries Society.)

appropriate measure for the toxic potential of subacute concentrations could not be attained in this way. As a consequence, for more than 10 yr, partial and full life-cycle tests with fishes including all developmental stages have been employed for estimating safe pollutant concentrations for aquatic life. In an early paper, MOUNT and STEPHAN (1967) reported on life-cycle studies with a freshwater species, the fathead minnow *Pimephales promelas,* using measurements of survival, growth and reproduction for estimating the maximum acceptable toxicant concentration (MATC). The MATC is defined as the toxic threshold concentration between (i) the highest toxicant concentration not yet leading to apparent negative effects and (ii) the next higher toxicant concentration causing toxic effects. MATC values for a number of organic compounds and various species of fish such as fathead minnows (*P. promelas*), bluegills (*Lepomis macrochirus*), brook trout (*Salvelinus fontinalis*) and flagfish (*Jordanella floridae*) are listed in Table 6-22.

In order to test chronic toxicity levels for large numbers of chemicals more economically, highly sensitive developmental stages may be used, such as embryo-larval and early juvenile stages (e.g. MCKIM, 1977). MCKIM showed that in 46 (82%) of 56 life-cycle tests the embryo-larval or early juvenile MATC was identical with the actual MATC, and that in all 56 tests this estimate was within a factor of 2 of the actual MATC. Life-cycle tests on marine species have been conducted, for example, on the opossum shrimp *Mysidopsis bahia* (NIMMO and co-authors, 1977) and the sheepshead minnow *Cyprinodon variegatus* (HANSEN and PARRISH, 1977; Table 6-22).

In order to provide a more general use of MATC data by extrapolation to other situations, an application factor (AF) may be calculated by using acute 96-h LC$_{50}$ data: 96-h LC$_{50}$ × AF = MATC and AF = MATC/96-h LC$_{50}$. In Table 6-23, a comparison of application factors is given for freshwater and estuarine fishes, exhibiting striking similarities.

Table 6-22

Maximum acceptable toxicant concentrations (MATC) of organic chemicals in freshwater and estuarine fishes (Based on McKim, 1977; Nimmo and co-authors, 1977; Hansen and Parrish, 1977; Hansen and co-authors, 1977)

No.[a]	Toxicant	Species	Type of life-cycle test[b]	Effect	MATC ($\mu g\ l^{-1}$)
1	Aroclor 1254	Fathead minnow	C	Mortality	1·8–4·6
2	Aroclor 1242	Fathead minnow	C	Mortality	5·4–15·0
3	Aroclor 1248	Fathead minnow	C	Mortality	1·1–3·0
4	Aroclor 1260	Fathead minnow	C	Mortality	2·1–4·0
5	Atrazine	Brook trout	P	Growth	60·0–120·0
6	Captan	Fathead minnow	P	Growth	16·5–39·5
7	Carbaryl	Fathead minnow	C	Mortality	210·0–680·0
8	Diazinon	Flagfish	C	Hatchability	54·0–88·0
9	Diazinon	Fathead minnow	P	Mortality; growth	6·8–13·5
10	Diazinon	Brook trout	P	Growth	<0·80
11	Endrin	Flagfish	C	Growth	0·22–0·30
12	Lindane	Fathead minnow	C	Growth	9·1–34·5
13	Malathion	Flagfish	C	Growth	8·6–10·9
14	Heptachlor	Fathead minnow	C	Mortality	0·86–1·84
15	Endosulfan	Fathead minnow	P	Hatchability	0·20–0·40
16	Toxaphene	Brook trout	P	Mortality; growth	<0·039
17	Endrin	Sheepshead minnow	C	Mortality; growth; hatchability; fertility of eggs	>0·12 <0·31
18	Heptachlor	Sheepshead minnow	P	Mortality; egg production	>0·97 <1·9
19	Methoxychlor	Sheepshead minnow	P	Hatchability	>12 <23
20	Malathion	Sheepshead minnow	P	Mortality	>4 <9
21	Kepone	Opossum shrimp	C	Growth	0·026

[a] 1–16, freshwater species; 17–21, estuarine species.
[b] C, complete life-cycle test; P, partial life-cycle test.

Table 6-23

Lower limits of application factors[a] for freshwater and estuarine fishes (After Hansen and Parrish, 1977. Copyright American Society for Testing and Materials, Philadelphia, PA. Reprinted with permission)

Chemical	Species	Habitat	Application factor
Endrin	*Cyprinodon variegatus*	Saltwater	0·35
	Jordanella floridae	Freshwater	0·25
Heptachlor	*Cyprinodon variegatus*	Saltwater	<0·07 or 0·09
	Pimephales promelas	Freshwater	0·07
Malathion	*Cyprinodon variegatus*	Saltwater	0·08
	Pimephales promelas	Freshwater	0·05
	Lepomis macrochirus	Freshwater	0·04
	Jordanella floridae	Freshwater	0·02

[a] Application factor = $\dfrac{\text{no-effect concentration}}{\text{96-h LC}_{50}}$

In addition to fishes which have already been established as test organisms in toxicity testing, other species may be used, such as algae. Marine algae exhibit a number of easily measurable responses when exposed to critical organic pollutant levels: inhibition of growth or photosynthesis, changes in species composition and cell-size distribution in mixed cultures, as well as modifications in species-specific physiological processes, such as calcification of cell-wall scales (ELDER and co-authors, 1971), inhibition of membrane-bound enzymes (FISHER, 1975) and effects on RNA levels (KEIL and co-authors, 1971). Marine algae are not only suitable test organisms, but also provide evidence from trophic interactions.

Inhibitions in rates of growth and photosynthesis are the most frequently used criteria for quantifying pollutant effects in algae (Table 6-24). FISHER (1975) investigated the relationship between growth and photosynthesis in *Thalassiosira pseudonana* (*Cyclotella nana*) exposed to PCBs (Aroclor 1254) at 10 and 50 μg l^{-1}, and stated that although rates of growth and photosynthesis were reduced, ^{14}C uptake per cell remained unaffected. These findings reveal that overall reduction of photosynthesis in the culture reflected growth inhibition rather than a lowering of photosynthetic rate as such. MACFARLANE and co-authors (1972), on the other hand, observed effects on gross morphology in *Nitzschia delicatissima* after exposure to 9·4 μg l^{-1}; compared with controls the chloroplasts were considerably reduced in size and their shape appeared spherical rather than ovate. These findings indicate that the decrease in rates of photosynthesis and the amounts of chlorophyll *a* could possibly be due to damage in the photosynthetic apparatus.

Mixed Phytoplankton—Compared with Single Species

Natural, mixed phytoplankton communities are more sensitive than single species to organochlorine compounds, such as PCBs and DDT (FISHER and co-authors, 1974; MOORE and HARRISS, 1974). Alterations in species composition have been observed after exposure of a mixed culture of *Thalassiosira pseudonana* and *Dunaliella tertiolecta* to 1 μg l^{-1} of PCBs and 10 μg l^{-1} of DDT (MOSSER and co-authors, 1972). FISHER and co-authors (1974) were able to demonstrate experimentally that continuous cultures of gnotobiotic communities were superior to batch cultures in eliciting such effects at concentrations as low as 0·1 μg l^{-1} of PCBs (Aroclor 1254). BIGGS and co-authors (1978) found PCBs to inhibit the growth of phytoplankters larger than 8 μm in mixed estuarine phytoplankton, causing the community composition to shift towards smaller sized algae. This, in turn, may affect zooplankters choosing their food on the basis of size and shape. Thus the decline of certain phytoplankton species might have deleterious effects on higher trophic levels. Consequently, the lower marine food web may be affected by qualitative alterations in phytoplankton communities, even if the total quantity of phytoplankton remains unchanged.

Geographic Differences

A correlation between organochlorine sensitivity and non-genetic adaptation of phytoplankters has been investigated in three species of diatoms exposed to 10 μg l^{-1} of PCBs (Aroclor 1254) by FISHER and co-authors (1973). Comparing the growth of *Thalassiosira pseudonana*, *Fragilania pinnata* and *Bellerochia* sp. from estuarine and Sargasso waters, they found the open-ocean clones to be more sensitive than clones from the

Table 6-24

Effects of organic pollutants on marine algae (Compiled from the sources indicated)

Toxicant	Algae	Effect tested[a]	Effective concentration (μg l^{-1})[b]	Source
DDT	*Skeletonema costatum*	P	<10	WURSTER (1968)
DDT	*Coccolithus huxleyi*	P		
DDT	*Pyramimonas* sp.	P		
DDT	*Peridinium trochoideum*	P		
DDT	Phytoplankton community (neritic)	P		
DDT	*Dunaliella tertiolecta*	P	No effect at 100	MENZEL and co-authors (1970)
DDT	*Skeletonema costatum*	P	10	
DDT	*Coccolithus huxleyi*	P	10	
DDT	*Cyclotella nana*	P	1	
DDT	*Dunaliella tertiolecta*	P	No effect at 80	BOWES (1972)
DDT	*Skeletonema costatum*	P	Slight effect at 80, but not quantified	
DDT	*Thalassiosira fluviabilis*	P		
DDT	*Cyclotella nana*	P		
DDT	*Coccolithus huxleyi*	P		
DDT	*Amphidinium carteri*	P		
DDT	*Porphyridium* sp.	P		
DDT	Phytoplankton community	P	5	MOORE and HARRISS (1972)
DDE	*Exuviella baltica*	G	0·2	POWERS and co-authors (1975)
DDE	*Exuviella baltica*	G (complete inhibition)	10	
DDE	*Exuviella baltica*	P	25	POWERS and co-authors (1979)
Dieldrin	*Exuviella baltica*	G (disintegration of cells)	10	POWERS and co-authors (1977)
Dieldrin, endrin	*Dunaliella tertiolecta*	P	No effect at 100	MENZEL and co-authors (1970)
Dieldrin, endrin	*Skeletonema costatum*	P	10	
Dieldrin, endrin	*Coccolithux huxleyi*	P	10	
Dieldrin, endrin	*Cyclotella nana*	P	1	
Chlordane	Mixed estuarine phytoplankton	P, G	10	BIGGS and co-authors (1978)
Mirex	*Chlorococcum* sp.	G, P	No effect at 0·2	HOLLISTER and co-authors (1975)
Mirex	*Dunaliella tertiolecta*	G, P		
Mirex	*Chlamydomonas* sp.	G, P		

Table 6-24—continued

Toxicant	Algae	Effect tested[a]	Effective concentration ($\mu g\ l^{-1}$)[b]	Source
Mirex	*Nitzschia* sp.	G, P		
Mirex	*Thalassiosira pseudonana*	G, P		
Mirex	*Porphyridium cruentum*	G, P		WALSH and co-authors (1977)
Kepone	*Chlorococcum* sp.	G (EC$_{50}$)	350	
Kepone	*Dunaliella* sp.	G (EC$_{50}$)	580	
Kepone	*Nitzschia* sp.	G (EC$_{50}$)	600	
Kepone	*Thalassiosira pseudomona*	G (EC$_{50}$)	600	
EPN, O-ethyl-O-(4-nitrophenyl)phenyl phosphorothioate	*Skeletonema costatum*	G (96 h EC$_{50}$)	340	WALSH and ALEXANDER (1980)
Methylparathion,O,O-dimethyl-O-(nitrophenyl) phosphorothioate	*Skeletonema costatum*	G (96 h EC$_{50}$)	5300	
Carbophenothion, S-[[(4-chlorophenyl)thio]methyl]-O,O-diethyl phosphorodithioate	*Skeletonema costatum*	G (96 h EC$_{50}$)	109	
Ametryne	*Chlorococcum* sp.	G (EC$_{50}$)	10	WALSH (1972)
		P (EC$_{50}$)	20	
Ametryne	*Dunaliella tertiolecta*	G (EC$_{50}$)	40	
		P (EC$_{50}$)	40	
Atrazine	*Chlorococcum* sp.	G (EC$_{50}$)	100	WALSH (1972)
		P (EC$_{50}$)	100	
Atrazine	*Dunaliella tertiolecta*	G (EC$_{50}$)	300	
		P (EC$_{50}$)	300	
2,4-D	*Chlorococcum* sp.	G(EC$_{50}$)	50 000	WALSH (1972)
		P (EC$_{50}$)	60 000	
2,4-D	*Dunaliella tertiolecta*	G (EC$_{50}$)	75 000	
		P (EC$_{50}$)	50 000	
Diquat (dibromide)	*Chlorococcum* sp.	G (EC$_{50}$)	200 000	WALSH (1972)
		P (EC$_{50}$)	>5 000 000	

Compound	Organism	Effect[a]	Concentration[b]	Reference
Diquat (dibromide)	Dunaliella tertiolecta	G (EC$_{50}$) / P (EC$_{50}$)	30 000 / >5 000 000	DEXTER and PAVLOU (1972)
PCBs				
Aroclor 1242	Thalassiosira fluviatilis and Skeletonema costatum	G	10^{-8} M[c]	
Aroclor 1254	Skeletonema costatum	P		
Aroclor 1260				
2,4'-Dichlorobiphenyl-	Phytoplankton communities	G	7	MOORE and HARRISS (1972)
Aroclor 1242	Phytoplankton communities	G	30 (LC$_{50}$) / 1–2	
Aroclor 1254	Phytoplankton communities	G	6·5 (LC$_{50}$) / 1–2	
Aroclor 1242	Estuarine phytoplankton communities	P	15 (LC$_{50}$)	MOORE and HARRISS (1972)
Aroclor 1254	Phytoplankton communities (natural)	P	10	HARDING (1976)
Aroclor 1254	Phytoplankton communities	P	10	BIGGS and co-authors (1978)
		G (inhibition of 50–80%)	5–10	
Chlorinated naphthalenes Halowax 1013 (26% chlorine)	Dunaliella tertiolecta	G	500–1000	
	Chlorococcum sp.	G	500–1000	
	Nitzschia sp.	G	500–1000	WALSH and co-authors (1977)
	Thalassiosira pseudomona	G	500–1000	
	Phaeodactylum tricornutum	P	8000	
Trichlorethylene, Perchlorethylene,	Phaeodactylum tricornutum	P	10 500	PEARSON and McCONNELL (1975)
1,2-Dichlorethane	Phaeodactylum tricornutum	P	350 000	

[a] P, inhibition of photosynthesis; G, inhibition of growth.
[b] If not stated otherwise, lowest reported concentrations being effective have been selected.
[c] Molecular weights used by authors: 256 (trichloro), 324 (pentachloro) and 392 (heptachloro) for Aroclor 1242, 1254 and 1260, respectively.

estuary. This finding supports the known fact that estuarine forms are more tolerant to environmental stress than their oceanic counterparts (Volume I).

Combined Effects

In nature, organisms are exposed to different toxicants acting simultaneously. This fact has thus far received insufficient attention in terms of experimental design (Volum I: ALDERDICE, 1972). Few papers have reported on the combined effects of pollutants. MOSSER and co-authors (1974) investigated interactions of PCBs, DDT and DDE in *Thalassiosira pseudonana* and found that DDE responded synergistic to PCBs. Growth inhibition at 10 μg l^{-1} of PCBs or 100 μg l^{-1} of DDE was only very slight, but increased substantially in a mixture of the compounds. Growth of the marine dinoflagellate *Exuviella baltica* was inhibited by DDE or PCBs, while both compounds in combination produced an approximately additive effect (POWERS and co-authors, 1975).

Miscellaneous Effects

Effects other than lethality have been observed with pesticides at low concentrations. COUCH and co-authors (1977) reported that scoliosis, curvature of the spine, occurred in the sheepshead minnow *Cyprinodon variegatus*, after 11 d of exposure to 0·0008 mg l^{-1} of Kepone in flowing sea water; in scoliotic fish the most striking effect was the breaking of the centra of the vertebrae at the epicentre of flexure in the spinal column. Other effects were, for example, darkening of the posterior one third of the body, fin rot, uncoordinated swimming and cessation of feeding. These symptoms were also produced after shorter exposure times, at elevated Kepone concentrations (HANSEN and co-authors, 1977).

There have been a variety of studies on the effects of pollutants on the biochemical composition of animals and on the influence of enzyme systems. Generally, experimentally used concentrations of the tested substances were higher than the actual concentrations found in sea water. However, more recent investigations have revealed that wild animals exhibit reactions on environmental pollution as indicated, for example, by enzyme inductions. These have been noticed in natural populations of *Nereis virens* in an oil-polluted area; the polychaetes had higher MFO activities and P-450 contents than those from a cleaner area (LEE, 1981), and in the polluted area the concentration of benzo(*a*)pyrene was elevated in the oiled sediment in relation to that in the clean area. Results of laboratory experiments were closely related to these findings, and also Aroclor 1254 has proved to be an inducing agent for P-450 contents in *Nereis virens* (LEE, 1981). The consequences of such enzyme inductions cannot fully be evaluated at present, but changes of MFO activities during moult cycles, e.g. in the blue crab (SINGER and LEE, 1977) and the possibility that MFO controls moulting hormone levels (LEE, 1981) stress the importance of these mechanisms as promising tools for measuring pollutant effects.

Effects of chronic exposure to organic chemicals other than pesticides have been reported for PCB mixtures. Aroclor 1254, at minimum concentrations of 0·00032 mg l^{-1}, killed fry of the sheepshead minnow *Cyprinodon variegatus* after 21 d; the maximum concentration with apparent effect was 0·0001 mg l^{-1}; juveniles and adults were less susceptible (SCHIMMEL and co-authors, 1974). Chronic exposure of spot *Leiostomus xanthurus* and pinfish *Lagodon rhomboides* to 0·0005 mg l^{-1} of Aroclor 1254 resulted in 50%

mortality within 38 and 12 d, respecitvely (HANSEN and co-authors, 1971). After 2 wk, changes in histological patterns were observed in spot; parenchymal cell vacuolation was 2 to 3 times more pronounced than in controls and hepatic cells exhibited fatty accumulations. After 3 wk extensive vacuolation in the liver had occurred, as well as degenerative changes in pancreatic tissues; the most striking changes in liver tissues were recorded in moribund fish (COUCH, 1975).

BENGTSSON (1980) studied growth, reproduction and swimming performance in the minnow *Phoxinus phoxinus*. After they had received a diet containing 20 (low), 200 (medium) and 2000 (high) μg g^{-1} of PCBs (Clophen A 50) for 40 d, they showed during the following 260-d observation period in PCB-free water increased growth in the high-level group; such an effect had also been observed at lower concentrations of Clophen A 50 (BENGTSSON, 1979). Hatchability was strongly reduced in the high-level group, and in the gonads PCBs were found at a level of 24 mg kg^{-1} fresh weight, compared with 0·51 to 6·2 mg kg^{-1} in the low-, and medium-level groups.

HOOFTMAN and VINK (1980) measured reproductive potentials in the worm *Ophryotrocha diadema* after long-term exposure to pentachlorophenol, dieldrin and 3, 4-dichloroaniline. It can be seen from Table 6-25 that these effects are already produced at low concentrations of the chemicals compared with the LC$_{50}$ values, which, again provide only poor indications of the toxic potentials of the substances involved.

In addition to facilitating laboratory work with such a sensitive toxicity parameter as the reproductive potential, this procedure represents a promising approach for examining relationships between levels of organic pollutants in wild animals and their reproductive success. Previous work in this field with salt-and freshwater fish suggests that DDT levels are responsible for the failure of eggs to develop in sea trout (BUTLER and co-authors, 1972) and in rainbow trout (HOPKINS and co-authors, 1969). More recently, VON WESTERNHAGEN and co-authors (1981) studied the reproductive success of wild Baltic flounder *Platichthys flesus*, by determining the numbers of viable larvae hatching from artificially inseminated and incubated eggs. They found 120 ng g^{-1} of PCBs in wet ovaries as the threshold level beyond which reduced survival of developing eggs can be expected. Oysters exposed for 30 wk to 0·001 mg l^{-1} of Aroclor 1254 showed only slight alterations in leucocytic infiltration around the gut. After 24 wk at 0·005 mg l^{-1} of Aroclor 1254, general tissue alterations in the parenchyma around the diverticula of the hepatopancreas became apparent; the normal, compact vesicular pattern was lost and

Table 6-25

Ophryotrocha diadema: acute and long-term toxicity of organic compounds (mg l^{-1}) to the polychaete (Based on HOOFTMAN and VINK, 1980)

| Compound tested | LC$_{50}$ (96 h) | | 50% inhibition of reproductive potential[*] | |
	Larvae	Adult	Larvae	Adult
Pentachlorophenol	0·6	1·2	0·025[a]	0·075[d]
Dieldrin	>0·1	>0·1	0·0009[b]	0·0025[d]
3,4-Dichloroaniline	4	15	0·01[c]	—

[*] Duration of tests: (a) 48 d; (b) 47 d; (c) 38 d; (d) 37 d.

replaced by infiltrating leucocytes. The most specific effects were confined to epithelia of the digestive diverticula, where atrophy occurred; recovery of these tissues could be achieved by depuration for 12 wk in PCB-free water (LOWE and co-authors, 1972).

Effects on Man

Residues of organic pollutants in or on marine seafood may pass over to man on consumption. This requires a careful hygienic toxicological evaluation of the chemicals involved. Such evaluations follow recommendations of the WHO (World Health Organization), which are based on extended toxicological work with pesticide chemicals conducted to extablish no-effect levels in mammals. These levels are multiplied by a safety factor to give the acceptable daily intake (ADI) for man, expressed in mg kg^{-1} body weight; ADI values have world-wide validity and are taken over into national legislation to establish maximum residue limits for food.

(c) Conclusions

In the evaluation of potential pollutant impacts on marine ecosystems, toxicological aspects play a major role. Acute toxicity data can provide a relative index of pollutant toxicity, but do not allow the evaluation of long-term effects on biota at the very low pollutant levels actually prevailing in oceans and coastal waters. Life-cycle studies have provided ecologically more meaningful and reliable information for estimating threshold concentrations. They revealed that embryo-larval and early juvenile stages are most sensitive to toxicity stress. There are also several other promising criteria for specific and sensitive tests, such as population dynamics, species composition, swimming performance, behaviour and histopatholgical properties. Effects on physiological performance and biochemical compositions have been determined, but mostly at relative high pollutant concentrations. A major difference between terrestrial and aquatic organisms is that the latter are more directly and more intimately exposed to ambient chemicals (Volume III: KINNE, 1976, p. 7). Effects on marine organisms due to mixtures of pollutants acting in concert have been studied in only a few cases. Here is fertile ground for the future. For methodological details consult Volume I: ALDERDICE (1972).

Currently prevailing *in situ* levels of pesticides and technical organic chemicals do not seem to pose an acute threat to most members of the marine biota. However, our knowledge of long-term sublethal effects, especially at the supra-individual level (populations, ecosystems), is still very limited. For a sound ecological assessment much more information is needed on pollutant effects on reproductive potentials and competition dynamics.

(7) Future Prospects

Organic chemicals are now required for so many facets of the life of man that it can hardly be imagined how we could exist without them unless serious consequences are accepted. The fact that numerous new chemicals are being produced every year means that we cannot expect a reduction in the overall pollution impact—at best qualitative changes in the pollution impact. Many chemicals are produced in small quantities and their pattern of usage does not imply wide distribution; others, however, produced on a large scale, are widely distributed and therefore require appropriate attention (Table 6-26).

Table 6-26

Estimated percentage of manufactured chemicals that reach
the production rates indicated (After SCHMIDT-BLEEK and
WAGENKNECHT, 1979; reproduced by permission of Pergamon
Press)

Chemicals (%)	Production (tons yr^{-1})
95	1
85	5
50	25
33	50
20	100
5	500
2	1000

Many pesticides, organic solvents and their synthetic intermediates—as well as their by-products—may exert biological effects at low concentrations, accumulate in living matter and non-living compounds of the ecosystem, exhibit high persistence to degradation and yield degradation products of toxicity similar to or higher than that of the parent compound. These chemicals contribute to environmental hazards. The ecologically undesirable properties of DDT and PCBs have initiated trends to reduce their use. In the case of PCBs, the lower chlorinated compounds which are more degradable and less bioconcentrated should be used. PCBs have been recommended for primary use in 'closed' systems. Attempts to find substitutes for PCBs, e.g. as a capacitor fluid, led to the development of a product the main component of which is monochloromonobutyl-diphenyl oxide; this chemical exhibits more favourable properties than PCB compounds with respect to fish toxicity, bioconcentration potential and degradation (BRANSON, 1977). In the marine environment the substitute is not very likely to pose a problem (ADDISON, 1979). DDT as a pesticide, although it has been banned in a number of countries, cannot yet be discounted since it is also used in the anti-malaria programme, and the predicted total requirement is 69 000 tons for 1981 (GOLDBERG, 1975). In India the use of DDT has increased about 4-fold in the last 10 yr and 25% of the pesticidal chemicals used are expected to reach the coastal marine environment. The future trend in a country such as India will be increased application of pesticides, estimated to be more than 1.8×10^5 tons yr^{-1} by 1988–89 (JALEES and VEMURI, 1980). Maintaining the technical standards of today and taking into account that food production and food preservation cannot be achieved on an economic basis at present, except with synthetic pesticides, it can be expected that the production of these chemicals is not going to be reduced in the next 10 yr.

Since organic chemicals may exert deleterious effects on life in the oceans and coastal waters and also on man via residues of pollutants in sea food, a careful and reliable assessment of their fates and effects in the sea is absolutely necessary. Newly produced chemicals should be thoroughly tested in order to assess possible undesirable effects before general application.

The ecological impact of a chemical cannot be judged on the basis of toxicity data alone. We must have sufficient information on the rates of bioconcentration and degradation. The complex structures of ecosystems and their dependence on numerous factors make the assessment of ecologically valid pollution effects extremely difficult. As has been pointed out repeatedly in previous volumes of this treatise, extrapolation from laboratory data to field conditions involves numerous uncertainties. Laboratory data should therefore exemplarily be tested for their reliability under natural conditions.

There are effects that might be easily overlooked, such as reduced resistance to parasitic infection and diseases under the influence of pollutants (e.g. KINNE, 1980; see also Chapter 1). Other effects may exist for which sufficiently sensitive tests have not yet been developed.

A number of indications point to an increasing, slow intoxication of estuarine and coastal areas. According to SINDERMANN's (1979) comprehensive review, several diseases and abnormalities of fish and shellfish seem to be associated with pollutant stress. These can be categorized as follows: (i) diseases caused by contaminant stress and related pathogens; (ii) stress-provoked latent infections; (iii) environmentally induced abnormalities; (iv) genetic abnormalities with mutagenic and other properties of contaminants; (v) contaminant effects on resistance and immune response; and (vi) pollutant–parasite interactions. For detailed comprehensive reviews on diseases in marine animals, consult KINNE (1980, 1983, in press).

In order to assess the overall possible impact on marine life of currently known organic chemicals, it seems appropriate to compare actual concentrations of pollutants with toxicological data (Fig. 6-19; ERNST, 1980). The concentrations of substances taken from Tables 6-10 to 6-13 relate to those in lower estuaries and the open sea. Estuarine concentrations may vary considerably and thus might sometimes be higher

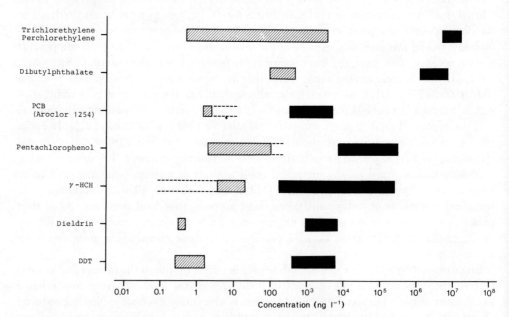

Fig. 6-19: Actual concentrations of chemicals in sea water (hatched bars) and their experimentally derived toxic concentrations (black bars). The most sensitive species were selected from Table 6-21. Left ends of bars refer to most sensitive species, and open-ocean concentrations respectively. (After ERNST, 1980; reproduced by permission of Biologische Anstalt Helgoland.)

than indicated here. Toxicological data were those obtained with fishes or, where possible, by employing the most sensitive criteria available. The differences between substance concentrations and toxic concentrations can be read directly on the abscissa as orders of magnitude and might be regarded as safety margins. While it seems that for the substances listed in Fig. 6-19 the safety margins will be high enough to preclude any toxic effects, at least in the open ocean, the following considerations tend to reduce these margins: toxic effects are usually ascribed to single compounds, but in the presence of many compounds, each will contribute a toxic potential (Chapter 1); this fact has not been taken into consideration in Fig. 6-19. Furthermore, biomagnification has not been considered. Pollution loads of sediments offer additional routes of entry into organisms. Additional stress factors, such as low oxygen concentrations, promote sensitivity at lower pollutant concentrations. At present established safety factors, as applied in toxicological evaluations for man, do not exist for ecosystems (Chapter 1.)

Future efforts in the abatement of marine pollution will depend on the status of our understanding of the complex interactions of pollutants and ecosystems. It is important to recognize that the fate of pollutants and the genesis of diseases is a multi-factorial problem and that further research must concentrate more on the fate and effects of pollutants under natural conditions. Investigations on bioconcentration and elimination have been shown to be feasible under natural conditions (ERNST and co-authors, 1982, in press). It should be possible to conduct toxicity experiments in a similar way under known exposure concentrations. However, such experiments are costly. They must be carefully planned and coordinated.

Literature Cited (Chapter 6)

ADAMSON, R. H. (1967). Drug metabolism in marine vertebrates. *Fedn Proc. Fedn Am. Socs exp. Biol.*, **26**, 1047–1055.

ADDISON, R. F. (1979). An assessment of the hazard to the marine environment of a PCB replacement based on butylated monochlorodiphenyl ethers. *C.M.-ICES E.*, **38**, 1–8.

ADDISON, R. F. and SMITH, T. G. (1974). Organochlorine residue levels in Arctic ringed seals: variation with age and sex. *Oikos*, **25**, 335–337.

ADDISON, R. F. and WILLIS, D. E. (1978). The metabolism by rainbow trout (*Salmo gairdnerii*) of p,p'-[^{14}C]DDT and some of its possible degradation products labelled with ^{14}C. *Toxic. appl. Pharmac.*, **43**, 303–315.

ADEMA, D. M. M. and VINK, G. J. (1981). A comparative study of the toxicity of 1,1,2-trichloroethane, dieldrin, pentachlorophenol and 3,4-dichloroaniline for marine and fresh water organisms. *Chemosphere*, **10**, 533–554.

AHNHOFF, M. and JOSEFSSON, B. (1974). Apparatus for on-site continuous liquid–liquid extraction of organic compounds from natural waters. *Analyt. Chem.*, **46**, 658–663.

AHOKAS, J. T. (1979). Cytochrome P-450 in fish liver microsomes and carcinogen activation. In M. A. Q. Khan, J. J. Lech and J. J. Menn (Eds), *Pesticide and Xenobiotic Metabolism in Aquatic Organisms*, ACS Symposium Series, No. 99. American Chemical Society, Washington, DC. pp. 279–296.

ALDERDICE, D. F. (1972). Factor combinations: responses of marine poikilotherms to environmental factors acting in concert. In O. Kinne (Ed.), *Marine Ecology, Vol. I, Environmental Factors*, Part 3. Wiley, London. pp. 1659–1722.

ALZIEU, C. (1976). Présence de diphenylpolychlores chez certains poissons de l'Atlantique et de la Méditerranée. *Sci. Pêche*, **258**, 1–11.

AMICO, V., IMPELLIZZERI, G., ORIENTE, G., PIATTELLI, M., SCIUTO, S. and TRINGALI, C. (1979). Levels of chlorinated hydrocarbons in marine animals from the central Mediterranean. *Mar. Pollut. Bull., N.S.*, **10**, 282–284.

ANDERSON, R. B. and EVERHART, W. H. (1966). Concentrations of DDT in landlocked salmon (*Salmo salar*) at Sebago lake, Maine. *Trans. Am. Fish. Soc.*, **95**, 160–164.

ANDERSON, R. B. and FENDERSON, O. C. (1970). An analysis of variation of insecticide residues in landlocked Atlantic salmon (*Salmo salar*). *J. Fish. Res. Bd Can.*, **27**, 1–11.

ANDERSSON, Ö. and BLOMKVIST, G. (1981). Polybrominated aromatic pollutants found in fish in Sweden. *Chemosphere*, **10**, 1051–1060.

ARGYLE, R. L., WILLIAMS, G. C. and DUPREE, H. K. (1973). Endrin uptake and release by fingerling Channel catfish (*Ictalurus punctatus*). *J. Fish. Res. Bd. Can.*, **30**, 1743–1744.

BACHE, C. A., SERUM, J. W., YOUNGS, W. D. and LISK, D. J. (1972). Polychlorinated biphenyl residues: accumulation in Cayuga lake trout with age. *Science, N.Y.*, **177**, 1191–1192.

BALLSCHMITER, K. and ZELL, M. (1980a). Analysis of polychlorinated biphenyls (PCB) by glass capillary gas chromatography. *Z. analyt. Chem.*, **302**, 20–31.

BALLSCHMITER, K. and ZELL, M. (1980b). Baseline studies of the global pollution. I. occurrence of organohalogens in pristine European and Antarctic aquatic environments. *Int. J. environ. analyt. Chem.*, **8**, 15–35.

BALLSCHMITER, K., ZELL, M. and NEU, H. J. (1978). Persistence of PCBs in the ecosphere: will some PCB components "never" degrade? *Chemosphere*, **7**, 173–176.

BAKER, R. W. R. (1978). Gel filtration of phthalate esters. *J. Chromat.*, **154**, 3–11.

BAUMANN OFSTAD, E., LUNDE, G. and MARTINSEN, K. (1978). Chlorinated aromatic hydrocarbons in fish from an area polluted by industrial effluents. *Sci. Total Environ.*, **10**, 219–230.

BEND, J. R., BALL, L. M., ELMAMLOUK, T. H., JAMES, M. O. and PHILPOT, R. M. (1979). Cytochrome P-450 in fish liver microsomes and carcinogen activation. In M. A. Q. Khan, J. J. Lech and J. J. Menn (Eds), *Pesticide and Xenobiotic Metabolism in Aquatic Organisms*, ACS Symposium Series, No. 99. American Chemical Society, Washington, DC. pp. 297–315.

BEND, J. R., BEND, S. G., GUARINO, A. M., RALL, D. P. and FOUTS, J. R. (1973). Distribution of ^{14}C-2,4,5,2′,5′-pentachlorobiphenyl in the lobster *Homarus americanus* at various times after a single injection into the pericardial sinus. *Bull. Mt Desert Isl. biol. Lab.*, **13**, 1–4.

BEND, J. R., JAMES, M. O. and DANSETTE, P. M. (1977). *In vitro* metabolism of xenobiotics in some marine animals. *Ann. N.Y. Acad. Sci.*, **298**, 505–521.

BENEZET, H. J. and MATSUMURA, F. (1973). Isomerization of γ-BHC to α-BHC in the environment. *Nature, Lond.*, **243**, 480–481.

BENGTSSON, B.-E. (1979). Increased growth in minnows exposed to PCBs. *Ambio*, **8**, 169–170.

BENGTSSON, B.-E. (1980). Long-term effects of PCB (Clophen A 50) on growth, reproduction and swimming performance in the minnow, *Phoxinus phoxinus*. *Wat. Res.*, **14**, 681–687.

BEVENUE, A., HYLIN, J. W., KAWANO, Y. and KELLERY, T. W. (1972). Organochlorine pollutants of the marine environments of Hawaii—1970–71. *Pestic. Monit. J.*, **6**, 56–64.

BIDLEMAN, T. F. and OLNEY, C. E. (1974). Chlorinated hydrocarbons in the Sargasso Sea atmosphere and surface water. *Science, N.Y.*, **183**, 516–518.

BIGGS, D. C., ROWLAND, R. G., O'CONNORS, H. B., JR, POWERS, C. D. and WURSTER, C. F. (1978). A comparison of the effects of chlordane and PCB on the growth, photosynthesis, and cell size of estuarine phytoplankton. *Environ. Pollut.*, **15**, 253–263.

BODMAN, R. H., SLABAUGH, L. V. and BOWEN, V. T. (1961). A multipurpose large volume seawater sampler. *J. mar. Res.*, **19**, 141–148.

BOWES, G. W. (1972). Uptake and metabolism of 2,2-bis-(*p*-chlorophenyl)-1,1,1-trichloroethane (DDT) by marine phytoplankton and its effect on growth and chloroplast electron transport. *Pl. Physiol.*, **49**, 172–176.

BOWES, G. W. and JONKEL, C. J. (1975). Presence and distribution of polychlorinated biphenyls (PCB) in Arctic and Subarctic marine food chain. *J. Fish. Res. Bd. Can.*, **32**, 2111–2123.

BRANSON, D. R. (1977). A new capacitor fluid—a case study in product stewardship. In F. L. Mayer and J. L. Hamelink (Eds.), *Aquatic Toxicology and Hazard Evaluation*, ASTM Special Technical Publ. 634. ASTM, Philadelphia. pp. 44–61.

BRANSON, D. R., BLAU, G. E., ALEXANDER, H. C. and NEELY, W. B. (1975). Bioconcentration of 2,2′,4,4′-tetrachlorobiphenyl in rainbow trout as measured by an accelerated test. *Trans. Am. Fish. Soc.*, **104**, 785–792.

BRETT, J. R. (1970). Temperature: animals. Fishes. Functional responses. In O. Kinne (Ed.), *Marine Ecology, Vol. I, Environmental Factors*, Part 1. Wiley, London. pp. 515–560.

BRODIE, B. B. and MAICKEL, R. P. (1962). Comparative biochemistry of drug metabolism. In *1st International Pharmacological Meeting* (Symp. Publ. div.), Pergamon Press, New York. **6**, 299–324.

BROOKS, C. J. W. and MIDDLEDITCH, B. S. (1979). Gas chromatography–mass spectrometry. In R. A. W. Johnstone (Ed.), *Mass Spectrometry*, Vol. 5. Chemical Society, London. pp. 172–185.

BUTLER, P. A. (1971). Influence of pesticides on marine ecosystems. *Proc. R. Soc. Lond., B.* **177**, 321–329.

BUTLER, P. A., CHILDRESS, R. and WILSON, A. J. (1972). The association of DDT residues with losses in marine productivity. In M. Ruivo (Ed.), *Marine Pollution and Sea Life*. Fishing News Books (Ltd.), London. pp. 262–266.

CAREY, A. E. and HARVEY, G. R. (1978). Metabolism of polychlorinated biphenyls by marine bacteria. *Bull. environ. Contam. Toxicol.*, **20**, 527–534.

CHIOU, C. T., FREED, V. H., SCHMEDDING, D. W. and KOHNERT, R. L. (1977). Partition coefficient and bioaccumulation of selected organic chemicals. *Environ. Sci. Technol.*, **11**, 475–478.

CLAEYS, R. R., CALDWELL, R. S., CUTSHALL, N. H. and HOLTON, R. (1975). Residues in fish, wildlife and estuaries. Chlorinated pesticides and polychlorinated biphenyls in marine species, Oregon/Washington coast, 1972. *Pestic. Monit. J.*, **9**, 2–10.

CLAUSEN, J., BRAESTRUP, L. and BERG, O. (1974). The content of polychlorinated hydrocarbons in Arctic mammals. *Bull. environ. Contam. Toxicol.*, **12**, 529–534.

CLAYES, R. C. (1979). In C. E. van Hall (Ed.), *Measurement of Organic Pollutants in Water and Waste-water* (ASTM Special Technical Publ. 685), ASTM, Philadelphia. pp. 20–21.

CONTARDI, V., CAPELLI, R., PELLACANI, T. and ZANICCHI, G. (1979). PCBs and chlorinated pesticides from the Ligurian Sea. *Mar. Pollut. Bull., N. S.*, **10**, 307–311.

COOK, G. H., MOORE, J. C. and COPPAGE, D. L. (1976). The relationship of malathion and its metabolites to fish poisoning. *Bull. environ. Contam. Toxicol.*, **16**, 283–290.

COOKE, M., NICKLESS, G., POVEY, A. and ROBERTS, D. J. (1979). Polychlorinated biphenyls, polychlorinated naphthalenes and polynuclear aromatic hydrocarbons in Severn Estuary (UK) sediments. *Sci. Total Environ.*, **13**, 17–26.

COUCH, J. A. (1975). Histopathological effects of pesticides and related chemicals on the liver of fishes. In W. E. Ribelin and G. Mikaki (Eds), *The Pathology of Fishes*. University of Wisconsin Press, Madison, pp. 559–584.

COUCH, J. A., WINSTEAD, J. T. and GOODMAN, L. R. (1977). Kepone-induced scoliosis and its histological consequences in fish. *Science, N.Y.*, **197**, 585–587.

COURTNEY, W. A. M. and LANGSTON, W. J. (1978). Uptake of polychlorinated biphenyl (Aroclor 1254) from sediment and from seawater in two intertidal polychaetes. *Environ. Pollut.*, **15**, 303–309.

COX, J. L. (1972). DDT residues in marine phytoplankton. *Residue Rev.*, **44**, 23–38.

CURTIS, M. W., COPELAND, T. L. and WARD, C. H. (1979). Acute toxicity of 12 industrial chemicals to freshwater and saltwater organisms. *Wat. Res.*, **13**, 137–141.

DAMICO, J. N. (1972). Pesticides. In G. R. Waller (Ed.), *Biochemical Applications of Mass Spectrometry*. Wiley, New York. pp. 623–653.

DAWSON, R. (1976). Determination of chlorinated hydrocarbons in seawater. In K. Grasshoff (Ed.), *Methods of Analysis*. Verlag Chemie, Weinheim, New York. pp. 234–255.

DAWSON, R. and RILEY, J. P. (1977). Chlorine-containing pesticides and polychlorinated biphenyls in British coastal waters. *Estuar. coast. mar. Sci.*, **4**, 55–69.

DeLAPPE, B. W., RISEBROUGH, R. W., MENDOLA, J. T., BOWES, G. W. and MONOD, J.-L. (1973). Distribution of polychlorinated biphenyls on the Mediterranean coast of France. *Journées Étud. Pollut., C.I.E.S.M.*, **1973**, 43–45.

DeLAPPE, B. W., RISEBROUGH, R. W., SPRINGER, A. M. SCHMIDT, T. T., SHROPSHIRE, J. C., LETTERMANN, E. F. and PAYNE, J. R. (1980). The sampling and measurement of hydrocarbons in natural waters. In B. K. Afghan and D. Mackay (Eds), *International Symposium on the Analysis of Hydrocarbons and Halogenated Hydrocarbons*. Plenum Press, New York. pp. 29–68.

DELEON, J. R., WARREN, V. and LASETER, J. L. (1978). Quantitative analysis of trace levels of mirex in edible fish by mass spectrometry. In A. P. de Leenheer, R. R. Roncucci and C. van Petegham (Eds), *Quantitative Mass Spectrometry in Life Sciences*, Vol. II. Elsevier, Amsterdam. pp. 483–492.

DEMETER, J., VAN PETEGHAM, C. and HEYNDRICKX, A. (1978). Determination of endosulfan in biological samples by offline high-pressure liquid chromatography–mass spectrometry. In A. P. de Leenheer, R. R. Roncucci and C. van Petegham (Eds), *Quantitative Mass Spectrometry in Life Sciences*, Vol. II. Elsevier Amsterdam. pp. 471–481.

DEXTER, R. N. and PAVLOU, S. P. (1972). Chemical inhibition of phytoplankton growth dynamics by synthetic organic compounds. *Journées Étud. Pollut., C.I.E.S.M.*, **1972**, 155–157.

DEXTER, R. N. and PAVLOU, S. P. (1973). Chlorinated hydrocarbons in sediments from Southern Greece. *Mar. Pollut. Bull., N.S.*, **4**, 188–190.

DICKSON, A. G. and RILEY, J. P. (1976). The distribution of short-chain halogenated aliphatic hydrocarbons in some marine organisms. *Mar. Pollut. Bull., N.S.*, **7**, 167–169.

VAN DIJK, J. J., VAN DER MER, C. and WIJNANS, M. (1977). The toxicity ot sodium penta-chlorophenolate to three species of decapod crustaceans and their larvae. *Bull. environ. Contam. Toxicol.*, **17**, 622–630.

DOUGHERTY, R. C. (1980). Toxic residues and pollutants. In G. R. Waller and O. C. Dermer (Eds), *Biochemical Applications of Mass Spectrometry* (1st Suppl. Ed.). Wiley, New York. pp. 951–968.

DRESCHER, H. E., HARMS, U. and HUSCHENBETH, E. (1977). Organochlorines and heavy metals in the harbour seal *Phoca vitulina* from the German North Sea coast. *Mar. Biol.*, **41**, 99–106.

DUCE, R. A., QUINN, J. G., ONLEY, C. E., PIOTROWICZ, S. R., RAY, B. J. and WADE, T. L. (1972). Enrichment of heavy metals and organic compounds in the surface microlayer of Narragansett Bay, Rhode Island. *Science, N.Y.*, **176**, 161–163.

DYBERN, B. I. and JENSEN, S. (1978). DDT and PCB in fish and mussels in the Kattegat–Skagerrak area. *Meddn Havsfisk. Lab., Lysekil*, **232**, 1–17.

EDER, G. (1976). Polychlorinated biphenyls and compounds of the DDT group in sediments of the central North Sea and the Norwegian Depression. *Chemosphere*, **5**, 101–106.

EDER, G. (1980). The formation of chlorophenols from the corresponding chlorooxyacetic acids in estuarine sediment under anaerob conditions. *Veröff. Inst. Meeresforsch. Bremerh.*, **18**, 217–221.

EDER, G. and WEBER, K, (1980). Chlorinated phenols in sediments and suspended matter of the Weser Estuary. *Chemosphere*, **9**, 111–118.

EDER, G., SCHAEFER, R. G., GOERKE, H. and ERNST, W. (1976). Chlorinated hydrocarbons in animals of the Skagerrak. *Veröff. Inst. Meeresforsch. Bremerh.*, **16**, 1–9.

EHRHARDT, M. (1976). A versatile system for the accumulation of dissolved, nonpolar organic compounds from seawater. *'Meteor' Forschungsergeb.*, A, **18**, 9–12.

EHRHARDT, M. (1978). An automatic sampling buoy for the accumulation of dissolved and particulate organic material from seawater. *Deep Sea Res.*, **25**, 119–126.

EHRHARDT, M. (1980). Preliminary results of a novel sampling technique for dissolved and particulate organic material in deeper water layers. In *Conference of Baltic Oceanographers, April 1980, Leningrad*, pp. 1–23.

EHRHARDT, M. and DERENBACH, J. (1980). Phthalate esters in the Kiel Bight. *Mar. Chem.*, **8**, 339–346.

EISLER, R. (1970a). Factors affecting pesticide-induced toxicity in an estuarine fish. *Tech. Pap. Fish Wildl. Serv. U. S.*, **45**, 1–20.

EISLER, R. (1970b). Acute toxicities of organochlorine and organophosphorus insecticides to estuarine fishes. *Tech. Pap. Fish Wildl. Serv. U.S.*, **46**, 1–12.

ELDER, D. L. (1976). PCBs in N.W. Mediterranean coastal waters. *Mar. Pollut. Bull., N.S.*, **7**, 63–64.

ELDER, D. L. and FOWLER, S. W. (1977). Polychlorinated biphenyls: penetration into deep ocean by zooplankton faecal pellet transport. *Science, N.Y.*, **197**, 459–461.

ELDER, D. L., FOWLER, S. W. and POLIKARPOV, G. G. (1979). Remobilization of sediment-associated PCBs by the worm *Nereis diversicolor*. *Bull. environ. Contam. Toxicol.*, **21**, 448–452.

ELDER, D. L. and VILLENEUVE, J.-P. (1977). Polychlorinated biphenyls in the Mediterranean Sea. *Mar. Pollut. Bull. N.S.*, **8**, 19–22.

ELDER J. H., LEMBI, C. A., ANDERSON, L. and MOORE, D. J. (1971). Scale calcification in a chrysophycean alga: a test system for the effects of DDT on biological calcification. *Proc. Indiana Acad. Sci.*, **81**, 106–113.

ELMAMLOUK, T. H. and GESSNER, T. (1976). Mixed function oxidases and nitroreductases in hepatopancreas of *Homarus americanus*. *Comp. Biochem. Physiol.* **53c**, 57–62.

ELMAMLOUK, T. H., GESSNER, T. and BROWNIE, A. C. (1974). Occurrence of cytochrome P-450 in hepatopancreas of *Homarus americanus*. *Comp. Biochem. Physiol.*, **48B**, 419–425.

ENVIRONMENTAL PROTECTION AGENCY (EPA), (1975), Review and evaluation of available techniques for determining persistence and routes of degradation of chemical substances in the environment. Washington D.C. *Final Report*, PB 243 825, 1–549.

ENVIRONMENTAL PROTECTION AGENCY (EPA), (1979). Water-related environmental fate of 129 priority pollutants. Volumes I and II. EPA-440/4–79–029a. Washington D.C.

ERNST, W. (1970). Stoffwechsel von Pestiziden in marinen Organismen. II. Biotransformation und Akkumulation von DDT-^{14}C in Plattfischen: *Platichthys flesus*. *Veröff. Inst. Meeresforsch. Bremerh.*, **12**, 353–360.

ERNST, W. (1973). Pesticides as marine pollutants—their distribution and fate. *Atti 5° Coll. Int. Oceanogr. Med., Messina*, 1973, 495–502.

ERNST, W. (1977). Determination of the bioconcentration potential of marine organisms—a steady state approach. 1. Bioconcentration data for seven chlorinated pesticides in mussels (*Mytilus edulis*) and their relation to solubility data. *Chemosphere*, **6**, 731–740.

ERNST, W. (1979a). Factors affecting the evaluation of chemicals in laboratory experiments using marine organisms. *Ecotoxical. Environ. Saf.*, **3**, 90–98.

ERNST, W. (1979b). Metabolic transformation of 1-[1-^{14}C]naphthol in bioconcentration studies with the common mussel, *Mytilus edulis*. *Veröff. Inst. Meeresforsch. Bremerh.*, **17**, 233–240.

ERNST, W. (1980). Effects of pesticides and related organic compounds in the sea. *Helgoländer Meeresunters.*, **33**, 301–312.

ERNST, W. (1982). Tiere als Monitororganismen für organische Schadstoffe. *Decheniana-Beihefte*, **26**, 55–66.

ERNST, W. and GOERKE, H. (1974). Anreicherung, Verteilung, Umwandlung und Ausscheidung von DDT-^{14}C bei *Solea solea* (Pisces: Soleidae). *Mar. Biol.*, **24**, 287–304.

ERNST, W. and GOERKE, H. Identification of metabolites of γ-hexachlorocyclohexane in *Nereis virens* (Polychaeta). Unpublished work.

ERNST, W., GOERKE, H., EDER, G. and SCHAEFFER, R. G. (1976). Residues of chlorinated hydrocarbons in marine organisms in relation to size and ecological parameters. 1. PCB, DDT, DDE and DDD in fishes and molluscs from the English Channel. *Bull. environ. Contam. Toxicol.*, **15**, 55–65.

ERNST, W., GOERKE, H., EDER, G. and WEBER, K. (1982). Comparison of bioconcentration data obtained under laboratory and natural conditions. In: *Environment and Quality of Life*. EEC Brussels, Luxembourg 1982, ISBN 92-825-3057-4. pp. 249–255.

ERNST, W., GOERKE, H. and WEBER, K. (1977). Fate of ^{14}C-labelled di-, tri- and penta-chlorobiphenyl in the marine annelid *Nereis virens*. II. Degradation and faecal elimination. *Chemosphere*, **6**, 559–568.

ERNST, W., SCHAEFER, R. G., GOERKE, H. and EDER, G. (1974). Aufarbeitung von Meerestieren für die Bestimmung von PCB, DDT, DDE, DDD, γ-HCH und HCB. *Z. Analyt. Chem.*, **272**, 358–363.

ERNST, W. and WEBER, K. (1978). The fate of pentachlorophenol in the Weser Estuary and the German Bight. *Veröff. Inst. Meeresforsch. Bremerh.*, **17**, 45–53.

FAULKNER, D. J. and ANDERSEN, R. J. (1974). Natural products chemistry of the marine environment. In E. D. Goldberg (Ed.), *The Sea*. Wiley, London. pp. 679–715.

FISHBEIN, L. (1975). *Chromatography of Environmental Hazards. Vol. II: Pesticides. Vol. III: Metals, Gaseous and Industrial Pollutants*, Elsevier, Amsterdam, London, New York.

FISHER, N. S. (1975). Chlorinated hydrocarbon pollutants and photosynthesis of marine phytoplankton: A reassessment. *Science, N.Y.*, **189**, 463–464.

FISHER, N. S., CARPENTER, E. J., REMSEN, C. C. and WURSTER, C. F. (1974). Effects of PCB on interspecific competition in natural and gnotobiotic phytoplankton communities in continuous and batch cultures. *Microb. Ecol.*, **1**, 39–50.

FISHER, N. S., GRAHAM, L. B., CARPENTER, E. J. and WURSTER, C. F. (1973). Geographic differences in phytoplankton sensitivity to PCBs. *Nature, Lond.*, **241**, 548–549.

FOWLER, S. W., POLIKARPOV, G. G., ELDER, D. L., PARSI, P. and VILLENEUVE, J.-P. (1978). Polychlorinated biphenyls: accumulation from contaminated sediments and water by the polychaete *Nereis diversicolor*. *Mar. Biol.*, **48**, 303–309.

GARNAS, R. L. and CROSBY, D. G. (1979). Comparative metabolism of parathion by intertidal invertebrates. In W. B. Vernberg, F. P. Thurber, A. Calabrese and F. J. Vernberg (Eds), *Marine Pollution: functional responses*. Academic Press, New York. pp. 291–305.

GARRETT, W. D. (1965). Collection of slick-forming materials from the sea surface. *Limnol. Oceanogr.*, **10**, 602–605.

GARRETT, W. D. (1967). The organic chemical composition of the ocean surface. *Deep Sea Res.*, **14**, 221–227.

GARSIDE, E. T. (1970). Temperature: animals. Fishes. Structural responses. In O. Kinne (Ed.), *Marine Ecology, Vol. 1, Environmental Factors*, Part 1. Wiley, London. pp. 561–573.

GERLACH, S. A. (1981). *Marine Pollution. Diagnosis and Therapy*, Springer-Verlag, Berlin, Heidelberg, New York.

GESSNER, F. (1970). Temperature: plants. In O. Kinne (Ed.), *Marine Ecology, Vol. 1. Environmental Factors*, Part 1. Wiley, London. pp. 363–406.

GESSNER, F. and SCHRAMM, W. (1971). Salinity: plants. In O. Kinne (Ed.), *Marine Ecology, Vol. 1, Environmental Factors*, Part 2. Wiley, London. pp. 705–820.

GIAM, G. S., CHAN, H. S. and NEFF, G. S. (1975). Sensitive method for the determination of phthalate ester plasticizers in open-ocean biota samples. *Analyt. Chem.*, **47**, 2225–2229.

GIAM, G. S., CHAN, H. S. and NEFF, G. S. (1976). Concentrations and fluxes of phthalates, DDTs and PCBs to the Gulf of Mexico. In H. L. Windom, R. A. Duce and D. C Heath (Eds), *Marine Pollutant Transfer*. Lexington Books, Lexington, Mass. pp. 375–386.

GIAM, G. S., CHAN, H. S., and NEFF, G. S. (1978a). Phthalate ester plasticizers, DDT, DDE and polychlorinated biphenyls in biota from the Gulf of Mexico. *Mar. Pollut. Bull. N.S.*, **9**, 249–251.

GIAM, G. S., CHAN, H. S., NEFF, G. S. and ATLAS, E. L. (1978b). Phthalate ester plasticizers: a new class of marine pollutants. *Science, N.Y.*, **199**, 419–421.

GLEDHILL, W. E., KALEY, R. G., ADAMS, W. J., HICKS, O., MICHAEL, P. R., SAEGER, V. W. and LeBLANC, G. A. (1980). An environmental safety assessment of butyl benzyl phthalate. *Environ. Sci. Technol.*, **14**, 301–305.

GLICKMAN, A. H., STATHAM, C. N., WU, A. and LECH, J. L. (1977). Studies on the uptake, metabolism, and disposition of pentachlorophenol and pentachloroanisole in rainbow trout. *Toxicol. appl. Pharmac.*, **41**, 649–658.

GOERKE, H. and ERNST, W. (1977). Fate of ^{14}C-labelled di-, tri- and pentachlorobiphenyl in the marine annelid *Nereis virens*. 1. Accumulation and elimination after oral administration. *Chemosphere*, **6**, 551–558.

GOERKE, H. and ERNST, W. (1980). Accumulation and elimination of ^{14}C-γ-HCH (lindane) in *Nereis virens* (Polychaeta) with consideration of metabolites. *Helgoländer Meeresunters.*, **33**, 313–326.

GOERKE, H., EDER, G., WEBER, K. and ERNST, W. (1979). Patterns of organochlorine residues in animals of different trophic levels from the Weser Estuary. *Mar. Pollut. Bull. N.S.*, **10**, 127–133.

GOERLITZ, D. F. and LAW, L. M. (1971). Note on removal of sulfur interferences from sediment extracts for pesticide analysis. *Bull. environ. Contam. Toxicol.*, **6**, 9–10.

GOERLITZ, D. F. and LAW, L. M. (1974). Determination of chlorinated insecticides in suspended sediment and bottom material. *J. Assoc. off. anal. Chem.*, **57**, 176–181.

GOLDBERG, E. D. (1975). Synthetic organohalides in the sea. *Proc. R. Soc. Lond., B*, **189**, 277–289.

GOLDBERG, E. D. (1976). *Strategies for Marine Pollution Monitoring*, Wiley, New York.

GRASSHOFF, K. (1976). *Methods of Seawater Analysis*, Verlag Chemie, Weinhein.

GRICE, G. D., HARVEY, G. R., BOWEN, V. T. and BACKUS, R. H. (1972). The collection and preservation of open ocean marine organisms for pollutant analysis. *Bull. environ. Contam. Toxicol.*, **7**, 125–132.

GRUGER, E. H., HRUBY, T. and KARRICK, N. L. (1976). Sublethal effects of structurally related tetrachloro-, pentachloro- and hexachlorobiphenyl on juvenile coho salmon. *Environ. Sci. Technol.*, **10**, 1033–1037.

GRZENDA, A. R., PARIS, D. F. and TAYLOR, W. J. (1970). The uptake, metabolism and elimination of chlorinated residues by goldfish (*Carassius auratus*) fed a ^{14}C-DDT contaminated diet. *Trans. Am. Fish. Soc.*, **99**, 385–396.

GRZENDA, A. R., TAYLOR, W. J. and PARIS, D. F. (1971). The uptake and distribution of chlorinated residues by goldfish (*Carassius auratus*) fed a ^{14}C-dieldrin contaminated diet. *Trans Am. Fish. Soc.*, **100**, 215–221.

GUARINO, A. M., JAMES, M. O. and BEND, J. R. (1977). Fate and distribution of the herbicides 2,4-dichlorophenoxyacetic acid (2,4-D) and 2,4,5-trichlorophenoxyacetic acid (2,4,5-T) in the dogfish shark. *Xenobiotica*, **7**, 1–9.

HALCROW, W., MACKAY, O. W. and BOGAN, J. (1974). PCB levels in Clyde marine sediments and fauna. *Mar. Pollut. Bull. N.S.*, **5**, 134–136.

HALL, C. E. VAN (1979). *Measurement of Organic Pollutants in Water and Wastewater*, Special Technical Publication, No 686, American Society for Testing and Materials, Philadelphia.

HAMILTON, E. I. and CLIFTON, R. J. (1979). Techniques for sampling the air–sea interface for estuarine and coastal waters. *Limnol. Oceanogr.*, **24**, 188–193.

HANSEN, D. J., GOODMAN, L. R. and WILSON, A. J., JR, (1977). Kepone: chronic effects on embryo, fry, juvenile, and adult sheepshead minnows (*Cyprinodon variegatus*). *Chesapeake Sci.*, **18**, 227–232.

HANSEN, D. J. and PARRISH, P. R. (1977). Suitability of sheepshead minnows (*Cyprinodon variegatus*) for life-cycle toxicity tests. In F. L. Mayer and J. L. Hamelink (Eds), *Aquatic Toxicity and Hazard Evaluation*. American Society for Testing and Materials, Philadelphia, pp. 117–126.

HANSEN, D. J., PARRISH, P. R. and FORESTER, J. (1974). Aroclor 1016: toxicity to and uptake by estuarine animals. *Environ. Res.*, **7**, 363–373.

HANSEN, D. J., PARRISH, P. R., LOWE, J. I., WILSON, JR., A. J. and WILSON, P. D. (1971). Chronic toxicity, uptake, and retention of Aroclor 1254 in two estuarine fishes. *Bull. environ. Contam. Toxicol.*, **6**, 113–119.

HANSEN, D. J. and WILSON, A. J. (1970). Residues in fish, wildlife and estuaries. *Pestic. Monit. J.*, **4**, 51–56.

HARDING, L. W. (1976). Polychlorinated biphenyl inhibition of marine phytoplankton photosynthesis in the northern Adriatic Sea. *Bull. environ. Contam. Toxicol.*, **16**, 559–566.

HARDING, L. W., JR., and PHILLIPS, J. H., JR. (1978). Polychlorinated biphenyls (PCB) uptake by marine phytoplankton. *Mar. Biol.*, **49**, 103–111.

HARVEY, G. R. (1972). Adsorption of chlorinated hydrocarbons from sea water by a cross-linked polymer. *U.S. Natn. Tech. Inform. Serv. Publ. Rep.*, 213954/5, 1–37.

HARVEY, G. R., BOWEN, V. T., BACKUS, R. H. and GRICE, G. D. (1972). Chlorinated hydrocarbons in open-ocean Atlantic organisms. In D. Dryssen and D. Jagner (Eds.), *The Changing Chemistry of the Oceans* (Nobel Symposium 20). Wiley, New York, London, Sydney, pp. 177–186.

HARVEY, G. R. and GIAM, G. S. (1976). Polychlorobiphenyls and DDT compounds. In E. D. Goldberg (Ed.), *Strategies for Marine Pollution Monitoring*. Wiley, New York. pp. 35–46.

HARVEY, G. R., MIKLAS, H. P., BOWEN, V. T. and STEINHAUER, W. G. (1974a). Observations on the distribution of chlorinated hydrocarbons in the Atlantic Ocean ogranisms *J. mar. Res.*, **32**, 103–118.

HARVEY, G. R., STEINHAUER, W. G. and MIKLAS, H. P. (1974b). Decline of PCB concentrations in North Atlantic surface water. *Nature, Lond.*, **252**, 387–388.

HATCHER, R. F. and PARKER, B. C. (1974). Laboratory comparisons of four surface microlayer samplers. *Limnol. Oceanogr.*, **19**, 162–165.

HELZ, G. R. and HSU, R. Y. (1978). Volatile chloro- and bromocarbons in coastal waters. *Limnol. Oceanogr.*, **23**, 858–869.

HERBST, E., WEISGERBER, I., KLEIN, W. and KORTE, F. (1976). Beiträge zur ökologischen Chemie. CXVIII. Bilanz, Bioakkumulierung und Umwandlung von 2,2'-Dichlorbiphenyl-^{14}C in Goldfischen. *Chemosphere*, **5**, 127–130.

HIRAIZUMI, Y., MANABE, T., TAKAHASHI, M., NISHIDA, K. JOH, H. and NISHIMURA, H. (1975). Analysis of PCBs contamination of sediment in the coastal waters of Seta Inland Sea. *La Mer*, **13**, 163–170,

HOLDEN, A. V. (1973a). Effects of pesticides on fish. In G. A. Edwards (Ed.), *Environmental Pollution by Pesticides*, Plenum Press, London, New York. pp. 213–253.

HOLDEN, A. V. (1973b). Monitoring PCB in water and wildlife. *Nat. Swed. Environ. Prot. Bd.*, PCB Conf. II Publ. 4E. pp. 23–33.

HOLDEN, A. V. and MARSDEN, K. (1969). Single-stage clean up of animal tissue extracts for organochlorine residue analysis. *J. Chromat.*, **44**, 481–492.

HOLDEN, A. V. and TOPPING, G. (1971/1972). Occurrence of specific pollutants in fish in the Forth and Tay estuaries. *Proc. R. Soc. Edinb., B*, **71**, 189–194.

HOLLIDAY, F. G. T. (1971). Salinity; animals. Fishes. In O. Kinne (Ed.), *Marine Ecology, Vol. 1. Environmental Factors,* Part 2. Wiley, London. pp. 997–1033.

HOLLISTER, T. A., WALSH, G. E. and FORESTER, G. E. (1975). Mirex and marine unicellular algae: accumulation, population growth and oxygen evolution. *Bull. environ. Contam. Toxicol.,* **14**, 753–759.

HOOFTMAN, R. N. and VINK, G. J. (1980). The determination of toxic effects of pollutants with the marine polychaete *Ophryotrocha diadema. Ecotoxicol. environ. Saf.,* **4**, 252–262.

HOPKINS, C. L., SOLLY, S. R. B. and RITCHIE, A. R. (1969). DDT in trout and its possible effect on reproductive potential. *N.Z. J mar. Freshwat. Res.,* **3**, 220–229.

HUTZINGER, O., SAFE, S. and ZITKO, V. (1974). *The Chemistry of PCBs* CRC Press, Cleveland, Ohio.

INTERNATIONAL COUNCIL FOR THE EXPLORATION OF THE SEA, CHARLOTTENLUND (1977a). The ices coordinated monitoring programmes, 1975–1976. *Co-op Res. Rep.* **72**, 1–26.

INTERNATIONAL COUNCIL FOR THE EXPLORATION OF THE SEA, CHARLOTTENLUND (1977b). Studies of the pollution of the Baltic Sea. *Co-op. Res. Rep.,* **63**, 1–97.

INTERNATIONAL COUNCIL FOR THE EXPLORATION OF THE SEA, CHARLOTTENLUND (1977c). A baseline study of the level of contaminating substances in living resources of the North Atlantic. *Co-op. Res. Rep. 1,* **69**, 1–82.

JACKSON, G. A. (1976). Biologic half-life of endrin in Channel catfish tissues. *Bull. environ. Contam. Toxicol.,* **16**, 505–507.

JALEES, K. and VEMURI, R. (1980). Pesticide pollution in India. *Int. J. environ. Stud.,* **15**, 49–54.

JANSSON, B., VAZ, R., BLOMKVIST, G., JENSEN, S. and OLSSON, M. (1979). Chlorinated terpenes and chlordane components found in fish, guillemot and seal from Swedish waters. *Chemosphere,* **8**, 181–190.

JARVINEN, A. W., HOFFMAN, M. J. and THORSLUND, T. W. (1977). Long-term toxic effects of DDT food and water exposure on fathead minnows (*Pimephales promelas*). *J. Fish. Res. Bd Can.,* **34**, 2089–2103.

JENSEN, S. (1966). Report of a new chemical hazard. *New Scient.,* **32**, 612.

JENSEN, S. (1972). The PCB story. *Ambio,* **1**, 123–131.

JENSEN, S. and JANSSON, B. (1976). Methyl sulfone metabolites of PCB and DDE. *Ambio,* **5**, 257–260.

JENSEN, S., JOHNELS, A. G., OLSSON, M. and OTTERLIND, G. (1972a). DDT and PCB in herring and cod from the Baltic, the Kattegat and the Skagerrak. *Ambio* (Special Rep.), **1**, 71–85.

JENSEN, S., JERNELOV, A., LANGE, R. and PALMORK, K. H. (1972b). Chlorinated by-products from vinylchloride production: a new source of marine pollution. In M. Ruivo (Ed.), *Marine Pollution on Sea Life.* Fishing News (Books) Ltd., London. pp. 242–244.

JENSEN, S., LANGE, R., BERGE, G., PALMORK, K. H. and RENBERG, L. (1975). On the chemistry of EDC-tar and its biological significance in the sea. *Proc. R. Soc. Lond., B,* **189**, 333–346.

JENSEN, S. and SUNDSTRÖM, G. (1974). Structures and levels of most chlorobiophenyls in two technical PCB products and in human adipose tissue. *Ambio,* **3**, 70–76.

JOHNSTON, R. (1976). *Marine Pollution,* Academic Press, London.

JONES, R. (1978). Estimates of the food consumption of haddock (*Melanogrammus aeglefinus*) and cod (*Gadus morhua*). *J. Cons. int. Explor. Mer,* **38**, 18–27.

JUENGST, F. W. and ALEXANDER, M. (1976). Conversion of 1,1,1-trichloro-2,2-bis-(*p*-chlorophenyl)ethane (DDT) to water-soluble products by microorganisms. *J. Fd Chem.,* **24**, 111–115.

KARINEN, J.F., LAMBERTON, J. G., STEWARD, N. E. and TERRIERE, L. C. (1967). Persistence of carbaryl in marine estuarine environment. Chemical and biological stability in aquarium systems. *J. agric. Fd Chem.,* **15**, 148–156.

KEIL, J. E. and PRIESTER, L. E. (1969). DDT uptake and metabolism by a marine diatom. *Bull. environ. Contam. Toxicol.,* **4**, 169–173.

KEIL, J. E., PRIESTER, L. E. and SANDIFER, S. H. (1971). Polychlorinated biphenyl (Aroclor 1242): effects of uptake on growth, nucleic acids, and chlorophyll of a marine diatom. *Bull. environ. Contam. Toxicol.,* **6**, 156–159.

KENAGA, E. E. (1975). Partitioning and uptake of pesticides in biological systems. In R. Haque and V. H. Freed (Eds), *Environmental Dynamics of Pesticides*. Plenum Press, New York. pp. 217–273.

KHAN, M. A. Q., STANTON, R. H., SUTHERLAND, D. J., ROSEN, J. D. and MAITRA, N. (1973). Toxicity–metabolism relationships of the photo-isomers of certain chlorinated cyclodiene insecticide chemicals. *Archs environ. Contam. Toxicol.*, **1**, 159–169.

KHORRAM, S. and KNIGHT, A. W. (1977). The toxicity of kelthane to the grass shrimp (*Crangon franciscorum*). *Bull. environ. Contam. Toxicol.*, **18**, 674–682.

KINNE, O. (1970). Temperature: animals. Invertebrates. In O. Kinne (Ed.), *Marine Ecology, Vol. I. Environmental Factors*, Part 1. Wiley, London. pp. 407–514.

KINNE, O. (1971). Salinity: animals. Invertebrates. In O. Kinne (Ed.), *Marine Ecology, Vol. I, Environmental Factors*, Part 2. Wiley, London. pp. 821–995.

KINNE, O (1976). Introduction to Volume III. In O. Kinne (Ed.), *Marine Ecology, Vol. III, Cultivation*, Part 1. Wiley, London. pp. 1–17.

KINNE, O. (1980). Diseases of marine animals: general aspects. In O. Kinne (Ed.), *Diseases of Marine Animals, Vol. I. General Aspects. Protozoa to Gastropoda*. Wiley, Chichester. pp. 13–64.

KINNE, O. and BULNHEIM. H.-P. (Eds) (1980). *Protection of life in the sea*. 14th European Marine Biology Symposium. Helgoländer *Meeresunters*, **33**, 1–772.

KINTER, W. B., MERKENS, L. S., JANICKI, R. H. and GUARINO, A. M. (1972). Studies on the mechanism of toxicity of DDT and polychlorinated biphenyls (PCBs): disruption of osmoregulation in marine fish. *Environ. Hlth Perspect.*, 1 DHEW Publ. No. (NIH) pp. 72–218, pp. 169–173.

KOBAYASHI, K., AKITAKE, H. and TOMIYAMA, T. (1970). Studies on the metabolism of pentachlorophenate, a herbicide, in aquatic organisms—III. Isolation and identification of a conjugated PCP yielded by a shellfish, *Tapes philippinarum. Bull. Jap. Soc. scient. Fish.*, **36**, 103–108.

KORTE, F. (Ed.) (1980). *Ökologische Chemie*, Georg Thieme Verlag, Stuttgart, New York.

LAMBERTON, J. G. and CLAEYS, R. R. (1970). Degradation of 1-naphthol in sea water. *J. agric. Fd Chem.*, **18**, 92–96.

LANE, C. E. and LIVINGSTON, R. J. (1970). Some acute and chronic effects of dieldrin on the sailfin molly, *Poecilia latipinna. Trans. Am. Fish. Soc.*, **99**, 489–495.

LARSSON, K., ODHAM, G. and SÖDERGREN, A. (1974). On lipid surface films on the sea. I. A simple method for sampling and studies of composition. *Mar. Chem.*, **2**, 49–57.

LAUGHLIN, R. B., JR., NEFF, J. M. and GIAM, G. S. (1977). Effects of polychlorinated biphenyls, polychlorinated naphthalenes, and phthalate esters on larval development of the mud crab *Rhithropanopeus harrisii*. In C. S. Giam (Ed.), *Pollutant Effects on Marine Organisms* Lexington Books, D. C. Heath and Company, Lexington, Massachusetts, Toronto. pp. 95–110.

LEE, R. F. (1981). Mixed function oxygenases (MFO) in marine invertebrates. *Mar. biol. Lett.*, **2**, 87–105.

LEO, A. J. (1975). Calculation of partition coefficients useful in the evaluation of the relative hazards of various chemicals in the environment. In G. D. Veith and D. E. Konasewich (Eds), *Structure–Activity Correlations in Studies of Toxicity and Bioconcentration with Aquatic Organisms*. Great Lakes Research Advisory Board, Windsor, Ontario. pp. 151–176.

LEO, A., HANSCH, C, and ELKINS, D. (1971). Partition coefficients and their uses. *Chem. Rev.*, **71**, 525–616.

LINDEN, E., BENGTSSON, B. E., SVANBERG, O. and SUNDSTROM, G. (1979). The acute toxicity of 78 chemicals and pesticides formulations against two brackish water organisms, the bleak (Alburnus alburnus and the harpacticoid *Nitocra spinipes*). *Chemosphere*, **8**, 843–851.

LINKO, R. R., RANTAMÄKI, P., RAINO, K. and URPO, K. (1979). Polychlorinated biphenyls in plankton from the Turku Archipelago. *Bull. environ. Contam. Toxicol.*, **23**, 145–152.

LOVELOCK, J. E. (1975). Natural halocarbons in the air and in the sea. *Nature, Lond.*, **256**, 193–194.

LOVELOCK, J. E., MAGGS, R. J. and WADE, R. J. (1973). Halogenated hydrocarbons in and over the Atlantic. *Nature, Lond.*, **241**, 194–196.

LOWE, J. I., PARRISH, P. R., PATRICK, J. M., and FORESTER, J. (1972). Effects of the polychlorinated biphenyl Aroclor 1254 on the american oyster *Crassostrea virginica. Mar. Biol.*, **17**, 209–214.

LU, P.-Y. and METCALF, R. L. (1975). Environmental fate and biodegradability of benzene derivatives as studied in a model aquatic ecosystem. *Environ. Hlth Perspect.*, **10**, 269–284.

LUCKAS, B., BERNER, M. and PSCHEIDL, H. (1978). Zur Kontamination von Dorschlebern aus Ostseedorschfängen mit chlorierten Kohlenwasserstoffen in den Jahren 1976/77. *Fisch.-Forsch.*, **16**, 77–81.

LUNDE, G. and BAUMANN OFSTAD, E. (1976). Determination of fat-soluble chlorinated compounds in fish. *Z. Analyt. Chem.*, **282**, 395–399.

LUNDE, G. GETHER, J. and JOSEFSSON, B. (1975). The sum of chlorinated and of brominated non-polar hydrocarbons in water. *Bull. environ. Contam. Toxicol.*, **13**, 656–661.

LUNDE, G., GETHER, J. and STEINESS, E. (1976). Determination of volatility and chemical persistence of lipid-soluble halogenated organic substances in marine organisms. *Ambio*, **5**, 180–182.

LUNDE, G. and STEINNES, E. (1975). Presence of lipid-soluble chlorinated hydrocarbons in marine oils. *Environ. Sci. Technol.*, **9**, 155–157.

MACEK, K. J. and KORN, S. (1970). Significance of the food chain in DDT accumulation by fish. *J. Fish. Res. Bd. Can.*, **27**, 1496–1498.

MACEK, K. J., RODGERS, C. R. STALLING, D. L. and KORN, S. (1970). The uptake and elimination of dietary ^{14}C-DDT and ^{14}C-dieldrin in rainbow trout. *Trans. Am. Fish. Soc.*, **99**, 689–695.

MACFARLANE, R. B., GLOOSCHENKO, W. A. and HARRIS, R. C. (1972). The interaction of light intensity and DDT concentration upon the marine diatom *Nitzschia delicatissima* Cleve. *Hydrobiologia*, **39**, 373–382.

MACGREGOR, J. S. (1974). Changes in the amount and proportion of DDT and its metabolites, DDE and DDD, in the marine environment off Southern California, 1949–72. *Fish. Bull. U.S.*, **72**, 275–293.

MACGREGOR, J. S. (1976). DDT and its metabolites in the sediments off Southern California. *Fish. Bull. U.S.*, **74**, 27–35.

MACLEOD, R. A. (1971). Salinity: bacteria, fungi and blue-green algae. In O. Kinne (Ed.), *Marine Ecology, Vol. I, Environmental Factors*, Part 2. Wiley, Chichester pp. 683–688.

MASON, J. W. and ROWE, D. R. (1976). The accumulation and loss of dieldrin and endrin in the eastern oyster. *Archs environ. Contam. Toxicol.*, **4**, 349–360.

MAUGH, T. H. (1978). Chemicals: how many are there? *Science, N.Y.*, **199**, 162.

MAYER, F. L. MEHRLE, P. M. and SANDERS, H. O. (1977). Residue dynamics and biological effects of polychlorinated biphenyls in aquatic organisms. *Archs environ. Contam. Toxicol.*, **5**, 501–511.

MCCONNELL, O. J. and FENICAL, W. (1977). Halogen chemistry of the red alga *asparagospis*. *Phytochemistry*, **16**, 367–374.

MCFADEN, W. H. (1973). *Techniques in Combined Gas Chromatography–Mass Spectrometry*, Wiley, New York.

MCKIM, J. M. (1977). Evaluation of tests with life stages of fish for predicting long-term toxicity. *J. Fish. Res. Bd Can.*, **34**, 1148–1154.

MCLEESE, D. W. and METCALFE, C. D. (1979). Toxicity of mixtures of phoshamidon and methidathion to lobsters (*Homarus americanus*). *Chemosphere*, **8**, 59–62.

MCLEESE, D. W., ZITKO, V. and SERGEANT, D. B. (1979). Uptake and excretion of fenithrothion by clams and mussels. *Bull. environ. Contam. Toxicol.*, **22**, 800–806.

MCLEESE, D. W., ZITKO, V., SERGEANT, D. B., BURRIDGE, L. and METCALFE, C. D. (1981). Lethality and accumulation of alkylphenols in aquatic fauna. *Chemosphere*, **10**, 723–730.

MELANCON, M. J. and LECH, J. J. (1976). Isolation and identification of a polar metabolite of tetrachlorobiphenyls from bile of rainbow trout exposed to ^{14}C-tetrachlorobiphenyl. *Bull. environ. Contam. Toxicol.* **15**, 181–188.

MENZEL, D. W., ANDERSON, J. and RANDTKE, A (1970). Marine phytoplankton vary in their response to chlorinated hydrocarbons. *Science, N.Y.* **167**, 1724–1726.

MESTRES, R., DUBOUL-RAZAVET, C. and LONG, B. (1974). Presence de micropollutants de la classe des organochlorés dans les sediments d'une vasière littorale: Le Fier d'Ars. *Trav. Soc. Pharm. Montpellier*, **34**, 79–90.

MESTRES, R., PAGNON, M. and DUBOUL-RAZAVET, C. (1975). Étude de résidue de pesticides et d'hydrocarbures organochlorés dans les sédiments du plateau continental languedocien en Méditerranée. *Trav. Soc. Pharm. Montpellier*, **35**, 181–194.

METCALF, R. L. (1977). Biological fate and transformation of pollutants in water. In I. H. Suffet (Ed.), *Fate of Pollutants in the Air and Water Environments*, Part 2. Wiley, New York. pp. 195–221.

MIGET, R., KATOR, H., OPPENHEIMER, C., LASETER, J. L. and LEDET, E. J. (1974). New sampling device for the recovery of petroleum hydrocarbons and fatty acids from aqueous surface films. *Analyt. Chem.*, **46**, 1154–1157.

MILLS, P. A. (1959). Detection and semiquantitatvie estimation of chlorinated organic pesticide residues in foods by paper chromatography. *J. Ass. off. agric. Chem.*, **42**, 734–740.

MITCHELL, A. I., PLACK, P. A. and THOMSON, I. M. (1977). Relative concentrations of ¹⁴C-DDT and of two polychlorinated biphenyls in the lipids of cod tissues after a single oral dose. *Archs. environ. Contam. Toxicol.*, **6**, 525–532.

MIYAZAKI, T., AKIYAMA, K., KANEKO, S., HORII, S. and YAMAGISHI, T. (1980). Identification of chlordanes and related compounds in goby-fish from Tokyo Bay. *Bull. environ. Contam. Toxicol.*, **24**, 1–8.

MOORE, R. E. (1978). Algal nonisoprenoids. In P. J. Scheuer (Ed.), *Marine Natural Products*, Vol. I. Academic Press, New York, San Francisco, London. pp. 43–124.

MOORE, S. A. and HARRISS, R. C. (1972). Effects of polychlorinated biphenyl on marine phytoplankton communities. *Nature, Lond.*, **240**, 356–358.

MOORE, S. A. and HARRISS, R. C. (1974). Differential sensitivity to PCB by phytoplankton. *Mar. Pollut. Bull., N.S.*, **5**, 174–176.

MOSSER, J. L., FISHER, N. S. and WURSTER, C. F. (1972). Polychlorinated biphenyls and DDT alter species composition in mixed cultures of algae. *Science, N.Y.*, **176**, 533–535.

MOSSER, J. L., TENG, T.-C., WALTHER, W. G. and WURSTER, C. F. (1974). Interactions of PCBs, DDT and DDE in a marine diatom. *Bull. environ. Contam. Toxicol.*, **12**, 665–668.

MOUNT, D. I. and STEPHAN, C. E. (1967). A method for establishing acceptable toxicant limits for fish—malathion and the butoxyethanol ester of 2,4-D. *Trans. Am. Fish. Soc.*, **96**, 185–193.

MÜLLER, G., DOMINIK, J., REUTHER, R., MALISCH, R., SCHULTE, E., ACKER, L. and IRION, G. (1980). Sedimentary record of environmental pollution in the Western Baltic Sea. *Naturwissenschaften*, **67**, 595–600.

MURRAY, A. J. and RILEY, J. P. (1973a). The determination of chlorinated aliphatic hydrocarbons in air, natural waters, marine organisms, and sediments. *Analyt. chim. Acta*, **65**, 261–270.

MURRAY, A. J and RILEY, J. P. (1973b). Occurrence of some chlorinated aliphatic hydrocarbons in the environment. *Nature, Lond.*, **242**, 37–38.

MURRAY, H. E., RAY, L. E. and GIAM, G. S. (1981). Analysis of marine sediment, water and biota for selected organic pollutants. *Chemosphere*, **10**, 1327–1334.

MUSTY, P. R. and NICKLESS, G. (1974). The extraction and recovery of chlorinated insecticides and polychlorinated biphenyls from water using porous polyurethane foams. *J. Chromat.*, **100**, 83–93.

MUSTY, P. R. and NICKLESS, G. (1975). Foams as extractants for organochlorine insecticides, polychlorobiphenyls and heavy metals from water. *Proc. Analyt. Div. Chem. Soc.*, **12**, 295–296.

NATIONAL RESEARCH COUNCIL (NRC), (1975). *Assessing Potential Ocean Pollutants*, Publication No. 2325, National Academy of Science, Washington DC.

NEELY, W. B., BRANSON, D. R. and BLAU, G. E. (1974). Partition coefficient to measure bioconcentration potential of organic chemicals in fish. *Environ. Sci. Technol.*, **8**, 1113–1115.

NESTEL, H. and BUDD, J. (1975). Chronic oral exposure of rainbow trout (*Salmo gairdneri*) to a polychlorinated biphenyl (Aroclor 1254): pathological effects, *Can. J. comp. Med.*, **39**, 208–215.

NEU, H. J., ZELL, M. and BALLSCHMITER, K. (1978). Identifizierung von Einzelkomponenten in komplexen Gemischen durch Retentions-Index-Vergleich nach Capillar-Gas-Chromatographie mit Elektroneneinfangdetektor (ECD). *Z. analyt. Chem.*, **293**, 193–200.

NILSSON, C.-A., NORDTRÖM, A., ANDERSSON, K. and RAPPE, C. (1978). Impurities in commercial products related to pentachlorophenol. In K. R. Rao (Ed.), *Pentachlorophenol, Chemistry, Pharmacology and Environmental Toxicology*. Plenum Press, New York. London. pp. 313–324.

NIMMO, D. R., BAHNER, L. H., RIGBY, R. A., SHEPPARD, J. M. and WILSON, A. J., JR. (1977). *Mysidopsis bahia*: An estuarine species suitable for life-cycle toxicity tests to determine the effects of a pollutant. In F. L. Mayer and J. L. Hamelink (Eds), *Aquatic Toxicology and Hazard Evaluation*. American Society for Testing and Materials, Philadelphia. pp. 109–116.

NIMMO, D. R., WILSON, P. D., BLACKMAN, R. R. and WILSON, A. J., JR. (1971). Polychlorinated biphenyl absorbed from sediments by fiddler crabs and pink shrimp. *Nature, Lond.*, **231**, 50–52.

OPPENHEIMER, C. H. (1970). Temperature: bacteria, fungi and blue-green algae. In O. Kinne (Ed.), *Marine Ecology, Vol. I. Environmental Factors*, Part 1. Wiley, London. pp. 347–361.

OSTERROHT, C. (1974). Development of a method for the extraction and determination of nonpolar, dissolved organic substances in sea water. *J. Chromat.*, **101**, 289–298.

OSTERROHT, C. (1977). Dissolved PCBs and chlorinated hydrocarbon insecticides in the Baltic, determined by two different sampling methods. *Mar. Chem.*, **5**, 113–121.

OSTERROHT, C. and SMETACEK, V. (1980). Vertical transport of chlorinated hydrocarbons by sedimentation of particulate matter in Kiel Bight. *Mar. Ecol. Progr. Ser.* **2**, 27–34.

PARKER, B. C., LEEPER, G. and HURNI, W. (1968). Sampler of studies of thin horizontal layers. *Limnol. Oceanogr.*, **13**, 172–175.

PARRISH, P. R., COUCH, J. A., FORESTER, J., PATRICK, J. M., JR, and COOK, G. H. (1974). Dieldrin: effects on several estuarine organisms. In A. L. Mitchell (Ed.), *Proc. 27th Ann. Conf. Southeastern Ass. Game and Fish Commissioners, Oct. 1973, Hot Springs, Arkansas, USA*. pp. 427–434.

PARRISH, P. R., SCHIMMEL, S. C., HANSEN, D. J., PATRICK, J. M. and FORESTER, J. (1976). Chlordane: effects on several estuarine organisms. *J. Toxicol. environ. Hlth*, **1**, 485–494.

PATIL, K. C., MATSUMURA, F. and BOUSH, G. M. (1972). Metabolic transformation of DDT, dieldrin, aldrin and endrin by marine microorganisms. *Environ. Sci. Technol.*, **6**, 629–632.

PEAKALL, D. B. (1975). PCBs and their environmental effects. In C. P. Straub (Ed.), *CRC Critical Reviews in Environmental Control*, Vol. IX. CRC Press, Inc., Boca Raton, Florida, pp. 469–508.

PEARSON, C. R. and McCONNELL, G. (1975). Chlorinated C_1 and C_2 hydrocarbons in the marine environment. *Proc. R. Soc. Lond. B*, **189**, 305–332.

PETROCELLI, S. R., ANDERSON, J. W. and HANKS, A. R. (1975a). Biomagnification of dieldrin residues by food-chain transfer from clams to blue crabs under controlled conditions. *Bull. environ. Contam. Toxicol.*, **13**, 108–116.

PETROCELLI, S. R. ANDERSON, J. W. and HANKS, A. R. (1975b). Controlled food-chain transfer of dieldrin residues from phytoplankters to clams. *Mar. Biol.*, **31**, 215–218.

POHL, R. J., BEND, J. R., GUARINO, A. M. and FOUTS, J. R. (1974). Hepatic microsomal mixed-function oxidase activity of several marine species from coastal Maine. *Drug Metab. Disposit.*, **2**, 545–555.

PORTMANN, J. E. and WILSON, K. W. (1971). The toxicity of 140 substances to the brown shrimp and other marine animals. *Shellfish Inform. Leafl.*, No. 22, p. 11.

POWERS, C. D., ROWLAND, R. G., MICHAELIS, R., FISHER, N. S. and WURSTER, C. F. (1975). The toxicity of DDE to a marine dinoflagellate. *Environ. Pollut.*, **9**, 253–262.

POWERS, C. D., ROWLAND, R. G. and WURSTER, C. F. (1977). Dieldrin-induced destruction of marine algal cells with concomitant decrease in size of survivors and their progeny. *Environ. Pollut.*, **12**, 18–25.

POWERS, C. D., WURSTER, C. F. and ROWLAND, R. G. (1979). DDE inhibition of marine algal cell division and photosynthesis per cell. *Pestic. Biochem. Physiol.*, **10**, 306–312.

PUCCETTI, G. and LEONI, V. (1980). PCB and HCB in the sediments and waters of the Tiber Estuary. *Mar. Pollut. Bull., N.S.*, **11**, 22–25.

REINERT, R. E. (1970). Pesticide concentrations in Great Lake fish. *Pestic. Monit. J.*, **3**, 233–240.

REYNOLDS, L. M. (1969). Polychlorinated biphenyls (PCBs) and their interference with pesticide residue analysis. *Bull. environ. Contam. Toxicol.*, **4**, 128–143.

RICE, C. P. and SIKKA, H. C. (1973). Uptake and metabolism of DDT by 6 species of marine algae. *J. agric. Chem.*, **21**, 148–152.

RILEY, J. P. and TAYLOR, D. (1969). Analytical concentration of traces of dissolved organic materials from sea water with Amberlite XAD-1 resin. *Analyt. chim. Acta*, **46**, 307–309.

RINEHART, K. L., JOHNSON, R. D., SIUDA, J. F., KREJCAREK, G. E., SHAW, P. D., McMILLAN, J. A. and PAUL, I. C. (1975). Structures of halogenated and antimicrobial organic compounds from marine sources. In E. D. Goldberg (Ed.), *The Nature of Seawater*. Dahlem Konferenzen, Abakon Verlagsgesellschaft, Berlin. pp. 651–665.

RISEBROUGH, R. W., DeLAPPE, B. W. and WALKER, W. (1976). Transfer of higher molecular weight chlorinated hydrocarbons to the marine environment. In H. L. Windom and R. A. Duce (Eds), *Marine Pollutant Transfer*. Lexington Books, Lexington, MA. pp. 261–321.

ROESIJADI, G., PETROCELLI, S. R., ANDERSON, J. W., GIAM, G. S. and NEFF, G. E. (1976). Toxicity of polychlorinated biphenyls (Aroclor 1254) to adult, juvenile and larval stages of the shrimp *Palaemonetes pugio. Bull. environ. Contam. Toxicol.*, **15**, 297–304.

RUDLING, L. (1970). Determination of pentachlorophenol in organic tissues and water. *Wat. Res.*, **4**, 533–537.

RUIVO, M. (Ed.) (1972). *Marine Pollution and Sea Life,* Fishing News Books (Ltd.), London.

SAFE , S. (1979). Environmental applications of mass spectrometry. In R. A. W. Johnstone (Ed.), *Mass Spectrometry*, Vol. 5. Chemical Society, London. pp. 172–185.

SAFE, S. and HUTZINGER, O. (1973). *Mass Spectrometry of Pesticides and Pollutants.* CRC Press, Cleveland, Ohio. p. 220.

SAYLER, G. S., THOMAS, R. and COLWELL, R. R. (1978). Polychlorinated biphenyl (PCB) degrading bacteria and PCB in estuarine and marine-environments. *Estuar. Coast Mar. Sci.*, **6**, 553–567.

SCHAEFER, R. G. (1974). Mass-spectrometric identification and quantitation of chlorinated hydrocarbons in fish. *Chem.-Ztg.*, **98**, 241–247.

SCHAEFER, R. G., ERNST, W., GOERKE, H. and EDER, G. (1976). Residues of chlorinated hydrocarbons in North Sea animals in relation to biological parameters. *Ber. dt. wiss. Kommn Meeresforsch.*, **24**, 225–233.

SCHEUER, P. J. (1978). *Marine Natural Products. Chemical and Biological Perspectives,* Vol. I. Academic Press, New York.

SCHIMMEL, S. C., HANSEN, D. J. and FORESTER, J. (1974). Effects of Aroclor 1254 on laboratory-reared embryos and fry of sheepshead minnows (*Cyprinodon variegatus*). *Trans. Am. Fish. Soc.*, **103**, 583–586.

SCHIMMEL, S. C. PATRICK, J. M. and FAAS, L. (1978). Effects of sodium pentachlorophenate on several estuarine animals: toxicity, uptake, and depuration. In K. Ranga Rao (Ed.), *Pentachlorophenol. Chemistry, Pharmacology and Environmental Toxicology.* Plenum Press, New York, London. pp. 147–155.

SCHIMMEL, S. C., PATRICK, J. M. and FORESTER, J. (1977a). Toxicity and bioconcentration of BHC and lindane in selected estuarine animals. *Arch environ. Contam. Toxicol.*, **6**, 355–363.

SCHIMMEL, S. C., PATRICK, J. M. and WILSON, A. J. (1977b). Acute toxicity to bioconcentration of endosulfan by estuarine animals. In F. L. Mayer and J. L. Hamelink (Eds). *Aquatic Toxicology and Hazard Evaluation.* American Society for Testing and Materials, Philadelphia. pp. 241–252.

SCHIMMEL, S. C. and WILSON, A. J., JR (1977). Acute toxicity of kepone to four estuarine animals. *Chesapeake Sci.*, **18**, 224–227.

SCHMIDT-BLEEK, F. and WAGENKNECHT, P. (1979). Umweltchemikalien. *Chemosphere*, **8**, 583–721.

SCHNEIDER, R. and OSTERROHT, C. (1977). Residues of chlorinated hydrocarbons in cod livers from the Kiel Bight in relation to some biological parameters. *Ber. dt. wiss. Kommn Meeresforsch.*, **25**, 105–114.

SCHULTE, E. and ACKER, L. (1974). Identifizierung und Metabolisierbarkeit von polychlorierten Biphenylen. *Naturwissenschaften*, **61**, 79–80.

SCHULTZ, D. M. and QUINN, J. G. (1972). Fatty acids in surface particulate matter from the North Atlantic. *J. Fish. Res. Bd Can.*, **29**, 1482–1486.

SCHWARTZ, H. E., ANZION, C. J. M., VAN VLIET, H. P. M., COPIUS PEEREBOOMS, J. W. and BRINKMANN, U. A. Th. (1979). Analysis of phthalate esters in sediments from Dutch rivers by means of high performance liquid chromatography. *Int. J. environ. analyt. Chem.*, **6**, 133–144.

SCURA, E. D. and MCCLURE, V. E. (1975). Chlorinated hydrocarbons in seawater. Analytical methods and levels in North Eastern Pacific. *Mar..Chem.*, **3**, 337–346.

SINDERMANN, C. J. (1979). Pollution-associated diseases and abnormalities of fish and shellfish: a review. *Fish. Bull., U.S.*, **76**, 717–749.

SINGER, S. C. and LEE, R. F. (1977). Mixed function oxygenase activity in blue crab, *Callinectes sapidus:* tissue distribution and correlation with changes during molting and development. *Biol. Bull.*, **153**, 377–386.

SIUDA, J. F. and DeBERNARDIS, J. F. (1973). Naturally occurring halogenated organic compounds. *Lloydia*, **36**, 107–143.

SMOKLER, P. E., YOUNG, D. R. and GARD, K. L. (1979). DDTs in marine fishes following termination of dominant California input: 1970–77. *Mar. Pollut. Bull. N.S.*, **10**, 331–334.

SÖRENSEN, O. (1973). Gaschromatographischer Pestizidnachweis mit gekoppelten Detektoren. *Gas-Wasserfach (Wasser)*, **114**, 224–227.

STADLER, D. (1977). Chlorinated hydrocarbons in the seawater of the German Bight and the Western Baltic in 1975. *Dt. hydrogr. Z.*, **30**, 189–215.

STADLER, D. and SCHOMAKER, K. (1976). Beschreibung eines Schöpfers zur kontaminationsfreien Entnahme von Seewasser unter der Oberfläche für die Analyse auf chlorierte Kohlenwasserstoffe. *Dt. hydrogr. Z.*, **29**, 83–86.

STADLER, D. and SCHOMAKER, K. (1977a). Ein Glaskugelschöpfer zur kontaminationsfreien Entnahme von Seewasser unter der Oberfläche für die Analyse von Kohlenwasserstoffen und halogenierten Kohlenwasserstoffen. *Dt. hydrogr. Z.*, **30**, 20–25.

STADLER, D. and SCHOMAKER, K. (1977b). Beschreibung zweier Probennehmer für die Entnahme von Mikroschichtproben aus der Meeresoberfläche für die Analyse auf chlorierte Kohlenwasserstoffe. *Dt. hydrogr. Z.*, **30**, 60–67.

STADLER, D. and ZIEBARTH, U. (1975). Beschreibung einer Methode zur Bestimmung von Dieldrin, *p,p'*-DDT und PCBs in Seewasser und Werte für die Deutsche Bucht, 1974. *Dt. hydrogr. Z.*, **28**, 263–273.

STADLER, D. and ZIEBARTH, U. (1976). *p,p'*-DDT, Dieldrin und polychlorierte Biphenyle (PCB) im Oberflächenwasser der westlichen Ostsee (1974). *Dt. hydrogr. Z.*, **29**, 25–31.

STAINKEN, D. and ROLLWAGEN, J. (1979). PCB residues in bivalves and sediments of Raritan Bay. *Bull. environ. Contam. Toxicol.*, **23**, 690–697.

STALLING, D. L., TINDLE, R. C. and JOHNSON, J. L. (1972). Clean-up of pesticide and chlorinated biphenyl residues in fish extracts by gel permeation chromatography. *J. Ass. off. analyt. Chem.*, **55**, 32–38.

STALLING, D. C., HOGAN, J. W. and JOHNSON, J. L. (1973). Phthalate ester residues—their metabolism and analysis in fish. *Environ. Hlth Perspect.*, **3**, 159–173.

STATHAM, C. N., PEPPLE, S. K. and LECH, J. J. (1975). Biliary excretion products of 1-(1-^{14}C)-naphthyl-*N*-methylcarbamate (carbaryl) in rainbow trout (Salmo gairdneri). *Drug Metab. Disposit.* **3**, 400–406.

STENERSEN, J. and KVALVAG, J. (1972). Residues of DDT and its degradation products in cod liver from two Norwegian fjords. *Bull. environ. Contam. Toxicol.*, **8**, 120–121.

STOUT, V. F. (1980). Organochlorine residues in fishes from the Northwest Atlantic Ocean and Gulf of Mexico. *Fish. Bull., U.S.*, **78**, 51.

SUNDSTRÖM, G., JANSSON, B. and JENSEN, S. (1975). Structure of phenolic metabolities of *p,p'*-DDT in rat, wild seal and guillemot. *Nature, Lond.*, **255**, 627–628.

TAGATZ, M. E., BORTHWICK, P. W. and FORESTER, J. (1975). Seasonal effects of leached mirex on selected estuarine animals. *Archs environ Contam. Toxicol.*, **3**, 371–383.

TAKESHITA, R., TAKABATAKE, E., MINAGAWA, K. and TAKIZAWE, Y. (1977). Microdetermination of total phthalate esters in biological samples by gas–liquid chromatography. *J. Chromat.*, **133**, 303–310.

TEICHMAN, J., BEVENUE, A. and HYLIN, J. W. (1978). Separation of polychlorobiphenyls from chlorinated pesticides in sediment and oyster samples for analysis by gas chromatography. *J. Chromat.*, **151**, 155–161.

TULP, M. Th. M. and HUTZINGER, O. (1978). Some thoughts on aqueous solubilities and partition coefficients of PCB, and the mathematical correlation between bioaccumulation and physico-chemical properties. *Chemosphere*, **7**, 849–860.

UTHE, J. F., REINKE, J. and GESSER, H. (1972). Extraction of organochlorine pesticides from water by porous polyurethane coated with selective absorbent. *Environ. Lett.*, **3**, 117–135.

UTHE, J. F., REINKE, J. and O'BRODOVICH, H. (1974). Field studies on the use of coated porous polyurethane plugs as indwelling monitors of organochlorine pesticides and polychlorinated biphenyl contents of streams. *Environ. Lett.*, **6**, 103–115.

VEITH, G. D., DE FOE, D. L. and BERGSTEDT, B. V. (1979). Measuring and estimating the bioconcentration factors of chemicals in fish. *J. Fish. Res. Bd Can.*, **36**, 1040–1048.

VILICIC, D., PICER, N., PICER, M. and NAZANSKY, B. (1979). Monitoring of chlorinated hydrocarbons in biota and sediments of South Adriatic coastal waters. International Commission for the Scientific Exploration of the Mediterranean Sea, Monaco. *Workshop on Pollution of the Mediterranean*, Nov. 1978. pp. 143–146.

VREELAND, V. (1974). Uptake of chlorobiphenyls by oysters. *Environ. Pollut.*, **6**, 135–140.

WALSH, G. E. (1972). Effects of herbicides on photosynthesis and growth of marine unicellular algae. *Hyacinth Control J.*, **10**, 45–48.

WALSH, G. E., AINSWORTH, K. A. and FAAS, L. (1977). Effects and uptake of chlorinated naphthalenes in marine unicellular algae. *Bull. environ. Contam. Toxicol.*, **18**, 297–302.

WALSH, G. E. and ALEXANDER, S. V. (1980). A marine algal bioassay method: results with pesticides and industrial wastes. *Wat. Air Soil Pollut.*, **13**, 45–55.

WARE, D. M. and ADDISON, R. F. (1973). PCB residues in plankton from the Gulf of St. Lawrence. *Nature, Lond.*, **246**, 519–521.

WEBER, K. (1976). Degradation of parathion in seawater. *Wat. Res.*, **10**, 237–241.

WEBER, K. and ERNST, W. (1978). Levels and pattern of chlorophenols in water of the Weser Estuary and the German Bight. *Chemosphere*, **7**, 873–879.

WESTERNHAGEN, H. VON., ROSENTHAL, H., DETHLEFSEN, V., ERNST, W., HARMS, U. and HANSEN, P.-D. (1981). Bioaccumulating substances and reproductive success in Baltic flounder *Platichthys flesus*. *Aquat. Toxicol.*, **1**, 85–99.

WHITE, R. H. and HAGER, L. P. (1975). A biogenetic sequence of halogenated sesquiterpenes. In E. D. Goldberg (Ed.), *The Nature of Seawater*. Dahlem Konferenzen, Abakon Verlagsgesellschaft, Berlin. pp. 633–650.

WICKSTRÖM, K., PYYSALO, H. and PERTTILÄ, M. (1981). Organochlorine compounds in the liver of cod (*Gadus morhua*) in the northern Baltic. *Chemosphere*, **10**, 999–1004.

WILLIAMS, P. M. (1967). Sea surface chemistry: organic carbon and organic and inorganic nitrogen and phosphorus in surface films and subsurface waters. *Deep Sea Res.*, **14**, 791–800.

WILLIAMS, P. M. and ROBERTSON, K. J. (1975). Chlorinated hydrocarbons in sea-surface films and subsurface waters at nearshore stations and in the North Central Pacific Gyre. *Fish. Bull. U.S.*, **73**, 445–447.

WILLIAMS, R. and HOLDEN, A. W. (1973). Organochlorine residue from plankton. *Mar. Pollut. Bull., N.S.*, **4**, 109–111.

WILLIS, D. E. and ADDISON, R. F. (1974). Hydroxylation of biphenyl *in vitro* by tissue preparations of some marine organisms. *Comp. Gen. Pharmac.*, **5**, 77–81.

WOFFORD, H. W., WILSEY, C. D., NEFF, G. S., GIAM, G. S. and NEFF, J. M. (1982). Bioaccumulation and metabolism of phthalate esters by oysters, brown shrimp, and sheepshead minnows. *Ecotoxicol. environ. Saf.*, **5**, 202–210.

WOLFE, I. N. I., ZEPP, R. G., BAUGHMAN, G. I. and GORDON, J. A. (1975). Kinetic investigation of malathion degradation in water. *Bull. environ. Contam. Toxicol.*, **13**, 707–713.

WURSTER, C. F. (1968). DDT reduces photosynthesis by marine phytoplankton. *Science, N.Y.*, **159**, 1474–1475.

YAMAGISHI, T., AKIYAMA, K., MORITA, M. TAKAHASHI, R. and MURAKAMI, H. (1978). Isolation and identification of 1,3,5-trichloro-2-(4-nitrophenoxy)benzene (CNP) in shellfish. *J. environ. Sci. Hlth*, **B13**, 417–424.

YOUNG, D. R., HEESEN, T. C. and McDERMOTT, D. J(1976). An offshore biomonitoring system for chlorinated hydrocarbons. *Mar. Pollut. Bull., N.S.*, **7**, 156–159.

YOUNGS, W. D., GUTEMANN, W. H. and LISK, D. J. (1972). Residues of DDT in lake trout as a function of age. *Environ. Sci. Technol.*, **6**, 451–452.

ZELL, M. and BALLSCHMITER, K. (1980). Baseline studies of the global pollution. II. Global occurrence of hexachlorobenzene (HCB) and polychlorocamphenes (toxaphene) (PCC) in biological samples. *Z. analyt. Chem.*, **300**, 387–402.

ZITKO, V. (1971). Polychlorinated biphenyls and organochlorine pesticides in some freshwater and marine fishes. *Bull. environ. Contam. Toxicol.*, **6**, 464–470.

ZITKO, V., HUTZINGER, O. and CHOI, P. M. K. (1972). Contamination of the Bay of Fundy—Gulf of Maine area with polychlorinated biphenyls, polychlorinated terphenyls, chlorinated dibenzodioxins and dibenzofurans. *Environ. Hlth Perspect.*, **1**, 47–50.

ZITKO, V., HUTZINGER, O. and CHOI, P. M. K. (1974). Determination of pentachlorophenol and chlorobiphenylols in biological samples. *Bull environ. Contam. Toxicol.*, **12**, 649–653.

Marine Ecology Vol. V, Part 4
Edited by Otto Kinne
© 1984 John Wiley & Sons Ltd

7. DOMESTIC WASTES

D. J. Reish

(1) Introduction

It was recognized only in the twentieth century that large amounts of domestic wastes could alter and degrade the marine environment. Although water pollution laws have existed in England for over 500 yr (Fair, 1964), it was not until the early nineteenth century that water was used to convey domestic wastes to a river or estuary for disposal. The Industrial Revolution lead to urbanization which resulted in ever larger quantities of waste being discharged into the aquatic environment. In the latter part of the nineteenth and early twentieth centuries, the pioneering work of Kolkwitz and Marsson (1908, 1909) in Germany and of Forbes and Richardson (1913) in the USA documented the deterioration of the aquatic biota in rivers receiving domestic waste discharges. Kolkwitz and Marsson (1908, 1909) proposed the saprobic system zones of organic enrichment on which the majority of later studies were based. They, as well as Forbes and Richardson, found that different populations of plants and animals inhabited different zones, based on the degree of contamination by sewage and the distance from the discharge. From their system of environmental classification evolved the indicator-organism concept: the presence of a species, or group of species, is taken to be indicative of a particular environmental condition, while the absence of the organisms concerned indicates the absence of those particular conditions.

That domestic wastes could alter the marine environment was first noted by Wilhelmi (1916). He found several polychaetes to be indicative of areas contaminated with domestic wastes. One of these species, *Capitella capitata*, played a similar role in polluted marine waters as did the oligochaete *Tubifex* sp. in fresh waters. Blegvad (1932) studied the benthic fauna in the vicinity of domestic outfall sewers in Copenhagen Harbour. He divided the region surrounding an outfall into three zones: an inner zone lacking animals, an intermediate zone containing only a few animal species and an outer zone unaffected by the discharge. In the latter part of the 1920s, the Water Pollution Research Board initiated comprehensive studies of many estuaries in the UK, including both polluted and unpolluted bodies of water, such as the Tees, Tay, Tamar, Dee and Mersey estuaries (Alexander and co-authors, 1935; Bassindale, 1938; Stopford, 1951). They found that the polluted Tees estuary was characterized by a smaller number of species than the unpolluted estuaries.

The studies on macrofauna and flora were paralleled by microbiological investigations which attempted to characterize marine waters in terms of their significance to public health (reviewed by Greenberg, 1956) or contamination of shellfish (reviewed by Baughman, 1948).

Characterization of the marine environment with respect to the effect of all types of pollutants began in earnest in the late 1960s. Sanitation districts, charged with treatment and disposal of domestic wastes, began to initiate surveys and monitoring programmes in order to analyse the environmental conditions around the area of the discharge. Perhaps the longest and most comprehensive monitoring programmes are those conducted at Hyperion Sewage Treatment Plant (City of Los Angeles) and Whites Point Treatment Plant (County of Los Angeles). Here, monitoring programmes were initiated in 1956 and 1971, respectively. Each of the plant operations discharged over 10^9 l d^{-1} of domestic wastes into the marine environment (SCHAFER, 1978). The results of much of this ecological work have appeared in various reports and publications of the South California Coastal Waters Research Project (SCCWRP) in 1973, 1975, 1976, 1977 and 1978.

(a) Sewage Treatment

With urbanization it became evident that the discharge of raw sewage into rivers and streams was no longer acceptable as a method of waste disposal. The source of disease agents (such as those of typhoid fever) was traced to domestic wastes. Treatment of sewage began in Europe in the late 1800s, followed by the USA and several other countries. However, sewage-treatment facilities are still not available in many parts of the world today. Sewage treatment varies according to local geography, population and whether or not storm waters and industrial wastes are included in the discharge. For convenience, sewage treatment may be categorized in terms of primary, secondary and tertiary methods. The flow of domestic waters through these treatment processes is illustrated in Fig. 7-1.

Primary treatment involves the removal of grit with screens, with the liquid-containing sediments flowing into a grit chamber, where sand and other heavy particles settle out and are removed. The liquid is passed into a primary sedimentation tank where about 30–55% of the solid organic wastes and sludge settle out. The liquid is then disposed into estuarine or oceanic waters through a pipe which discharges the wastes either intertidally or subtidally. These wastes may receive chlorination before discharge, which depends on variables such as the geography of the locality, season or local laws or practices. The sludge may be burned, buried, used as fertilizer or disposed of at sea—in the last case, either by barging or through a separate pipe.

Secondary treatment involves the foregoing steps followed by transferring the liquid from the primary sedimentation tank to either an activated-sludge process tank provided with aeration or to a trickling-filter process tank provided with a spraying mechanism. Each process involves the use of bacteria which act on the organic material in the sludge and break it down to less harmful substances. The fluid is next transferred to a final sedimentation tank where additional solid particles settle down. Both the liquid and the sludge are disposed of in the same manner as in primary treatment. Secondary treatment removes about 85–95% of the suspended solids. The biological oxygen demand (BOD) of the resulting liquid has been reduced by a similar amount.

The number of tertiary treatment plants (advanced treatment) has increased rapidly in the USA from 10 in 1968 to 992 in 1974 (METCALF and EDDY, Inc., 1979). Much of the impetus for the rise in tertiary treatment has been due to public demands for cleaner waters. Many different processes are employed in the tertiary treatment of domestic wastes. The particular process used depends on the nature of the waste and the intended

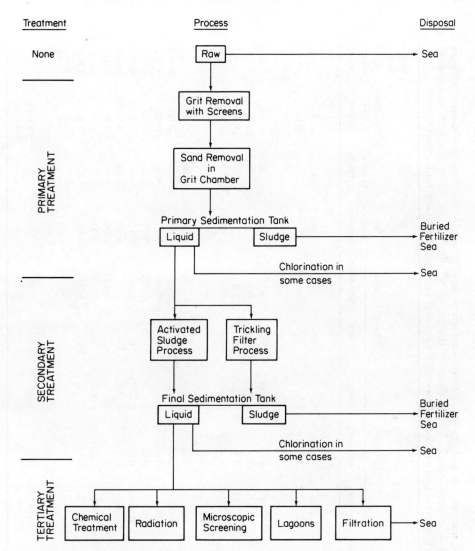

Fig. 7-1: Processing and disposal of domestic sewage via different treatment processes. Diagrammatic representation. (Original.)

use of the effluent. The objective of tertiary treatment is to reduce further the quantity of suspended solids, BOD and nutrients. Processes used in tertiary treatment include mechanical flocculation, microstrainers, sand filtration, tank filtration, oxidation ponds, chemical treatment and radiation (Fig. 7-1).

Comparisons of the various treatment processes in the reduction of BOD, nutrients, trace materials and synthetic organic compounds are given in Table 7-1. The concentration of copper, as an example, was 0.01 mg l^{-1} in the effluent from a tertiary treatment plant which was 1 order of magnitude greater than natural waters but was less than that measured from primary and secondary treatment wastes. Copper was over 3 orders of magnitude less in tertiary wastes than in the sludge from the Hyperion treatment plant.

Table 7-1

Concentrations of general constituents, trace metals and chlorinated hydrocarbons in final effluent of six municipal waste discharges in southern California (USA) compared with open oceanic waters (After SCHAFER, 1979, and YOUNG, 1979; modified; reproduced by permission of the Southern California Coastal Water Research Project and Los Angeles County Sanitation District)

Parameter	Whites Point		Hyperion			Orange County		San Diego: primary	Oxnard: primary	San Jose Creek: tertiary	Surface water: open oceanic waters
	Primary	Secondary	Primary	Secondary	Sludge	Primary	Secondary				
Flow rate (1 d^{-1} × 10^6)	1290	1·9	1225	378	17·5	723	109	466	42·4	—	—
pH	7·3	7·3	7·2	7·4	7·3	7·3	7·3	6·9	—	7·1	—
General constituents (mg l^{-1})											
Suspended solids	131	18	82	4	12,000	54	7·1	165	98	<4	—
Total nitrogen	49	—	31	9·5	740	48	8·3	36	—	20·2	—
Total phosphate-P	20	—	8	0·7	220	1·7	2·4	7·4	—	15·1	—
BOD	200	9	200	9	4700	160	8·8	230	258	5·8	—
Trace metals (mg l^{-1})											
Arsenic	0·004	0·004	0·01	0·01	0·2	—	—	0·014	0·006	0·004	—
Cadmium	0·018	0·008	0·02	0·01	1·3	0·02	0·025	0·004	0·009	0·003	0·0001
Chromium	0·22	0·09	0·18	0·03	12·8	0·045	0·049	0·15	0·05	<0·02	0·005
Copper	0·14	0·05	0·25	0·03	15·5	0·042	0·082	0·19	0·13	0·01	0·003
Lead	0·11	0·003	0·03	0·02	2·2	0·02	0·03	0·11	0·05	0·03	0·00003
Mercury	0·0003	0·0001	0·003	0·002	0·13	—	—	0·002	0·002	0·00004	0·0002
Nickel	0·2	0·22	0·2	0·15	4·1	0·3	0·18	0·08	0·2	<0·02	0·007
Selenium	0·007	0·007	0·002	0·003	0·069	—	—	0·003	—	0·004	—
Zinc	0·37	0·26	0·4	0·17	28	0·6	0·09	0·03	0·2	0·118	0·01
Chlorinated hydrocarbons (µg l^{-1})											
DDT	1·58[a]	1·58[a]	0·2[a]	0·2[a]	3·65	0·05[a]	0·05[a]	—	—	<0·00006	—
PCB	1·81[a]	1·81[a]	2·13[a]	2·13[a]	35·4	0·73[a]	0·73[a]	—	—	<0·0002	—

[a] Analyses were made on combined primary and secondary effluent

Tertiary treatment can reduce the concentration of DDT and PCBs more effectively than the other parameters measured; the levels of these synthetic compounds were about 5 orders of magnitude less in the tertiary effluent than some primary effluents and sludge. These data show the effectiveness in tertiary treatment of further reducing the concentrations of many potential pollutants. However, an equally important consideration, which always must be taken into account, is the increase in capital investment and operating costs as one proceeds from primary to tertiary treatment.

(2) Identification and Quantification

For many years it was a common belief that the oceans were large and could receive an infinite amount of wastes without any damage to this environment. Only recently it was realized that the marine environment is fragile and can become detrimentally affected by waste discharges. The emergence of the ecological movement in the 1960s led to the initiation of baseline studies prior to the installation of a new discharge. Since degradation of the marine environment had occurred around many discharges, environmental studies became necessary. Surveys were conducted with available equipment and known techniques. Environmental studies either followed those used in fresh water pollution surveys as outlined in Standard Methods (APHA, 1975) or oceanographic procedures; however, BARNARD and JONES (1960) had described the procedures to follow in a large-scale oceanographic survey several years earlier.

Standardization of techniques and procedures for biological studies has lagged behind those of a chemical or geological nature, probably because of the greater complexity of biological systems. Although chemical studies have been extended in recent years to include additional parameters with the advent of more sophisticated instrumentation, most of the procedures are already accepted. Standardization of procedures is attainable in chemical and geological surveys and monitoring, but less likely in biological programmes owing to the fact that the extreme variability of the habitat necessitates different collection procedures.

(a) Chemical Analyses

The extent of chemical analyses of sewage effluent, sludge and the receiving waters will depend on the size of the discharge and the type of sewage treatment. The greater the quantity of the discharge, the larger is the number of parameters analysed. The number of parameters analysed has increased considerably in the past decade or so, because of greater public concern for preserving the marine environment. Improvement of scientific instruments, especially the atomic-absorption spectrophotometer, has made it possible to obtain more accurate determinations in a shorter period of time. Metal and pesticide standards available from the US Environmental Protection Agency (EPA) aid in ensuring more precise chemical determinations. Table 7-2 indicates many of the parameters measured by the different sanitation districts in southern California, USA. Included are pH, suspended solids, BOD, nutrients, trace metals and chlorinated hydrocarbons. More thorough analyses were made of the 120 trace organics, 15 metallic constituents and 15 waste water parameters present in the waste waters of sewage treatment plants in southern California (YOUNG, 1979). These parameters constitute the list of priority pollutants compiled by the EPA. The results of this analysis indicated

Table 7-2

Mass emission rates of general constituents, trace metals and chlorinated hydrocarbons in final effluent of two municipal waste-water discharges in southern California (USA). (After McDERMOTT, 1974, and SCHAFER, 1976, 1978; modified; reproduced by permission of the Southern California Coastal Water Research Project)

Parameter	Whites Point				Hyperion			
	1971	1973	1975	1977	1971	1973	1975	1977
Flow (l d^{-1} × 10^9)	5·12	496	471	462·8	463	468	477	441
General constituents (metric tons yr^{-1})								
Suspended solids	167 000	128 000	130 966	102 000	38 000	36 000	40 460	27 300
Ammonia nitrogen	41 000	29 500	17 713	18 000	7010	6·080	6·625	7·590
Total phosphate-P	22 700	23 700	6912	5500	9·030	11 600	4289	3350
Trace metals (metric tons yr^{-1})								
Arsenic	Not measured	7·4	5·2	4·16	3·0	5·6	4·8	4·4
Cadmium	15·4	13·4	17	11·6	22·1	8·4	9·5	8·8
Chromium	462	357	377	176	134	136	62	57·3
Copper	267	233	198	116	106	65·5	91	88
Lead	144	104	118	87·8	29	21·1	14·2	13·2
Mercury	0·7	0·6	0·5	0·46	1·3	1·2	1·0	0·9
Selenium	8·2	5	6·1	7·4	3·8	8	9·5	4·4
Silver	10·8	5·5	6·1	3·7	0·7	14	9·5	13·3
Zinc	1400	882	683	388	212	117	110	141
Chlorinated hydrocarbons (kg yr^{-1})								
DDT	21 500	3720	<1098	730	73·7	304	777	88
PCB	5180	1280	<1427	838	364	936	1868	939

that only a few of these priority pollutants were above the 10 mg l^{-1} limit. Los Angeles County discharge at Whites Point had 15% of the priority pollutants above the 10 mg l^{-1} level. Effluents from secondary treatment plants had lower concentrations of primary pollutants than primary treatment wastes.

Chemical analyses of effluents and solid wastes from sewage treatment plants will continue to remain an important aspect of marine environmental studies. Not only will we see further improvement and sophistication of the scientific instruments used in chemical analysis, but also an increase in the number of parameters measured for identifying potentially hazardous or toxic chemicals.

(b) Geological Analysis

Particle-size determination of the sediments may indicate whether or not the particles have settled on the bottom in sufficient quantity to alter its composition. The data will be more meaningful if the sediment characteristics are measured prior to waste-disposal initiation. Samples are taken with a coring device measuring 5 cm in diameter; the upper 10 cm of the sample is placed in a 500 ml screw-top jar for transport to the laboratory (HOLME, 1964). Generally, a 25 g sample is used for making a particle-size analysis. The larger particles (>0·0625 mm) are separated by graded sieves. Sediments with particle size <0·0625 mm can be measured by a variety of techniques based on the sinking rates of particles through water (BARNES, 1959).

Most of the sediments from estuaries and inshore waters where municipal wastes are discharged fall within the sand–silt–clay fractions (Table 7-3). The ϕ scale is used as a convenience to denote the size of a particle by number rather than by its diameter in millimetres. A particle with a 1·0 mm diameter is defined as zero. A negative scale is employed (Table 7-3) for sediment particles larger than 1·0 mm and a positive scale for particles smaller than 1·0 mm. Data for percentages of sand, silt or clay, may be given in tabular form, plotted on a triangular diagram or presented pictorially (Fig. 7-2). Biologists use the median particle diameter for plotting sediment distribution as a method of analysing benthic communities. The distribution of many communities is directly related to the sediment characteristics (JONES, 1969; Fig. 7-3).

Table 7-3

Classification of marine sediments (Original)

Size of particle (mm)	ϕ scale	Name
2–1	0 to −1	Very coarse sand
1–0·5	0–1	Coarse sand
0·5–0·25	1–2	Medium sand
0·25–0·125	2–3	Fine sand
0·125–0·0625	3–4	Very fine sand
0·0625–0·0039	4–8	Silt
<0·0039	8–14	Clay

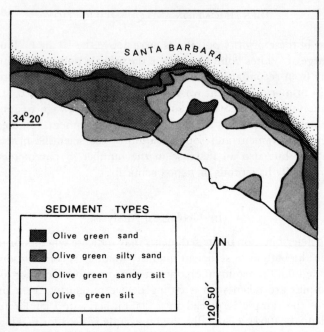

Fig. 7-2: Distribution of various sediment types off the coast of
Santa Barbara, California, USA. (After JONES, 1969; modified.)

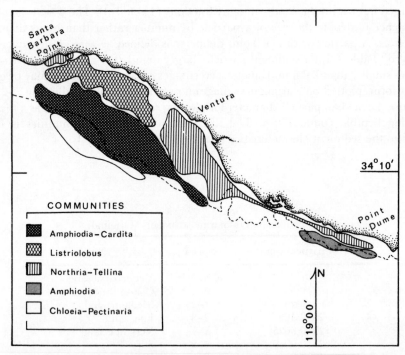

Fig. 7-3: Distribution of benthic communities off Santa Barbara and Ventura
Counties, California, USA. *Amphiodia*, Ophiuroidea; *Cardita* and *Tellina*,
Pelecypoda; *Listriolobus*, Echiura; *Choleia* and *Pectinaria*, Polychaeta. (After
JONES, 1969; modified.)

(c) Biological Surveys and Monitoring

Biological techniques of sampling the water column, epibenthos, infauna and intertidal environment in the vicinity of marine discharges have not yet been standardized. There is a lack of standardization also in the treatment of the sample once it has been taken. Probably complete standardization will never be achieved; in fact, it does not seem absolutely essential that we achieve that goal. The marine environment, itself, is not 'standardized', and the methods employed depend to a considerable degree upon the geography of the area, as well as on the type of research vessel available. Techniques are different for sampling within estuaries and offshore waters. Sampling intertidal sandy beaches or mud flats is different from sampling rocky shores. Nevertheless, manuals or reviews have been written which list the different methods for sampling various marine environments (e.g. BARNARD and JONES, 1960; HOLME, 1964). 'Standard Methods'—for many years an important reference in the examination of freshwater—has included the marine environment to a limited extent in its recent edition (APHA, 1975). Some attempts have been made to evaluate and compare various sampling devices. WORD (1976), for example, compared the effectiveness of 7 benthic grabs (Table 7-4).

Over the past 2 decades, as more and more investigators have become involved in marine pollution studies, research has become increasingly sophisticated. Earlier studies simply recorded the scientific name of the species along with the number of individuals present (REISH, 1959a). Later, a species diversity numerical value was introduced which divided the number of specimens within a sample by the number of species present (BARNARD and JONES, 1960). Further refinement in the statistical treatment of data came with the Shannon–Wiener index of diversity and Sanders rarefaction technique (PEARSON and ROSENBERG, 1978). These indices are especially useful in comparing stations within an area. However, comparisons of numerical values from one study to another should be made with caution because of the possibility that different samplers and methods have been employed (PEARSON and ROSENBERG, 1978). Station data can be compared by the Moriseta–Ono index of similarity or by computer cluster analyses (PEARSON and ROSENBERG, 1978; BENDER and co-authors, 1979).

Other refinements which add greater mathematical validity to the data include replication of sample collection (BENDER and co-authors, 1980) and the use of finer sieves in processing the sample (REISH, 1959b). Agreement as to the number of replicates to be taken or the mesh size to be employed does not yet exist. However, one must not forget that the time and cost of sample processing at a given station increase with each additional replicate and use of a finer mesh size.

During the initial stages of planning a monitoring programme, the computer facilities and programs available must be considered so that the data obtained can be analysed quickly and efficiently at the lowest possible cost. It may be feasible, after data accumulation for 2–3 yr, to reduce the frequency and number of samples, depending on the degree of natural variability. By analysing existing benthic data with the aid of computer-assisted programmes, SHARP and co-authors (1979) found that the monthly monitoring programme in Lavaca Bay (Texas, USA) at 11 stations could have been reduced to sampling 3 stations twice a year. It is therefore important that the monitoring programme be appraised periodically in order to determine if the results obtained are yielding meaningful results. Monitoring programmes should not become exercises for collecting data for the data's sake.

Table 7-4

Comparison of the effectiveness of benthic samplers (After WORD, 1976; modified; reproduced by permission of Southern California Coastal Water Research Project)

Criterion	Sampling device						
	Box corer	Van Veen 1	Van Veen 2	Smith–McIntyre	Ponar	Shipek	Orange-peel
Uniformity in area sampled	Uniform	Uniform	Uniform	Uniform	Uniform	Variable	Variable
Penetration to 10 cm	Yes	Yes	Yes	Yes	Variable	Variable	Yes
Leakage from sampler	Minimal	Yes	Minimal	Minimal	Minimal	Yes	Yes
No descending pressure waves	Minimal	Yes	Minimal	Minimal	Minimal	Yes	?
Variation in area or depth sampled	Minimal	Can vary	Minimal	Minimal	Variable	Variable	Variable
Percentage of obtaining samples	100	100	100	63–86	77–100	20–100	48–100
Number of operators required	3	2	2	2–3	1	1	1

Water column

Numerous methods are employed for characterizing the water mass near municipal waste discharges. Some of the field techniques involve the specific identification of phytoplankton and/or zooplankton, others measure holistic properties, such as quantities of chlorophyll *a* or ATP, productivity or biomass. Each of these involves a different method of sampling and yields different numerical results.

For species identification, plankton is collected with various types and sizes of plankton nets. These may be towed horizontally at a specific depth or vertically. Quantitative data on plankton populations have also been obtained through the use of a plankton pump or plankton recorder (e.g. LENZ, 1972). Identification of plankton, especially phytoplankton, is extremely important in the vicinity of municipal waste discharges because of possible eutrophication effects resulting from introduction of nutrients with the domestic wastes. Whereas in rivers and lakes the species composition of phytoplankton in the vicinity of domestic outfall sewers is usually well known (HOLLAND, 1968; PALMER, 1969), related knowledge in marine waters is limited.

Quantification of phytoplankton is facilitated by counting cells in which the various species contained in a 1·0 ml aliquot are identified and the number of individuals counted. The number of species and individuals per unit volume are determined. Small planktonic organisms are enumerated by passing the fluid through a Coulter Counter. Unfortunately, this technique does not distinguish between species, but the number of cells per unit volume is recorded.

Plankton biomass is measured by drying the sample of 90 °C overnight and determining the weight of the sample per m^3. Approximate measurements of phyto- and zooplankton can be made by the use of an appropriate sized mesh. Most phytoplankters readily pass through a 350 μm mesh, which largely separates these 2 major components of the plankton. The data are recorded as mg m^{-3}.

Chlorophyll *a* is found in all organisms which carry out phytosynthesis. Its concentration is a good measurement of phytoplankton activity. The chlorophyll *a* concentration can be determined either spectrophotometrically or fluorimetrically (APHA, 1975). A known volume of water is passed through a Millipore filter which retains the plankton. Pigments are extracted with 90% acetone and the optical density is measured in the red portion of the spectrum with a spectrophotometer. The fluorimetric procedure is similar, the optical density being determined at wavelengths of 430 and 663 nm. With both methods the data are recorded as mg m^{-1} or μg l^{-1} of chlorophyll *a*.

The adenosine triphosphate (ATP) concentration in the water is an accurate measure of the amount of plankton present, since ATP occurs only in living cells, i.e. it is not associated with non-living matter. A sample of known volume is filtered through a net (183–250 μm mesh) to remove the larger zooplankton. The net is then immersed in boiling Tris buffer (100 °C) and the concentration of ATP determined with a scintillation counter using the bioluminescent reaction with luciferin–luciferase. The data are recorded as mg l^{-1} (EPPLEY and co-authors, 1972).

Productivity is defined as the amount of organic carbon synthesized through photosynthetic activity in a given volume of water per unit time. It can be measured as the amount of oxygen released through photosynthesis or as the amount of ^{14}C fixed in organic matter. In both cases water samples are placed in light and dark bottles for a defined period of time after which either the dissolved oxygen is determined or the ^{14}C measured. Results are expressed as g C m^{-2} (time)$^{-1}$ (APHA, 1975).

In laboratory studies, pure cultures of a species of phytoplankton are inoculated into a water sample and the number of cells or the rate of productivity is determined after a defined period of time. This type of study is particularly useful in comparing the productivity of receiving waters with natural waters located some distance from the discharge. Further refinements are obtained by enriching sea-water samples in order to overcome possible effects of nutrient depletion so that any inhibitory reaction can be distinguished. The data are recorded as cells ml^{-1} or as mg C m^{-3} (YOUNG and BARBER, 1973; THOMAS and co-authors, 1974).

Zooplankton is usually a heterogeneous assemblage of animals. Nearly every animal phylum contains species which spend at least part of their life in the water column. The techniques for collecting zooplankton for biomass measurements or species enumeration are identical with those used for phytoplankton, except for the mesh size of the net. Much effort can be saved by consulting a manual of zooplankton sampling procedures (UNESCO, 1968; STEEDMAN, 1976) and by carefully working out a monitoring programme prior to practical research. If the purpose of the sampling programme is to identify both the phyto- and zooplankton species, the mesh aperture should be 0·063 mm. If, however, phytoplankters are to be studied by some other method—such as the measurement of chlorophyll a—then the mesh aperture may be increased to 0·093 mm. The larger mesh size will greatly facilitate identification of zooplankton by excluding most phytoplankters and some debris.

Zooplankton data are quantified in terms of volume, dry weight or species identification and enumeration. The volume of the plankton is measured by placing the preserved sample in a graduated cylinder and allowing the organisms to settle overnight; thereafter the volume can be read from the scale. Zooplankton volume can also be measured by displacement. In this case, the volume of the organisms and the preservative is measured, the sample is filtered and the preservative which passes through is measured and subtracted from the initial volume. Biomass weight is obtained by splitting the initial sample at the time of collection into two equal halves. One half of the sample is preserved for species enumeration and the other is dried at 50 °C, then weighed. Macrozooplankton is usually identified with the unaided eye. Microplankton is generally identified in aliquot samples with the use of dissecting and/or compound microscopes. The data are recorded as ml, g or number of individual species m^{-3}.

Benthos

The subtidal benthic environment is rich and diversified. Epibenthos organisms live on top of the substratum; they vary in size from smaller forms, such as harpacticoid copepods and small algae, to larger crustaceans, echinoderms, kelp and fish. The infauna or benthic fauna are subdivided according to size into macrofauna, meiofauna, microfauna or algae. To the macrofauna belong organisms which are too large to pass through a 1·0 mm sieve, including conspicuous animals such as polychaetes, crustaceans, molluscs and echinoderms. Members of the meiofauna readily pass through a 1·0 mm sieve but are retained on a 0·1 mm sieve; this group includes copepods, nematodes, flatworms, gastrotrichs and larval stages of macrofauna. The microfauna largely consist of protozoans which pass through a 0·1 mm sieve (MARE, 1942). Bacteria are usually considered as a separate category since they require different techniques in sampling and identification.

Macrofauna

Methods for sampling the benthic macrofauna have not been standardized, and it is likely that they never will be. Factors contributing to the lack of standardization include variations in the type of sampler, and substratum, in the techniques employed for processing the sample, as well as in the capabilities of the research vessel used. Sampling a shallow estuary requires very different methods than sampling an offshore locality. With the invention and widespread use of SCUBA, it is now possible to obtain samples of nearshore subtidal waters which heretofore have been difficult or impossible to collect from land or a research vessel.

WORD (1976) compared the efficiency of seven benthic grabs, including ponar, box corer, shipek, two types of Van Veen, orangepeel and Smith–McIntyre. He found the box corer to be an effective device followed by the Van Veen 2 and Smith–McIntyre (Table 7-4). However, the box corer requires three persons for operation, the highest personnel demand of the seven devices tested. Collection of samples from rocky substrates is always difficult, regardless of the type of sampler used. The best samples from this habitat could probably be obtained by SCUBA divers, unless the water is too deep.

Replication of samples at a particular site is made in order to insure an adequate statistical base for an accurate description of the benthos at that locality. No consensus exists as to what constitutes adequate replication. WORD (1976) found that even after 10 replicates the asymptotic point for the addition of new species had not been reached. However, most of the additional species after the first replication were rare species so that only about 10% of the specimens in the second replicate were members of the new species group. The percentage is less for the third sample, and so on. BENDER and co-authors (1980) found that it took 10 replicates in order to encounter 90% of the species present in 16 replicates; however, when species present less than 6 times were excluded, the 90% recovery was reached by the third replication. A solution to the problem of replication of samples is difficult, in most instances it will be directed by the resources available. The reviewer questions the advisability of exceeding three replicates per station, especially in monitoring programmes. Further, the shorter the period of time between two successive sampling periods the smaller is the number of replicates necessary to describe adequately benthic communities.

It is better to wash and preserve the specimens collected on board a ship in order to ensure adequate penetration of the formalin. A 10% solution of buffered formalin is generally used to preserve the material retained on the sieve. However, it is not always possible to wash the samples when they are taken in an estuary from a skiff. Either periodic trips should be made to a dock where washing and preserving the samples can be accomplished, or the entire sample should be placed in a plastic bag, and full-strength, buffered formalin added. The samples should be washed within a day or two with the material retained on the sieves transferred to either 70% ethanol or isopropyl alcohol.

Since marine benthos studies began with the pioneering work of PETERSEN (1913) in the early part of this century, the mesh size of the sieve has been progressively reduced. PETERSEN used a 1·5 mm mesh size to wash his collections free of sediment. HARTMAN (1955) employed a 0·7 mm mesh sieve in her study of California benthos. Currently, the 0·5 mm mesh sieve is used in many benthic studies, including those taken around domestic outfall sewers. Since each progression to a smaller mesh size leads to an increasing amount of specimens and debris to sort through, it is unlikely that an even smaller mesh size sieve will be introduced at a later date. In a study to emphasize the importance of sieve size in washing quantitative samples, the reviewer (1959b) demonstrated that all species of macro-invertebrates were retained on the 0·5 mm sieve, and 68% of the specimens. If, however, the nematodes are excluded from the totals—they are generally not included in the macro-invertebrates—the proportion of specimens recovered on the 0·5 mm sieve increases to 88%. Clearly, the selection of an appropriate-sized sieve to wash quantitative samples is as critical as the type of sampler employed.

The processing of samples in the laboratory is the most time-consuming task in any quantitative study. If sorting into major animal groups is not done carefully, much of the time, effort and precision of the sampling programme are lost. In some laboratories a 10% solution of rose bengal is added to the sample at the time of initial preservation. Living organisms take up the dye and are coloured rose; this facilitates sorting. It is important that all residues be examined to make certain that the sorter did not overlook some specimens.

The proper identification of benthic species by competent specialists is another critical step in the treatment of the sample. Without trained individuals previous work in sample processing is wasted. Requirements for the training of a specialist in this field have been discussed in an earlier review (REISH, 1979).

Meiofauna

Many marine phyla have representatives in the meiofauna, either as larvae or adults. A number of authors include the foraminiferans in this group; others prefer to treat them separately. Thus far, the meiofauna has hardly been used as an indicator of domestic-waste pollution. Undoubtedly this group holds much promise for the future. The identification of meiofauna species is no easy task, since the group comprises so many different phyla. A manual was edited by HULINGS and GRAY (1971) and it should be consulted if a meiofaunal programme is contemplated.

Microfauna

This group is largely composed of protozoans. Especially the foraminiferans have been used extensively in marine pollution studies, more so than other protozoans. Detailed procedures for the collection, staining, preservation and sample sorting have been described for foraminiferans in the meiofaunal manual edited by HULINGS and GRAY (1971).

Macrophyta

Included under macrophytes are Chlorophyta, Cyanophyta, Phaeophyta and Rhodophyta, as well as the Spermatophyta. Since most of these macrophytes require a solid substratum for attachment, they are frequently absent from the vicinity of sewer outfalls. Little is known about the responses of macrophytes to stress from domestic wastes. An extensive study was conducted by DAWSON (1959) in southern California,

USA). SCUBA gear has greatly facilitated careful and selected collection of organisms from inshore waters.

Quantification of algal distribution may be difficult because it is frequently impossible to determine how many individuals of a given species are present within a defined area. The most commonly used method involves a transect, e.g. from the intertidal zone into the subtidal zone. For quantification, quadrats along this line are selected at random. Photographs of the quadrats are taken from which the larger species can be identified and the percentage of cover determined (MURRAY and LITTLER, 1974). Smaller forms, including epiphytes, are collected. For each species the data are recorded as the percentage of surface area covered. At the same tidal level, individual species from quadrats of two transects can be compared for statistical significance of distribution problems by using the one-tailed Wilcoxen single-rank test (MURRAY and LITTLER, 1974). The quadrat method along a transect line has also been used to measure the amounts of biomass as either for each species or for the algae occupying the entire sample area.

Micro-organisms

This group consists of bacteria, viruses and protozoans (*Entamoeba histolytica*), as well as metazoan eggs (i.e. of nematodes, trematodes or cestodes). Only bacteria and viruses are of major concern for the marine microbiologist; human disease caused by protozoans or by metazoan ova and transmitted in marine waters are unknown. Enteric micro-organisms, including potentially pathogenic species, enter marine waters via sewage discharges.

Since detection and measurement of pathogenic bacteria are difficult and time consuming and since non-pathogenic coliform bacteria are generally found in all warm-blooded animals, the presence and concentration of coliforms in receiving waters is considered presumptive evidence of the potential presence of human disease agents. Measurement of coliforms is a standardized laboratory procedure for public health laboratories (APHA, 1975). The data are recorded as the most probable number (MPN) per 100 ml (Table 7-5). The bacteriological standards for marine waters for the State of California (USA) are summarized as follows (SCCWRP, 1972):

'In areas used for body contact sports, the samples of water from each station shall have a MPN of coliforms less than 1000/100 ml, provided that not more than 20% of the samples at any station in a 30-day period exceeds 1000/100 ml, and furthermore, no repeat sample taken within 48 hours shall exceed 10 000/100 ml. In an area where shellfish are harvested for human consumption, the median coliform concentration shall not exceed 70/100 ml and not more than 10% of the samples shall exceed 230/100 ml'.

Enteroviruses are not normally found in intestinal tracts of invertebrates but many viruses have been identified in human faeces. Viruses present in sewage include adenovirus, coxsackievirus, echovirus, virus of infectious hepatitis, poliovirus and reovirus (SCCWRP, 1971). Not all of these viruses were found in sewage discharged into estuarine or oceanic waters, but the ingestion of raw shellfish has been recognized as a significant mode of transmission of infectious hepatitis in the past 2 decades. Techniques

Table 7-5

Concentration (mean) of coliform bacteria in sewage effluents in southern California, USA (After
YOUNG, 1979; reproduced by permission of Southern California Coastal Water Research Project)

Locality	Effluent characteristics (MPN per 100 ml)		
	Primary	Secondary	Sludge
Whites Point	$7 \cdot 5 \times 10^6$	$7 \cdot 5 \times 10^4$	No data
Hyperion	21×10^6	$9 \cdot 3 \times 10^4$	$3 \cdot 9 \times 10^6$
Orange County	34×10^6	$1 \cdot 1 \times 10^4$	No data
Pt. Loma	80×10^6	No data	No data

used to determine virus concentrations are listed as 'tentative' in Standard Methods
(APHA, 1975). The data are recorded as plaque forming units (PFU) per litre. Studies
on the occurrences of viruses in the vicinity of sewage discharges are limited (SCCWRP,
1973).

Fish

A knowledge of what species of fish are present is important in any survey or monitoring programme dealing with domestic waste discharges because of their importance as
commercial and recreational resources. Fish can also be important to human health
since they can serve as agents in the biomagnification and transfer of contaminants to
mankind. Since the concentrations of contaminants are higher in sediments than in
other environments, monitoring of fish populations has emphasized the collection and
enumeration of demersal species.

Demersal fish are mostly caught by trawling; however, where the bottom topography
is rocky, other means of collection are required. Methods of trawling vary according to
the dimensions of net openings and mesh sizes of the netting, day/night collections,
speed and the duration of trawling. Data comparisons between two studies must be
made with caution because of these variabilities. An attempt was made to evaluate
different trawling nets and techniques as a first step towards standardization of collecting methods (MEARNS, 1974). The efficiency of each method can be measured addition-
ally by cameras or SCUBA diver observations (ALLEN, 1975a). Since 1974 many of the
trawlings in southern California inshore waters have been made with 7·6 m otter trawl
with a 1·25 cm cod-end mesh and towed at 3·7 km h^{-1} for 10 min over the bottom. It
was necessary to double trawling duration in depths greater than 200 m because of the
low fish population density (ALLEN and MEARNS, 1977).

The data collected consisted of enumeration, weight and length of species. Each
specimen was examined for diseases or abnormalities. These data can be analysed as to
species diversity, relationship of length, weight or diseases to distance from a domestic
waste discharge or water depth. Since natural variations in fish populations may occur,
comparisons of species from contaminated and non-contaminated areas must be made
with caution (ALLEN, 1975b).

Data Analysis

In the past, long species lists were published which had not necessarily been subjected to critical analysis. The species were more or less subjectively grouped into certain community structures (e.g. THORSON, 1956; REISH, 1959a). The publication of long species lists with the number of specimens per sample is important but expensive. This reviewer laments that exclusion of such data from publication results in the loss of information on which many of the conclusions are based. Data from species list were plotted on maps as number of species per sample or as number of specimens per m² (SMITH and GREENE, 1976).

The first attempt at a numerical reduction of the data obtained consisted in simply dividing the number of specimens by the number of species. Comparisons of these indices were unsatisfactory since a high number could be obtained near a stressed environment as well as a low one. The analysis became more sophisticated through the use of computer-assisted programmes involving, for example, species diversity and faunal similarity indices which can be represented in the form of a trellis diagram (Fig. 7-4) or dendrogram (Fig. 7-5). The advantages of some of these analyses, using the same data base, were the subject of a critical appraisal by SMITH and GREENE (1976). They concluded that different indices were designed to measure different ecological quantities and each had their merits; however, they favoured the use of either the Brillion (PIELOU, 1969) or Shannon–Wiener's index of diversity. Both of these indices emphasize the number of species and distribution of individuals among the species.

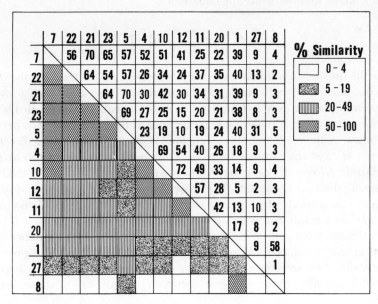

Fig. 7-4: Percentage similarity between pairs of stations on Byfjord Estuary, Sweden, in October 1971. (After ROSENBERG, 1977; modified; reproduced by permission of Elsevier/North Holland Biomedical Press.)

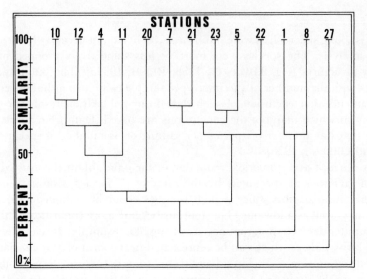

Fig. 7-5: Dendrogram showing similarities between benthic stations
in Byfjord Estuary, Sweden, in October 1971. (After ROSENBERG,
1977; modified; reproduced by permission of Elsevier/North Holland Biomedical Press.)

The Shannon–Wiener index of diversity (H') is commonly employed. The data are
reduced to a single number using the equation

$$H' = -\sum_{j=1}^{s} \frac{n_j}{N} \ln\left(\frac{n_j}{N}\right)$$

where n_j is the number of individuals in the jth species, s is the total number of species
and N is the number of individuals. The numerical value varies from 0 in a highly
stressed environment to over 5–6 in a clean environment. However, BOESCH (1977)
cautions the use of H' as a pollution scale because of the variations both between and
within habitats. While the H' numbers were low, WARE (1979) showed a direct relationship between H' and the distance from the Terminal Island sewage discharge in Los
Angeles Harbor. H' was 0·3 near the terminus with values of 0·56, 0·73, 0·82 and 1·05
with increasing distance. PEARSON (1975) documented a steady decrease in H' results
over a decade as a result of pulp mill wastes. ROSENBERG (1976) demonstrated an
increase in H' as a result of a pollution abatement programme.

The rarefaction technique of SANDERS (1968) interpolates the number of species
distributed among a reduced number of specimens from a sample and presents these
data graphically. This technique reduces the data in such a way that it is possible to
discern readily the effect of increasing or decreasing the amount of waste discharge on
marine life.

The use of the index of faunal similarity or index of affinity is useful in comparing the
fauna of several stations along a polluted gradient. The data derived can be presented in
the form of a trellis diagram (Fig. 7-4) or dendrogram (Fig. 7-5). The index of similarity
can be determined with a hand calculator and the dendrogram drawn or the entire

operation done with the use of a computer. One such index is that described by MORISETA (1959); it is defined as

$$C_\lambda = \frac{2 \sum x_i y_i}{(\lambda_A + \lambda_B) N_B N_A}$$

where A and B are the numbers of specimens in the two samples, x_i and y_i are the species found in both samples and λ_A and λ_B are Simpson's index of diversity, which is defined as

$$\lambda = \frac{\sum n_i (n_i - 1)}{N(N - 1)}$$

where n_i is the number of specimens in species i and N is the total number of specimens within the sample. A numerical value is obtained which ranges from 0 to 1, with 1 indicating complete identity and 0 complete dissimilarity. The index of similarity for several stations is represented in and by a trellis diagram in Fig. 7-4 with the same data represented in a dendrogram in Fig. 7-5. Variable shadings in the trellis diagram indicate stations with similar populations. Similarities and differences are more apparent with the dendritic presentation of data (Fig. 7-5); this accounts for its wide acceptance in the past few years.

BELLAN (1979a) proposed an index of pollution based on the benthic polychaete population and defined as the ratio of the number of specimens of polluted water indicators, (i.e. *Platynereis dumerili, Theostoma oerstedi, Dorvillea rudolphi*) to the number of clean water specimens (Table 7-6). The ratio increases with increasing intensity of pollution and the figure obtained always exceeds 1 in areas under environmental stress. This particular treatment of data was based on offshore data from southern France; its usefulness is probably limited to highly polluted, protected waters where 1 or 2 pollution indicator species are present. Table 7-6 gives a wide range of numerical values from less than 1·0 in clean waters to 398 in polluted waters. A good correlation is shown between H' and Bellan's index.

WORD (1979) described an infaunal index based on the feeding strategies of the dominant benthic species. The impetus, in part, was the potential reduction of time spent in identifying hundreds of different species of benthic invertebrates in monitoring programmes in southern California. The 47 selected species were grouped into 4 categories: (i) dominated by suspension feeders (primarily amphipods and ophiuroids); (ii) dominated by a combination of suspension and surface detritus feeders (equal numbers of amphipods, ostracods, polychaetes and pelecypods); (iii) dominated by surface deposit feeders (pelecypods, a gastropod and a polychaete); and (iv) dominated by subsurface feeders (primarily polychaetes).

The index number was calculated for each sample according to the equation

$$\text{Infaunal index} = 100 - \left[\frac{33\frac{1}{3}(0_{n_1} + 1_{n_2} + 2_{n_3} + 3_{n_4})}{n_1 + n_2 + n_3 + n_4} \right]$$

where n_1 is the number of specimens in Group 1, n_2 in Group 2, etc., and $33\frac{1}{3}$ is a scaling factor dimensioned so that the number will range from 1 to 100. The higher the number

Table 7-6

Index of pollution and species diversity (H') from vicinity of Marseille, France (After BELLAN,
1979; modified; reproduced by permission of Station Marine d'Endoume, Marseille)

Station	Index of pollution	H'
Cortiou (clean to polluted)		
C3	0·2	2·7
C4	0·5	2·79
C5	2·49	2·48
C6	5·17	1·98
C7	9·75	1·82
Golfe de Fos (clean to polluted)		
F1	0·23	3·22
F2	0·42	2·51
F3	1·03	1·60
F4	2·39	1·45
F5	4·36	1·01
F6	34·78	0·48
Vieux Port (polluted)		
P3	2·59	1·83
P1	136·94	1·30
P2	398·0	0·79

the more natural the environment and, conversely, the lower the number, the greater is
the degradation. An example of the data obtained by these calculations for southern
California (USA) is represented graphically in Fig. 7-6. The lowest numbers were noted
in the Los Angeles–Orange Counties regions which corresponded to the location of the
three largest outfall sewers. Other dips in the curve occurred at Oxnard and San Diego,
the locations of the next largest discharges.

Many indices for presenting biological data have been devised; some of these have
been presented here. Some have been applied more widely than others, and two new
indices by WORD (1979) and BELLAN (1979a) are presented which have thus far only
been used at the initial locality. The underlying motivation of these techniques is to
reduce the vast amounts of biological data into a readily assimilable form that can be
understood not only by biologists but also by members of other disciplines concerned
with environmental matters. Of course, the usefulness of any of these indices depends on
the accuracy of the data base.

(3) Fate of Domestic Waters Discharged into the Marine Environment

(a) Dispersion

Disposal of domestic wastes into marine waters is accomplished in different ways. In
most cases the treated or untreated wastes are discharged through pipes which release
the sewage on the water surface or in submarine waters. Subtidal pipes may be provided
with many holes to facilitate dispersion, bent at an angle to direct the discharge
upwards, or they may terminate in a Y to aid dispersion. The type of pipeline depends
on variables such as the physical features of the area, the location of the thermocline and

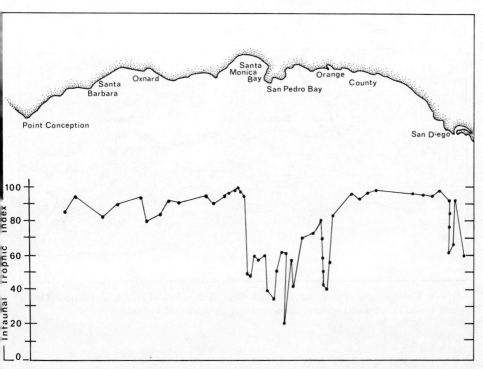

Fig. 7-6: Infaunal trophic index values for southern California, USA, for 1977. (After WORD, 1977; modified; reproduced by permission of Southern California Coastal Water Research Project.)

water currents. Cities located near areas with a broad offshore continental shelf, such as New York, or within a semi-enclosed sea, such as the North or Baltic Sea, are faced with different problems of sewage disposal than cities located near deep inshore waters, such as Los Angeles. A pipeline to convey the sewage of New York City to deeper waters would have to be too long and costly; the material is transported by barge and discharged at a dumpsite located offshore. Los Angeles disposes of its solid wastes separately from the liquid wastes into the head of Santa Monica Canyon, a submarine canyon which is located 11 km offshore at a depth of 100 m. As a general rule, smaller cities tend to discharge their wastes either intertidally or in shallow inshore waters. Marseille (France), however, discharges much of its untreated wastes into intertidal waters at Cortiou (Fig. 7-7).

A final and important consideration in siting pipes for discharging sewage into the sea is the availability of funds. The longer the pipes and the deeper the water where the terminus is located, the greater is the cost of domestic waste disposal. As towns grow into cities, outfall sewers are relocated from shallow waters or estuaries to offshore submarine waters. In fact, it has been suggested that the metropolitan Los Angeles area extend its pipelines into even deeper waters and discharge the buoyant sludge into depths of 300–400 m (JACKSON and co-authors, 1979).

Discharge of sewage into intertidal waters results in unsightly inshore waters and contamination of the beaches. The extent of the affected area depends on the volume and

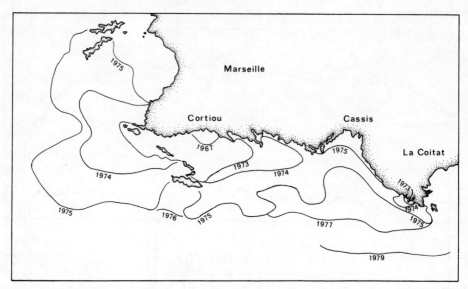

Fig. 7-7: Enlargement of polluted zone with time off the coast of Marseille, France. (Original, based on personal communications from BELLAN and PICARD.)

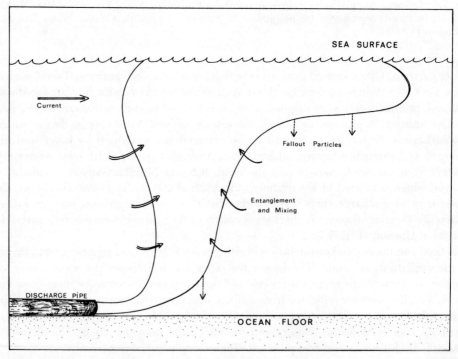

Fig. 7-8: Fresh water plume arising from discharge of liquid sewage wastes into the marine environment. (Original.)

type of treatment, tidal fluctuation, currents and local physiographic features. Dispersion of domestic waters into intertidal waters is an unsatisfactory method of disposal because of its unsightly appearance and poor dispersal of wastes. Siting a pipeline into submarine waters results in more thorough dispersion of the wastes with sea water. The rate of dispersion can be increased through the use of multiple outlets along the pipes. Dispersion effectiveness is directly related to the depth of the pipe. An open-ended pipe in 6·1 m of water would provide an initial dilution of 10%, but the dilution would be 0·5–1% at a depth of 30·5 m (LUDWIG and STORRS, 1970).

A plume from a discharge pipe may be compared to smoke arising from a chimney (Fig. 7-8). Since the density of the discharge is primarily fresh water containing various sized particles, the waste is buoyant and will rise. The larger particles settle down and the remainder are entrained with the plume as it moves upward. The plume continues to rise until its density becomes equal to that of sea water or a thermocline exists which prevents a further rise. Particles continue to settle down with the liquid phase being subjected to diffusion and transport by the currents.

Frequently it is impossible to see any surface indication of the discharge when the pipe is sited in deep waters because of the thermocline or density characteristics (PEARSON, 1956; BROOKS, 1960; LUDWIG and STORRS, 1970). The sewage plume from Hyperion treatment plant rises from a discharge depth of 60 m to about 20 m below the surface (KOLPACK, 1979).

(b) Water Column

In attempting to assess whether or not a particular discharge has altered the environment, it is important to know the natural variability of the parameters measured. This is particularly true with water-column data, since currents can transport water into an area with entirely different chemical characteristics. The chemical composition of the natural water mass will vary according to depth, distance from shore, amount of runoff, composition of the terrestrial environment, density of the population, etc. Table 7-1 summarizes the concentration of some of the trace metals in the surface waters of the world. In determining whether or not an elevated concentration is the result of some natural or man-caused event, it is important to have adequate data characterization for statistical validity. Comparison of the concentration of trace elements in the effluent of the different municipal wastes in southern California (USA) to national ocean waters indicate that they are above background levels at the point of discharge (Table 7-1). Trace elements were highest in the sludge disposed at the head of Santa Monica Canyon and lowest in the tertiary wastes from the San Jose Creek treatment plant.

Data on the spatial distribution of nutrients, trace elements, etc., in the waste discharge plume are limited in number because the dilution of the effluent, generally 0·5–1% (LUDWIG and STORRS, 1970), soon dissipates the liquid, making it impossible to be detected above background levels.

BALCH and co-authors (1976) compared 13 water-quality parameters before and after the construction of a new discharge pipeline on Vancouver Island (Canada). The 5·3 mg l^{-1} of untreated municipal wastes were discharged at a depth of 61 m. Five statistically significant changes were noted from the pre-to-post-discharge period including nitrates, phosphates, coliforms, water transparency and colour. Differences were noted within 0·8 km of the terminus. The 1959/1960 data for chlorophyll a, nitrate and phosphate concentrations were compared for Kaneohe Bay, Oahu, to determine whether or not eutrophication had occurred during a decade when the amount of waste disposal discharged into the bay had more than doubled. All parameters were highest near the sewer outfall. The chlorophyll a concentration increased from 50 to over 100% during this period. The authors concluded that during the decade the bay changed from an oligotrophic one to an eutrophic body of water (CAPERON and co-authors, 1971).

ROGERS (1977) analysed the water mass at varying levels at 11 stations around Whites Point discharge for temperature, salinity, density, transparency, dissolved oxygen, chlorophyll a, nitrates, nitrites, ammonia, phosphate and silicates. A cline existed between the 20 and 25 m levels in September 1974 which prevented the waste discharge from mixing with the upper levels. Fig. 7-9 presents the data for ammonia at the surface and at a depth of 25 m. High ammonia concentrations were noted at a depth of 25 m within about a 2-km radius of the discharges, with intermediate amounts upcoast and downcoast from the discharges. Lowest concentrations were measured offshore from the terminus of the pipes. Surface concentrations of ammonia were lowest in the vicinity of the discharge with higher values upcoast and downcoast. From these data it appears as if the plume was trapped by a cline at about 25 m. Currents then transported the effluent in both directions with a large quantity transported upcoast from the discharge point. Similar findings were reported by BEERS and co-authors (1979) for the Pt. Loma (USA) outfall. Maximum ammonia amounts were measured at a depth of 40 m, approximately 4 km upstream and 2 km downstream from the discharge point. However, in August 1972 no differences in ammonia concentration were noted, regardless of station or depth. KAYSER (1968) was able to trace nitrogen and phosphorus originating from the nutrient-rich Elbe River (F.R. Germany) from Cuxhaven to Helgoland, a distance of 70 km. An estimated 1·5 h after a sludge-transport barge had discharged its sludge load in the New York Bight dumpsite, ammonia levels were found to be normal. Apparently, intermittent sludge dumping increases the ammonia content in the water column, but not for any length of time (DUEDALL and co-authors, 1975). In a related study, DRAXLER (1979) found that most of the ammonia and phosphorus remained dissolved in surface waters for at least 6·5 h following a sludge dumping.

Distributions of heavy metals (Chapter 3) were used to trace the fate of sewage in the water column in Chesapeake Bay, USA (HELZ and co-authors, 1975). Decreases in metal concentrations varied, indicating that factors other than simple dilution were playing a role in the reduction. Comparison of the total metal input from sewage sources in Chesapeake Bay suggested that the wastewater input may be within one order of magnitude from the fluvial input for Cr, Cd, Cu, Pb and Zn. Dispersion of heavy metals in sewage apparently occurs more rapidly when the wastes are emptied into open coastal waters. The Tel Aviv–Tafo sewage plant discharges 63·4 to 95·1 × 10^6 l d^{-1} at a point 800 m off the coast of Israel. Significant quantities of Ag, Co, Cr, Cu, Hg, Ni, Pb and Zn were found within 400 m of the pipe terminus, especially during summer, but the concentrations were normal at a distance of 800 m (AMIEL and NAVROT, 1978).

Dispersion of sewage effluent presents unique problems in coral atolls. Water exchange between the lagoon and open coast through narrow channels may be as low as 15% within a tidal cycle (HARDY and HARDY, 1972). Domestic sewage, generally untreated, is discharged directly into the lagoon usually through many widely separated small pipelines. Coliform counts may run as high as 3·4 × 10^7 MPN per 100 ml. Treated

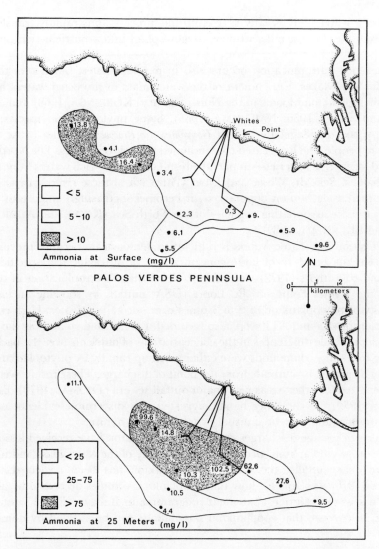

Fig. 7-9: Distribution of ammonia at the surface and at 25 m depth off
the Whites Point discharge, Palos Verdes Peninsula, California,
USA. (Original, based on data in ROGERS, 1977.)

effluents discharged into the lagoon could lead to eutrophication. The most practical
and lowest cost solution of sewage disposal is apparently to collect the sewage and
discharge it into deep coastal waters where circulation would lead to a rapid dispersion
(HARDY and HARDY, 1972; MABBETT, 1975).

Eutrophication, i.e. the process by which nitrogen, phosphorus and other plant nutrients increase significantly in concentration in a body of water with time, is well known
in fresh water lakes, such as Lake Zurich (Switzerland) and Lake Erie (USA–Canada)
(HASLER, 1947). In principle, eutrophication is a natural process during which the
nutrient concentration is increased by geological processes. However, the introduction of

sewage into a water body greatly speeds up this process, thus rapidly transforming a lake, for example, from a nutrient-poor (oligotrophic) into a nutrient-rich (eutrophic) one.

Man-hastened eutrophication occurs also in marine wastes, but not to the extent reported for some lakes. Enrichment of the water mass in protected waters has led to massive phytoplankton blooms in the North Sea (e.g. KORRINGA, 1968) and elsewhere (PINTO and SILVA, 1956; NAKAZIMA, 1965). Some phytoplankton species, such as members of the dinoflagellate genera *Gonyaulax* or *Prorocentrum*, can cause paralytic poisoning to people who have eaten contaminated mussels or oysters. The North Sea has been noted for its rich fisheries—in part the result of nutrient-rich waters from the rivers Rhine, Thames, Scheldt, Weser and Elbe. With each of these rivers carrying sewage from ever increasing human populations, it becomes increasingly important to determine the ultimate assimilating capacity of the North Sea (KINNE and AURICH, 1964; KINNE and BULNHEIM, 1980).

Eutrophication of oceanic waters is difficult to demonstrate because prevailing currents prevent the build-up of large concentrations of nutrients at a specific locality. EPPLEY and co-authors (1972) were able to demonstrate eutrophication in water collected off the Whites Point and Pt. Loma (USA) outfalls by showing an increase in productivity of phytoplankton of 2 to 3 times over normal sea water. High concentrations of chlorophyll *a* and ATP were also recorded. EPPLEY and co-authors did not find, however, appreciable differences in the concentrations of nutrients over the background, suggesting that these compounds were either taken up rapidly by phytoplankton and/or are rapidly dispersed by currents from the point of discharge. These results were similar to those reported for other areas near major outfall sewers (THOMAS, 1972). BEERS and co-authors (1979), on the other hand, suggested that discernible evidence of oceanic outfall eutrophication may be unusual or transitory in nature.

The effects of sewage discharges on marine zooplankton have received less attention than those on phytoplankton. Because of the mobility of the water mass and its transitory nature near outfalls, the study of zooplankton has been neglected. However McNULTY (1970) was able to show that zooplankton volumes decreased by about half following a sewage pollution abatement programme in Biscayne Bay, Florida (USA), indicating, in reverse, that zooplankton populations can be affected by domestic discharges.

(c) Interaction with Sediments

Physical and chemical characteristics play a significant role in determining the composition of the benthic biota. The discharge of primary wastes, for example, on the ocean floor results in the deposition of solid wastes over a large area, the extent of which depends on prevailing currents, amount of discharge and depth of the pipe. Particle size and the chemical nature of the sediments can be altered by the discharge of domestic wastes. Of particular importance is the enrichment of the sediments by the deposition of organic solids. Dissolved organic matter and other contaminants, such as heavy metals, flocculate and precipitate out and settle to the bottom, where they may become attached to individual sediment particles. If present in sufficient concentration, the organic solids and flocculants will alter the physical characteristics of the sediment surface; this in turn, may play an important role in determining the composition of benthic communities (e.g. LUDWIG and STORRS, 1970).

The chemical composition of the organic solids undoubtedly varies considerably. Table 7-1 characterizes sludge discharged from the 7-mile Hyperion pipeline. As expected, in every parameter measured its composition greatly exceeds, often by several orders of magnitude, that of the liquid discharge. Since these solids contain high concentrations of contaminants, it is readily apparent that the benthic surface can be altered more extensively than other oceanic environments.

Sediment particles generally exceed 62 μm in diameter in depths at which domestic wastes are discharged into open oceanic waters. Where finer solids from a sewer outfall accumulate, the sediments receive a higher percentage of particles smaller than 62 μm. Since data on sediment characteristics are lacking prior to pipeline construction, GREENE (1976) compared sediments of the Palos Verdes Peninsula (USA) in 1973 with data collected in 1954–59. During this interval the volume of discharge nearly doubled at the nearby sewage outfall. Sediments within 9 km of the discharge contained more than 40% 4 μm particles in 1973, compared to 20% in the 1950s. In contrast, sludge dumping in New York Bight did not alter the sediment characteristics; no evidence of significant accumulation was found (FOLGER and co-authors, 1979; WILLIAMS, 1979) Sewage sludge is nearly neutrally buoyant and therefore more mobile. It remains in the water column hours after dumping and concentrates at density discontinuities from where it may be transported (FREELAND and co-authors, 1979).

Sewage particulates which reach the benthos in appreciable amounts alter this environment not only physically but also chemically; the BOD increases because of the deposition of organics, which tend to decrease the concentration of dissolved oxygen, especially near the bottom. Elevated BOD values are a problem, especially in estuaries with a limited water circulation; here they may result in dissolved-oxygen depletion (REISH, 1959a).

Trace metals are associated with solid particulates and are generally 2 orders of magnitude greater than when present in the liquid discharge (Table 7-1). The concentration of some of the heavy metals present at Whites Point and at the terminus of the 7-mile Hyperion sludge line are compared with background levels for southern California sediments in Table 7-7. All figures are higher for the sediments around the Whites Point discharge than Hyperion; this can be attributed to the presence of industrial wastes in the effluent. The enrichment ratio for both Whites Point and Hyperion sediments are given in Table 7-7. This ratio was calculated by dividing the mean of the concentration level of a trace metal present within a 1·5-km radius of the discharge by the background levels for those elements in southern California sediments (ROHATGI and CHEN, 1976). The ratios varied from a low of 1·3 for nickel at Hyperion to a high of 54 for lead at Whites Point. At both outfalls lead had the highest enrichment ratio and nickel the lowest. For southern California these figures are generally higher than in most other localities receiving domestic wastes. HALCROW and co-authors (1973) reported lower values for Cd, Cr and Cu but intermediate values for Ni, Pb and Zn in a sewage dump area in the Firth of Clyde (UK).

The measurement of heavy metals present within the sediments represents an accurate method of determining the distribution of sewage. HERSHELMAN and co-authors (1977) plotted the distribution of several trace metals at 44 stations from the Palos Verdes Peninsula. The results of their study are presented in Fig. 7-10 in the form of concentration isopleths (mg kg^{-1} dry weight). The background level for each metal is indicated. As indicated by the distribution of these metals, the prevailing current is upcoast in a northwesterly direction. The highest concentration was always present near

Table 7-7

Mean concentrations and enrichment factors of trace metals in surface sediments in southern California (USA). All values expressed in mg kg^{-1} dry weight (Compiled from the sources indicated)

| Metal | Whites Point | | Hyperion[b] | | |
	Sediments[a]	Enrichment ratio[c]	Sediments	Enrichment ratio	Natural sediments
Cadmium	60	23	15	5·8	2·6
Chromium	1170	17·5	380	5·7	67
Copper	810	27	190	6·3	30
Lead	540	54	91	9·1	10
Nickel	80	4	26	1·3	20
Zinc	2720	24·7	315	2·9	110

[a] Data condensed from HERSCHELMAN and co-authors (1977).

[b] Data condensed from ROHATGI and CHEN (1976).

[c] Calculated by the reviewer using data from HERSCHELMAN and co-authors (1977) and ROHATGI and CHEN (1976).

the Y-shaped pipeline. Since Hyperion pipelines are located only 20 km upcoast from Whites Point, it is possible that the influence of these two large discharges may overlap at some later date; however, a submarine canyon is located between the two outfalls which may trap any sediments moving along the bottom.

In an attempt to obtain an indication of what percentage of total trace metals present in the waste effluent is deposited within the sediment, ROHATGI and CHEN (1976) calculated from Hyperion data that only about 10% of the total Cd, Cu, Ni, Pb and Zn would be deposited within a 2-km radius of the outfall. The remainder of the metals as particulates is transported elsewhere or subjected to chemical mobilization processes, or both. In the latter case, soluble complexes form and enter the solution phase. It was estimated by GALLOWAY (1979) that the introduction of metals into the ocean by municipal wastes in southern California is now of the same order of magnitude as those entering the system via natural weathering processes. However, while the natural rate will remain constant, it is expected that the anthropogenic contribution will continue to rise.

Other chemicals in addition to metals may occur in sediments in the vicinity of the outfall sewer terminus. These compounds include—but are not necessarily limited to—DDT and its metabolites and PCBs. The concentrations of these two synthetic organic compounds in waste waters of southern California are included in Table 7-1 and the mass emission rates for several years in Table 7-2. The largest quantity of DDT was discharged at Whites Point in 1971; however, the point source control at the manufacturer of this compound has resulted in a significant drop in the amount entering the marine environment at this locality (Table 7-2; YOUNG and HEESEN, 1978a). The concentrations of PCBs also dropped during this period. The highest concentrations of these compounds are now in Hyperion sludge where 3·65 and 35·4 μg l^{-1} of DDT and PCBs, respectively, were measured.

In summary, less than 10% of the solid particulates to which trace metals and organic

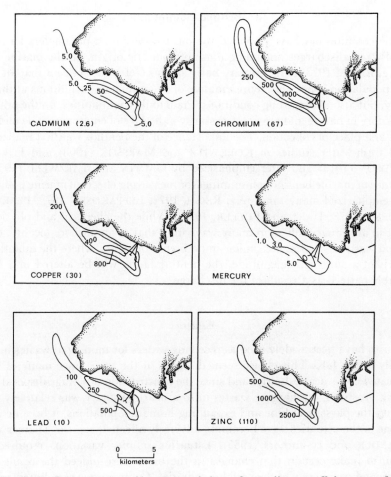

Fig. 7-10: Concentration of heavy metals in surface sediments off the coast of
Palos Verdes Peninsula, California, USA. (After HERSHELMAN and co-
authors, 1977; reproduced by permission of the South California Coastal
Water Research Project.)

compounds may attach settle in the vicinity of municipal outfalls. The remaining 90%
are carried elsewhere and remain undetected. If the quantity is sufficient then the
substratum in the vicinity of the discharge will be composed of finer particulates. These
could alter the composition of the benthic community. Of greater immediate concern to
mankind is the fate of trace metals and synthetic organic compounds. Will these chemicals enter the food chain and possibly affect man? Such was the case with methylmer-
cury, which resulted in Minimata disease. Elevated levels of DDT and PCBs have been
reported from fish in southern California (YOUNG and co-authors, 1978). The levels
were sufficiently high to affect egg production in pelicans and to cause death in zoo
animals fed contaminated fish (YOUNG and HEESEN, 1978b). Fin erosion in Dover sole
was apparently caused by contact with contaminated sediments (see below). Disposal of
sludge at sea by barging releases the material at the surface, which apparently allows
sufficient time for dispersion before the sludge reaches the bottom.

(d) Benthic Communities

Benthic communities have provided the most widely used parameters for assessing effects of waste discharges, including those of domestic origin, on the marine environment (e.g. REISH, 1973). The primary basis for this fact is the concept that an infaunal benthic population is not only representative of environmental conditions at the time of sampling, but also of preceding conditions. Planktonic communities, on the other hand, are transitory in nature and are not necessarily indicative of environmental conditions at the time and place of collection. The importance of the benthos was first realized in the classical fresh-water studies of KOLKWITZ and MARSSON (1908) and FORBES and RICHARDSON (1913) and later emphasized by GAUFIN and TARZWELL (1952). The significance of marine benthic communities for measuring effects of marine pollution has been re-emphasized many times (e.g. REISH, 1973; LEPPÄKOSKI, 1975; PEARSON and ROSENBERG, 1978; HART and FULLER, 1979). While the chemical and physical environment in an estuary differs in many ways from that of an offshore locality, effects of municipal waste discharge are similar in both environments; where the quantity of the effluent is of a sufficient magnitude, the benthic fauna will be altered in a basically predictable pattern as a result of the discharge.

Estuaries

Estuaries have been widely used as receiving waters for municipal wastes in Europe, especially in the UK. This choice seemed logical at the time, since many of the cities were located on or near estuaries and since the construction of sewage disposal systems, including pipelines, discharging wastes in to a nearby estuary was relatively inexpensive. With the passage of time and expanding human populations, it became apparent that some of the estuaries were becoming adversely affected, e.g. the Thames and Tees (ALEXANDER and co-authors, 1935). Estuarine salinity variations required special attention to make certain that changes in the biota were indeed the result of waste discharge and not merely of the variable salinities. Comparisons of polluted and unpolluted estuaries by the Water Pollution Research Board in the UK demonstrated that it was possible to distinguish a salinity effect from that of a waste discharge effect. The results of the surveys are shown in Fig. 7-11. In the region between Stockton-on-Tees and Cargo Fleet, both domestic and industrial wastes are emptied into the Tees estuary. The benthic fauna in the Tees estuary was reduced in the region receiving waste discharges in comparison with the unpolluted Tay estuary (for Station locations, see Fig. 7-11). The pollution effect becomes particularly striking when one considers that the benthic species are more numerous in the Tees than in the Tay outside the region affected by the waste discharges.

GRAY (1976) sampled the benthos near the mouth of the River Tees and compared his results with the conditions present in the 1930s (ALEXANDER and co-authors, 1935). The human population had increased by one-third (to 400000) during the ensuing 4 decades and untreated sewage was still being discharged into the river. GRAY took his samples near the mouth of the river in the vicinity of Stations XII and XIII (Fig. 7-11) of the earlier study. As can be seen in Fig. 7-11, benthic invertebrates were very numerous in the 1930s, but in 1976 GRAY found a reduction in the number of species, especially in pelecypods. Only *Macoma balthica* and an occasional *Mytilus edulis* were found.

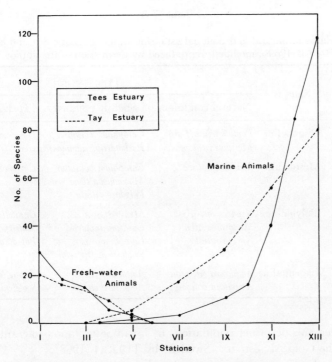

Fig. 7-11: Number of fresh water and marine species from the river to the sea in the polluted Tees Estuary, UK, and unpolluted Tay Estuary, UK. (After ALEXANDER and co-authors, 1935; modified; reproduced by permission of the Controller of Her Majesty's Stationery Office, London.)

Absent were the previously common *Acrobicularia plana, Tellina tenuis, Venerupis pollustra, Cerastoderma edulis* and *Mya arenaria*.

In an attempt to assess the causes of pollution, BAGGE (1969b) compared data from five estuaries in the Baltic–Skagerak region (Table 7-8). While no two estuaries are identical with regard to abiotic factors, the amount of dissolved oxygen near the bottom, the salinity and the type of substratum are the three most important ecological parameters governing the faunal composition of the estuarine benthos. The dissolved oxygen concentration and the substrate type are modified by the amount and type of sewage disposal. In natural, unpolluted areas, the benthos is dominated by animal groups such as ophiuroids, isopods, pelecypods and polychaetes. In heavily polluted, oligohaline or mesohaline estuaries, the benthos is dominated by *Chironomus* larvae; in the most saline environment species of the polychaetes *Capitella, Polydora* and *Scolelepis* are present. The intermediate zone supports a mixture of species from both environmental extremes and also some additional ones (Table 7-8). BAGGE's conclusions are most significant, especially since benthic biological background data are available for the past 40 yr for one of the sites, Gullmar Fjord (BAGGE, 1969a). Using historical data, for comparison, BAGGE (1969b) was able to document the replacement of the original ophiuroid community by a heavily pollution-stressed community dominated by species of *Capitella* and *Scolelepis*.

Table 7-8

Type species groups in natural and polluted estuarine waters of Baltic Sea and Skagerak (After
BAGGE, 1969b; modified; reproduced by permission of the author)

Salinity (‰S)	Salinity zone	Type species		
		Natural conditions	Slightly polluted	Heavily polluted
0·5 to ±5	Oligohaline	*Pontoporeia affinis* *Mesidotea entomon*	*Corophium volutator* *Euilyodrilus hammoniensis*	*Chironomus plumosus*
±5 to ±18	Mesohaline	*Macoma baltica*	*Corophium volutator* *Macoma baltica* *Polydora ciliata*	*Chironomus plumosus*
±18 to ±30	Polyhaline	*Nephtys incisa* *Syndosmya alba* *Nucula nitidu*	*Mya arenaria* *Cardium lamarcki* *Corophium* spp. *Syndosmya filiformis*	*Capitella capitata* *Scolelepis fuliginosa* *Polydora ciliata*
±30	Mixoeuhaline	*Amphiura filiformis* *Amphiura chiajei*	*Amphiura filliformis* *Amphiura chiajei*	*Capitella capitata* *Scolelepis fuliginosa*

Similar replacements of former populations by invading pollution-tolerant species were
recorded for the Gota Alv estuary (Sweden) by TULKKI (1968).

In summary, the assessment of relationships between municipal-waste discharges and
benthic animal populations in an estuarine environment is complicated by the inten-
sively varying salinities, and usually by lack of knowledge regarding the original, natural
conditions in the area. Nevertheless, some generalities emerge: in natural areas, an
intermediate zone occurs where the strength of freshwater and marine animal populations is minimal (Volume 1: KINNE, 1971). The population density in this zone can be
further reduced if wastes are discharged nearby, as shown for the River Tees (ALEXAN-
DER and co-authors, 1935). If the wastes are released in the upper reaches of the estuary,
tolerant oligochaete species will flourish (MCLUSKY and co-authors, 1980). Natural
populations of more sensitive species will be replaced by more pollution-tolerant ones
near the source of the waste discharge, and will be restricted to cleaner waters (TULKKI,
1968). At a specific locality, populations of benthic organisms respond within a relatively short period of time either to degradation (TULKKI, 1968; MCLUSKY and co-
authors, 1980) or improvement (REISH, 1971; LEPPÄKOSKI, 1977) of environmental
conditions. For a critical assessment, historical data must be available. This fact was
particularly well documented by GRAY (1976) and TULKKI (1968).

Intertidal

Studies dealing with the effects of domestic pollution on the intertidal flora and fauna
are limited in number and significance since pipelines are generally sited some distance
from the shore in subtidal waters. Sewer outflows into the intertidal environment generally release small amounts of waste. Interpretation of intertidal dynamics in the vicinity
of domestic discharges must take into consideration the high natural variability of
populations (HARTNOLL and HAWKINS, 1980). In fact, the variability of intertidal

natural populations may exceed that caused by man-made pollutant stress (SHARP and co-authors, 1979).

San Clemente Island, the southernmost island of the southern California Channel Island group, is located about 75·6 km from Long Beach. A small naval base is situated near the northern end, and the base is serviced by a sewage outfall which discharges about 95 000 l d^{-1} of untreated wastes into the rocky intertidal environment. Extensive studies have been made of the intertidal macrophytes and macro-invertebrates (MURRAY and LITTLER, 1974), the intertidal and subtidal polychaetous annelids (DORSEY, 1975; ROWE, 1975) and algal succession and energetics (LITTLER and MURRAY, 1978; MURRAY and LITTLER, 1978). Flora and fauna comparisons were made to a control area located some distance from the discharge. A total of 30 species of algae were observed at the control area, compared with 13 at the discharge site (MURRAY and LITTLER, 1974). Intertidal polychaete species were reduced in the vicinity of the discharge (ROWE, 1975), but the subtidal species remained unaffected (DORSEY, 1975). Blue–green algae were more plentiful in the outfall area than elsewhere (LITTLER and MURRAY, 1978). Denuded quadrats in the vicinity of the discharge area re-populated with the same species within a month, but similar plots in the control area had not recovered to former levels by 30 months (MURRAY and LITTLER, 1978).

Perhaps the most ambitious intertidal study dealing with the effects of marine pollution was that undertaken by DAWSON (1959, 1965) to compare the algal flora at 42 mainland localities in southern California (USA) in the 1956–59 period with those species present prior to 1912. Later, others (FOSTER and co-authors, 1971; NICHOLSON and CIMBERG, 1971; WIDDOWSON, 1971; CIMBERG and co-authors, 1971) revisited some of the same stations, and finally all 42 stations were sampled again in 1973–74 (THOM and WIDDOWSON, 1978). DAWSON noted two distinct differences with regard to algal composition: (i) he was unable to locate at least 50% of the species reported more than 40 yr ago; (ii) he observed a marked reduction in the number of species near the sewage discharge site at Whites Point and a decided increase in the coralline algae population. Further species reductions were recorded in samples made in the late 1960s, but in the most recent study (THOM and WIDDOWSON, 1978) a general increase in the number of species since 1955–58 was noted. A shift from massive species to tuft and crustose-type algal growth forms was observed. The reasons for the recent increase in number of species were not stated; possibly it was related to pollution abatement programmes, the establishment of intertidal preserves or natural variability. The brown alga *Cystoseira stricta* disappeared from an area shortly after the region, in the vicinity of Marseille (France), became contaminated with sewage (BELLAN-SANTINI, 1968). Coralline algae and mussels replaced the brown alga until conditions deteriorated further, at which time no macroscopic life existed around the Marseille discharge at Cortiou.

In summary, domestic sewage discharged into a rocky shore intertidal environment can alter the species composition even if the discharge is small. Brown algae and/or massive species disappear first; they are replaced by corallines and tufted species. Where the discharge is large, as is the case near Marseille, all macroscopic algae are excluded.

Subtidal

It has been well established that substantial amounts of municipal wastes affect the benthic community. Bioenhancement of the benthic community apparently occurs if the primary treated waste discharge is small $(7\cdot5 \times 10^6 \text{ l d}^{-1})$ (REISH, 1980). However, with an increase in the amount of discharge, the benthos is no longer able to assimilate the waste; the benthic community becomes stressed and altered (Fig. 7-6). According to data collected in southern California (USA), the environment becomes stressed when the amount of waste discharged ranges between $7\cdot5$ and $42\cdot2 \times 10^6 \text{ l d}^{-1}$ (REISH, 1980). MEARNS and WORD (1981) attempted to forecast ecological changes in the benthos based on the amount of solid wastes discharged. They calculated the amount of excess standing crop around each outfall above background levels and related it to the amount of solid wastes discharged (Fig. 7-12). However, the faunal diversity—as measured by the infaunal trophic index (WORD, 1979)—was depressed most at localities where the amount of solid wastes was greatest (Fig. 7-6). Excess standing crop presumably represents bioenhancement as a result of the increase in organic matter accompanying the discharge. A direct relationship was found between solid-mass emission rates and excess benthic faunal standing crop. While predicting faunal changes on the basis of either the amount of liquid or solid wastes discharged may involve a simplification of a benthic community's response to municipal wastes, it does provide a first step in forecasting

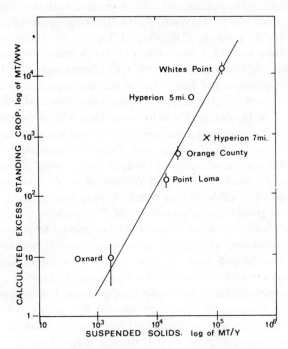

Fig. 7-12: Relationship between suspended solids, mass emission rates and total excess standing crop of benthic infauna surrounding 6 outfalls in southern California, USA. Hyperion 5-mile discharges of liquid waste containing suspended solids, and Hyperion 7-mile discharges of sludge. (After MEARNS and WORD, 1981; modified; reproduced by permission of Stan Hitts Graphics Inc.)

possible ecological effects of these discharges on the benthic environment prior to construction of a marine outfall.

Subjective and objective evaluations of data have been made to determine the extent of offshore benthic community alterations as a result of the discharge of municipal wastes. In either type of analysis it has been shown that from 2 to 6 distinct benthic communities become established around an outfall sewer (REISH, 1979). Generally, 3 are observed: a stressed or polluted community surrounding the end of the pipe, an intermediate degree or semi-healthy community, and an outer normal or healthy community unaffected by the discharge (Fig. 7-7). The stressed community is characterized by a few species—one of which was invariably *Capitella capitata*—present in large numbers and high biomass. The benthic fauna in the degraded community is more diverse, but a few species are present in large numbers; the biomass is elevated but to a lesser extent. The normal community is characterized by a high faunal diversity consisting of many feeding types. The discharge of sewage effluents into offshore waters apparently tends to affect first the suspension feeders such as ophiuroids and amphipods; amphipods; the next group to be affected are surface detrital feeders. Subsurface detrital feeders constitute the most tolerant feeding group (i.e. *Capitella capitata*; WORD, 1979). While these generalities are based primarily on data from southern California, they have been found to apply whenever comparable studies have been completed (i.e. Germany: WILHELMI, 1916; Denmark: BLEGVAD, 1932; Finland: TULKKI, 1965; South Africa: OLIFF and co-authors, 1967; Adriatic Sea: GHIRADELLI and PIGNATTI, 1968; France: BELLAN and PÉRÈS, 1970; Italy: COGNETTI, 1972; Chile: GALLARDO and co-authors, 1972; Japan: KITAMORI, 1972; Jamaica: WADE, 1972, 1976; UK: HALCROW and co-authors, 1973; Baja California (Mexico): LIZARRAGA-PARTIDA, 1974; Sweden: LEP-PÄKOSKI, 1975; British Columbia: BALCH and co-authors, 1976; New York: RISTICH and co-authors, 1977; Australia: POORE and KUDINOV, 1978; Turkey: KOCATAŞ and GELDIAY, 1980.

On the basis of the literature quoted above, benthic communities in the vicinity of outfall sewers tend to attain a steady state in the absence of changes in the quality or treatment of the wastes. However, this natural phenomena can alter the community structure and thus lead to misinterpretations. THOM and CHEW (1979) noted a *Capitella capitata* dominated community around the combined sewage and storm water discharge in Puget Sound (USA) during winter which was replaced by a *Nebalia pugettensis* dominated community during summer when there was little storm water runoff. Apparently, *N. pugettensis* is not tolerant of fresh water, whereas *C. capitata* can withstand considerable salinity fluctuations (LEPPÄKOSKI, 1975). *C. capitata* dominated the benthos in the vicinity of the Whites Point discharge in the 1971–72 period; in the ensuing years to 1977 this stressed community essentially disappeared. However, in 1978–79 it reappeared, during a period when no significant alterations in natural environmental factors or the quality of discharge had occurred. One noteworthy biological change, however, was observed during this period: the echiuroid worm *Listriolobus pelodes*, which measures several cm in length, first appeared in the area in mid-1973, and by 1975 had established itself near the outfall. It burrowed into the sediment and apparently this activity allowed the penetration of dissolved oxygen into the sediments. Increased sediment oxidation improved the living conditions for the benthic populations. The population of *L. pelodes* began to diminish in 1977, probably owing to natural causes. With the decrease in sediment aeration organic material accumulated, once again leading to a stressed condition (REISH, 1980). Unfortunately, no reliable data are available on cyclic occurrences of the majority of subtidal invertebrates. An extensive population of *L. pelodes* occurred off Santa Barbara in 1955–58 (BARNARD and HARTMAN, 1959) but was virtually absent 10 yr later when the next samples were taken. The cyclic occurrence of *L. pelodes* indicates the importance of analysing the data from both biological and statisti- cal–mathematical points of view.

Oceanic disposal of municipal wastes through pipelines is not practical for cities such as London, New York or Philadelphia because of the wide ocean shelf. In these cases, the sludge is loaded on to specially constructed barges for transport and dumped into offshore waters. A total of $8·7 \times 10^6$ metric tons of sludge from New York City was dumped in the New York Bight during the 1964–68 period (PEARCE, 1972); a similar amount was barged off the mouth of the Thames estuary by the London authorities (SHELTON, 1971). Philadelphia discharges a much smaller quantity off the coast of Delaware (GUARINO and co-authors, 1979). Additional materials, such as dredge spoils, are also discharged at the two US sites mentioned. Levels of sludge from municipal wastes were highest at the New York dumpsite where amounts of organic material in the sediment measured greater than 20% (PEARCE, 1972) or about 10 times above back- ground levels. Benthic populations were impoverished at stations with a high organic carbon content in the sediments. Organic carbon values in the sediment consisted of up to 5·7% in the Thames estuary; however, the benthic fauna showed little effect of the sludge dumping, probably because of the strong tidal currents in the area (SHEL- TON, 1971). Organic carbon values in the sediment measured up to 3·7% off Delaware and the benthic fauna is apparently changing as a result of accumulation of organic particles (WATLING and co-authors, 1975).

Indicator-organism concept

The use of one or more key species which could provide a convenient, quick assess- ment of prevailing environmental conditions grew out of and developed from the saprobic-system approach of KOLKWITZ and MANSSON (1908). The recent application of the indicator-organism concept in marine waters began with the studies in protected waters of California (ANONYMOUS, 1952; FILICE, 1954a, b; REISH, 1955). This con- cept was based on the premise that a pollutant entering the system will kill and exclude the sensitive species from the area. If the amount of pollutants is increased, the next level of less sensitive species will be eliminated. Elimination of these species from the com- munity makes it possible for the surviving species to flourish in the absence of their competitors. If, however, the amount of polluting material is too large for the environ- ment to assimilate, eventually all surviving species will be killed, leaving the area devoid of macroscopic life. One therefore needs to know only what species are indicators of the particular condition to assess the environment state at a locality. Presumably, by iden- tifying a few key species within a given sample, one could determine the general ecological conditions at that station; such a procedure could save a considerable amount of time, effort and money.

How valid is the indicator-organism concept? The use of indicator or assay organisms in marine environments has been reviewed by PEARSON and ROSENBERG (1978) and by several authors in HART and FULLER (1979); see also Volume III: KINNE (1977). According to PEARSON and ROSENBERG, pollution-indicator species by themselves do not indicate anything about the environmental condition except their own presence or absence, i.e. they witness changes in community structure as a function of time. This is a useful criterion for assessing ecological conditions. It is possible that subtle structural changes could be overlooked if only a few indicator species are identified rather than the complete sample. Since all species had been identified in the samples, it was possible to document marked community changes off Marseille (France) with time (Fig. 7-7) and relocation of the pipeline in Los Angeles Harbor (USA) (BELLAN, 1979b; REISH, 1980).

PEARSON and ROSENBERG (1978) listed various marine invertebrates which have been used as indicators of organic pollution. Polychaetes have been the most frequently used group (especially species of *Capitella, Polydora, Neanthes* and *Scolelepis*) followed by molluscs (principally species of *Macoma, Mya* and *Mytilus*), oligochaetes (*Peloscolex*), amphipods (*Corophium*) and sea anemones (*Cerianthus*). The case of amphipod crustaceans as indicators of varying degrees of water quality was strongly presented by BELLAN-SANTINI (1980), based on her 18-yr study of this group in Marseille. An inverse relationship was found to exist between the number of species and the degree of pollution, a finding supported by PARKER (1980) in Belfast. *Caprella actifrons, Podocerus variegatus* and *Jassa falcata* are suspension feeders and prefer the polluted waters. The group of amphipods characteristic of unpolluted waters were primarily algal eaters (BELLAN-SANTINI, 1980). Apparently these clean-water inhabitants were excluded by the absence of algae in polluted waters.

Foraminifera have shown promise as indicators of varying degrees of sewage pollution. RESIG (1960) reported that the number of specimens per gram of sediment decreased towards the point of discharge but the ratio of arenaceous to calcareous forms increased towards the sewage field. *Eggerella advena* and *Trochammina pacifica* showed strong affinities to the outfall areas in southern California. However, SCHAFER (1973) documented that the calcareous species *Elphidium incertum* and *E. clavatum* dominated the foraminiferan population near sewage outfalls in Nova Scotia (Canada).

Polychaetes have been the most widely used pollution indicators thus far, but undoubtedly other groups, such as molluscs, crustaceans and foraminiferans, will be equally important as additional information becomes available (KINNE, 1977). It is only logical to assume that some species within a group are more sensitive than others. Possibly additional sibling species—such as *Capitella capitata* in oil-polluted environments (GRASSLE and GRASSLE, 1976—will be found to exist in one or more of these groups. In the final analysis, the usefulness of the indicator-organism concept depends not only on good biological data but also on a competent ecologist to interpret them.

(4) Effects of Domestic Wastes on Marine Organisms

(a) Liquid Wastes

A limited number of bioassays have been undertaken which compare the toxicity of municipal wastes to marine organisms. Many of the sanitation districts routinely conduct bioassays with fresh water fish, i.e. the fathead minnow *Pimephalas promelas*, to measure the toxicity of sewage effluents. OSHIDA and co-authors (1981) reported on the effect of a mixture of primary and secondary effluents on the fathead minnow (Table 7-9). The 96-h LC 50 was nearly 100% effluent of the Hyperion discharge but only 26·6% of the Whites Point discharge. They compared these results with the concentra-

7. DOMESTIC WASTES (D. J. REISH)

Table 7-9

Effects of waste water on fathead minnows *Pimephalas promelas* (calculated 96-h LC_{50} values) and sea urchin *Strongylocentrotus purpuratus* eggs (effluent concentration causing 50% reduction in fertilization success) (After OSHIDA and co-authors, 1981; reproduced by permission of Academic Press, Inc.)

Effluents tested (Sanitation District, California, USA)[a]	Fathead minnow 96-h LC_{50} (% effluent)	50% reduction in sea urchin fertilization success occurred between these effluent concentrations
Hyperion	98·6	2·6–20
Whites Point	26·6	0·92–2·6
Orange County	77·1	0·97–7·1

[a] Mixture of primary and secondary treatment liquid wastes.

tion causing a 50% reduction in the fertilization rate of sea urchin eggs (*Strongylocentrotus purpuratus*). Sea urchin egg fertilization rates were found to be a more sensitive criterion than the survival rate of the fathead minnow. The pollution levels (Table 7-9) were generally 1 to 2 orders of magnitude less than for the fish.

Table 7-10

Neanthes arenaceodentata: concentration of metals in individuals exposed to primary and secondary treated wastewater (After MARTIN, 1981; modified; reproduced by permission of the author)

Experimental conditions	Dry wt. (g)	Metal concentration (μg g^{-1} dry wt.)						
		Ag	Cd	Cr	Cu	Ni	Pb	Zn
Controls								
Laboratory colony	0·14	0·8	0	0·3	0	2·3	11	88
Experiment with sea water	0·23	0·90	0·2	0·7	0·4	—	13	75
Experiment with sea salts	0·21	1·1	0·4	1·1	13	26	21	140
Primary treated waste water								
18%	0·23	1·0	2·7	3·9	16	2·7	21	150
56%	0·18	2·7	8·9	15	83	61	37	210
75%	0·08	8·2	11·0	120	130	85	21	240
87%	0·08	14·0	13·0	220	180	98	32	290
Secondary treated waste water								
18%	0·22	1·1	1·3	8·8	9·8	35	28	140
56%	0·20	4·6	2·9	9·0	20·0	96	25	120
100%	0·20	0·6	5·4	6·0	20·0	140	18	140
Sea water analysis (mg^{-1})								
Sea water		0·002	0·0004	0·001	0·001	0·001	0·001	0·001
Sea salts		0·002	0·003	0·003	0·004	0·001	0·001	0·002
Primary treated waste water		0·02	0·09	0·2	0·5	0·2	0·2	0·6
Secondary treated waste water		0	0·02	0·04	0·02	0·2	0·03	0·08

KINDIG and LITTLER (1980) exposed untreated, primary and secondary effluents to different macrophytes, including coralline algae, and measured their growth rates. Three species of corallines, *Bossiella orbigniana*, *Lithothrix aspergillum* and *Corallina officialis var. chilensis*, revealed enhanced growth rates when exposed to primary wastes. Chlorination of secondary wastes had no effect after 2 wk. Populations of *C. officialis* collected near sewer outfalls were more tolerant than those collected from cleaner waters, indicating that this species is able to acclimate to sewage stress. These data tend to support DAWSON's (1959, 1965) hypothesis that certain intertidal algae, especially corallines, are extremely tolerant of domestic waste waters.

The toxicity of primary and secondary treated wastes to laboratory populations of the polychaete *Neanthes arenaceodentata* was determined and the body burden of metals measured (MARTIN, 1981). Sea salts were added to the municipal wastes to bring the salinity to that of normal sea water. The 96-h LC 50 in primary treated wastes was 92%, compared with the 82% value measured for the fathead minnow (MARTIN, 1981). Individuals exposed to higher concentrations of primary treated wastes failed to feed, or to construct mucoid tubes; they exhibited jerky movements and lost weight during the 28-d test period (Table 7-10). After the 28-d experimental period, all surviving test organisms were analysed for Ag, Cd, Cr, Cu, Ni, Pb and Zn (Table 7-10). There was a direct relationship between the concentration of both wastes and the body-burden level of these metals in the tissue, except for lead. The concentration of the metals in the tissue ranged from slightly less than 1 to 4 magnitudes greater than the concentration of the medium, regardless of the source. Apparently, *N. arenaceodentata* is unable to metabolize these metals, with the possible exception of lead.

(b) Diseases of Marine Organisms

Diseases of marine organisms, especially fish and shellfish have been the subject of reviews (MAWDESLEY-THOMAS, 1974; SHERWOOD, 1978; SINDERMANN, 1966, 1979; KINNE, 1980) and symposia (KRAYBILL and co-authors, 1977; KINNE and BULNHEIM, 1984). Interest has focused on these particular groups since many of their representatives are used as food by humans. Public health officials have been concerned about the possibility of diseased organisms transmitting the causative agent to man by way of food. Diseases or diseased conditions have been reported from abalones (FITCH, 1956), pelecypods (BROWN and co-authors, 1977), decapods (YOUNG and PEARCE, 1975) and fish (AUBERT and co-authors, 1979). Whereas diseased organisms, especially fish, are often easily detected, the causative agent of the disease is much more difficult to determine.

Diseases of marine fish (for the latest reviews consult CONROY, in press; KABATA, in press; LOM, in press; PETERS, in press; LAUCKNER, in press; ROHDE, in press; WEDEMEYER, in press; WOLF, in press) have been more thoroughly studied than those of invertebrates (LAUCKNER, 1980 and in press). External diseases include fin erosion (or fin rot; Fig. 7-13a), skin ulcers, skin tumours (Fig. 7-13b), lip papillomas, epidermal papillomas (Fig. 7-13b), abnormal pigmentation patterns (Fig. 7-13c) and exophthalmos. Enlarged livers have been noted which exhibit structural disorganization in addition to histological changes (SHERWOOD, 1978) and skeletal deformaties (SINDERMANN, 1979). While at this stage of our knowledge it is not possible to implicate that a specific contaminant is a cause of a specific disease, circumstantial evidence is overwhelming that at least for some diseases a relationship exists between a polluting agent in the domestic waste and the disease. The problem is complicated by a lack of adequate

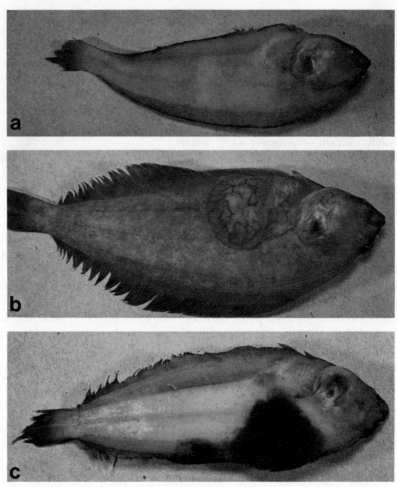

Fig. 7-13: *Microstomus pacificus*. External diseases in Dover sole: (a) fin erosion; (b) skin tumours; (c) abnormal pigmentation pattern and fin rot. (Original photographs: courtesy of MARJORIE SHERWOOD, Southern California Coastal Water Research Project.)

baseline data; in addition, the contaminants are usually not from a single source but consist of both municipal and industrial wastes.

Fin erosion is more prevalent in demersal fish than in pelagic species. The identifiable features of fin erosion is the irregular appearance of the fins (Fig. 7-13a) as a result of the wasting away of the fins and the reabsorption of fin rays. Histological examination of the diseased areas is characterized by epidermal hyperplasia, papillary folding of the epidermis, increase in fibroblasts, reduction or absence of mucocytes and eosinophilic granular cells and abnormal pigment-cell distribution (SHERWOOD, 1979). The disease does not appear to be related to an invasion by a micro-organism. No pathogenic bacteria have been isolated from the site of the lesions (MANFREDI, 1976). Infected Dover soles *Microstomus pacificus* have larger livers, the calcium content of the kidneys is higher and severely affected fish produce less slime than normal fish (MEARNS and SHERWOOD, 1977b).

Fin erosion has been reported in demersal fish from many parts of the world and always near metropolitan areas including southern California (SHERWOOD, 1979), Washington (WELLINGS and co-authors, 1976), Massachusetts (SHERWOOD, 1979), New York Bight (MAHONEY and co-authors, 1973; MURCHELANO and ZISKOWSKI, 1976), Florida (COUCH amd NIMMO, 1974; SINDERMANN and co-authors, 1978), Gulf of Mexico (OVERSTREET and HOWSE, 1977), Irish Sea (PERKINS and co-authors, 1972), French Mediterranean Sea (AUBERT and co-authors, 1979) and Japan (NAKAI and co-authors, 1973). The prevalence of fin erosion in marine fish has been well documented in southern California (MEARNS and SHERWOOD, 1977a,b; SHERWOOD, 1979). A total of 33 species of fish have been observed to be infected with this disease (Table 7-11) in which the incidence varies from 0·02 to 30%. At least 30% of the Dover soles examined during a 5-yr period were infected with fin erosion. High incidences were also noted in the greenstriped rockfish *Sebastes elongatus*, rex sole *Glyptocephalus zachirus* and barred sand bass *Paralabrax nebulifer*. The incidence of fin erosion was not uniformly distributed in southern California waters (Table 7-11); the highest occurrence for these three species was observed in the Palos Verdes shelf area, which is near the Whites Point discharge. No diseased fish were noted in the vicinity of the discharges from San Diego or Ventura (SHERWOOD, 1979). Up to 38% of the trawled marine fish in the New York Bight were diseased; here, a total of 22 species are known to be affected (MAHONEY and co-authors, 1973).

While the cause of fin erosion is unknown, the data indicate that the incidence of the disease is related to municipal waste discharges which also include industrial effluents. The prevalence of the disease was highest at the Palos Verdes locality, which is near the domestic and industrial outfall at Whites Point (Table 7-12). Fin erosion did not occur at Corteau, the site of the municipal waste outfall for Marseille, but diseased fish were noted on the west side of the city, which receives both industrial and municipal wastes (BELLAN, personal communication).

A working hypothesis for the cause of fin erosion in southern California was proposed by SHERWOOD (1979); it states that the disease is apparently related to high levels of PCBs in the sediment off the Palos Verdes Peninsula. Evidence in support of this hypothesis comes from (i) the fact that fins in maximum contact with the sediment are most affected; (ii) a higher incidence of fin erosion in demersal rather than in pelagic fish; (iii) a higher incidence of the disease in areas with the maximum PCB concentrations in the sediments; (iv) lack of abnormal counts of micro-organisms from affected areas; and (v) induction of early fin erosion stages in demersal fish exposed to sewage sludge under laboratory conditions (MEARNS and SHERWOOD, 1977b). Elevated PCB levels were also noted in diseased flounder from Puget Sound and New Jersey (SHERWOOD, 1978). High levels of heavy metals and petroleum, not necessarily associated with municipal wastes, are also thought to be related to fin erosion disease in the New York Bight (ZISKOWSKI and MURCHELANO, 1975). However, MAHONEY and co-authors (1973) implicated bacteria of three genera—*Aeromonas*, *Vibrio* and *Pseudomonas*—which were always present within diseased fins. They were able to induce fin rot disease under laboratory conditions with these bacteria, but only after the fins had been lightly abraded and the bacteria rubbed on. They concluded that two conditions were necessary for fin rot disease to occur: a dense bacterial population and environmental stress. Both of these conditions occur in the New York Bight.

Other externally manifested diseases of fish—including skin ulcers, skin tumours, lip papillomas, abnormal pigmentation patterns and exhophthalmos—are less prevalent

Table 7-11

Prevalence of fin erosion in demersal fishes from southern California (USA) captured by otter trawl, May 1972 to May 1976 (After SHERWOOD and MEARNS, 1977; reproduced by permission of the South California Coastal Water Research Project)

Family, species	Common name	Total	Number of individuals with fin erosion	%
Pleuronectidae				
Microstomus pacificus	Dover sole	27 991	8318	30
Glyptocephalus zachirus	Rex sole	2666	349	13
Lyopsetta exilis	Slender sole	3277	158	4·8
Pleuronichthys decurrens	Curlfin sole	3656	84	2·3
Parophrys vetuls	English sole	2924	12	0·41
Pieuronichthys verticalis	Hornyhead turbot	1311	6	0·46
Eopsetta jordani	Petrale sole	18	1	5·6
Pothidae				
Citharichthys sordidus	Pacific sanddab	17 879	4	0·02
Citharichthys fragilis	Gulf sanddab	28	1	3·6
Hippoglossina stomata	Bigmouth sole	696	1	0·14
Xystreurys liolepis	Fantail sole	66	1	1·5
Cynoglossidae				
Symphurus atricauda	California tonguefish	4762	27	0·57
Scorpaenidae				
Sebastes miniatus	Vermilion rockfish	555	30	5·4
Sebastes elongatus	Greenstriped rockfish	138	20	14
Sebastes jordani	Shortbelly rockfish	2803	12	0·43
Sebastes rosenblatti	Greenblotched rockfish	193	11	5·7
Sebastes dalli	Calico rockfish	6459	8	0·12
Sebastes rubrivinctus	Flag rockfish	75	2	2·7
Sebastolobus alascanus	Shortspine thornyhead	99	2	2·0
Sebastes semincinctus	Halfbanded rockfish	4027	1	0·02
Sciaenidae				
Genyonemus lineatus	White croaker	7224	183	2·5
Embiotocidae				
Cymatogaster aggregata	Shiner perch	7498	10	0·13
Phanerodon furcatus	White seaperch	966	1	0·10
Zalembius rosaceous	Pink seaperch	5913	1	0·02
Ophidiidae				
Chilara taylori	Spotted cusk-eel	254	6	2·4
Anoplopomatidae				
Anoplopoma fimbria	Sablefish	560	5	0·89
Hexagrammidae				
Zaniolepis latipinnis	Longspine combfish	2239	4	0·18
Zaniolepis frenata	Shortspine combfish	581	2	0·34
Batrachoididae				
Porichthys notatus	Plainfin midshipman	5782	2	0·03
Serranidae				
Paratabrax nebulifer	Barred sand bass	22	2	9·1
Agonidae				
Xeneretmus latifrons	Blacktip poacher	1186	1	0·08
Engraulidae				
Engraulis mordax	Northern anchovy	3304	1	0·03
Zoarcidae				
Lycodopsis pacifica	Blackbelly eelpout	2184	1	0·04

Table 7-12

Geographical distribution of fin erosion disease in the most commonly affected fish in southern California (USA) (1972–1976) (After SHERWOOD, 1979; modified; reproduced by permission of the Southern California Coastal Water Research Project)

Species	Common name	Santa Monica Bay		Palos Verdes		South San Pedro Bay		Dana Point	
		Number of fish	% incidence	Number of fish	% incidence	Number of fish	% incidence	Number of fish	% incidence
Microstomus pacificus	Dover sole	894	3·5	20 854	39	5354	2	889	067
Glyptocephalus zachirus	Rex sole	194	2·1	1661	21	758	0	53	0
Sebastes elongatus	Greenstriped rockfish	11	0	111	18	9	0	7	0
Waste discharger		Hyperion		Whites Point		Orange County		No major discharges	

than fin erosion and have been subjected to less thorough investigation. While evidence for a relationship between pollution and these diseases is circumstantial at best, it appears that some factor(s) cause(s) a stress which in turn may cause the particular disease.

Internal diseases are more difficult to detect in field collections and their causative agent are more difficult to determine. Skeletal deformities of various types have been reported, but it is possible that these might be genetically induced. VALENTINE (1975) found a significantly higher incidence of abnormal gill rakers in barred sand bass *Paralabrax nebulifer* collected from southern California waters (USA) than from Baja California (Mexico). Of the 487 fish examined, nearly 50% had either mildly or severely deformed gill rakers in southern California and 1% in Baja California collections. VALENTINE theorized that these deformities may be related to elevated environmental levels of chlorinated hydrocarbons and heavy metals, both of which are high in southern California marine waters, and are known to interfere with calcium metabolism.

BROWN and co-authors, (1977) examined populations of the soft shell clam *Mya arenaria* from 10 localities in New England (USA) for neoplasia. The incidence ranged from 0 to 65%, with the higher values from populations taken from polluted areas especially with larger amounts of oil contamination. In contrast, FARLY (1976) did not find any relationship between environmental pollution and neoplastic disease.

Diseases of decapod crustaceans have been noted in collections from polluted estuaries and offshore waters. Skeletal erosion, gill erosion and 'black-spot disease' were reported in the USA from the New York Bight (YOUNG and PEARCE, 1975) and Florida (IVERSEN and BEARDSLEY, 1976), from the Gulf of Mexico (COUCH, 1978) and from Germany (SCHLOTFELDT, 1972). Crabs and lobsters collected from the sludge disposal site in New York Bight and other areas where contaminated sediments could collect showed skeletal erosions of the tips of their walking legs and the ventral half of their chelipeds. YOUNG and PEARCE (1975) were able to induce diseased skeletal tips under laboratory conditions in specimens exposed to contaminated sediments for 6 wk. The exoskeleton was pitted and cracks permitted an invasion of agglutinated blood cells. Gill filaments were also found to be eroded. The black spot disease involves erosion of the carapace and appendages of the shrimp *Crangon crangon* resulting in an appearance of a blackened area in the region of the infection (SCHLOTFELDT, 1972). The cause of these different crustacean diseases apparently is related to the presence of contaminated sediments of domestic origin; whether the agent is fungal or bacterial is unknown (YOUNG and PEARCE, 1975). Chitinoclastic bacteria have been found associated with shell disease in the American lobster (HESS, 1937). It is possible that the breakdown of the exoskeleton could lead to invasion by other micro-organisms from domestic sludge.

The knowledge of the diseases of fish and invertebrates is fragmentary and only beginning to be understood (KINNE, 1980; LAUCKNER, 1980 and in press), but at least in some cases a relationship between polluted sediments and disease cause has been established. Organisms which come in direct contact with the contaminated sediments seem to be more susceptible to attack. Micro-organisms have been implicated as primary causative agents in fin rot fish diseases but not in crustacean diseases. The possible effect on man of eating diseased organism is unknown.

(c) Movement of Toxicants Through a Marine Food Chain

The potential dangers of biomagnification of toxic compounds in food chains became a public issue with the discovery that the concentration of DDT was highest in animals at the top of the food chain. The concern was intensified when the cause of Minimata disease linked the disposal of industrial wastes containing mercury to human disease and death. With the knowledge that these two toxicants were magnified at each step of the food chain, intensive research was undertaken throughout the world to measure the body-burden levels of trace metals, pesticides and hydrocarbons in all types of marine plants and animals (e.g. REISH and co-authors, 1980). The mussel watch, a world-wide monitoring effort, is an attempt to coordinate these studies and to detect possible danger sites (BAYNE, 1978).

In most studies dealing with the body burden of one or more chemicals the source of the contaminant is either unknown or not stated. As indicated in Table 7-1, domestic wastes are characterized by a complex array of chemicals, any one of which may vary in concentration and biological effect according to the nature of discharge and treatment. Trace metals were higher in the discharges from Whites Point and Hyperion than from the other sewers in southern California (Table 7-1). Liquids following secondary treatment contained lower concentrations of trace metals than primary wastes, except for some elements in the Orange County discharge. The Hyperion sludge contained the highest concentration of chemicals.

In most instances it is not known if the chemicals present in the waste waters enter into a marine food chain. Perhaps the best documented case is the movement of DDT in the discharge at Whites Point through the food chain to the California brown pelican (YOUNG and HEESEN, 1978). In these birds thin-shelled eggs were laid as a result of metabolic disorder of the shell gland brought about by high concentrations of DDT. Only one nestling out of 552 nests was successful in 1970 on Anacapa Island, the breeding site of the pelican. Prior to 1970, 272·2 kg of DDT were discharged daily into the ocean at Whites Point. Point source control was accomplished in 1970; it resulted in a reduction in the discharge to 4·5 kg d^{-1} since that date. With the reduction of the principle source of DDT in the discharge, breeding of the pelican has continued to become more successful again in southern California.

Since no other instance has been documented where a toxicant in a domestic discharge has been biomagnified during movement through the marine food chain, YOUNG and MEARNS (1979) used the ratio of caesium to potassium to measure toxicant increase in tissues from one trophic level to the next. Their rationale was based on the premise that while potassium must be maintained at a more or less constant level within the cells, caesium must not. The ratio of Cs to K is particularly useful in assigning a certain trophic-level status to larger marine animals. For example, the Cs to K ratio near Whites Point was $6·4 \times 10^{-6}$ for abalones and scallops, 11·2 for decapods and sand dabs *Citharichthus* spp. and 15·1 for scorpionfish *Scorpaena guttato* and bocaccii *Sebastes paucipinis*; the trophic levels assigned were 2–3, 3–4 and 4–5, respectively. With the trophic level determined, it is possible to examine the body-burden levels for these organisms and to determine whether or not a particular toxicant is being magnified when passing through a food chain. In this type of analysis there was no indication that Ag, Cd, Cr, Cu, Fe, Mg, Ni, Pb and Zn increased in concentration at the next trophic level. On the other hand, mercury (primarily in the organic form), total DDT and PCB 1254 concentrations increased with each trophic level. Although YOUNG and MEARNS were not able to relate quantitatively the increases in Cs to K ratios with corresponding increases in toxicant concentrations, their method of analysis shows promise as a means of interpreting body-burden data in areas where the trophic level of the animals involved is unknown.

(5) Management of Municipal Waste Disposal

The question of where to dispose municipal wastes from coastal communities is difficult to answer. Human population growth and migrations continue in many areas of the world; in coastal areas, especially in the USA, human populations increase rapidly. Hence sewage treatment facilities must be enlarged in order to expand and to handle the increasing quantities of waste. Storm drains, often combined with sewage treatment facilities, place an additional burden on plant operations during periods of storm. At the same time, population growth in coastal areas had increased the demand for more recreational areas and for cleaner waters and beaches. As a response to this, the US Congress enacted the Clean Water Act, which requires mandatory secondary treatment of all sewage discharged into marine waters. Since most US coastal communities discharge either untreated or primary treated wastes into marine waters, the Clean Water Act necessitates upgrading of almost all US sewage-treatment facilities. Taking into account the testimony of many experts, Congress amended the Clean Water Act, adding Section 301 (h), which allows a marine discharger to acquire a waiver from secondary treatment if it could be shown that the effluent does not harm the oceanic environment. In terms of ecology, the applicant is required to demonstrate, among other things, that the discharge protects a balanced indigenous population of fish, shellfish and other wildlife, and to submit a plan for monitoring the environment. A total of 70 coastal communities have submitted waiver applications, the majority of which are from the Pacific coast. Action on these is pending (KISH, 1980).

The Clean Water Act represents the most ambitious attempt to date by any governmental agency to control municipal waste effluents which empty into the ocean. If a waiver is denied, in addition to the cost factors, the question remains of what should be done with the sludge. Although a certain percentage can be dried and sold as fertilizer, for the remainder the only alternative is land disposal. However, this not only becomes increasingly expensive but it also creates more and more difficulties, especially for metropolitan communities, to find and set aside suitable disposal sites. In southern California, for example, it would soon be necessary to transport the sludge 50–100 km from the treatment plant to a suitable disposal site. However, most people residing in the vicinity of the proposed dumpsite strenuously object to the disposal of sludge near their community. The 70 sanitation districts seeking a waiver are faced with a dilemma which it seems impossible to solve. Compounding the problem is the lack of sufficient funds for modifying existing facilities to secondary treatment if the waiver application is denied (KISH, 1980).

Since the primary concern for upgrading traditional facilities to a secondary sewage treatment level is to protect or improve the environment, it is of paramount importance to determine whether or not upgrading does, in fact, lead to that aim. Ecologically valid data for assessing consequences of changes in sewage treatment are limited. Nevertheless, the large capital investment required for upgrading plants to secondary treatment capabilities requires careful and critical evaluation of all biological data available. Unfortunately, in most cases known to the reviewer, such a study has not yet been completed. Some indications of improvement in benthic communities have been observed after a change from primary to secondary treatment in Los Angeles Harbor (WARE, 1979; REISH and BIKSEY, unpubl.). Wastes from tuna canneries and primary sewage effluents were discharged less than 0·5 km from each other. A seasonal quantitative study was made in the benthic area prior to re-routing the fish cannery waste to the sewage treatment plant and its change into secondary treatment facility (WARE, 1979). The cannery wastes were transported and mixed with domestic sewage, subjected to secondary treatment and discharged into outer Los Angeles Harbor. A second quantitative study has been initiated at the same station localities 3 yr after change-over to secondary treatment. Only preliminary results are available at this time (REISH and BIKSEY, unpubl.). The change produced dramatic alterations in the benthic fauna. Station 1, located near the fish cannery outfall, initially had a median species diversity of $H' = 0·28$ and was characterized by the presence of *Capitella capitata*. Following secondary treatment the species diversity increased to $H' = 2·4$ with no single species dominating. *C. capitata* was still present but no longer dominant. Similar results were obtained at Station 2, located near the exit of the outfall sewer. Here, the species diversity increased from $H' = 0·58$ to $2·2$ after secondary treatment. Since no toxic industrial wastes were discharged in the vicinity, the benthic-fauna depression was presumably due to the inability of the fauna to assimilate all the organic wastes received. Once nearly all the solid wastes had been removed, numerous benthic animals could settle and establish themselves in the area. Other dramatic changes in the area were noted (REISH and co-authors, 1981): the fish and bird populations decreased following a cessation of cannery discharges in late 1977. While the species composition remained the same, populations of birds and fish decreased in individual numbers. The populations of white croakers *Genyonemus lineatus* and anchovy *Engraulis mordax* decreased from 1 to 10% of the former levels. The birds no longer aggregated for feeding at the boil of the sewage plant discharge. It is assumed that the fish cannery wastes contributed to a large percentage of the food for the white croaker and other fish; once this source of food was no longer available, the fish populations presumably dispersed. The bird populations probably decreased because of the fewer fish present and the absence of solid food particles from the discharge.

Diversion of untreated sewage from Cannes (France) in 1973 resulted in the complete recovery of the local benthic fauna by 1977, but increased the degree of pollution at the western end of the Golfe de la Napoule where a new, longer discharge pipe now released these plus additional wastes from other communities (BELLAN, 1979b). Similar changes were noted off Orange County (USA) when a new and longer pipeline was constructed and the use of the shorter one discontinued. Recovery near the old discharge was completed in 1 yr; degradation near the new pipeline outlet took 4–5 yr (REISH, 1980).

As noted above, partial or complete recovery resulting in a normal benthic environment apparently is rapid once the source of the pollutant has been removed from the system. Similar fast recovery rates of the benthos have also been observed with other types of pollutants. The ecological conditions began to improve in less than 1 yr in the inner harbour of Los Angeles, once oil refinery wastes ceased to be discharged into the area (REISH, 1971). In a Swedish estuary recovery began 3 yr after closure and was completed in 8 yr, following the closure of a sulphite pulp mill (ROSENBERG, 1972, 1976).

There is no concensus of opinion with regard to the resolution for municipal waste disposal into the sea. CARTER (1973), for example, states that sewage disposal into the sea can be considered as sewage treatment provided that any toxic substances from industrial wastes, if present, are eliminated prior to the discharge. OFFICER and RYTHER (1977) believe that eutrophication of marine waters would be greatly enhanced by secondary treatment, because the nutrients would be in a more effective chemical form, especially if the toxic material has a long decomposition time.

Is secondary treatment of sewage necessary for wastes destined to be discharged at sea? Unfortunately, we do not know the assimilative capacity of the sea (DALLAIRE, 1971). Neither British nor Australian sewage discharges into the sea have caused major environmental problems (KISH, 1980). Any solution must take into account the entire sewage treatment process. Not only must the ocean environment be carefully assessed, but also the disposal of sludge. Will sludge disposal cause greater problems to the terrestrial than to the marine environment? Will trucking of sludge cause serious air pollution problems in metropolitan areas? Clearly there is no simple, all-inclusive solution. Many countries have followed the USA in the construction of larger and longer sewer outfalls. Undoubtedly, they will be observing how the USA solves this issue.

Clearly, we need to know more about ecological processes, especially how they relate to sewage disposal. We have seen increasing evidence of fish diseases near discharges, we have seen the deterioration of many km^2 of benthos around waste discharges and we have seen indications of toxicity of the effluent to such important biological processes as fertilization. On the other hand, we have been encouraged by the rapid recovery of the environment once the waste discharge has been eliminated or altered or its toxic components removed. It becomes apparent that any decision as to the nature and type of waste treatment must consider the particular receiving environment. Control of toxic wastes at the sources should be attempted whenever feasible. The removal of a toxic substance or substances from the waste would make it less detrimental to the environment. If estuaries or tropical lagoons are the site for a sewer outfall, secondary treatment should be the minimum requirement. Small, isolated sewers, such as the one located on San Clemente Island, do little or no damage to the biota. Construction of secondary treatment facilities at such localities seems unnecessary. Discharge of wastes from metropolitan areas such as southern California presents the most difficult problem to solve.

It has been suggested that a separate sludge line—such as that present today at the Hyperion treatment plant—be constructed near other large discharges in order to eliminate solid wastes at greater depths (300—400 m) than currently practiced (JACKSON and co-authors, 1979). Never before has sludge been disposed of through a pipe at depths this great, but Orange County Sanitation District (USA) is planning such a project on a more or less experimental basis. If such a method of sludge disposal has little or no effect on the marine environment, then this offers to be an alternative to land disposal. Municipal wastes would then undergo secondary treatment with control at the source to eliminate or reduce toxic substances with the liquid wastes being discharged through one pipe in shallow waters and the sludge being discharged at 300 m. If, on the other hand, it is found that sludge discharged at 300 m does affect a significant amount of the benthos, then some other method of sludge disposal would be necessary. It is hoped that it will not, because this method of disposal seems to offer the greatest promise as to the best possible compromise to protect all elements of the environment.

Literature Cited (Chapter 7)

ALEXANDER, W. B., SOUTHGATE, B. A. and BASSINDALE, R. (1935). Survey of the River Tees. Part 2. The estuary. Chemical and biological. *Tech. Pap. Wat. Pollut. Res. D.S.I.R.*, **5**, 1–171.

ALLEN, M. J. (1975a). Alternate methods for assessing fish populations. *Sth. Calif. Coast. Wat. Res. Proj., Annual Rep.*, El Segundo, pp. 95–98.

ALLEN, M. J. (1975b). Regional variation in the structure of fish communities. *Sth. Calif. Coast. Wat. Res. Proj., Annual Rep.*, El Segundo, pp. 99–102.

ALLEN, M. J. (1977). Bottom fish populations below 200 meters. *Sth. Calif. Coast. Wat. Res. Proj., Annual Rep.*, El Segundo, pp. 109–116.

AMERICAN PUBLIC HEALTH ASSOCIATION (1975). *Standard Methods for the Examination of Water and Wastewater*, 14th ed., American Public Health Association, American Water Works Association and Water Pollution Control Federation, Washington, DC.

AMIEL, A. J. and NAVROT, J. (1978). Nearshore sediment pollution in Israel by trace methods derived from sewage effluent. *Mar. Pollut. Bull., N.S.*, **9**, 10–14.

ANONYMOUS (1952). *Los Angeles–Long Beach Harbor Pollution Survey*, Regional Water Pollution Control Board, Los Angeles.

AUBERT, M., AUBERT, J. and ORCEL, L. (1979. Étude sur l'extension geographique et l'origine des necroses des poissons sur le littoral mediterraneen. *Rev. int. Océanogr. méd.*, **53–54**, 3–21.

BAGGE, P. (1969a). Effects of pollution on estuarine ecosystems. 1. Effects of effluents from wood processing industries on the hydrography, bottom and fauna of Saltkallefjord (W. Sweden). *Merentutkimuslait. Julk.*, **228**, 1–118.

BAGGE, P. (1969b). The succession of the bottom fauna communities in polluted estuarine habitats in the Baltic–Skagerak region. *Merentutkimuslait. Julk.*, **228**, 119–130.

BALCH, N., ELLIS, D., LITTLEPAGE, J., MARLES, E. and PYM, R. (1976). Monitoring a deep marine wastewater outfall. *J. Wat. Pollut. Control Fed.*, **48**, 429–457.

BARNARD, J. L. and HARTMAN, O. (1959). The sea bottom of Santa Barbara, California. Biomass and Community structure. *Pacif. Nat.*, **1** (6), 1–16.

BARNARD, J. L. and JONES, G. F. (1960). Techniques in a large scale survey of marine benthic biology. In E. A. Pearson (Ed.), *Waste Disposal in the Marine Environment*. Pergamon Press, New York. pp. 413–447.

BARNES, H. (1959). *Apparatus and Methods in Oceanography, Part 1. Chemical*, Interscience, New York.

BASSINDALE, R. (1938). The intertidal fauna of the Mersey estuary. *J. mar. biol. Ass. U.K.*, **23**, 83–98.

BAUGHMAN, J. L. (1948). *An Annotated Bibliography of Oysters with Pertinent Material on Mussels and other Shell-fish and an Appendix on Pollution*, Texas A & M Res. Fd., College Station, Texas.

BAYNE, B. L. (1978). Mussel watch. *Nature Lond.*, **275**, 87.

BEERS, J. R., REED, F. M. H. and STEWART, G. L. (1979). Microplankton and other seston components in coastal waters near the Point Loma (San Diego) sewage outfall. *Calif. Mar. Res. Comm. CALCOFI Dept.*, **20**, 125–134.

BELLAN, G. (1979a). Annélides polychètes des substrats solides de trois millieux pollués sur la côtes de provence (France): Cortiou, Golfe de Fos, Vieux Port de Marseille. *Téthys*, **9**, 267–277.

BELLAN, G. (1979b). An attempted pollution abatement in the Gulf of La Napoule (Cannes, France). *Mar. Pollut. Bull., N.S.*, **10**, 163–166.

BELLAN, G. and PÉRÈS, J. M. (1970). Etat général des pollutions sur les Côtes méditerranéenes de France. *Quad. Civ. Staz. Idrobiol.*, **1**, 35–65.

BELLAN-SANTINI, D. (1968). Influence de la pollution sur les peuplementos benthiques. *Revue int. Océanogr. méd.*, **10**, 27–53.

BELLAN-SANTINI, D. (1980). Relationship between populations of amphipods and pollution. *Mar. Pollut. Bull., N.S.*, **11**, 224–227.

BENDER, M. E., REISH, D. J. and WARD, C. H. (1980). Re-examination of the offshore ecology investigation. In C. H. Ward, M. E. Bender and D. J. Reish (Eds), *The Offshore Ecology Investigation*. Rice Univ. Studies, 65 (4–5), 35–102.

BENDER, M. E., SHARP, J. M., REISH, D. J., APPAN, S. G. and WARD, C. H. (1979). An independent appraisal of the offshore ecology investigation. *Offshore Tech. Conf. Houston*. Paper No. 3506, 2163–2172.

BLEGVAD, H. (1932). Investigations of the bottom fauna at outfalls of drains in the Sound. *Rep. Dan. biol. Stn*, **37**, 1–20.

BOESCH, D. F. (1977). Application of numerical classification in ecological investigations of water pollution. *Ecological Res. Sec.*, EPA-600/3–77–033.

BROOKS, N. H. (1960). Diffusion of sewage effluent in an ocean current. In E. A. Pearson (Ed.), *Waste Disposal in the Marine Environment*. Pergamon Press, New York. pp. 246–267.

BROWN, R. S., WOLKE, R. E., SAILA, S. B. and BROWN, E. W. (1977). Prevalence of neoplasia in 10 New England populations of the soft-shell clam (*Mya arenaria*) *Ann. N.Y. Acad. Sci.*, **298**, 522–534.

CALIFORNIA STATE WATER RESOURCES CONTROL BOARD (1972). *Water Quality Control Plan for Ocean Waters of California* (Mimeogr. Rep.), State Water Resources Control Board, Sacramento.

CAPERON, J., CATTELL, S. A. and KRASNICK, G. (1971). Phytoplankton kinetics in a subtropical estuary. Eutrophication. *Limnol. Oceanogr.*, **16**, 599–607.

CARTER, L. (1973). Disposal to the sea is sewage treatment. *Effluent Wat. Treat. J.*, **13**, 647–649.

CIMBERG, R., MANN, S. and STRAUGHAN, D. (1973). A re-investigation of southern California rocky intertidal beaches three and one-half years after the 1969 Santa Barbara oil spill. *Proc. Joint Conf. Prevention and Control of Oil Spills* (Prelim. Rep.). Washington, D.C. pp. 697–702.

COGNETTI, G. (1972). Distribution of polychaeta in polluted waters. *Rev. int. Océanogr. méd.*, **25**, 23–24.

CONROY, D. A. (in press). Diseases of fishes: micro-organisms. Bacteria. In O. Kinne (Ed.), *Diseases of Marine Animals*, Vol. IV. Biologische Anstalt Helgoland, Hamburg.

COUCH, J. A. (1978). Diseases, parasites and toxic responses of commercial penaid shrimps of the Gulf of Mexico and South Atlantic coasts of North America. *Fish. Bull., U.S.*, **76**, 1–44.

COUCH, J. A. and NIMMO, D. R. (1974). Detection of interactions between natural pathogens and pollutants in aquatic animals. In R. L. Amborski, M. A. Hood and R. R. Miller (Eds), *Diseases of Aquatic Animals*. Centre Wetland Resources Publ., No. LSU-SG-74-05. Louisiana State Univ., Baton Rouge. pp. 261–268.

DALLAIRE, E. E. (1971). Ocean dumping: what and where if at all? *American Society of Civil Engineering* (ASCE), **Nov. 1971**, 58–62.

DAWSON, E. Y. (1959). A preliminary report on the marine benthic flora of southern California. In *Oceanographic Survey of the Continental Shelf Area of Southern California*, 20. Calif. State Water Pollution Control Bd, Sacramento. pp. 169–264.

DAWSON, E. Y. (1965). Intertidal algae. In *Oceanographic and Biological Survey of the Southern California Mainland Shelf*, 27. Calif. State Water Pollution Control Bd, Sacramento. pp. 220–231 and 351–438.

DORSEY, J. H. (1975). *Effects of Sewage Discharge on the Subtidal Polychaetous Annelids of Wilson Cove, San Clemente Island, California*, Master's Thesis, California State Univ., Long Beach.

DRAXLER, A. F. J. (1979). Transient effects of ocean wastewater dumping. *J. Wat. Pollut. Control Fed.*, **51**, 741–748.

DUEDALL, I. W., BOWMAN, M. J. and O'CONNORS, H. B., JR. (1975). Sewage sludge and ammonium concentrations in the New York Bight. *Estuar. coast. mar. Sci.*, **3**, 457–463.

EPPLEY, R. W., CARLUCCI, A. F., HOLM-HANSEN, O., KIEFER, D., McCARTHY, J. J. and WILLIAMS, P. M. (1972). Evidence for eutrophication in the sea near southern California coastal sewage outfalls, July 1970. *Calif. Mar. Res. Comm., CALCOFI Rept.*, **16**, 74–83.

FAIR, G. M. (1964). Water pollution research and the revolution in science and technology. In B. A. Southgate (Ed.), *Advances in Water Pollution Research*, Vol. I. Pergamon Press, Oxford. pp. xi–xxv.

FARLY, C. A. (1976). Proliferative disorders in bivalve molluscs. *Mar. Fish. Rev.*, **38**(10), 30–33.

FILICE, F. P. (1954a). An ecological survey of the Castro Creek area in the San Pablo Bay. *Wasmann J. Biol.*, **12**, 1–24.

FILICE, F. P. (1954b). A study of some factors affecting the bottom fauna of a portion of the San Francisco Bay estuary. *Wasmann J. Biol.*, **112**, 257–292.

FITCH, J. E. (1956). The effect of White's Point River on marine life. *Am. Malacol. Union Bull.*, **23**, 20–21.

FOLGER, D. W., PALMER, H. D. and SLATER, R. H. (1979). Two waste disposal sites on the continental shelf off the middle Atlantic states: observations made from submersibles. In H. D. Palmer and M. G. Gross (Eds), *Ocean Dumping and Marine Pollution*. Dowden, Hutchinson and Ross, Inc., Stroudsberg, Pa. pp. 163–184.

FORBES, S. A. and RICHARDSON, R. E. (1913). Studies on the biology of the Upper Illinois River. *Bull. Ill. St. Lab. nat. Hist.*, **9**, 481–574.

FOSTER, M., NEUSHUL, M. and ZIRGMARH, R. (1971). The Santa Barbara oil spill. Part 2. Initial effects on intertidal and kelp bed organisms. *Environ. Pollut.*, **2**, 97–113.

FREELAND, G. L., SWIFT, D. P. and YOUNG, R. A. (1979). Mud deposits near the New York Bight dumpsites: origin and behaviour. In H. D. Palmer and M. G. Gross (Eds), *Ocean Dumping and Marine pollution*. Dowden, Hutchinson and Ross, Inc., Stroudsberg, Pa. pp. 73–95.

GALLARDO, V. A., CASTILLO, J. G. and YANEZ, L. A. (1972). Algunas consideraciones preliminares abre la ecologia benthonica de los fondas sublittordes blandos en la Bahia de Concepcion. *Boln Soc. Biol. Concépcion*, **44**, 169–190.

GALLOWAY, J. M. (1979). Alteration of trace metal geochemical cycles due to marine discharge of wastewater. *Geochim. cosmochim. Acta*, **43**, 207–218.

GAUFIN, A. R. and TARZWELL, C. M. (1952). Aquatic macro-invertebrate communities as indicators of organic pollution in Lytle Creek. *Sewage ind. Wastes*, **28**, 906–924.

GHIRADELLI, E. and PIGNATTI, S. (1968). Conséquences de la pollution sur les peuplements du "Vallone de Muggia" près se Trieste. *Revue int. Océanogr. Méd.*, **10**, 111–122.

GRASSLE, J. P. and GRASSLE, J. F. (1976). Sibling species in marine pollution indicator *Capitella* (Polychaeta). *Science, N.Y.*, **192**, 567–569.

GRAY, J. S. (1976). The fauna of the River Tees estuary. *Estuar. coast. mar. Sci.*, **4**, 653–676.

GREENBERG, A. E. (1956). Survival of entire organisms in seawater. *Publ. Hlth Rep., Wash.*, **71**, 77–86.

GREENE, C. S. (1976). Changes in grain size of sediments on the Palos Verdes Shelf. *Sth. Calif. Coast. Wat. Res. Proj., Annual Rep.*, El Segundo, pp. 91–93.

GUARINO, C. F., NELSON, M. D. and ALMEIDA, S. S. (1979). Ocean disposal as an alternate disposal method. *J. Wat. Pollut. Control Fed.*, **51**, 773–782.

HALCROW, W., MACKAY, D. W. and THORNTON, I. (1973). The distribution of trace metals and fauna in the Firth of Clyde in relation to the disposal of sewage sludge. *J. mar. biol. Ass. U.K.*, **53**, 721–739.

HARDY, J. T. and HARDY, S. A. (1972). Tidal circulation and sewage pollution in a tropical lagoon. *Environ. Pollut.*, **3**, 195–203.

HART, C. W., JR and FULLER, S. L. H. (Eds) (1979). *Pollution Ecology of Estuarine Invertebrates*, Academic Press, New York.

HARTNOLL, R. G. and HAWKINS, S. J. (1980). Monitoring rocky-shore communities: a critical look at spatial and temporal variations. *Helgoländer Meeresunters.*, **33**, 484–494.

HARTMAN, O. (1955). Quantitative survey of the benthos of San Pedro Basin, southern California. 1. Preliminary results. *Allan Hancock Foundation Pacific Exped.*, **19**, 1–185.

HASLER, A. D. (1947). Eutrophication of lakes by domestic sewage. *Ecology*, **28**, 383–395.

HELZ, G. R., HUGGETT, K. J. and HILL, J. M. (1975). Behaviour of Mn, Fe, Cu, Zn, Cd and Pb discharged from a wastewater treatment plant into an estuarine environment. *Wat. Res.*, **9**, 631–636.

HERSHELMAN, G. P., JAN, T. K. and SCHAFER, H. A. (1977). Pollutants in sediments of Palos Verdes. *Sth. Calif. Coast. Wat. Res. Proj., Annual Rep.*, El Segundo, pp. 63–68.

HESS, E. (1937). A shell disease in lobsters (*Homarus americanus*) caused by chitinovorous bacteria. *J. biol. Bd Can.*, **3**, 358–362.

HOLLAND, R. E. (1968). Correlation of *Melosira* species with trochic conditions in Lake Michigan. *Limnol. Oceanogr.*, **13**, 555–557.

HOLME, N. A. (1964). Methods of sampling the benthos. *Adv. mar. Biol.*, **2**, 171–260.

HULINGS, N. C. and GRAY, J. S. (Eds) (1971). A manual for the study of meiofauna. *Smithson. Contr. Zool.*, **78**, 1–84.

IVERSEN, E. J. and BEARDSLEY, G. L. (1976). Shell disease in crustaceans indigenous to South Florida. *Prog. Fish Cult.*, **38**, 195–276.

JACKSON, G. A., KOH, R. C. Y., BROOKS, N. H. and MORGAN, J. J. (1979). Assessment of alternative strategies for sludge disposal into deep ocean basins off southern California. Environmental Quality Laboratory, Calif. Inst. Tech., Pasadena. EOL Rep. 14, pp. 1–121.

JONES, G. F. (1969). The benthic macrofauna of the mainland shelf of Southern California. *Allan Hancock Monogr. mar. Biol.*, **4**, 1–219.

KABATA, Z. (in press). Diseases of fishes: metazoans. Arthropoda. In O. Kinne (Ed.), *Diseases of Marine Animals*, Vol. IV. Biologische Anstalt Helgoland, Hamburg.

KAYSER, M. (1968). Hauptquellen häuslicher Abwässer und deren Bedeutung für die Verunreinigung der Nordsee. *Helgoländer wiss. Meeresunters.*, **17**, 25–43.

KINDIG, A. C. and LITTLER, M. M. (1980). Growth and primary productivity of marine macrophytes exposed to domestic sewage effluents. *Mar. environ. Res.*, **3**, 81–100.

KINNE, O. (1971). Salinity: animals. Invertebrates. In O. Kinne (Ed.), *Marine Ecology, Vol. 1, Environmental Factors*, Part 2. Wiley, London. pp. 821–995.

KINNE, O. (1977). Cultivation of animals: research cultivation. In O. Kinne (Ed.), *Marine Ecology, Vol. III, Cultivation*, Part 2. Wiley, Chichester. pp. 579–1124.

KINNE, O. (1980). Diseases of marine animals: general aspects. In O. Kinne (Ed.), *Diseases of Marine Animals, Vol. I, General Aspects. Protozoa to Gastropoda*. Wiley, Chichester. pp. 13–64.

KINNE, O. and AURICH, H. (Eds) (1968). Internationales Symposium "Biologische und hydrographische Probleme der Wasserverunreinigung in der Nordsee und angrenzenden Gewässern". *Helgoländer wiss. Meeresunters.*, **17**, 1–530.

KINNE, O. and BULNHEIM, H.-P. (Eds) (1980). 'Protection of Life in the Sea'. Proceedings of the International Helgoland Symposium, September 1979. *Helgoländer Meeresunters.*, **33**, 1–769.

KINNE, O. and BULNHEIM, H.-P. (Eds) (1984). *Diseases of Marine Organisms*. Proceedings of the International Helgoland Symposium, September 1983. *Helgoländer Meeresunters.*, **38**.

KISH, T. (1980). And the winners are . . . . *J. Wat. Pollut. Control Fed.*, **52**, 2076–2081.

KITAMORI, R. (1972). Fauna and flora changes by pollution in coastal waters of Japan (2nd). *Int. Ocean Dev. Conf.*, pp. 71–77.

KOCATAS, A. and GELDIAY, R. (1980). Effects of domestic pollution on Izmir Bay (Turkey). *Helgoländer Meeresunters.*, **33**, 393–400.

KOLKWITZ, R. and MARSSON, M. (1980). Oekologie der pflanzliken Saprabein. *Ber. dt. bot. Ges.*, **26**, 505–519.

KOLKWITZ, R. and MARSSON, M. (1909). Oekologie der tierischen Saprobien. Beiträge Lehre von der biologischen Gewässerbeurteilung. *Int. Rev. ges. Hydrobiol. Hydrogr.*, **2**, 126–152.

KOLPACK, R. L. (1979). Distribution of suspended particulate matter near sewage outfalls in Santa Monica Bay, California. In H. D. Palmer and M. G. Gross (Eds), *Ocean Dumping and Marine Pollution*. Dowden, Hutchinson and Ross, Inc., Stroudsberg, Pa. pp. 205–239.

KORRINGA, P. (1968). Biological consequences of marine pollution with special reference to the North Sea fisheries. *Helgoländer wiss. Meeresunters.*, **17**, 126–140.

KRAYBILL, H. F., DAWE, C. J., HARSHBARGER, J. C. and TARDIFF, R. G. (1977). Aquatic pollutants and biologic effects with emphasis on neoplasia. *Ann. N.Y. Acad. Sci.*, **298**, 1–604.

LAUCKNER, G. (1980). Diseases of marine animals: protozoa to gastropoda. In O. Kinne (Ed.), *Diseases of Marine Animals, Vol. I, General Aspects. Protozoa to Gastropoda*. Wiley, Chichester. pp. 75–400.

LAUCKNER, G. (in press). In O. Kinne (Ed.), *Diseases of Marine Animals*, Vol. IV. Biologische Anstalt Helgoland, Hamburg.

LENZ, J. (1972). Analysis of the stock. A: plankton. In C. Schlieper (Ed.), *Research Methods in Marine Biology*. Univ. Wash. Press, Seattle. pp. 46–63.

LEPPÄKOSKI, E. (1975). Assessment of degree of pollution on the basis of macrozoobenthos in marine and brackish water environments. *Acta Acad. åbo. (Ser. B)*, **35**(2), 1–90.

LEPPÄKOSKI, E. (1977). Monitoring the benthic environment of organically polluted river mouths. In J. S. Alabaster (Ed.), *Biological Monitoring of Inland Fisheries*. Applied Science, Barking. pp. 125–132.

LITTLER, M. M. and MURRAY, S. N. (1978). Influence of domestic wastes on energetic pathways in rocky tidal communities. *J. appl. Ecol.*, **15**, 583–595.

LIZARRAGA-PARTIDA, M. L. (1974). Organic pollution in Ensenada Bay, Mexico, *Mar. Pollut. Bull., N.S.*, **5**, 109–112.

LOM, J. (in press). Diseases of fishes: protistans. In O. Kinne (Ed.), *Diseases of Marine Animals*, Vol. IV. Biologische Anstalt Helgoland, Hamburg.

LOS ANGELES COUNTY SANITATION DISTRICT. (1980). *Annual Report, 1979* (Mimeogr. Rep), Whittier, California.

LUDWIG, H. F. and STORRS, P. N. (1970). Effects of waste disposal into marine waters. A survey of studies carried out in the last ten years. *Wat. Res.*, **4**, 709–720.

MABBETT, A. N. (1975). Water quality of a micronesian atoll. *J. environ. Hlth*, **37**, 332–341.

MAHONEY, J. B. MIDLIGE, F. H. and DEVEL, D. G. (1973). A fin rot disease of marine and euryhaline fishes in the New York Bight. *Trans. Am. Res. Soc.*, **102**, 596–605.

MANFREDI, E. (1976). *A Bacteriological Study of Fin Erosion Disease in Dover Sole (Microstomus pacificus)*, Masters Thesis, California State University, Long Beach, Calif.

MARE, M. F. (1942). A marine benthic community, with special reference to the microorganisms. *J. mar. biol. Ass. U.K.*, **25**, 517–534.

MARTIN, A. (1981). *The Effects of Primary and Secondary Municipal Wastewater on the Polychaetous Annelid Neanthes arenaceodentata*, Masters Thesis, California State University, Long Beach, Calif.

MAWDESLEY-THOMAS, L. E. (1974). Some aspects of neoplasia in marine mammals. *Adv. mar. Biol.*, **12**, 151–231.

MCDERMOTT, D. J. (1974). Characteristics of municipal wastewaters, 1971–1973. *Sth. Calif. Coast. Wat. Res. Proj., Annual Rep.*, El Segundo, pp. 89–96.

MCLUSKY, D. S., TEARE, M. and PHIZACKLEA, P. (1980). Effects on domestic and industrial pollution on distribution and abundance of aquatic oligochaetes in the Forth estuary. *Helgoländer Meeresunters.*, **33**, 113–121.

MCNULTY, J. K. (1970). *Effects of Abatement of Domestic Sewage Pollution in Biscayne Bay*, Univ. of Miami Press, Coral Gables.

MEARNS, A. J. (1974). Standardizing sampling procedures. *Sth. Calif. Coast. Wat. Res. Proj., Annual Rep.*, El Segundo, pp. 63–68.

MEARNS, A. J. and SHERWOOD, M. J. (1977a). Fin erosion prevalence and environmental changes. *Sth. Calif. Coast. Wat. Res. Proj., Annual Rep.*, El Segundo, pp. 139–142.

MEARNS, A. J. and SHERWOOD, M. J. (1977b). Distribution of neoplasia and other diseases in marine fishes relative to the discharge of wastewater. *Ann. N.Y. Acad. Sci.*, **298**, 210–224.

MEARNS, A. J. and WORD, J. Q. (1981). Forecasting effects of sewage solids on marine benthic communities. *Proc. Ecological Effects of Environmental Stress. NOAA–MESA*.

METCALF and EDDY, Inc. (1979). *Wastewater Engineering. Treatment, Disposal and Reuse*, 2nd ed., McGraw-Hill, New York.

MORISETA, M. (1959). Measuring interspecific association and similarity between communities. *Mem., Fac. Sci. Kyushu Univ. (Series E)*, **3**, 65–80.

MURCHELANO, R. A. and ZISKOWSKI, J. (1976). Fin rot disease studies in the New York Bight. *Limnol. Oceanogr. (Special Symposium)*, **2**, 329–336.

MURRAY, S. N. and LITTLER, M. M. (1974). Analysis of standing stock and community structure of macro-organisms. In S. N. Murray and M. M. Littler (Eds), *Biological Features of Intertidal Communities near the U.S. Navy Sewage Outfall, Wilson Cove, San Clemente Island, California*. U.S. Navy, NUC Tech Paper No. 396, pp. 23–85.

MURRAY, S. N. and LITTLER, M. M. (1978). Patterns of algae succession in a perturbed marine intertidal community. *J. Phycol.*, **14**, 506–512.

NAKAI, A., KOSAKA, M., KUDOH, S., NAGAI, A., HAYASHIDA, F., KUBOTA, T., OGURA, M., MIZUSHIMA, T. and UOTANI, I. (1973). Summary report on marine biological studies of Suruga Bay accomplished by Tokai University 1964–1972. *Fac. Mar. Sci. Technol. Tokai Univ.*, **7**, 63–117.

NAKAZIMA, M. (1965). Studies on the source of shellfish poison in Lake Hamana. *Bull. Jap. Soc. scient. Fish.*, **31**, 198–207.

NICHOLSON, N. and CIMBERG, R. (1971). The Santa Barbara Oil Spill of 1969: a post-spill survey of the rocky intertidal. In D. S. Straughan (Ed.), *Biological and Oceanographic Survey of the Santa Barbara Channel Oil Spill of 1969–1970*, Vol. 1. Allan Hancock Foundation, University of South California. pp. 325–402.

OFFICER, C. B. and RYTHER, J. H. (1977). Secondary sewage treatment versus ocean outfalls: an assessment. *Science, N.Y.*, **197**, 1056–1060.

OLIFF, W. D., BERRISFORD, C. D., TURNER, W. D., BALLARD, A. J. and MCWILLIAM, D. C. (1967). The ecology and chemistry of sandy beaches and nearshore submarine sediments of Natal. II. Pollution criteria for nearshore sediments of the Natal coast. *Wat. Res.*, **1**, 131–146.

OSHIDA, P. S., GOOCHEY, T. K. and MEARNS, A. J. (1981). Effects of municipal wastewater on fertilization, survival and development of the sea urchin, *Strongylocentrotus purpuratus*. In F. S. Vernberg, A. Calabrese, F. P. Thurberg and W. B. Vernberg (Eds), *Biological Monitoring of Marine Organisms*. Academic Press, New York.

OVERSTREET, R. M. and HOWSE, H. D. (1977). Some parasites and diseases of estuarine fishes in polluted habitats of Mississippi. *Ann. N.Y. Acad. Sci.*, **298**, 427–462.

PALMER, C. M. (1969). A composite rating of algae tolerating organic pollution. *J. Phycol.*, **5**, 78–85.

PARKER, J. G. (1980). Effects of pollution upon the benthos of Belfast Lough. *Mar. Pollut. Bull.*, *N.S.*, **11**, 80–83.

PEARCE, J. B. (1972). *The Effects of Solid Waste Disposal on Benthic Communities in the New York Bight. Marine Pollution and Sea Life*, Fishing News (Books) Ltd., Survey.

PEARSON, E. A. (1956). An investigation into the efficiency of submarine outfall disposal of sewage and sludge. *Calif. State Water Pollution Control Bd, Sacramento, Calif. Publ.*, **14**, 1–154.

PEARSON, T. H. (1975). The benthic ecology of Lock Limhe and Lock Eil, a sea-lock system on the west coast of scotland. IV. Changes in the benthic fauna attributed to organic enrichment. *J. exp. mar. Biol. Ecol.*, **20**, 1–41.

PEARSON, T. H. and ROSENBERG, R. (1978). Macrobenthic succession in relation to organic enrichment and pollution of the marine environment. In H. Barnes (Ed.), *Oceanography and Marine Biology Annual Reviews*, 16. Aberdeen Univ. Press. pp. 229–311.

PERKINS, E. J., GILCHRIST, J. R. S. and ABBOTT, O. J. (1972). Incidence of epidermal lesions in fish of the North-East Irish Sea area, 1971. *Nature, Lond.*, **238**, 101–103.

PETERS, N. (in press). Diseases of fishes: neoplasia. In O. Kinne (Ed.), *Diseases of Marine Animals*, Vol. IV. Biologische Anstalt Helgoland, Hamburg.

PETERSEN, C. G. J. (1913). Valuation of the sea. II. The animal communities of the sea bottom and their importance for marine zoogeography. *Rep. Dan. biol. Stn*, **22**, 1–111.

PIELOU, E. C. (1969). *An Introduction to Mathematical Ecology*, Wiley–Interscience, New York.

PINTO, J. S. and SILVA, E. S. (1956). The toxicity of *Cadium edule* L. and its possible relation to the dinoflagellate *Prorocentrum micans* Ehr. *Notas Estud. Inst. Biol. mar., Lis.* **12**, 1–12.

POORE, G. C. B. and KUDINOV. J. D. (1978). Benthos around an outfall of the Werribee sewage-treatment farm, Port Phillip Bay, Victoria. *Aust. J. mar. Freshwat. Res.*, **29**, 157–167.

REISH, D. J. (1955). The relation of polychaetous annelids to harbor pollution. *Publ. Hlth Rep., Wash.*, **70**, 1168–1174.

REISH, D. J. (1959a). An ecological study of pollution in Los Angeles–Long Beach Harbors, California. *Occ. Pap. Allan Hancock Fdn*, **22**, 1–119.

REISH, D. J. (1959b). A discussion of the importance of the screen size in washing quantitative marine bottom samples. *Ecology*, **40**, 307–309.

REISH, D. J. (1971). Effect of pollution abatement in Los Angeles Harbors. *Mar. Pollut. Bull., N.S.*, **2**, 71–74.

REISH, D. J. (1973). The use of benthic animals in monitoring the marine environment. *J. environ. Plann. Pollut. Control*, **1**, 32–38.

REISH, D. J. (1979). Bristle-worms (Annelida: Polychaeta). In C. W. Hart, Jr., and S. L. H. Fuller (Eds), *Pollution Ecology of Marine Invertebrates*. Academic Press, New York. pp. 77–125.

REISH, D. J. (1980). Effect of domestic wastes on the benthic marine communities of Southern California. *Helgoländer Meeresunters.*, **33**, 377–383.

REISH, D. J., GEESEY, G. G., KAUWLING, T. J., WILKES, F. G., MEARNS, A. J., OSHIDA, P. S. and ROSSI, S. S. (1980). Marine Estuarine Pollution. *J. Wat. Pollut. Control Fed.*, **52**, 1533–1574.

REISH, D. J., SOULE, D. F. and SOULE, J. D. (1981). The benthic biological conditions of Los Angeles–Long Beach Harbors: the results of 28 years of investigations and monitoring. *Helgoländer Meeresunters.*, **34**, 193–205.

RESIG, J. M. (1960). Forminiferal ecology around ocean outfalls off Southern California. In E. A. Pearson (Ed.), *Waste Disposal in the Marine Environment*. Pergamon Press, Oxford. pp. 104–121.

RISTICH, S. S., CRANDALL, M., and FORTIER, J. (1977). Benthic and epibenthic macroinvertebrates of the Hudson River. I. Distribution, natural history and community structure. *Estuar. coast. mar. Sci.*, **5**, 255–266.

ROGERS, T. C. (1977). *A Study of Phytoplankton Growth near a Submarine Sewage Outfall*, Masters Thesis, California State Univ., Long Beach, Calif.

ROHATGI, N. K. and CHEN, K. Y. (1976). Fate of metals in wastewater discharge to ocean. *Proc. Am. Soc. civ. Engrs*, **102**(EE3), 675–685.

ROHDE, K. (in press). Diseases of fishes: helminthes. In O. Kinne (Ed.), *Diseases of Marine Animals*, Vol. IV. Biologische Anstalt Helgoland, Hamburg.

ROSENBERG, R. (1972). Benthic faunal recovery in a Swedish fjord following the closure of a sulphite pulp mill. *Oikos*, **23**, 92–108.

ROSENBERG, R. (1976). Benthic faunal dynamics during succession following pollution abatement in a Swedish estuary. *Oikos*, **27**, 414–427.

ROSENBERG, R. (1977). Benthic macrofaunal dynamics, production, and dispersion in an oxygen-deficient estuary of west Sweden. *J. exp. mar. Biol. Ecol.*, **26**, 107–133.

ROWE, R. C. (1975). *The Effects of Sewage Discharge on Intertidal Polychaetous Annelid Assemblages at San Clemente Island*, Masters Thesis, California State Univ., Long Beach, Calif.

SANDERS, H. L. (1968). Marine benthic diversity: a comparative study. *Am. Nat.*, **102**, 243–282.

SCHAFER, C. T. (1973). Distribution of foraminifera near pollution sources in Chaleur Bay. *Water, Air and Soil Pollut.*, **2**, 219–233.

SCHAFER, H. A. (1976). Characteristics of municipal wastewater discharges, 1975. *Sth. Calif. Coast. Wat. Res. Proj.*, *Annual Rep.*, El Segundo, pp. 57–60.

SCHAFER, H. A. (1978). Characteristics of municipal wastewater discharges, 1976. *Sth. Calif. Coast. Wat. Res. Proj.*, *Annual Rep.*, El Segundo, pp. 19–23.

SCHAFER, H. A. (1979). Characteristics of municipal wastewater discharges, 1977. *Sth. Calif. Coast. Wat. Res. Proj.*, *Annual Rep.*, El Segundo, pp. 97–101.

SCHLOTFELDT, H. J. (1972). Jahreszeitliche Abhängigkeit der 'Schwarzfleckenkrankheit' bei der Garnele, *Crangon crangon* L. *Ber. dt. wiss. Kommn Meeresforsch.*, **22**, 397–399.

SHARP, J. M., APPAN, S. G., BENDER, M. E., LINTON, T. G., REISH, D. J. and WARD, C. H. (1979). Natural variability of biological community structure as a quantitative basis for ecology impact assessment. *Proc. Soc. Petroleum Biol.*, *Arlington, Va.*, **1979**, 257–284.

SHELTON, R. G. J. (1971). Sludge Dumping in the Thames estuary. *Mar. Pollut. Bull.*, *N.S.*, **2**, 24–27.

SHERWOOD, M. J. (1978). Fin erosion disease and liver chemistry: Los Angeles and Seattle. *Sth. Calif. Coast. Wat. Res. Proj.*, *Annual Rep.*, El Segundo, pp. 213–219.

SHERWOOD, M. J. (1979). The fin erosion syndrome. *Sth. Calif. Coast. Wat. Res. Proj.*, *Annual Rep.*, El Segundo, pp. 203–221.

SHERWOOD, M. J. and MEARNS, A. J. (1977). Environmental significance of fin erosion in Southern California demersal fishes. *Ann. N.Y. Acad. Sci.*, **298**, 177–189.

SINDERMANN, C. J. (1966). Diseases of marine fishes. *Adv. mar. Biol.*, **4**, 1–80.

SINDERMANN, C. J. (1979). Pollution-associated diseases and abnormalities of fish and shellfish: a review. *Fish. Bull.*, *U.S.*, **76**, 717–749.

SINDERMANN, C. J., ZISKOWSKI, J. J. and ANDERSON, V. T., JR. (1978). A guide for the recognition of some disease conditions and abnormalities in marine fish. *Tech. Serv. Rep.*, *Natl. Mar. Fish. Serv.*, *Northwest Fish. Center*, **14**, 1–65.

SMITH, R. W. and GREENE, C. S. (1976). Biological communities near submarine outfalls. *J. Wat. Pollut. Control Fed.*, **48**, 1894–1912.

SOUTH CALIFORNIA COASTAL WATER RESEARCH PROJECT (1973). The ecology of the southern California Bight: implications for water quality management. *Sth. Calif. Coast. Wat. Res. Proj.*, TR104, 531 pp.

SOUTH CALIFORNIA COASTAL WATER RESEARCH PROJECT (1975). *Annual Rep.*, El Segundo, California, 197 pp.

SOUTH CALIFORNIA COASTAL WATER RESEARCH PROJECT (1976). *Annual Rep.*, El Segundo, California. 211 pp.

SOUTH CALIFORNIA COASTAL WATER RESEARCH PROJECT (1977). *Annual Rep.*, El Segundo, California. 263 pp.

SOUTH CALIFORNIA COASTAL WATER RESEARCH PROJECT (1978). *Annual Rep.*, El Segundo, California. 253 pp.

STEEDMAN, H. F. (Ed.) (1976). Zooplankton fixation and preservation. *Monographs on Oceanographic Methodology*, No. 4. UNESCO Press, Paris.

STOPFORD, S. (1951). An ecological survey of the Cheshire foreshore of the Dee Estuary. *J. Anim. Ecol.*, **20**, 103–122.

THOM, R. M. and CHEW, K. K. (1979). The response of subtidal infaunal communities to a change in wastewater discharge. In *Urban Stormwater and Combined Sewers Overflow Impact on Receiving Water Bodies*. Orlando, Florida, Nov. 26–28, 1977. pp. 174–191.

THOM, R. M. and WIDDOWSON, T. B. (1978). A re-survey of E. Yale Dawson's 42 intertidal algae transects on the southern California mainland after 15 years. *Bull. S. Calif. Acad. Sci.*, **77**, 1–13.

THOMAS, W. H. (1972). Nutrients, chlorophyll and phytoplankton productivity near the southern California sewage outfalls. *Inst. Mar. Res., La Jolla, California*, Ref. No. 72–19, 1–77.

THOMAS, W. H., SIEBERT, D. L. R. and DODSON, A. N. (1974). Phytoplankton experiments and bioassays in natural coastal sea water in sewage outfall receiving wastes off Southern California. *Estuar. coastal marine Sci.*, **2**, 191–206.

THORSON, G. (1956). Marine level-bottom communities of recent seas, their temperature adaptation, and their 'balance' between predators and food animals. *Trans. N.Y. Acad. Sci. (Series 2)*, **18**, 683–700.

TULKKI, P. (1968). Effect of pollution on the benthos off Gothenburg. *Helgoländer Meeresunters.*, **17**, 209–215.

TULKKI, P. (1965). Disappearance of the benthic fauna from the Basin of Bornholm (Southern Baltic) due to oxygen deficiency. *Cah. Biol. Mar.*, **6**, 455–463.

UNESCO (1968). Zooplankton sampling. *Monographs on Oceanographic Methodology*, No. 2. UNESCO Press, Geneva.

VALENTINE, D. W. (1975). Skeletal abnormalities in marine teleosts. In W. E. Ribelin and G. Mikaki (Eds), *The Pathology of Fishes*. Univ. Wisconsin Press, Madison. pp. 695–718.

WADE, B. A. (1972). A description of a highly diverse soft-bottom community in Kingston Harbor, Jamaica. *Mar. Biol.*, **13**, 57–69.

WADE, B. A. (1976). The pollution ecology of Kingston Harbor, Jamaica. Part 4. Benthic ecology. *Res. Rep. Zool. Dept., Univ. West Indies*, **2**(5), 1–104.

WARE, R. R. (1979). *The Food Habits of the White Croaker* Genyonemus lineatus *and an Infaunal Analysis near Areas of Waste Discharge in Outer Los Angeles Harbor*, Masters Thesis, California State Univ., Long Beach, Calif.

WATLING, L., LEATHEM, W., KINNER, P., WETHE, C. and MAURER, D. (1974). Evaluation of sludge dumping off Delaware Bay. *Mar. Pollut. Bull., N.S.*, **5**, 39–42.

WEDEMEYER, G. A. (in press). Diseases of fishes: stress effects at individual and population levels. In O. Kinne (Ed.), *Diseases of Marine Animals*, Vol. IV. Biologische Anstalt Helgoland, Hamburg.

WELLINGS, S. R., ALPERS, C. E., McCAIN, B. B. and MILLER, B. S. (1976). Fin erosion disease of starry flounder (*Platichtys stellatus*) and English sole (*Parohyrys vetulus*) in the estuary of the Duivamish River, Seattle. *J. Fish. Res. Bd Can.*, **33**, 2577–2586.

WIDDOWSON, T. B. (1971). Changes in the intertidal algae flora of the Los Angeles area since the survey by E. Yale Dawson in 1956–1959. *Bull. Sth. Calif. Acad. Sci.*, **70**, 2–16.

WILHELMI, J. (1916). Übersicht über die biologische Beurteilung des Wasser. *Sber. Ges. naturf. Freunde Berl.*, **1916**, 297–306.

WILLIAMS, S. J. (1979). Geologic effects of ocean dumping on the New York Bight inner shelf. In H. D. Palmer and M. G. Gross (Eds), *Ocean Dumping and Marine Pollution*. Dowden, Hutchinson and Ross, Inc. Stroudsberg, Pa. pp. 51–72.

WOLF, K. E. (in press). Diseases of fishes: Micro-organisms. Virales. In O. Kinne (Ed.), *Diseases of Marine Animals*, Vol. IV. Biologische Anstalt Helgoland, Hamburg.

WORD, J. Q. (1976). A comparison of grab samples. *Sth. Calif. Coast. Wat. Res. Proj., Annual Rep.*, El Segundo, pp. 63–66.

WORD, J. Q. (1979). The infaunal trophic index. *Sth. Calif. Coast. Wat. Res. Proj., Annual Rep.*, El Segundo, pp. 19–39.

YOUNG, D. L. K. and BARBER, R. T. (1973). Effects of waste dumping in New York Bight on the growth of natural populations of phytoplankton. *Environ. Pollut.*, **5**, 237–252.

YOUNG, D. K. and MEARNS, A. J. (1979). Pollutant flow through food webs. *Sth. Calif. Coast. Wat. Res. Proj., Annual Rep.*, El Segundo, pp. 185–202.

YOUNG, D. R. (1979). Priority pollutants in municipal waste waters. *Sth. Calif. Coast. Wat. Res. Proj., Annual Rep.*, El Segundo, pp. 103–112.

YOUNG, D. R. and HEESEN, T. C. (1978a). Chlorinated benzenes in Palos Verdes flatfish. *Sth. Calif. Coast. Wat. Res. Proj., Annual Rep.*, El Segundo, pp. 149–152.

YOUNG, D. R. and HEESEN, T. C. (1978b). Marine bird deaths at Los Angeles Zoo. *Sth. Calif. Coast. Wat. Res. Proj., Annual Rep.*, El Segundo, pp. 193–198.

YOUNG, D. R., JAN, T. K. and HEESEN, T. C. (1978). Cycling of trace metal and chlorinated hydrocarbon wastes in the southern California Bight. In M. L. Wiley (Ed.), *Estuarine Interactions*. Academic Press, New York. pp. 481–496.

YOUNG, J. S. and PEARCE, J. B. (1975). Shell disease in crabs and lobsters from New York Bight. *Mar. Pollut. Bull., N.S.*, **6**, 101–105.

ZISKOWSKI, J. and MURCHELANO, R. P. (1975). Fin erosion in the winter flounder. *Mar. Pollut. Bull., N.S.*, **6**, 26–28.

Marine Ecology Vol. V, Part 4
Edited by Otto Kinne
© 1984 John Wiley & Sons Ltd

8. THERMAL DEFORMATIONS

P. R. O. BARNETT and B. L. S. HARDY

(1) Introduction

There can be no doubts about the vitally important role of temperature as a factor in the marine environment. In Volume I of *Marine Ecology*, KINNE (1970a, p. 322) stated that, 'with regard to life on earth temperature is—next to light—the most potent environmental component', and BRETT (1970, p. 516) wrote that 'among the abiotic entities of a fish's environment there is no greater contender for the position of ecological master factor than temperature'. It is scarcely surprising, therefore, that biologists should take an interest in, and at times express concern at, various human activities that can result in different kinds of thermal deformations being imposed on the marine and estuarine environments.

In Volume I, BRETT (1970), GARSIDE (1970), GESSNER (1970), KINNE (1970b) and OPPENHEIMER (1970) gave very detailed reviews and discussions of the role and importance of temperature in the ecology of marine organisms generally and of how it affects all aspects of the biology of individual species which in turn affects the ecology of communities. The situation with regard to thermal deformations is probably best summarized by KINNE's (1970b, p. 510) comments that:

'In view of the known effects of temperature on gamete maturation, spawning, embryonic development, length of planktonic phase and larval settling, one must expect considerable disturbances of such delicate interrelations between species of an ecosystem in case of abnormal or changing temperature conditions, since most species tested thus far reveal differences in their respective thermal responses'.

Thus the ecological implications for the marine environment of thermal deformations of the waters of estuaries and coasts may be expected to be considerable. There has been much speculation on the subject in the literature in recent years, some of it substantiated but a large proportion refuted in the light of experience with real situations.

Many industrial uses for sea water involve thermal modifications and these have resulted in considerable concern expressed by various sections of society about the uses of marine and estuarine waters. By industrialists, particularly those responsible for generating electricity, these waters are regarded as a resource exploitable for economic benefit. Considered to be an unlimited supply of cooling water, the seas are often regarded as a valuable facility for the disposal of wastes, including waste heat, because of the rapidity of mixing and exchange of water in coastal regions (e.g. CARTER, 1976). The industries that thermally deform waters are mostly those for which the seas provide a dual function: firstly, as a source of either cooling water or heat, and secondly, as a sump into which waste heat from cooling processes may be disposed. Unlike many other polluting industries disposal to the sea is a transitory stage in the passage of waste heat to the atmosphere, assuming that all the heat is finally radiated or evaporated from the water surface. What is important is the length of time the heat remains in the water body and how much of it mixes with the ambient water before it finally passes to the atmosphere.

Whilst industrialists regard these water bodies as valuable facilities for the disposal of wastes, other sections of the community regard them as economic resources for other reasons such as fishing, mineral extraction, etc., or as natural resources which should be protected as much as possible. These different attitudes have resulted in various terms being introduced in the literature over the years to describe the effects of modifying water temperatures by heating. SORGE (1969) pointed out that the industrialist refers to waste-heat disposal as thermal enrichment whereas the conservationist regards it as thermal pollution. The neutral biologist will frequently call it thermal addition. MER-RIMAN (1970, 1971) used the term calefaction meaning the act of heating. It is, therefore, appropriate that man-made temperature modifications to the seas, consisting of both heating and cooling effects, should be described as 'thermal deformations'.

Deciding whether a temperature change is a deformation or not can sometimes be extremely difficult when considered against a background of naturally fluctuating temperature changes. It is well known that, in general, natural marine temperature changes are much less extreme than those of the terrestrial environment, although in certain habitats considerable fluctuations, sometimes as a result of tidal exposure, do occur. For example, in rocky intertidal areas (e.g. KINNE, 1963, 1970a,b; LEWIS, 1964) fluctuating terrestrial temperatures are imposed on the habitat twice daily, alternating with the less extreme temperature conditions of the sea and with the biota frequently subjected to considerable temperature shock when exposure to the atmosphere or inundation by water takes place. In subtidal regions natural temperature changes outside the normal seasonal range can take place on occasions as a result of a particularly cold winter (e.g. KINNE, 1963, 1970b; CRISP and SOUTHWARD, 1964) or a very warm summer in temperate regions of the world. Such changes are natural deformations and will not be considered in this chapter. The purpose here is to consider thermal deformations brought about as a result of man's activities and in particular those resulting from discharges into estuaries and open coasts.

Thermal deformations may be broadly classified into two groups: firstly, those in which heat is added to or removed from natural waters by some industrial process, and secondly, those in which human activities cause indirect temperature changes. For example, in freshwater environments extensive deforestation can reduce shading by trees to such an extent that stream temperatures become elevated and affect salmonid populations (TARZWELL, 1970). Some human activities cause reductions in the magnitude of natural temperature fluctuations, for example, the reclamation of extensive intertidal sand and mudflat areas in estuaries. Under normal circumstances the heat budget of these intertidal areas can play a very important role in the temperature regime of an estuary. Exposure to warm summer sun during low-tide periods results in considerable warming of the overlying, incoming body of water during flood tide, whilst during winter considerable cooling of the overlying water can take place (e.g. BARNETT, 1968). Thus, the extremes in the general temperature of the estuarine water may be lessened considerably following intertidal reclamation, in addition to other changes in the hydrography of the estuary. However, this chapter is devoted mainly to the effects of industrial activities in which the water is used for cooling or warming and is then discharged to the environment. More extensive thermal deformations due to man's activities are discussed in the final section.

(a) Cold Effluents

Although a variety of industries cause thermal deformations to the marine environment, by far the greatest effects are due to sea water use for cooling purposes, particularly electricity generation. In contrast, sea water is at present used very little as a source of heat, liquid natural gas plants being virtually the only such users. These are unloading facilities for natural gas imported by sea in the liquid state and fed into gas pipelines for distribution. At this stage the gas is vaporised by passing through a heat exchanger usually warmed by sea water.

According to PARKER and UHL (1975), a re-gasification plant at a receiving terminal, producing about $14 \times 10^6 \, m^3 \, d^{-1}$ of gas, uses about $22\,700 \, m^3 \, h^{-1}$ of sea water and reduces its temperature by about $5 \cdot 5 \, C°$. In some cases, some of the fuel (about $1 \cdot 5$ to 3% of the total) is burnt to provide the heat for vaporization, but sea water heating is a more economical alternative.

The cold effluent, or discharge, is denser than the ambient sea water and so sinks to the sea bed, where it spreads out as a thin layer, probably not mixing readily with the ambient water. This could have considerable effects on the distribution, spawning and growth of bottom fauna. By imposing unseasonally low temperatures such effluents might prevent some fish species from spawning on the sea bed; if they did spawn and bottom eggs were later inundated by cold effluent, this could have negative effects on their survival and development.

Further, cold effluents add to the extreme low temperatures occurring during winter in cooler regions. There appears to be little information on the effects of these types of effluent, although there is a great deal of information on the effects of natural cooling in estuaries and oceans as a result of cold winters (Volume I: BRETT, 1970; GARSIDE, 1970; GESSNER, 1970; KINNE, 1970b; OPPENHEIMER, 1970).

(b) Heated Effluents

The use of sea water for cooling purposes and its subsequent discharge as heated effluents into the sea are very common practice. Many different industries are responsible, including those concerned with electrical power generation, desalination and oil refining. HOCUTT (1980) quotes various examples for the USA (Table 8-1). Electricity generation, primary metals manufacturing and chemical plants are the three main users. Electricity generation is by far the largest user and in the USA, for example, 70 to 80% of all the water used for industrial cooling is for this purpose. In Japan, CHIBA (1981b) mentions heated discharges from industries such as paper making, silk reeling, pharmaceutical and cement manufacture. The steel industry is a particularly heavy user of cooling water (SCHEFFLER and LAMMERS, 1977).

All these industries use freshwater for cooling to a large extent but with some of the larger users, the pressures on limited inland freshwaters and the increasing development of these industries means that many are now being sited on estuaries and open coasts to take advantage of more plentiful supplies of cooling water. There are, however, engineer-

Table 8-1

Use of cooling water by various types of industry in the United States in 1967 (After US ENVIRONMENTAL PROTECTION AGENCY, 1976; modified; reproduced by permission of the US Department of Commerce, National Technical Information Service)

Type of industry	Total intake volume (m^3 yr^{-1} $\times 10^6$)	Number of plants	Average intake volume per plant (m^3 yr^{-1} $\times 10^6$)
Steam electricity power stations	151 416	1000	151·40
Primary metals manufacturing	13 741	840	16·36
Chemicals manufacturing	13 363	1130	11·83
Petroleum refineries	4656	260	17·91
Pulp and paper mills	2461	620	3·97
Food products manufacturing	1628	2350	0·69
Stone, clay and glass manufacturing	530	590	0·90
Rubber manufacturing	363	300	1·21
Wood products manufacturing	197	190	1·04
Textile mills	91	680	0·13
Leather manufacturing	3·8	90	0·04

ing problems associated with the use of sea water for cooling (YAMAZAKI, 1965; LANGFORD, 1977); corrosion and biological fouling of cooling systems are much more serious in marine and brackish waters than in freshwater. The elevated temperature conditions within the cooling system can also produce higher growth rates of fouling organisms, producing even more rapid blockage of the pipes. Nevertheless, these disadvantages are outweighed by the advantages (YAMAZAKI, 1965) so that salt waters are used extensively for industrial cooling purposes. The necessity for preventative treatments of the water to reduce fouling can mean the discharge of biocides such as chlorine with the warm water (p. 1834, p. 1888) thus complicating the effects of temperature elevation.

Table 8-2

Extent of various resource uses in the oceans (After HILL, 1977; reproduced by permission of the author)

Resource use	Volume of sea water per unit product	Exploitation of sea per 10^6 population per year	
		Volume of sea (km^3)	Area of coastal sea (km^2)
Desalination	2 m^3 m^{-3}	0·16	1·6
Power station cooling	10^9 m^3 $GW(e)^{-1}$ yr^{-1}	1·0	10·0
Fishing	4×10^5 m^3 t^{-1}	80·0	800·0
Plankton recovery			
(a) assuming 100 mg m^{-3}	10^7 m^3 t^{-1}	2000·0	20 000·0
(b) Antarctic krill	10^2 m^3 t^{-1}	0·2	
Uranium from sea water	10^9 m^3 t^{-1} or 10^{11} m^3 $GW(e)^{-1}$ yr^{-1}	200·0	2000·0

In industrialized countries the generation of electricity uses larger volumes of sea water for cooling purposes than other industries, but in many parts of the world desalination plants are now becoming important contributors. HILL (1977) has attempted to assess the scale of these two activities in the utilization of the oceans and to compare them with other uses (Table 8-2). Although power station cooling systems use comparatively small volumes compared with some other activities (e.g. fishing, uranium extraction), they take in large quantities of coastal and estuarine plankton and at times impinge on their intake screens large numbers of young fish, almost invariably causing mortalities (p. 1806 and p. 1836). They also discharge very considerable quantities of energy as waste heat. Thus, it may be that the scale of usage by this industry, as calculated by HILL (1977) and based on volumes of pumped water, may underestimate the significance of the impact on the biota of coastal and estuarine waters.

Power Station Effluents

As already pointed out, electricity generation is by far the largest user of sea water for cooling purposes. In the USA for example, of all the sea water and freshwater used for industrial cooling purposes, 80% was used by the electricity generating industry (BELTER, 1975). In a steam electricity generating station, steam is generated in boilers using fossil fuels or the heat from various designs of nuclear reactors. The steam is used to drive turbo-generators, the condensers of which are cooled by the circulation of very large volumes of fresh or sea water (Fig. 8-1). The simplest cooling system, the open-circuit type, discharges the heated cooling water directly back to the environment.

According to the laws of thermodynamics, all the energy stored in a fuel cannot be completely converted into electrical or mechanical energy using a heat engine such as a steam turbine (MARSHALL, 1979). The heat rejected by an electricity generating station using steam turbines is related to the thermodynamic efficiency of the power generation. For thermodynamic reasons the maximum ideal efficiency of a steam power station is described by the Carnot equation (SEARS, 1969):

$$\eta = \frac{T_B - T_{CO}}{T_B}$$

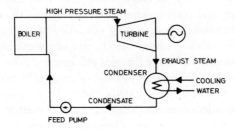

Fig. 8-1: General layout of a steam turbine electricity generating power station... (After MARSHALL, 1979; reproduced by permission of the Controller of Her Majesty's Stationery Office, London.)

where T_B = boiler temperature (K) and T_{co} = condenser temperature (K). Typically, the boiler will produce superheated steam at about 540 °C (813 K) (WIUFF, 1977), whilst the condenser cooling water may be about 20 °C (293 K). These temperatures would produce an ideal efficiency of 0·64, or 64%. In theory, the boiler steam temperature could be similar to the temperature of the burning fuel (about 2000 °C) but there is a metallurgical/economic pressure limit for pipework, etc., which imposes a level of about 540 °C (WIUFF, 1977). Advanced gas-cooled and fast reactor nuclear stations have overall efficiencies similar to those of fossilized fuel stations. However, with the water reactor type of nuclear power station, temperature/pressure restrictions of reactors mean that a lower steam temperature of about 315 °C must be maintained in the boiler. Using cooling water of 20 °C, this would give an ideal efficiency of only 0·5, or 50%. According to WIUFF, in practice the real efficiency is only about 60% of the ideal so that the examples of fossil-fuelled and water reactor nuclear power station efficiencies are reduced to 38% and 30%, respectively. Both types of station lose heat to the atmosphere; about 15% for fossil-fuelled stations mainly through chimney stacks and about 5% for nuclear. Thus, for fossil-fuelled stations 38% of the fuel generates electricity and 15% is lost to the atmosphere, both accounting for 53%. This leaves 47% of the fuel energy being discharged with the cooling water. With water reactor stations the comparable value is $100 - (30 + 5) = 65\%$. The fact that these two types of modern power stations discharge almost half and two thirds, respectively, of their fuel energy into the cooling water of their condensers emphasizes the enormous quantities of waste heat being discharged to the marine and estuarine environments. It also emphasizes what a small proportion of that fuel energy, 38% and 30%, respectively, is actually utilized to generate electricity. In practice, the average value for conventional fossil-fuelled stations is only about 33% (KARKHECK and POWELL, 1979), which raises the question of whether burning fuel in this way is a good use for these limited resources or whether they should be preserved for use as chemical feedstock or as liquid fuel for transport.

In the fossil-fuelled power station example mentioned above, it was shown that a 540 °C boiler temperature and 20 °C condenser temperature gave an ideal efficiency of 0·64, or 64% If the condenser temperature is reduced by 8 C°, to 12 °C, that ideal efficiency, according to the Carnot equation, is raised to 0·65, or 65%. An 8 C° reduction between these levels improves the ideal efficiency by 1%. In practical terms this would be about 0·6% and it would require a condenser temperature reduction of about 13 C° to achieve a practical improvement in efficiency of 1%. This emphasizes the very great importance of condenser and cooling water temperatures in the efficient operation of a power station. An improvement in operating efficiency of 1% can represent very great savings in fuel costs. For example, in southern Scotland it was reported that, over a 10-yr period, an improvement in the average thermal efficiency of power stations of about 3%, to 33·56%, represented annual fuel cost savings of about £21 million (US $42 million) (SOUTH OF SCOTLAND ELECTRICITY BOARD, 1980). There are, therefore, considerable incentives to maintain condenser temperatures as low as possible. This requires a plentiful supply of cold sea water and adequate volumes to be pumped through the condensers to maintain the lowest possible temperatures. It is important to use cooling water at the lowest possible temperature. For example, KÜTÜKÇÜOĞLU (1975) reported that in Turkey, by siting the intakes of power stations 5 to 20 m deeper than surface intakes, it was possible to reduce the cooling water intake temperature by about 5 C°.

The volumes of cooling water used by power stations are very large. They can be calculated from the ratio (η_H) of the heat discharged in the cooling water to the heat used to generate electricity (WIUFF, 1977). In WIUFF's example given above for conventional fossil-fuelled stations, $\eta_H = 47\%/38\% = 1\cdot24$ and for a water reactor nuclear station with a boiler temperature of 315 °C $\eta_H = 65\%/30\% = 2\cdot17$. In these two examples, the water reactor station would discharge 75% more heat to the cooling water than the conventional plant for the generation of a given amount of electricity. The volume of cooling water (Q) is calculated from

$$Q \ (\text{m}^3 \text{ s}^{-1}) = \frac{\eta_H E_e}{\Delta T_{co} \rho c}$$

where E_e = electrical power being generated (watts), ΔT_{co} = temperature elevation (C°) of the cooling water in the condensers, ρ = density of the cooling water (10^3 kg m^{-3}) and c = heat capacity of the cooling water ($4\cdot19 \times 10^3$ J kg^{-1} C$^{°-1}$). Thus a 1000 MW(e) station with a ΔT_{co} of 10 °C would use 30 m^3 s^{-1} (108 000 m^3 h^{-1}) for the conventional fossil-fuelled example and 53 m^3 s^{-1} (187 200 m^3 h^{-1}) for the water reactor nuclear station. These figures compare with HILL's (1977) average estimate of 10^9 m^3 GW(e)$^{-1}$ yr^{-1} [$\equiv$ 114 155 m^3 h^{-1} for a 1000 MW(e) station] used in Table 8-2. It should be emphasized that these are only examples of discharge rates to give some idea of their magnitude and that there is considerable variation in the quantities of water discharged by different types and location of power station. According to BELTER (1975), cooling-water flows for a 1000 MW(e) station in the USA usually range from about 45 to 60 m^3 s^{-1} (from 62 000 to 216 000 m^3 h^{-1}), depending on the temperature rise of the cooling water as it passes through the condensers.

The total volume of water discharged by a power station depends on the electrical capacity of the plant. Nowadays there is a tendency for power stations to be designed with increased electrical capacities because they are more economical to operate. In the USA, turbo-generator units with capacities up to 1300 MW(e) are being operated and units of about 650 MW(e) are becoming typical in industrial countries. Two or four 650 MW(e) units will provide large generating stations with capacities of 1300 and 2600 MW(e), respectively. In a fossil-fuelled station using 108 m^3 MW(e)$^{-1}$ h^{-1} of cooling sea water, discharge rates would be about 140 400 and 280 800 m^3 h^{-1}, respectively, with a temperature elevation of about 10 C°. In a water reactor nuclear station using 187 m^3 MW(e)$^{-1}$ h^{-1} of sea water the comparable volumes would be 243 100 and 486 200 m^3 h^{-1}. REID (1981) mentions a 3000 MW(e) Canadian CANDU nuclear plant under construction with a design cooling water discharge of 612 000 m^3 h^{-1}. To give some impression of the quantities, if this volume of water discharged in 1 h formed a 1 m deep layer without any mixing with ambient water it would cover an area of 61·2 ha.

Growth in the electricity supply industry has been enormous in recent years. For example, LEE and SENGUPTA (1979b) pointed out that in the USA in 1977 approximately 350 000 MW(e) of steam-generating plant was under design or construction. This represented approximately 80% of all the electricity generating capacity that existed in the country in 1973. In future, power generation could be concentrated more in specially constructed facilities known as energy parks. LEE and SENGUPTA (1979b) suggested that waste heat problems could be further compounded by the fact that in the next few years 5000 MW(e) [5 GW(e)] energy parks would come into existence. BELTER (1975) described the possibilities of energy parks in the USA eventually having com-

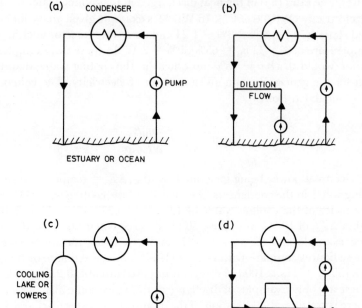

Fig. 8-2: (a) Open-circuit or once-through power station condenser cooling; (b) open-circuit cooling with dilution of heated
effluent by water at ambient temperature; (c) open-circuit cooling with cooling canal, lake or tower; (d) closed-circuit cooling
with cooling lakes or towers. (Original.)

bined capacities of 20 000 to 40 000 MW(e) [20 to 40 GW(e)] at a single site which will
require the discharge to the environment of heat concentrations of about 40 000 to
80 000 MW(th).

Basically, there are two types of cooling systems in power stations: open-circuit and
closed-circuit systems (Fig. 8-2). In open-circuit or open-cycle cooling the water is
pumped from the source into the power station where it cools the turbine condensers
before being discharged back to the aquatic environment. This system is sometimes
referred to as once-through cooling (MINER and WARRICK, 1975). In the simplest and
most commonly used type, all waste heat from the condensers is discharged to the
aquatic environment where some of it mixes with the ambient water and some is passed
to the atmosphere if the cooling water remains at the surface. In closed-circuit cooling
systems, water is recycled between the turbine condensers and some special cooling
device such as cooling towers or cooling lakes which act as evaporators and radiators
transferring the waste heat in the cooling water directly to the atmosphere. Although
termed closed circuits, these systems still take in water to replace that lost by evaporation and they also discharge some water, generally termed purge water or blow-down, in
order to maintain the correct cooling water quality. The volume used in this way is
much smaller—generally only about 2 to 4% of the water taken in by open-circuit

systems. Thus a 1000 MW(e) power station with closed-circuit cooling may discharge the equivalent of about 2000 to 3000 m³ h⁻¹, although this may not occur on a continuous basis. In addition, various chemicals may be added to the cooling water of a closed-circuit system to maintain the quality of that water; for example, anti-fouling and anti-corrosion chemicals (BECKER and THATCHER, 1973; MATSON, 1977; HART and DELANEY, 1978). Consult p. 1888 for a description of these treatments.

Closed-circuit cooling is mostly used at inland sites with restrictions on the quantities of available cooling water and on the amount of heat which may be permitted to be discharged to a river or lake. It is used to some extent in estuaries and in tropical and sub-tropical regions, particularly in the USA, where open-circuit systems can cause greater damage to aquatic organisms. However, most estuarine and open-coast power stations use open-circuit cooling which, in engineering terms, is much simpler, less expensive and produces electricity more efficiently. The implications for the environment may not be so advantageous and making compromises between the economic, environmental and engineering requirements has resulted in a great deal of research.

Power stations with open-circuit cooling systems can thermally affect the biota in a number of different ways. These may be grouped into two categories: (i) as a result of organisms being entrained in the water drawn into and passed through the cooling system, and (ii) as a result of heated effluent being discharged to the sea or estuary and affecting the organisms living there.

Considering the sequence of events as the water is utilized in the station, the first stage is when organisms may impinge on mesh screens usually placed at the intakes to prevent fish being drawn into the cooling system (Fig. 8-3). Young fish and other small organisms may be drawn on to these screens and killed as a result of mechanical damage. There are examples in the literature of large fish kills caused by this particular problem

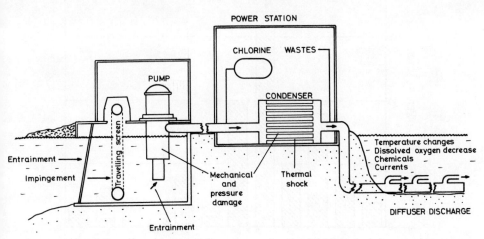

Fig. 8-3: Schematic representation of the sources of potential biological damage in an open-circuit power station cooling system. Variations on the vertical travelling band screens at the intake include inclined travelling screens, fixed screens, perforated pipe screens, horizontal travelling screens, revolving drum screens and rotating disc screens (US ENVIRONMENTAL PROTECTION AGENCY, 1976). The injection point for biocides such as chlorine may be at some earlier stage, in the pump house. (After CLARK and BROWNELL, 1973; modified; reproduced by permission of JOHN R. CLARK, Washington, DC.)

(p. 1836), especially in estuaries that are important breeding areas. The second (some-times third) stage affecting entrained organisms is when cooling sea water is chlorinated to reduce bio-fouling of the cooling system. Chlorination is discussed in some detail on pp. 1834 and 1888. In some parts of the world (e.g. California, USA), periodic back-flushing of heated cooling water is used to thermally control fouling organisms (CLARK and BROWNELL, 1973). One of the main disadvantages with marine waters as coolants is the degree of fouling which can take place. In some parts of the world mussels are a particular problem since they can both reduce the diameter of the large culvert pipes and block the small diameter condenser tubes in the turbines (e.g. HOLMES, 1970a,b). In addition, bacterial slimes on the surfaces of condenser tubes can significantly reduce the heat transfer between the condensing steam and cooling water, with subsequent loss of efficiency in electrical generation. Chlorination regimes can vary greatly but in the UK, for example, a fairly standard practice is to continuously inject chlorine gas into the cooling system, close to the intake, to produce a concentration of about 0·5 ppm (pp. 1797 and 1889). These apparently low concentrations are intended to irritate mussels settling on the surface of pipes and induce them to move right through the system.

The third (sometimes second) stage occurs when cooling water passes through the large pumps of the cooling system and organisms are subjected to great turbulence, pressure changes and the possibility of mechanical damage (p. 1836). The fourth stage consists in the water being pumped to the turbine condensers where the temperature elevation occurs (p. 1775); it produces a rapid thermal shock to the entrained organisms. The size of the temperature elevation may vary considerably between power stations but

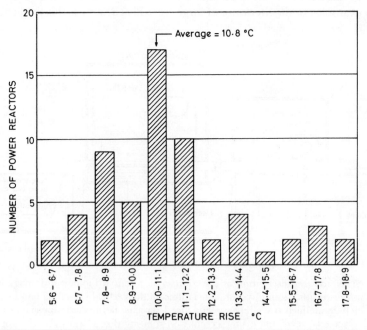

Fig. 8-4: Designed temperature elevations of cooling water in the con-densers of 61 nuclear power stations. (After COUTANT, 1970b; reproduced by permission of the US Department of Energy.)

is usually in the region of about 10 C°. COUTANT (1970b), for example, lists the design temperature elevations in the condensers of 67 nuclear power stations in the USA (Fig. 8-4); these ranged from 5·6 to 18·9 C° with an average of 10·8 C°.

The final stage occurs when the heated effluent is discharged to the environment and mixes with the ambient water. This may produce further biological changes in the surrounding environment (p. 1807).

Natural Gas Liquefaction

Another industry that discharges cooling waters with similar volumes to those of power stations is the liquefaction of natural gas. Liquefaction produces a 600:1 volume reduction allowing great economies in shipment (PARKER and UHL, 1975). A plant using feed gas at 28×10^6 m^3 d^{-1} requires between 91 and 181 m^3 h^{-1} of cooling water, which is discharged with a typical ΔT of 10 C°. There has been a big growth in this industry in recent years (PARKER and UHL, 1975) but the plants are far less numerous than power stations.

Desalination Plant Effluents

In recent years there have been considerable developments in design and construction of desalination plants as a means of providing large volumes of potable water in regions with shortages of natural water. The two major material crises facing the world today are shortages of energy and water (CHANNABASAPPA, 1977). Although about two thirds of the earth is covered with water, only 0·001% occurs in freshwater lakes and rivers. Since about 97% is in the oceans, the economic desalination of sea water to provide supplementary freshwater is an urgent problem in many parts of the world (HICKLING and BROWN, 1973).

Desalination is particularly important in countries with shortages of freshwater for drinking and industrial purposes. In particular, the increasing wealth of the oil-producing nations of the Middle East has seen the installation of large numbers of these plants. About one third of the existing sea water desalination plants in the world are situated in these regions (HARASHINA, 1977), most of them in Kuwait, and there is considerable concern about the effects of their effluents. Many plants produce heated effluents which also contain high salt concentrations and sometimes metal contaminants (DEPARTMENT OF THE ENVIRONMENT, 1977; ROMERIL, 1977). Pollution effects will increase with the increasing wealth and industrial development of oil-producing nations, particularly in the Middle East, in areas such as the Arabian Gulf and Red Sea. Even in countries where, until recently, adequate rainfall has ensured sufficient supplies of freshwater, increasing demands are beginning to cause problems. Although freshwater is often used several times between source and estuary of a river, increasing demands on supplies may mean that desalination will become more important as a method of supplementing natural supplies. HICKLING and BROWN (1973) have suggested that with increasing difficulties in finding sites for reservoirs and with increasing costs of construction, desalination may become a reasonably economical solution. The steel industry is a particularly heavy user of freshwater for cooling and process uses (SCHEFFLER and LAMMERS, 1977). In the Federal Republic of Germany, these authors show that despite the reduction in water consumption from 37 m^3 t^{-1} of steel in 1969 to only 2 m^3 t^{-1} in

1977, this greatly reduced requirement still causes water-supply problems for which desalination provides a good solution. In recent years more and more steel-work complexes in industrial countries are being set up close to the sea because of the availability of water.

There are various types of desalination plant (DEPARTMENT OF THE ENVIRONMENT, 1977): (i) distillation, (ii) reverse osmosis, (iii) electrodialysis and (iv) secondary refrigerant freezing. Distillation produces heated discharges with elevated salt concentrations.

Desalination effluent from distillation typically produces effluents with ΔT values of up to 15 C°, rather higher than those from most power stations, although the volumes are considerably less (DEPARTMENT OF THE ENVIRONMENT, 1977). The salinity may be higher than 50‰ S and the pH about 9 as a result of increased carbonate concentration. HARASHINA (1977) described one of the largest multiple-flash distillation plants in the world, in Kuwait, which produced about 22 730 m³ d⁻¹ of freshwater. The plant used sea water of 31 °C at a rate of about 7030 m³ h⁻¹, discharging effluent at temperatures ranging from 36 to 41 °C. Since the brine effluents have a high salinity they are denser than the surrounding water (Fig. 8-5) and the plume of discharged water usually sinks to the sea bed, where it spreads out as a hypersaline layer (p. 1796). In many cases, the heat for distillation plants is provided by power stations as thermal energy, and SPANHAAK and co-authors (1977) described a multiple-flash distillation plant in Texel, Holland, which uses bypass steam from a turbo-generator. However, the high costs of energy are resulting in the development of low temperature vapour compression distillation plants (HOFFMAN, 1977).

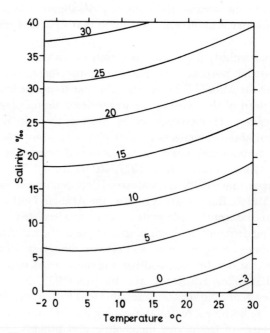

Fig. 8-5: Density of sea water (σ_t) as a function of salinity and temperature. (After KALLE, 1971; modified; reproduced by permission of Wiley, Chichester.)

Even alternative desalination systems such as reverse osmosis, electrodialysis and secondary refrigerant freezing still require considerable quantities of electrical power and so result, indirectly, in the production of heated effluents from the power stations supplying the power. Because of the high cost of energy, GLUECKSTERN and KANTOR (1977) evaluated the economics of using small-sized nuclear reactors for seawater desalination. The problem with most normal-sized reactors is that in most developing countries they produce too much electricity for the grid distribution networks, hence small reactors are better possibilities. It has been found in Israel that hybrid desalination/electricity generating nuclear reactors could provide a flexible and economic alternative for producing large amounts of desalinated water.

(c) Pertinent Literature

Most of the literature dealing with thermal deformations concerns temperature increases as a result of power station discharges. There is little information, by comparison, on the environmental effects of discharges from distillation desalination plants, although a recent conference was devoted to technical aspects of desalination and water reuse (BALABAN, 1977a,b; p. 1831). There is very little information in the literature concerning cold effluent discharges and their environmental effects (p. 1830).

Over the past 20 yr there has been an enormous amount of literature dealing with thermal discharges to the marine, estuarine and freshwater environments as a result of the great increase in the construction of electricity generating stations. This has been particularly true of the industrialized and more prosperous nations of the world owing to increased demand for electricity for both industrial and domestic purposes. Most of the literature has been produced in the USA where, in the 1960s and 1970s, there was a very vocal and effective conservation lobby giving rise to much research and many impact statements. A considerable amount of this work has been criticized, particularly by HEDGPETH (1977) and others (e.g. GREEN, 1979, for general pollution studies) as being of doubtful value in assessing the real impact of heated discharges to the environment, many of the conclusions being based on inadequate sampling and unproven assumptions. However, much good and useful work has also emerged over the years, gradually building up a picture of the significance and impact of heated discharges in various localities.

Some of the earlier work was carried out by MARKOWSKI (1959, 1960, 1962) in the UK and the early nature of his work may have been one reason for its subsequent criticism by HEDGPETH and GONOR (1969). NAYLOR (1965a) conducted careful observations on a power station discharge to an enclosed dock in South Wales that did much to establish some of the earliest reliable conclusions about the ecological effects of artificial heating. Subsequently, NAYLOR (1965b) produced the first, and what has now become a classical, review of the subject.

During the late 1960s the conservation lobby, particularly in the USA, was gathering momentum as a result of the great concern about increasing numbers of power stations. Conferences specializing in the problems of conventional and nuclear power stations and their environmental effects were convened and their proceedings reported in MIHURSKY and PEARCE (1969), KRENKEL and PARKER (1969a,b), INTERNATIONAL ATOMIC ENERGY AGENCY (1971, 1974, 1975), INSTITUTION OF ENGINEERS, AUSTRALIA (1972), GIBBONS and SHARITZ (1974), ESCH and MACFARLANE (1976), KARAM and MORGAN (1976), MAROIS (1977) and LEE and SENGUPTA (1979a). Meetings have specialized also in various aspects of power station operations such as entrainment (JENSEN, 1974, 1976,

1978; SAILA, 1975; SCHUBEL and MARCY, 1978), entrainment and cropping (VAN WINKLE, 1977), waste heat utilization (YAROSH, 1972), utilization and regulation (LEE and SENGUPTA, 1979a) and thermal aquaculture (DEVIK, 1976; TIEWS, 1981). Recent important papers on the environmental effects in the coastal zone are those by CLARK and BROWNELL (1973), in SCHUBEL and MARCY (1978) and in HOCUTT and co-authors (1980). In addition to NAYLOR's (1965a) paper, other important reviews on thermal effects of power stations have been presented by COUTANT (1968, 1969a, 1970a,b, 1972), COUTANT and GOODYEAR (1972), COUTANT and PFUDERER (1974), COUTANT and TALMAGE (1975), SYLVESTER (1975) and TALMAGE and COUTANT (1978). Bibliographies dealing particularly with fish have been published by RANEY and MENZEL (1967, 1969) and RANEY and co-authors (1973). A particularly useful general list of references was produced by MÖLLER (1978a) as an indexed bibliography.

(d) Scope of the Chapter

This chapter reviews our present knowledge on instrumentation and methods of monitoring (see below); fate and *in situ* distribution of thermal effluents (p. 1788); their biological effects as well as those of desalination discharges (p. 1796); management of waste-heat impacts (p. 1840); and management of wate-heat uses (p. 1872). The chapter concludes with a view on various aspects of possible future thermal deformations by man, particularly the possibilities of deformations on a global scale (p. 1911).

(2) Instrumentation and Methods of Monitoring

(a) Thermometers, Thermistors and Thermographs

The measurement of temperature changes in waters receiving thermal discharges typically use well known techniques developed for general hydrographic survey work. KINNE (1970a) has briefly reviewed some of the methods for temperature measurement in the marine environment. KJERFVE (1979) discussed measurement and analysis of physical factors in estuaries. Occasionally, temperature elevations due to power-station discharges can result in temperatures too high to be measured by some hydrographic instruments. The types of instrumentation employed vary from simple mercury thermometers and thermistor probes to infrared scanning techniques for direct observations; various thermographs have been developed for continuous monitoring work.

CHENEY and RICHARDS (1966) used mercury thermometers to measure surf temperature in the early stages of a programme in California (USA) when the worker simply waded into the surf to take readings. The authors have found horticultural mercury maximum–minimum thermometers and the HARTLEY and MACLAUCHLAN (1969) glass integrating thermometer to be a useful and cheap combination for ecological studies on heated effluents over long time intervals. Buried in exposed sandy beaches and examined about every 2 wk the two instruments between them provide readings of maximum, minimum and mean temperatures over the time interval. GODSHALL and co-authors (1974) measured temperature and oxygen in the marine environment using low-cost sensors, including the sucrose inversion method for mean temperatures. MACFADYEN and WEBB (1968) developed an improved temperature integrator for ecological studies which uses a mercury current-integrator (Curtis meter) in series with a thermistor and mercury battery.

Although very simple, the properly calibrated mercury thermometer has an inherent accuracy and stability that are sometimes better than those of some thermistor probe systems. Although it can only be used for spot temperatures, it has been incorporated into a successful recording thermograph designed and originally manufactured by the Braincon Corporation and subsequently by the Endeco Corporation, USA. This instrument records on film the shadow of the mercury column of a specially made glass thermometer, the stem of which is painted with a radioactive light strip to irradiate the film. The film is wound on at regular pre-set intervals, and since this is only 2–3 mm each time, a roll of film can last for up to a year. The authors have found these instruments very reliable over a period of 18 yr for sea water and sand temperature measurements for studies on long-term effects of heated effluents. The film records were recovered and the thermographs serviced only at 6-monthly intervals, with a film stepping rate of 1 record h^{-1}. Once calibrated, the thermometer requires no further attention and in the long-term the instruments have proved to be more reliable and stable than thermistor systems in the past. Film records have an accuracy of ± 0.25 C°. The main disadvantage is that they must be tediously read by eye and then translated to punched or magnetic tape. Another and older design of temperature recorder is the Cambridge Instruments mercury-in-steel thermograph, which has been in use for many years and records on circular clock-driven charts (± 0.1 C°). JARMAN and DE TURVILLE (1974) used 15 of these recorders to monitor temperatures of the flooded estuary of Southampton Water, UK, over a 7-yr period to observe the effects of heated effluents from a power station and an oil refinery.

Modern developments in electronics have resulted in the design of thermographs that have largely replaced these older instruments, although the mercury-in-glass recorder is so reliable that its continued use is probably assured for long-term studies where the recorders are checked at only infrequent intervals, despite the problems of reading the film records. Thermistors are more sophisticated and generally more convenient because they can be used for direct readings at the time or connected to various recording devices such as chart, tape or solid-state recorders. Some thermistors are now capable of measuring with a precision of 0.001 C° (KJERFVE, 1979). At the time of writing it is likely that temperature recorders using thermistors and solid-state electronics with memory banks will become commercially available and, with no moving mechanical parts, should be very reliable.

Thermistors may be towed from a boom mounted on the side of a small craft traversing a thermal plume to provide a continuous temperature record which can then be related to the vessel's position at particular time intervals, so that a three-dimensional picture may be built up. Various workers have used this technique very successfully (e.g. WYNESS, 1960). It requires accurate plotting of the boat's position and this may be done from the boat or, if sufficiently close to the shore, by 2 or 3 separately positioned operators with theodolites stationed on land and taking bearings of the boat at known intervals and maintaining radio contact with the boat (WYNESS, 1960). A variety of techniques for position-fixing of vessels during estuarine hydrographic surveys are described in INGHAM (1975) and HOOPER (1979). VAN LOON and co-authors (1977) developed a useful position-fixing system consisting of two shore-based transponders and a receiver/transmitter unit and range console on board a small boat which fixes the position to ± 3 m. More expensive alternatives to fixing the positions are radar techniques, either shore-based or on the boat, or one of the large-scale radio, navigation systems such as Decca in which shore-based transmitting stations may be specially established (e.g. HOOPER, 1979).

A sophisticated electronic system for measuring thermal plumes using thermistors has been employed by BOLUS and co-authors (1973). This system was mounted on a boat small enough to be towed by road to the power station site. Temperature was measured ($\pm 0 \cdot 3$ C°) simultaneously at 6 depths with a train of thermistors towed from the side of the boat and provided with a V-fin to maintain the lower end of the train at depth. Temperatures, positions and depths were all recorded on a data acquisition system which could also interface with *in situ* thermographs, current meters and meteorological instruments to provide a three-dimensional description of the thermal plume plus analyses of the changes with time. This particular method has been used successfully for a number of years in the Great Lakes (USA, Canada) for studies of shore-sited thermal discharges and subsequently for studies of plumes resulting from multiport diffuser discharges placed at some distance from the shore. Improvements to this system and its application are described by DITMARS and co-authors (1979). Presumably, care must be taken to ensure that turbulence from the boat's propeller does not mix the water column prior to measurements being made. The method should be useful for measuring negatively buoyant plumes but in the case of plumes that spread out over the seabed to form a comparatively thin layer 1–2 m thick, as with desalination effluents, it would not be so easy to tow a thermistor train close to the bottom unless it is mounted on some kind of sledge.

(b) Infrared Techniques

These have proved to be particularly useful, versatile and convenient for measuring distributions and temperatures of thermal effluents spreading over the surface of sea, estuary, river or lake. The technique measures infrared radiation from the sea surface and cannot be used to measure temperatures below the surface. It requires special scanners sensitive to infrared radiation. Despite a frequent misconception, infrared camera film cannot be used because it is sensitive to infrared at frequencies different from those emitted by heat radiation.

In recent years there have been considerable developments in remote sensing devices for oceanographic survey work to detect variations in back-scattering light from the sea surface within the range ultraviolet to near-infrared (about $0 \cdot 32$ to $0 \cdot 85$ μm) and including the visible spectrum (e.g. CRACKNELL, 1981). Multispectral scanners covering this range have been used to detect water-colour variations and to relate these to various factors such as turbidity from rivers, phytoplankton in upwelling areas, etc. (e.g. PEARCY and KEENE, 1974; SZEKIELDA, 1976; MONAHAN and PYBUS, 1978; DICKSON and co-authors, 1980). Multispectral scanners observe narrow spectral bands and can produce images in several bands simultaneously (KENDRICK, 1976). Alternatively, scanners can have single channels receiving one narrow band, such as thermal infrared scanners that detect thermal radiation in the far-infrared range of $8 \cdot 0$ to $14 \cdot 0$ μm.

There has been considerable progress in the remote sensing of oceans and atmospheres from satellites (see DEEPAK, 1980, and CRACKNELL, 1981, for recent developments) although this method is insufficiently detailed for surveying heated effluents. Much of this work is applied to measuring the exchanges between oceans and atmosphere for climatological and oceanographical research. For a number of years remote sensing of temperatures from ocean and land surfaces has been carried out, but their accurate interpretation presents problems because the outgoing spectral radiance of the earth's surface is affected by atmospheric composition and temperature structure and is now the subject of much research (e.g. SIMPSON, 1981). According to CHAHINE (1980) there are two spectral regions for remote sensing of sea surface temperatures, around the 3·7 and the 11 μm bands. The accuracy of the results obtained from these two regions depends on the variations in atmospheric conditions and surface emissions. Atmospheric differences can cause considerable variations with satellite observations but for localized studies of heated effluents low-flying aircraft have provided invaluable data.

Early observations were carried out using infrared thermometers or radiometers mounted beneath aircraft and plotting data on a suitable recorder (e.g. CHENEY and RICHARDS, 1966; SQUIRE, 1967). Thermometers required frequent calibration because atmospheric conditions such as fog, cloud and smoke could have significant effects on the readings. The great advantage is the ability to survey large areas in a short time; for example, CHENEY and RICHARDS (1966) refer to the completion of three flights over an area of 13 km^2 in one day. It is best to have some *in situ* measurements, as well, to check on the infrared technique. At any one time, the instrument averages the temperature for the area scanned and this is determined by the optical field of view. For example, at an altitude of 152 m, SQUIRE'S (1967) instrument sensed an area of about 64 m^2, whilst CHENEY and RICHARDS' (1966) covered about 16 m^2. The rate at which surface temperatures are averaged is determined by the speed of the aircraft (SQUIRE, 1967).

In modern infrared surveys, the radiation received by the radiometer can be shown on a cathode-ray tube and the display recorded photographically to show the surface temperatures as lighter and darker tones corresponding to the warmer and colder zones, respectively. If the signals are recorded on magnetic tape, computer techniques such as image enhancement may be used (KENDRICK, 1976). Examples of black-and-white photographs of thermal imagery studies of power station effluents are presented in MOORE and JAMES (1973), KHALANSKI (1975) and VAILLANCOURT and COUTURE (1975). The system used by BLAND and co-authors (1979), for coastal power station discharges in the USA recorded directly on to analogue magnetic tape from which data were transferred to colour film with different colours for specified temperature ranges. BURWITZ and TOBIAS (1977) also used colour film to record their studies of thermal plumes in German rivers. SAUNDERS (1967), PEARCY and MUELLER (1970), MADDING and co-authors (1975) and SIMPSON (1981) discuss the calibration of infrared data for sea surface temperatures.

Reasonable accuracy is claimed for these detectors. BOLUS and co-authors (1973), for example, give an accuracy of ±0·7 C° for a Barnes infrared thermometer compared with ±0·2 C° for their thermistors. MOORE and JAMES (1973) used a hand-held Barnes infrared thermometer with an accuracy of ±0·5 C°. However, they claimed their airborne scanning detector, under suitable operating conditions, was capable of detecting temperature differences of ±0·1 C°. BLAND and co-authors (1979) quote accuracies of ±0·5 C° for an airborne system recording on magnetic tape and colour-coded photographs. MADDING and co-authors (1975) compared the results of *in situ* surveys of Lake Michigan, USA, towing thermistor strings, with those obtained by remote aerial infrared observations and showed that during conditions of constant wind, current and discharge velocities the general plume shapes and characteristics were in general agreement. However, during very calm conditions, with no wind, the plume became very diffuse and it was difficult for the boat survey to cover sufficient area with detailed observations to produce a good comparison with the remote infrared survey. Infrared methods measure only the surface water-film temperatures and during very calm conditions these may not be representative of the surface layer as measured by thermistors. Close to the discharge there is probably sufficient mixing for them to be virtually the same but further away significant differences may arise when factors such as solar heating and evaporative cooling create temperature gradients in the surface layer of water and prevent good agreement between the two methods. Both methods have their advantages (MADDING and co-authors, 1975). While infrared surveys cover a large area and obtain a great deal of detailed data from the sea surface, a boat survey is bound to produce a picture distorted by time. Plotting temperature contours on calm days from inadequate boat data is difficult because of the surface temperature patchiness, whereas infrared data give a complete picture of the surface water film, although this might not be representative of the surface layer that would be measured by thermistors. *In situ* data would be required for three-dimensional studies and for temperature data for correlation with other physical, chemical and biological data.

(c) Other Methods

In the methods described so far, the thermal plume is characterized by the temperature difference with the surrounding ambient water and the plume's distribution can be plotted from temperature observations. However, methods not using temperature measurements have been employed on occasion, mainly before modern temperature methods such as infrared scanning were developed or as methods of predicting the fate of a thermal effluent before the industry causing it is located on a particular site. GUIZERIX and co-authors (1976), for example, discussed the use of tracers for designing and locating industrial sea outfalls where thorough knowledge was required of the hydrodynamics of the location. Aspects to be studied would be turbulent diffusion, convection, differential convection due to variables such as the topography and wind. Tracers for such studies are radioisotopes, chemical dyes such as rhodamine B and floats. Floats provide an economical means of measuring flows in shallow water but must be properly designed and have no wind resistance. BURTON (1974) used pieces of chart paper floating in the surface water layer as markers for sequences of aerial photographs. Dyes may be more effective in measuring local advective processes than drogues or current meters, which can introduce distortions (DOOLEY, 1974). Rhodamine B can be used to measure turbulent diffusion, and aerial photography is useful to follow the dye's progress although photographs will only show dye in the surface few metres with a reduction, with depth, due to the absorption of light rays. According to GUIZERIX and co-authors (1976), a correction can be applied but it is not universal since it can vary according to the optical properties of the sea water. The dye cannot be observed from the air during darkness, unlike infrared techniques, although water samples may be taken for dye concentrations. Samples of sea water marked with rhodamine B have been taken for studies of vertical diffusion, and the dye can be measured with a fluorimeter provided there is no oil contamination.

Various papers discussing the use of rhodamine dye for hydrographic studies are in KULLENBERG and TALBOT (1974), particularly those by EWART and BENDINER (1974), KARABASHEV and OZMIDOV (1974), KULLENBERG (1974a), OKUBA (1974) and TALBOT and TALBOT (1974). PRITCHARD and CARPENTER (1960) originally described the use of rhodamine B and its tracing using fluorescence techniques. Other related papers are by KULLENBERG (1968, 1974b).

(d) Modelling

Modelling has been widely used to predict the distribution of power-station effluents in estuaries and the open sea. Mathematical models have been developed to simulate the development and fate of thermal plumes in different situations and to assess the possible effects of heat discharged into coastal and estuarine waters. With physical or hydraulic models the proposed site is simulated by the construction of a model copying the physical characteristics and imitating the varying hydrographic conditions of the site. Simulated effluent may then be introduced into this model and its fate and dispersion under varying hydrographic and meteorological conditions observed using various measuring devices such as thermistors.

Disadvantages of physical models are high costs of construction and problems in calibration. Modelling the dispersion of heat, its transfer to the atmosphere and different scales of turbulence are difficult to achieve with physical models. However, turbulent stresses and transports can be modelled correctly in conditions with a high Reynolds number, as with jet flow at a discharge (RODENHUIS, 1979). This only applies close to the discharge, in the near-field (p. 1789), but further away, in the far-field, RODENHUIS advises the use of mathematical models. Sometimes the near-field situation can be so complicated that it cannot be incorporated into a mathematical model. LIGTERINGEN (1979) used a hydraulic model of this part in combination with a mathematical model for the far-field. The first had the advantage of providing adequate boundary conditions for the second model.

Mathematical models are more versatile than physical models and different situations for different sites may be more readily tested, provided the worker has access to good computing facilities. Although there is increasing evidence that well designed mathematical models are being validated by real situations, RODENHUIS (1979) emphasized the errors and problems that can be involved with numerical models. For example, RODENHUIS discussed the use of different models at various stages in the evolution of a thermal discharge. Firstly, a plume model for the near-field, and a hydrodynamic model and a transport dispersion model for the far-field. The problem is in coupling the various models and this is when large errors can be introduced. If an ecological mathematical model is needed, the errors become even greater. The numerical solution of the equations involved in the models can introduce even more errors. There are particular problems in selecting coefficients in dispersion and heat transfer for the models. One reason is that field studies before a station is constructed can only give a very limited picture of the mixing characteristics of the water body because discharge volume, momentum and buoyancy are not represented.

Despite these problems, WILLIAMS (1974) and RODENHUIS (1979) suggested that as long as the inaccuracies were borne in mind, useful predictions could be made, particularly about the possibility of recirculation and possible ecological consequences. Most mathematical models, however, do not give an adequate account of time-dependent three-dimensional flows, or the effects of bottom topography and surface waves, all of which affect the three-dimensional character of a buoyant plume. Thus, LEE and co-authors (1979) and SENGUPTA and co-authors (1979) emphasized the need for suitable three-dimensional models and described a model that predicted the three-dimensional flow and temperature field in the sea associated with a submerged discharge. Their model combines the effects of currents, wind, surface cooling and bottom topography. Application of the model to a power station on an island on the east coast of Florida showed good agreement with the real situation shown by infrared observations (LEE and co-authors, 1979).

A recent review of numerical mathematical models for thermal studies is given by DUNN and co-authors (1975). AUDUNSON and co-authors (1975) described a one-dimensional model for studies of heated discharges into a Norwegian fjord. DAUBERT (1975) discussed the use of mathematical models based on results from physical models or from *in situ* measurements for coastal sites in France, and NIHOUL and SMITZ (1976) developed mathematical models for studies on the Belgian coast. WILLIAMS (1974) described a numerical model of temperatures in the sea at the Sizewell Power Station, UK. Volume 3 of LEE and SENGUPTA (1979a) includes various papers on this type of modelling in the USA. RODENHUIS (1979) applied numerical modelling to a desalination plant/power station effluent in Saudi Arabia and to power station discharges in Denmark.

Examples of early papers using physical, hydraulic models to study the dispersal of power station heat from the Forth estuary in east Scotland are BARR (1958) and FRAZER and co-authors (1968). More recently, MIKKOLA (1975) studied proposed discharges to the sea in Finland with a hydraulic model. In California (USA), ISAACSON and co-authors (1979) modelled the discharge of waste heat to the sea following heat treatment of the cooling-water system to reduce fouling, when the roles of the discharge and intake pipes were reversed. OSTRACH and co-authors (1979) examined discharges of a submerged buoyant jet into a stratified environment using hydraulic models, and SUNDARAM and co-authors (1979) used them to investigate some of the basic turbulent transport processes relevant to the prediction of the behaviour of thermal plumes.

(3) *In Situ* Distribution

(a) Fate of Buoyant Thermal Effluents

The largest volumes of thermally modified water are discharged by power stations. Owing to the requirement for very large quantities of cooling water many new power stations are sited on estuaries and coasts with practically unlimited quantities of water. However, estuarine sites can produce serious problems, particularly when used for power stations with open-circuit cooling.

During passage through the power station the cooling-water density is reduced owing to temperature elevation (Fig. 8-5); it becomes buoyant relative to the ambient water. If discharged to the same body of water the plume will remain on the surface. The simplest type of discharge is from a large pipe, usually about 3 m in diameter, or from a discharge canal from which the heated water flows to form a buoyant surface plume. All buoyant discharges tend to form plumes (FISCHER and co-authors, 1979). In some cases, the culvert pipes may open vertically on the sea bed as one or more tunnels from which the heated water flows vertically to spread out on the sea surface. More sophisticated alternatives are jets and various types of diffusers at the end of the discharge pipe and designed to improve mixing of the effluent with the ambient water. These are discussed more fully under 'Management of Waste-Heat Impacts' p. 1854). Three phases are usually recognized in the dispersal of a heated effluent (MACQUEEN, 1979): (i) discharge as a jet when the effluent reaches the surface of the ambient water; (ii) spreading on the sea surface; (iii) final mixing with ambient water and loss of heat to the atmosphere. The

first stage, the discharge, is usually referred to as near-field (MACQUEEN, 1979; SUN-
DARAM and co-authors, 1979), and the second and third phases as far-field (SUNDARAM
and co-authors, 1979). MACQUEEN (1979), however, defines the second phase as mid-
field and the third phase as far-field.

In these three different phases there are numerous factors affecting the warm water
after it is discharged into the sea (e.g. DAUBERT, 1975; MACQUEEN, 1979; SUNDARAM
and co-authors, 1979). There are three main types: (i) plume, (ii) ambient water and
(iii) atmospheric characteristics. Most of the various factors within these three main
groups, and their interactions are summarized in Fig. 8-6. Important plume factors in
the near-field are depth, orientation, velocity and temperature of the discharge, while
ambient water factors—such as temperature, current speed and direction, stratification,
turbulence and salinity—all affect the mid- and far-fields. Atmospheric conditions such
as wind speed and direction, temperature and humidity affect the orientation and rate of
cooling of surface plumes in the mid- and far-fields.

In the **near-field phase** the dispersal of the discharge is controlled by momentum from
its velocity. Both ambient and atmospheric conditions are less important in the near-field
(AUDUNSON and co-authors 1975); the discharge characteristics of depth, design and
orientation of the outfall, and the temperature and velocity of discharge are of greatest
significance. The water flows from the discharge point entraining some of the ambient
water. Some mixing occurs in the vicinity of the discharge where the higher velocity

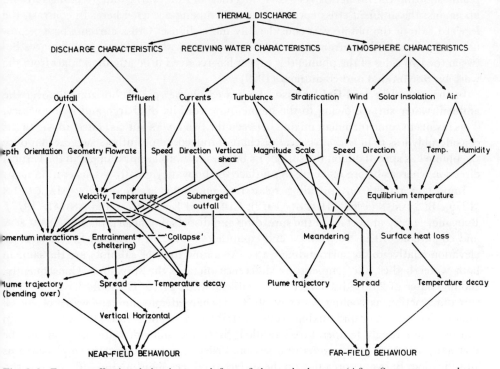

Fig. 8-6: Factors affecting behaviour and fate of thermal plumes. (After SUNDARAM and co-
authors 1979; reproduced by permission of Hemisphere Publishing Corporation, Washington,
DC.)

relative to the surrounding water produces turbulent mixing (e.g. SILL, 1979; SUN-DARAM and co-authors, 1979). If the water is discharged horizontally at the sea surface it spreads out as a buoyant plume, but if effluent is discharged vertically from the seabed it rises to the surface as a buoyant jet, then spreads out as a plume. Vertical discharges have been studied by MORTON and co-authors (1956) and SUNDARAM and co-authors (1979). ABRAHAM (1960, 1965) and JIJI and HOCH (1977) discuss both vertical and horizontal jets.

The turbulent entrainment of ambient water into a vertically rising plume is greatly influenced by plume buoyancy, unlike the entrainment that takes place in horizontally flowing plumes (SUNDARAM and co-authors 1979). The transfer of turbulence from a flowing plume into an adjacent static body of water occurs in various stages (TOWN-SEND, 1956); most of the entrainment is a result of the larger eddies. In a vertically rising buoyant discharge, large eddies create turbulence at the interface with ambient water, some of which is entrained. With a horizontal discharge, the interface between plume and ambient water is horizontal, with a stable density gradient at the interface, and the turbulence caused by larger eddies is considerably reduced.

The discharging effluent may also be carried horizontally by ambient tidal streams. The momentum of the discharged water will interact with ambient cross currents to determine the trajectory of the plume (SUNDARAM and co-authors, 1979). These authors investigated a sheltering effect as a result of discharging into an ambient cross-current. The upstream side of the plume had a very distinct boundary with a cross-section of the plume showing the isotherms very close together and the warm water tending to build up against the ambient cross current. Intensive mixing occurred here. In contrast, the leeward side of the plume was considerably more diffuse. Cross currents have a pro-nounced effect on the entrainment of ambient water with considerable differences be-tween the two sides of the plume; this aspect has received little attention apart from the work by SUNDARAM and co-authors (1979).

In the second or **mid-field phase**, the heated plume spreads horizontally over the ambient water surface owing to the momentum from its discharge and its buoyancy. This results in some turbulent mixing between the two layers but, as stated above, this is considerably less than with a vertical discharge. Indeed, some authors suggest that once the plume has spread out on the surface its buoyancy inhibits entrainment in the vertical direction whereas before reaching the surface the buoyancy greatly assists entrainment.

Important papers on spreading related to buoyancy are quoted by MACQUEEN (1979), BENJAMIN (1968), HOULT (1972), CHEN and LIST (1977) and ANWAR (1977). Buoyancy is very important in the spreading and dispersal of heated plumes (ISAACSON and co-authors, 1979; SUNDARAM and co-authors, 1979); it is related to temperature elevation relative to the surrounding water. Assuming that the salinities are the same in both water bodies, the temperature difference dictates the spreading. Consequently, cooling water of 20 °C discharged into ambient water of 10 °C should have the same spreading action as cooling water of 30 °C discharged into ambient water of 20 °C. Unless vigorous and rapid mixing occurs with the surrounding water, the water column can become strongly temperature stratified. Since the ambient water may already be thermally stratified owing to natural seasonal effects, a discharge may emphasize this stratification. Buoyant spreading on the surface produces some mixing with the underly-ing ambient water; turbulence from other causes amplifies the mixing, while stable stratification inhibits it (MACQUEEN, 1979); see also ELLISON and TURNER, 1959;

THORPE, 1973; KANTHA and co-authors, 1977; POSMENTIER, 1977; LONG, 1978; LUM-
LEY and co-authors, 1978.

In the mid-field phase, the heated surface plume is not only carried horizontally by a
tidal stream, but may also be carried horizontally by wind action which creates vertical
shear currents in the surface layers. In many cases this can also have a pronounced effect
on the trajectory of the plume (e.g. SUNDARAM and WU, 1973; DITMARS and co-authors,
1979). In the presence of winds and tidal currents the plume direction will be some
resultant of these two forces. Where tidal flows are fairly small, wind strength and
direction can have a pronounced effect on plume distribution. For example, BARNETT
(1971) has demonstrated the effect of an adjacent heated effluent on the temperatures of
an intertidal sand beach at Hunterston Power Station, UK (Fig. 8-7). Large-amplitude
temperature fluctuations of about 4 to 5 C° above ambient, and extending for very
variable periods of from 3 d to several weeks could be attributed mostly to wind effects.
Rapid declines in temperature occurred during easterly winds when the surface layer of
warm water was blown away from the shore. Indeed, there were occasions when the
beach temperature fell to about 2 C° below ambient sea water temperatures owing to
winds replacing heated water over the beach with cold intertidal water from nearby
sandflats during winter. During periods of calm or westerly onshore winds there was an
accumulation of warm water close to the beach with significant elevations of sediment
temperatures.

In the mid-field phase, the heated water normally becomes well mixed vertically with
ambient water, through the available water depth, in a period of less than one tidal cycle
(BURNETT and co-authors, 1975; TALBOT, 1976; MACQUEEN, 1978, 1979). At the same
time heat is being lost to the atmosphere.

In the third or **far-field phase**, heat continues to be lost to the atmosphere from the

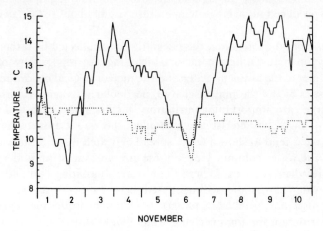

Fig. 8-7: Hourly temperature records at a sand depth of
5 cm in a beach near outfall from Hunterston 'A' generating station, Firth of Clyde, Scotland (continuous line) and
for ambient sea water at a depth of 10 m about 0·8 km
distant from the outfall (dotted line). November 1967.
(After BARNETT, 1971; reproduced by permission of the
Royal Society, London.)

surface owing to both sensible heat transfer and to transfer by latent heat of evaporation. At the same time, the warm water is gradually mixed with more ambient water. MCMILLAN (1973) investigated experimentally the rate at which heat is lost to the atmosphere, and SPURR and SCRIVEN (1975) and BURNETT and co-authors (1975) presented theoretical discussions. BURNETT and co-authors reported heat-transfer rates to the atmosphere of 30 to 40 W m^{-2} K^{-1} for an inland cooling lake in the UK. According to MACQUEEN (1978), greater air stability over the sea would reduce this to about 25 W m^{-2} K^{-1}; this again could be reduced to less than half during periods of light winds and prolonged sunshine.

The far-field phase can last for tens of tidal periods (MACQUEEN, 1979) and result in elevations of general water temperature. MACQUEEN developed theoretical models to assess the effects of various factors on the water temperature in this phase and discussed the differences between river/estuarine sites and open coastal sites. In the latter case he considered a water mass or 'tidal-plug' that moved to and fro along the coast next to the power station according to tidal currents. The power station discharged into this tidal plug, thus producing a temperature rise dependent on the various factors mentioned above. The drift or residual current past the power station site carries the heat further away, out of the 'tidal-plug'. MACQUEEN (1978) showed how even very small residual currents have a very significant effect by limiting the temperature rise in the tidal water plug more than the heat loss from the surface to the atmosphere. Thus, in enclosed areas with only small residual currents the general far-field temperature elevation can be expected to be more than that on an open coastal site with greater residual currents to transfer the heat away from the site. In Southampton Water, UK, JARMAN and DE TURVILLE (1974) studied the effects of heated effluents on this flooded estuary which has a fairly small residual current but a significant 'tidal-plug' effect. In warm water mixed with ambient estuarine water this causes movement to and fro, with only a slow drift towards the sea. Sited throughout the area, 15 thermographs showed that the cooling waters produced an average temperature rise of about 0·7 C° over much of the estuary.

As is obvious from the foregoing, the size of the areas affected by a thermal discharge depend greatly on many different factors including the topography of the adjacent land, hydrography of the ambient water and the meteorological conditions. In addition, the characteristics of the thermal discharge, particularly its temperature, volume and rate of discharge, are important contributory factors (EDINGER and POLK, 1971). Each site has its own characteristics which affect the way it absorbs heat. In general, shallow-water areas tend to show a wider spread of heated water if the warming extends to the base of the water column. Under these circumstances the underlying sediment becomes warmer than ambient. Where there is an alternating ebb and flow of heated and ambient water across the sediment, sediment temperatures integrate overlying water temperatures and produce a pattern of sediment isotherms representing mean values and distribution for the overlying water (GERCHAKOV and co-authors, 1973). The extent of the warming of underlying sediments depends on the degree of thermal stratification, particularly in relation to the rise and fall of any tide. For example, the reviewers have shown how the intertidal sediment temperatures of a sandy beach near the Hunterston Power Station, UK, can vary considerably according to the state of the tide (Fig. 8-8). The rising tide, warmed by effluent, inundates the beach and produces the highest sand temperatures of the tidal cycle. As the tide continues to rise, the

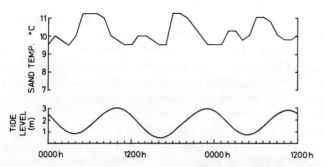

Fig. 8-8: Hourly temperature records at a sand depth of 5 cm
in a beach near outfall from Hunterston 'A' generating station, Firth of Clyde (Scotland) showing fluctuations relative
to the tidal cycle. (Original.)

temperatures start to fall as colder ambient water underlying the warm effluent moves in
over the sand. These cooler conditions prevail throughout most of the remainder of the
tidal cycle until the ebbing tide results in warmer surface water descending to the sand
surface prior to the sand being exposed to the atmosphere again.

This account is typical of the temperature cycle during calm conditions when thermal
stratification develops in the water column. However, during strong onshore winds, with
wave action mixing the thermal discharge with ambient water, the sediment temperature fluctuates less throughout the tidal cycle and tends to be smaller in magnitude.
Temperature changes caused by effluents are imposed, therefore, on intertidal areas in
addition to natural fluctuations in temperature already taking place. Natural fluctuations, however, occur throughout the tidal cycle, the most extreme being during the low
tide period. Effluent effects are imposed during the immersion periods. It is true that the
magnitude of the artificial temperature changes may be less than those frequently
imposed on intertidal organisms by atmospheric exposure (HAWES and co-authors,
1975), and it is sometimes assumed that, because of this, the artificial temperature
fluctuations are less important than the natural changes. However, this view should be
treated with caution because it is important to remember that the artificial fluctuations,
although often smaller, are imposed on intertidal organisms when these are most active
during their periods of immersion. Thermal tolerance levels for an intertidal organism
may be different during periods of immersion than when exposed to air. For example,
LEWIS (1964) cites work by SANDISON which showed such differences in intertidal
gastropods of the temperate rocky shores of Europe. In the dog whelk *Thais lapillus* the
difference was as high as 3 to 4 C°.

It is well known that intertidal areas can play a very important role in modifying the
temperature regimes of, particularly, shallow bays, estuaries and waddens (SMIDT,
1951) with extensive intertidal areas. The heat balances of these areas have special
characteristics that result from periodic flooding of extensive tidal flats resulting in
changes to surface areas through which solar radiation can penetrate and through which
heat exchange can take place with the atmosphere. VUGTS and ZIMMERMAN (1975)
discussed interactions between daily heat balances and tidal cycles and emphasized the
importance of understanding these heat balances in relation to the siting of power
stations in tidal flat areas. There are temperature changes brought about by (i) semi-

diurnal lunar tides; (ii) daily variation in solar radiation; (iii) interaction of (i) and (ii) which produces a variation in the daily mean temperature and the daily amplitude having a period of about 14·8 d (VUGTS and ZIMMERMAN, 1975); and (iv) seasonal climatic variations. The extent and magnitude of these natural variations vary considerably according to local topography, tidal range, hydrography and meteorological conditions. On a wider scale, climate and latitude are important general considerations in relation to thermal discharges.

The magnitude of seasonal variations in temperature can have an important bearing on the temperature elevations produced by thermal discharges. An example in temperate waters is shown in Fig. 8-9. High natural summer temperatures offer less scope for temperature elevations if mortalities are to be avoided. The annual variations in sea water temperatures in different latitudes are discussed in SVERDRUP and co-authors (1963) and Volume I: KINNE (1970a). In coastal and brackish waters, temperatures can range geographically, from about −2 to about 43 °C (KINNE, 1963). The annual variations in seawater temperatures are virtually negligible in the highest and lowest latitudes but reach their maximum in the middle latitudes (KINNE, 1970a). Surface water temperatures in coastal waters and estuaries will, however, vary greatly according to local conditions. For example, surface waters near the equator average about 27·5 °C whereas the average maximum in the Red Sea is about 30 °C. In shallow marginal waters the surface maximum temperatures may range from about 35 to 42 °C in tropical regions. Clearly, power station discharges must be considered carefully in relation to the conditions of a particular site, especially where effluents may impose ΔT values of about 10 C° on ambient temperatures as high as 30 to 40 °C.

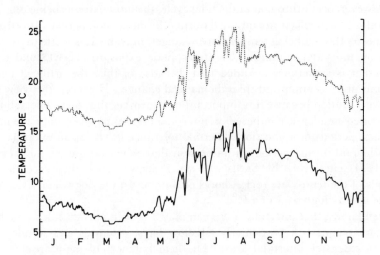

Fig. 8-9: Daily mean seawater temperatures in 1969, Firth of Clyde, Scotland. Continuous line, ambient temperature at 10 m depth near Hunterston 'A' power station; dotted line, temperatures due to passage through power station and a temperature elevation of 8·5 C°. (After BARNETT, 1972; reproduced by permission of the Royal Society, London.)

(b) Negatively Buoyant Effluents

Heated Effluents

Under certain conditions, the heated effluent from a power station may be discharged to ambient water which is less dense than the effluent. This is unusual since ambient water drawn into the power station becomes less dense owing to the heating. Discharge of heated effluent into less dense ambient water can occur if the intake water is different from that near the discharge. For example, a deep intake with dense cold water may produce water which, even with an elevation of temperature, may still be cooler and more dense than a naturally occurring warm surface layer. Such a discharge will be negatively buoyant relative to the surface layer and if discharged at the surface will sink below the surface layer and form a layer between the warm and cold ambient water masses. Similarly, in areas with salinity stratification, the heated discharge may be negatively buoyant relative to a less saline surface layer and will form a layer sandwiched between the ambient surface layer and the more saline deeper ambient waters. In these circumstances, the warm layer may be overlain by a cooler surface layer and in consequence there is no heat loss to the atmosphere. The only cooling that can take place is a result of heat transfer between the various water layers and by some turbulent mixing at the interfaces. However, the turbulent mixing is small between horizontal layers that are thermally stratified and the rate of cooling of discharged effluent is less than that of a surface plume.

An example of a heated layer becoming trapped beneath a less saline cold surface layer and a deeper dense layer was described by PANNELL and co-authors (1962) for Southampton Water, UK. In strongly stratified coastal and estuarine waters the effluent may sink irrespective of the time of year. In warmer waters of the southern USA estuarine power station discharges tend to be buoyant throughout the year but in the north they may be buoyant between the spring and autumn and sink during the winter months when heavy rainfall results in low surface salinities in upper estuaries (CLARK and BROWNELL, 1973).

In some localities the salinity/density variations of the ambient water may be so great as to make temperature/density variations insignificant by comparison (CARSTENS, 1975). In the 5 to 15 °C temperature range the density change brought about by a 1‰ S decrease would require a 5 C° ΔT (Fig. 8-5).

In a fjord-like inlet, with a sill at the entrance and deeper water behind, there is a strong possibility of discharged heated effluent being trapped behind the sill (AUDUNSON and co-authors, 1975; CARSTENS, 1975). The combination of a sill and vertical stratification due to freshwater run-off results in long residence times for the ambient high salinity water trapped behind the sill. The warm effluent may sink below the overlying and less saline water to a level that is deeper than the top of the sill and thus become trapped with the dense ambient water. Using numerical modelling, AUDUNSON and co-authors (1975) have shown that the expected temperature increases in the waters inside fjords caused by a 4000 MW(e) nuclear power station would be greatly dependent on the water exchange taking place across the sill, which may vary greatly according to the locality. CARSTENS (1975) emphasized that as a result of water exchange characteristics the waste heat present near a shallow sill may end up at the bottom of a deep basin and that it is important to mix warm effluent with the surface brackish layer during discharge.

Desalination Effluents

Although distillation desalination plants produce much smaller volumes of heated effluent than most electricity generating stations (p. 1780), the increased salinity results in a negatively buoyant plume spreading out on the sea bed beneath the overlying ambient water. Apart from the increased salinity of these discharges, there is no opportunity for heat exchange with the atmosphere to aid in cooling. The dense, warm saline water can spread across extensive areas of sea bed, filling hollows in the bottom topography and resulting in a stable thermally stratified layer, probably with little turbulent mixing with the overlying water. Aspects of the turbulent mixing of these dense plumes during discharge are discussed on p. 1857.

Cold Effluents

There is little information in the literature on cold effluents, but as with desalination-plant discharges cold effluents are negatively buoyant and sink to the sea bed, spreading out to form a thin cold layer next to the sediment. Again, there is little opportunity for turbulent mixing with the overlying water except that caused by the lateral spreading of the cold layer. Any differences in the directions and rates of flow of the two layers will increase such mixing but since the interface could be at some depth below the surface there will usually be little opportunity for mixing as a result of meteorological effects at the surface.

(4) Biological Effects

Extensive studies of the biological effects of thermal deformations on aquatic organisms have been carried out mainly with respect to elevated temperatures in power station cooling waters. There is very little information on the effects of desalination or cold effluents (pp. 1831 and 1830).

Marine and estuarine organisms may be affected by temperature increases at power stations in two distinct but related ways. Firstly, organisms present in the water column may be drawn into the intake culverts of a power station and thus become entrained in the station's cooling-water system, where they are subjected to temperature increases and other stresses. Secondly, organisms occurring near cooling-water outfalls may be affected by the warm water being discharged from the outfall. In the warm-water discharge area, the most pronounced temperature increases occur at the outfall, with decreasing temperature effects the greater the distance from the outfall. In addition to temperature, organisms present at the outfall may be affected by water flow and also by chemical residues present in the effluent water.

Entrained organisms, however, may be affected by a much more complex set of stresses than those occurring in the discharge area. For instance, they may be exposed to severe chemical as well as temperature stress, subjected to sudden pressure changes and may receive mechanical damage as a result of impingement against the intake screens, pumps and various baffles or corners in the cooling system.

(a) Entrainment Effects

In entrainment studies, therefore, because of the variety of possible non-thermal effects, it is sometimes difficult to separate thermal from non-thermal effects (CARPENTER and co-authors, 1974b; HEDGPETH, 1977), especially as more than one stress may affect the organism at the same time. In fact, several instances of observed damage to entrained organisms, formerly attributed to elevated temperatures, were later found to be caused by chlorine (LAUER and co-authors, 1974). Some authors simply refer to entrainment effects without attempting to identify the specific cause while others may refer to one or other of a number of possible causes (e.g. STEVENS and FINLAYSON, 1978).

For example, in entrainment studies carried out in Japan (ANRAKU and KOZASA, 1978), the amount of ATP in zooplankton sampled in the effluent canal was reduced compared with that in zooplankton in the intake. Although simulation experiments indicated that an intense decrease of ATP in zooplankton occurred when the temperature increased to 40 °C, the authors nevertheless regarded the field effects as being caused by a complex of physical and chemical effects experienced during entrainment. In addition, high mortalities of fish larvae observed in power station effluent canals (MARCY, 1971; GONZÁLEZ and YEVICH, 1977), were attributed by MARCY to a prolonged exposure to high temperatures, whereas GONZÁLEZ and YEVICH were of the opinion that the high mortalities they observed were probably a result of turbulence. Other studies suggested that mechanical damage (MARCY, 1973; SCHUBEL, 1974) or chemical stress, such as chlorine (SCHUBEL, 1974), were more damaging to entrained fish eggs and larvae than temperature elevation alone.

An additional complication is that increased temperatures are apparently synergistic with most other stresses (SAVAGE, 1969; CLARK and BROWNELL, 1973; SCHUBEL, 1974; YOUNG, 1974), thus enhancing the effects of other stresses and increasing the toxicity of trace substances (p. 1834).

Some entrainment stresses, however, are intermittent or variable in occurrence. Antifouling agents such as chlorine or chlorine producing compounds (e.g. sodium hypochlorite) are frequently added to the cooling system in order to prevent marine growths (LAUER and co-authors, 1974; see also pp. 1834 and 1888; their addition, however, may occur either intermittently or continuously at a low level, intermittent chlorination being common in the USA (ADAMS, 1969; CLARK and BROWNELL, 1973; BRIAND, 1975; CAIRNS, 1977) and low-level continuous chlorination—0·02 to 0·05 ppm (JAMES, 1967; BEAUCHAMP, 1969; BRUNGS, 1973) or 0·2 to 0·5 ppm (ASTON, 1981)—common in British coastal power stations.

Where intermittent chlorination occurs, therefore, the effects of chlorine and non-chlorine stress can be investigated separately by timing the entrainment study to coincide with, or to avoid the period of, chlorination. Also, the effects of raised temperatures are apparently dependent on ambient water temperatures, which vary seasonally, and as a result were most pronounced during the hottest months and least during the coldest months (p. 1803). In addition, entrainment studies can be supplemented by entrainment simulation studies in the laboratory by which one or more stress(es) can be studied independently. Although some authorities (e.g. MIHURSKY, 1967) criticize the usefulness of simulation studies, these studies at least give some indication of what might occur during entrainment.

In entrainment studies, therefore, a number of techniques have been used to try and identify the effects of specific stresses on entrained organisms. In particular, the stress effects of increased temperatures have been the subject of numerous studies with respect to bacteria, phytoplankton, zooplankton and fish.

Bacteria

Bacteria are tolerant of relatively large increases in temperature (Volume I: OPPENHEIMER, 1970; CAIRNS and LANZA, 1972; LAUER and co-authors, 1974; INTERNATIONAL ATOMIC ENERGY AGENCY, 1974). In entrainment simulation experiments for a Hudson River Power Station, USA, LAUER and co-authors found that bacteria could tolerate a temperature increase of 11·2 C° above an ambient temperature of 24·4 °C, a larger increase than that projected for the power station. In addition, quantitative bacterial samples from above and below a power station on the Connecticut river produced no numerical differences in bacterial counts during the course of a 4-yr study (INTERNATIONAL ATOMIC ENERGY AGENCY, 1974).

Temperature increases during entrainment may even be advantageous to bacteria. Bacterial populations have a reproductive potential that enables them to exploit quickly favourable changes in the environment or to make a rapid recovery if a portion of the population is killed (e.g. LAUER and co-authors, 1974). Post-entrainment increases in bacterial numbers have been reported by CAIRNS (1971) and FOX and MOYER (1973). In a simulated study of power station effects, CASPERS (1975) found that a temperature elevation (ΔT) of 10 C° stimulated decomposition by bacteria, the response being relatively quick in summer-acclimated bacteria, whereas winter-acclimated bacteria required a minimum of 2 d to be activated sufficiently to cause an increase in decomposition. An increase in bacterial decomposition may also occur in response to an increase in dead plankton killed during entrainment (CLARK and BROWNELL, 1973), this effect being most pronounced in summer (p. 1803). Increases in bacterial decomposition will probably result in an increase in BOD (CAIRNS, 1971; CLARK and BROWNELL, 1973; CASPERS, 1975). CASPERS, however, suggests that deoxygenation in summer may be compensated by the photosynthesis of phytoplankton. In addition, turbulence in marine and coastal locations will oxygenate the water and thus compensate for any increase in BOD caused by thermal increases. For a more detailed discussion on concentrations of dissolved oxygen in heated effluents, consult p. 1893.

Thermal increases do not necessarily have a uniform effect on all types of bacteria. For example, in a population of estuarine bacteria in Narragansett Bay, Rhode Island, USA, SIEBURTH (1967) found that natural seasonal changes in temperature produced changes in the relative abundance of bacterial types, with thermophilic bacteria becoming relatively more abundant during the warmest months of the year, whereas psychrophilic types were commoner during the winter months. Similar trends may occur at some power stations, provided the entrained bacteria are subjected to elevated temperatures for a sufficiently long period (a few hours instead of a few minutes) for changes to occur in the bacterial population.

Shifts in bacterial community structure as a result of thermal increases have been recorded by BUCK and RANKIN (1972) in field and laboratory studies at a power station on a low-salinity area of the Connecticut river estuary. Although no drastic bacterial changes occurred, there appeared to be some indications that thermal effluents enhanced the growth of certain types of bacteria, namely the *Achromobacter–Alcaligenes* group and coryneform bacteria, while suppressing the growth of the *Vibrio–Aeromonas* group and enteric bacteria. Differential temperature effects have also been recorded by HEFFNER and co-authors (1973) in entrainment simulation studies. Bacterial counts decreased slightly with increasing temperatures from 22 to 39 °C, with one type of bacteria, the coliforms, showing a more noticeable decrease than others.

Thermal stress at the levels encountered during entrainment, therefore, did not appear to have any serious effects on bacterial populations and, in fact, sometimes resulted in an increase in bacterial density. However, certain types of bacteria may undergo a reduction in density, one of the most sensitive to thermal increases being the enteric coliforms.

Phytoplankton

A variety of effects have also been observed in studies of entrained phytoplankton; various authorities report negligible, depressing or stimulating effects in response to thermal stress. Thermal stress during entrainment apparently had little or no effect on the phytoplankton populations studied by SAVAGE (1969), CARPENTER and co-authors (1972), HEFFNER and co-authors (1973), LAUER and co-authors (1974) and ISAAC (1979). However, these studies were carried out in northern temperate areas such as England and the north-eastern USA where ambient temperatures were relatively low.

Several studies in warmer waters (ambient water temperature in excess of 23 °C) have shown that primary productivity may decrease as a result of entrainment. Decreases ranged from 5 to 15% (BROOK and BAKER, 1972) and 2 to 37% (FOX and MOYER, 1973). MORGAN (1969) and HEINLE (1969) referred to a decrease in photosynthetic capacity following entrainment at a power station on the Patuxent estuary (USA), the concentration of chlorophyll *a* being significantly lower in the effluent compared with that in the intake. A considerable reduction in chlorophyll *a*, as a result of temperature increases, was also observed in field and laboratory studies at a power station in Japan (ANRAKU and KOZASA, 1978), whereas virtually no change in chlorophyll *a* concentration was recorded at a power station in Hawaii (BIENFANG and JOHNSON, 1980), even though there was a reduction of about 30% in carbon assimilation.

Reductions in the carbon assimilation rate, measured at coastal power stations in Florida and Hawaii (FOX and MOYER, 1973; BIENFANG and JOHNSON, 1980) occurred immediately following heat stress. In FOX and MOYER's study the severity of the effect was found to be proportional to the intake water temperature, a 30% reduction in primary productivity being recorded for an ambient water temperature of 27 °C and a ΔT of 6 C°.

Other workers have also shown that intake water temperature, and therefore the time of year, may influence the effect of thermal stress during entrainment. For instance, the results of studies by WARINNER and BREHMER (1966), MORGAN and STROSS (1969), WILLIAMS and co-authors (1971), MARSHALL and TILLY (1973) and BRIAND (1975) suggest that the amount of heating during entrainment may inhibit or stimulate photosynthesis depending on ambient water temperature.

Field and laboratory studies of entrained phytoplankton were carried out by WARINNER and BREHMER (1966) and MORGAN and STROSS (1969) at estuarine power stations in Virginia and Maryland (USA). They found that relatively small temperature increases of 3·5 to 8 C° during entrainment may cause significant reductions in carbon assimilation during the summer months, when intake water temperatures were 20 °C or warmer. In contrast, temperature increases of 2·5 to 14 C° during the winter months when intake water temperatures were 16 °C or cooler, stimulated primary production. At intermediate intake water temperatures, a ΔT of 3 C° was found to stimulate carbon assimilation while a ΔT of over 5·5 C° inhibited it. Similar results were reported by WILLIAMS and co-authors (1971) and BRIAND (1975), working in Long Island Sound and southern California, respectively. BRIAND suggested that the degree of thermal damage to phytoplankton during entrainment depended on two interacting factors intake water temperature and magnitude of temperature increase.

Some investigators, however, made no mention of seasonality or did not observe any seasonality in the effects of entrainment. For instance, HAMILTON and co-authors (1970) noted a slight stimulation of primary production during entrainment but did not mention if the effect varied with the time of year. In addition, HEFFNER and co-authors (1973) and BIENFANG and JOHNSON (1980) observed no seasonality of effects in their studies at power stations on the Hudson river estuary and in Hawaii, respectively.

Where entrainment resulted in a decrease in primary production, the effects observed may be caused by impairment of some metabolic function, such as photosynthesis, or by death of phytoplankton cells. For instance, FOX and MOYER (1973) considered that the reduction in primary productivity (average, 18·5%; range 2 to 37%), which they observed at a coastal power station in Florida, was caused by a loss of efficiency of production rather than by the death of phytoplankton cells. According to BIENFANG and JOHNSON (1980), phytoplankton populations were stressed but not destroyed by thermal shock during entrainment. However, WILLIAMS and co-authors (1971) and BRIAND (1975), working in Long Island Sound and southern California, respectively, thought that the decreases in primary production during entrainment were caused by phytoplankton cell deaths. BRIAND recorded a 41·7% decrease in phytoplankton numbers and a 33·7% decrease in volume following entrainment in September.

Diatoms, in particular, appeared to be especially sensitive to temperature shock. For instance, laboratory studies by LANZA and CAIRNS (1972), simulating sudden temperature stresses during entrainment, resulted in severe internal destruction of cells of the freshwater diatom *Navicula seminulum* var. *Hustedtii* following a ΔT of 7 C°. However, experiments by CAIRNS (1956) indicated that, although diatoms did not grow at relatively high entrainment temperatures, not all individuals were destroyed and normal population levels could later be re-established.

Phytoplankton populations in general seem to have a large regenerative capacity; recovery of primary productivity following entrainment has been observed by a number of researchers (CLARK and BROWNELL, 1973; FOX and MOYER, 1973; CARPENTER and co-authors, 1974a; LAUER and co-authors, 1974; BRIAND, 1975; ISAAC, 1979; BIENFANG and JOHNSON, 1980). While BIENFANG and JOHNSON found that phytoplankton recovered from an approximately 30% loss in carbon assimilation after returning to ambient temperatures, FOX and MOYER, and LAUER and co-authors showed that recovery from minor reductions in carbon uptake may occur by the time the phytoplankton reaches the end of the discharge canal. Recovery of primary production, however, apparently also varied seasonally. For instance, in a study of entrained phytoplankton in southern California, phytoplankton cells that survived entrainment at temperatures of 25 or 26·5 °C (during the cooler part of the year) grew 3 times faster than phytoplankton at ambient temperature (BRIAND, 1975). According to BRIAND, these phytoplankton cells would have returned to their original density within 1 d following entrainment. However, when the effluent temperature increased above 31 °C (in September) no increase in phytoplankton growth rates was recorded among the surviving phytoplankton, an indication that 31 °C was above the optimum temperature for photosynthesis. As a result, the phytoplankton biomass in the effluent area would have become greatly reduced at a time of year when it was normally at its highest level. BRIAND concluded, therefore, that enhancement of photosynthesis may occur as a result of entrainment, provided that the intake water temperature does not exceed a certain limit.

The time of year, therefore, influenced both the degree of effect of thermal stress during entrainment and the recovery of the thermally stressed phytoplankton populations. Nevertheless, even taking season and latitude of the site into account, there does appear to be wide variation in reports on the effects of raised temperatures on entrained phytoplankton. In fact, HAWES and co-authors (1975), referring to the widely varying and often conflicting data on entrainment, suggested that each power station site should be considered independently.

Natural phytoplankton populations are usually composed of a large number of species, and different species or types of phytoplankton may be affected to varying extents by thermal change (Volume I: GESSNER, 1970; Volume III: UKELES, 1976; see also SAVAGE, 1969; RAYMONT, 1972). HIRAYAMA and HIRANO (1970b) found that *Skeletonema costatum* was more sensitive to raised temperatures than *Chlamydomonas* sp., growth being inhibited at 35 and 43 °C, respectively. Likewise, photosynthetic activity was completely inhibited in *S. costatum* after exposure for 10 min at 37 °C, while photosynthesis in *Chlamydomonas* sp. began to show temperature effects of 40 °C and was completely inhibited at 42 °C.

Even in experiments with single species, some variability was recorded, possibly because of differences in the physiological condition of individual phytoplankton cultures (HIRAYAMA and HIRANO 1970b). The duration of exposure was also important. Laboratory studies on two species of phytoplankton, *Isochrysis galbana* and *Dunaliella euchlora*, showed that exposure to 36 °C for 10 min had little effect on carbon assimilation, whereas a 22-min exposure at the same temperature resulted in a pronounced growth reduction (SAVAGE, 1969). Therefore, where entrainment temperatures are sufficiently high and of sufficient duration to cause differential mortalities, a shift in community structure may occur. There are several reports on shifts in algal dominance, from diatoms to green algae to blue–green algae, as a result of increased temperatures (both natural and man-made). Accounts of these changes in composition are given with respect to thermal effluents in CAIRNS (1956), CLARK and BROWNELL (1973) and INTERNATIONAL ATOMIC ENERGY AGENCY (1974) (mainly with reference to BEER and PIPES, 1969, and BUCK, 1970). These reports were largely based on freshwater studies but are presumably also applicable to estuarine and marine conditions. HEFFNER and co-authors (1973) found a seasonal change in algal composition at a power station on the Hudson river estuary. This area was dominated by diatoms for most of the year, with green and blue–green algae becoming more abundant in late summer and early autumn.

Some entrainment times, however, may be too short to permit changes in community structure. According to LAUER and co-authors (1974), the entrainment time at a power station on the Hudson river estuary was 5·9 to 33·3 min compared with an algal generation time of 8 h. They found no evidence of any alteration in species composition caused by thermal entrainment and no visible sign of damage or change in growth as a result of passing through a power station's cooling system.

In contrast, investigations by BRIAND (1975) revealed that although thermally stressed phytoplankton populations returned to their original density following entrainment, they nevertheless underwent a change in community structure. Thermal stress during entrainment reduced species diversity, killing diatoms in greater numbers (45·7%) than dinoflagellates (32·8%) and thereby increasing the percentage dominance of the dominant species (which were also the most tolerant of thermal stress) in the phytoplankton population. Reinforcement of the dominance of certain species, e.g. *Gonyaulax polyedra*, he suggested, may have been responsible for a 'red tide' observed near a thermal discharge (p. 1808).

Zooplankton

Responses of zooplankton to thermal stress have been reviewed in Volume I: KINNE 1970b (see also Volume III: KINNE, 1977). Compared with phytoplankton, zooplankton apparently suffer higher mortalities during passage through a cooling system because of a greater susceptibility to entrainment stresses (CLARK and BROWNELL, 1973). Individual studies of zooplankton, however, also produced very variable results, some authors (e.g. MARKOWSKI, 1959; HEINLE, 1976) claiming that the increases in temperature that occurred during entrainment had virtually no effect, while others (MITCHELL and NORTH, 1971; CARPENTER and co-authors, 1974b; LANGFORD, 1975; ICANBERRY and ADAMS, 1974; DAVIES and JENSEN, 1974, 1975; DAWSON, 1979) reported that temperature elevations caused only small mortalities among entrained zooplankton. In contrast, very large zooplankton mortalities, attributable to thermal stress during entrainment, have been reported by a number of investigators (MIHURSKY and KENNEDY, 1967; HEINLE, 1969; REEVE, 1970; SUCHANEK and GROSSMAN, 1971; REEVE and COSPER, 1972; CLARK and BROWNELL, 1973; LAUER and co-authors, 1974; MARCY, 1975; ALDEN and co-authors, 1976; YOUNGBLUTH, 1976; GONZALEZ and YEVICH, 1977; BARKER and STEWART, 1978; ALDEN, 1979; DAWSON, 1979).

The degree of effect appears to depend mainly on the magnitude of the temperature rise ΔT during entrainment. Most power stations have ΔT values in the range 8 to 12 C° (US ENVIRONMENTAL PROTECTION AGENCY, 1974) but both lower and higher ΔTs have been recorded (COUTANT, 1970b, 1971a). BURTON and co-authors (1976) carried out entrainment simulation studies for power stations with low ΔT values (p. 1867) with a maximum of 5 C°. No immediate or latent mortalities were recorded with several planktonic species of estuarine macrocrustaceans and the authors suggested that the effects of entrainment in low ΔT power stations in Chesapeake Bay (USA) would be minimal.

The size of entrainment mortalities were also related to the length of time the organisms spent in the cooling system, the longer the period of entrainment the less the chance of surviving entrainment stresses (YOUNG, 1974; BARKER and STEWART, 1978). Entrainment time varied considerably according to the design of the power station; depending on the rate of flow and the length of the system, it might extend from a few minutes to several hours (GONZÁLEZ and YEVICH, 1977). For instance, many power stations discharge directly into the environment while others have effluent canals of various lengths. Transit times of 2·7 h, for an effluent canal on the Patuxent estuary

(USA), was reported by HIDU and co-authors (1974). DAVIES and JENSEN (1975) suggested that the low zooplankton mortalities they observed may have been caused by a combination of a short transit time of 2 min and a small temperature increase of 6 C° at a Delaware (USA) power station.

The percentage of zooplankton reported killed at any one power station, however, also varied according to a number of other factors, ranging from the method of assessment to the time of year. REEVE (1970) and CARPENTER and co-authors (1974b) showed that the time of assessment was important. While only 15% of the entrained copepods died immediately after discharge, 50% died by $3\frac{1}{2}$ d and 70% by 5 d after entrainment (CARPENTER and co-authors, 1974b). Entrainment, therefore, may exert delayed as well as immediate effects on zooplankton. As the assessment of dead zooplankton is normally carried out immediately after entrainment, some investigators may consequently have underestimated the numbers killed. Additional assessment complications arise as a result of motility loss in stressed zooplankton, which may sink to the bottom of the water column (CARPENTER and co-authors, 1974b), mixing of entrained and non-entrained plankton following discharge and before the samples are collected (DAVIES and JENSEN, 1975) and patchiness in zooplankton populations (REEVE, 1970; CARPENTER and co-authors, 1974a; HEINLE, 1976). In addition, overestimates of the numbers killed during entrainment may occur as a result of mortalities produced by the sampling technique (YOUNGBLUTH, 1976).

Zooplankton mortalities during entrainment also varied throughout the year, apparently in response to variations in ambient water temperatures (SUCHANEK and GROSSMAN, 1971; CLARK and BROWNELL, 1973; LAUER and co-authors, 1974; ALDEN and co-authors, 1976; GONZALEZ and YEVICH, 1977; ALDEN, 1979), being negligible in winter and highest in summer. In summer, when ambient temperatures prevail, an increase in temperature during entrainment may raise the thermal environment of entrained organisms to levels above that of the thermal death point and result in high mortalities.

Lethal temperatures may also vary to a limited extent with respect to acclimation temperatures (Volume I: KINNE, 1970b), as was shown in entrainment simulation studies (HEINLE, 1969; REEVE and COSPER, 1972). However, in tropical and subtropical regions (e.g. Florida, USA) organisms are living so close to their lethal temperature level in summer that even a relatively small increase in temperature during entrainment may have catastrophic results (HEINLE, 1969; TARZWELL, 1970; CLARK and BROWNELL, 1973; YOUNGBLUTH, 1976; GONZÁLEZ and YEVICH, 1977; THORHAUG and co-authors, 1978; ALDEN, 1979). ALDEN'S studies at the Crystal River Power Station, on the Florida coast, showed that a temperature rise of only 5 to 7 C° was sufficient to raise the entrainment temperature into the mid-thirties and above the upper limits of all copepods occurring in the cooling system. The extent to which entrained organisms, therefore, are affected by raised temperatures depends on the geographical latitude of the power station site, being relatively more severe in tropical and subtropical compared with cooler temperate regions.

The degree of effect was also dependent on the species or ecological group of species entrained, some species being more susceptible to raised temperatures than others (MIHURSKY and KENNEDY, 1967; CLARK and BROWNELL, 1973; LAUER and co-authors, 1974; GONZÁLEZ and YEVICH, 1977; ALDEN, 1979). For instance, ALDEN'S studies at a coastal power station in Florida (p. 1809) indicated that estuarine copepods were relatively more resistant than coastal species to temperature increases during entrainment, presumably because the estuarine species were more adapted to relatively greater natural ranges in temperature than coastal species (Volume I: KINNE, 1970b). Studies by SUCHANEK and GROSSMAN (1971) at a power station on Long Island Sound showed that raised temperatures caused 100% mortalities in entrained copepods during the July to early October period. Other studies, reviewed by CLARK and BROWNELL (1973), revealed high mortalities more frequently in mysids, which had the lowest lethal temperature of the groups studied, and least often in copepods, where the lethal temperature was relatively high (*ca.* 34 °C compared with 27 °C in mysids and 32 °C in gammarids). In observations reported by MIHURSKY and KENNEDY (1967) and LAUER and co-authors (1974), working in Chesapeake Bay and the Hudson river estuary (USA), respectively, the mysid *Neomysis americana* was found to be the most temperature sensitive of the species studied and suffered very high mortalities. MIHURSKY and KENNEDY attributed this to the mysid being a northern species at the southern limit of its distribution. The results of studies by LAUER and co-authors, however, indicated that some of the zooplankton survived entrainment unharmed.

Differential effects of temperature stress on zooplankton species have also been reported in entrainment simulation studies. For instance, laboratory studies by GON-ZÁLEZ and YEVICH (1977) for a power station on Narragansett Bay (Rhode Island, USA) showed that the copepod *Acartia tonsa* tolerated elevated temperatures better than *A. clausi*, the latter apparently being affected by temperatures of 25 °C and above. This sensitivity, they suggested, may explain the occasional large zooplankton kills as a result of entrainment. Important planktonic organisms, therefore, can be damaged severely by entrainment stresses but the effects vary between species.

Differential mortalities during entrainment were probably responsible for changes in population structure observed when comparing zooplankton from effluent and intake (REEVE, 1970; CARPENTER and co-authors, 1974a,b; YOUNGBLUTH, 1976). MAR-KOWSKI (1959), in this study at Barrow-in-Furness, UK, found no differences in species composition as a result of entrainment but HEDGPETH and GONOR (1969) criticized some of his techniques.

In addition to species composition, some workers have observed that entrainment stresses had a differential effect on different stages of development both in entirely planktonic species and in the planktonic larvae of benthic species. For instance, HEINLE (1969, 1976) reported that temperature increases during entrainment had a lethal effect on some copepod eggs and early larvae even though virtually no adults were affected, while HEINLE (1969) and MORGAN (1969), working in the Patuxent estuary, reported a 90% decrease in the hatching success of zooplankton eggs following entrainment. Early developmental stages of bivalves also appeared to be more sensitive to raised temperatures than later stages.

In entrainment simulation studies with eggs and larval stages of the coot clam *Mulinia lateralis* (KENNEDY and co-authors, 1974a) and the hard-shell clam *Mercenaria mercenaria* (KENNEDY and co-authors, 1974b), cleavage stages of egg development were most sensitive to raised temperatures, trochophore larvae less so, while straight-hinge larvae were the least sensitive of the stages studied. Similarly, in laboratory studies by HIDU and co-authors (1974), the fertilized eggs, early cleavage stages and ciliated gastrulae of the oyster *Crassostrea virginica* were considerably more temperature sensitive than later developmental stages. In contrast, in studies at Bradwell Power Station (UK), entrainment of oyster larvae (*Ostrea edulis*) had little effect on viability except for a suggestion that the older larvae may have been slightly more affected than younger larvae (LANGFORD, 1975).

Eggs and larvae of many other benthic species, however, apparently passed unharmed through the cooling water system of Bradwell Power Station (UK) to settle in tanks fed by the discharge and where colonies of benthic organisms developed (LANGFORD, 1975). Similarly, the successful settlement in the discharge canal of a California Power Station of a number of bivalve species, including the oyster *Ostrea lurida* and cockle *Cardium corbis*, indicated that some of the free-swimming stages of these bivalves must have passed unharmed through the cooling water system of the power station (ADAMS, 1969). This was also indicated by laboratory studies. For example, planktonic larvae of the bivalve *Tellina tenuis* at Hunterston Power Station in Scotland (BARNETT and HARDY, 1969; BARNETT, 1972), when acclimated at 14 to 15 °C, had a 50% lethal death point (LD$_{50}$) of 32·6 °C for prolonged exposure. As maximum entrainment temperatures at Hunterston Power Station were only 24 °C, even during the warmest months of the year, it was suggested that these larvae were unlikely to be adversely affected by heat during passage through the cooling system of this station.

Thermal increases during entrainment, therefore, may have little or no effect on zooplankton (including larvae of macrobenthic organisms), or may have differential effects on different growth stages, species or ecological groups, or may cause mass mortalities in the entrained zooplankton. In addition, occasional very large zooplankton kills may also occur as a result of cold shock. In entrainment simulation experiments (LAUER and co-authors, 1974), *Gammarus* sp. became hyperactive when the temperature was raised close to the thermal death point but, after being returned to ambient water temperatures, immediately lost mobility and orientation for periods of up to 0·5 h and, therefore, became more vulnerable to predation (see below).

Some of the larvae that survived entrainment, however, may have been affected by elevated temperatures. HEINLE (1969) reported a slower growth rate in surviving copepod larvae subsequent to entrainment. Other cases of sublethal effects have been reported for adult copepods; for example, GONZÁLEZ and YEVICH (1977) found, as a result of entrainment simulation experiments with *Acartia clausi*, that reproduction did not occur at temperatures in excess of 25 °C. Reference has already been made (p. 1803) to studies by CARPENTER and co-authors (1974b) in the Patuxent estuary, in which copepods lost motility after entrainment and sank to the bottom of the water column. With time, this apparently sublethal effect became a delayed mortality, with the animals dying up to 5 d after entrainment. In other cases, animals affected by sublethal stresses died as a result of external factors. For instance, some zooplankton organisms may be completely disorientated by entrainment with the result that their ability to avoid capture by predatory fish at the discharge may be significantly reduced (NEILL, 1969; COUTANT 1969b, 1971a; CLARK and BROWNELL, 1973).

For zooplankters that survive entrainment, delayed mortalities and predation, a return to pre-entrainment population densities may occur, depending on the reproductive capacity of the species concerned (LAUER and co-authors, 1974; ISAAC, 1979). However, zooplankton recovery is unlikely to be as fast or as complete as that of phytoplankters that survive entrainment, population regeneration of zooplankton being significantly less than that of phytoplankton (CLARK and BROWNELL, 1973) where compensation for entrainment losses may occur within 1 day of being returned to ambient temperature (p. 1801).

Fish

Apparently, ichthyoplankton is even more susceptible to entrainment damage than zooplankton in general (CLARK and BROWNELL, 1973; SCHUBEL and co-authors, 1978)—although, as with other groups studied, the results of investigations vary considerably. While some early studies (KERR, 1953; MARKOWSKI, 1962) suggested that entrainment had little or no harmful effects on planktonic animals, including small fish, they were carried out on organisms entrained for less than 4 min in temperatures of less than 28 °C. Most other entrainment studies report high fish mortalities.

Studies of entrained fish stocks include both field and laboratory investigations of entrainment effects on eggs, larvae, juvenile and small fish of a size that could pass through the intake screens of a power station's cooling system (p. 1836). Very large numbers of the planktonic larvae and juvenile stages of inshore fish have been killed during entrainment in estuarine and coastal power stations (CLARK and BROWNELL, 1973; HOSS and co-authors, 1974; YOUNG, 1974; ENRIGHT, 1977) and also in power stations in freshwater location (MARCY, 1971).

In studies carried out in the absence of chlorination (p. 1797), MARCY (1971) found that no larval fish survived passage through a power station in Connecticut (USA) when the discharge temperature exceeded 30 °C. Larval fish apparently suffered two temperature shocks during entrainment (HOSS and co-authors, 1974), the first when passing from an ambient temperature to the elevated entrainment temperature, the second when passing from the high temperature back to the ambient. HOSS and co-authors also noted that the effect of the second shock was more pronounced than that of the first and that the degree of effect also depended on the acclimation (i.e. ambient) temperature.

However, because of the complexity of events occurring during entrainment several investigators have used simulation experiments in the laboratory to study the effects of entrainment on early fish stages. By this means the effects of the various stresses occurring during entrainment were analysed separately.

SCHUBEL (1974) studied the hatching success of eggs of a number of fish species, after subjecting them to temperature elevations and high-temperature exposures comparable to those experienced during entrainment. The results were not significantly different from those of the controls. SCHUBEL, therefore, suggested that the destruction of fish eggs that occurred during entrainment were not caused by elevated temperatures or, at least, not by this factor alone.

Simulation experiments were also carried out with fish eggs and larvae (LAUER and co-authors, 1974; HETTLER and CLEMENTS, 1978) and post-larval and juvenile fish (HOSS and co-authors, 1972, 1975; COUTANT and KEDL, 1975) and the effects of exposure to temperature elevations on the different developmental stages investigated. Pregastrula eggs of spot *Leiostomus xanthurus* suffered higher mortalities than later larvae when subjected to thermal shock, ΔT 8 to 14 C° (HETTLER and CLEMENTS, 1978). According to LAUER and co-authors, older larvae of tomcod *Microgadus tomcod* and late-stage eggs and older larvae of striped bass *Morone saxatilis* can withstand the temperature increases and durations projected for an estuarine power station (Hudson River, New York State,

USA), whereas the younger larval stages of both species may suffer mortalities if entrained during January to March. Simulation experiments by COUTANT and KEDL (1975) showed that larval striped bass, acclimated to 22 °C, could tolerate exposure to 29 °C for 30 min (longer than the entrainment time in most power stations) but survived only 5 to 6 min at 31 and 33 °C. However, the effects of thermal increases could be reduced by cycling acclimation temperatures instead of keeping them constant (HOSS and co-authors, 1975).

Temperature tolerance depends on the thermal history of the fish concerned (e.g. LAUER and co-authors, 1974). According to HOSS and co-authors (1974), fish larvae showed little or no distress with a temperature increase of 12 C° but temperature increases of 15 to 18 C° resulted immediately in loss of equilibrium, convulsions and possibly death. In general, fish mortalities during entrainment apparently depended mainly on the temperature increase relative to ambient temperatures and exposure duration (SCHUBEL and co-authors, 1978; see also p. 1864). Further details on the effects of entrainment on fish, especially with respect to fish eggs and larvae, are given in a review by MARCY (1975).

Exposure to heat shock during entrainment can lead to a number of debilitating sublethal effects which may result in delayed mortalities or otherwise reduce the chance of survival of fish when returned to ambient temperature conditions. Entrained fish larvae and juvenile fish (as with zooplankton, p. 1805) may sometimes be disorientated sufficiently to reduce their ability to avoid predators at the discharge (pp. 1811 and 1815).

In general, physiological responses of fish to entrainment stresses, such as abrupt temperature change, apparently occur rapidly and involve complex changes in ionic balance, red and white blood cell counts and increased corticosteroid inter-renal secretion (REAVES and co-authors, 1968; WEDEMEYER, 1969, 1973; PICKFORD and co-authors, 1971; summarized by SCHUBEL and co-authors, 1978). These changes may persist long after the fish has been returned to ambient temperature conditions. In addition, changes in relative abundance of various hormones, such as corticosteroids, could affect morphogenesis to such an extent as to cause morphological deformities. BERGAN (1960) and HOPKINS and DEAN (1975) have shown that even short thermal shocks to fish eggs, such as experienced during entrainment, can result in the development of gross deformities.

Simulation experiments (HOSS and co-authors, 1972, 1974) also revealed that oxygen consumption of larval fish was quickly affected by thermal shock and that survival of fish depended on the temperature increase, duration of the raised temperature, salinity and the species of fish concerned. One of the most temperature sensitive species was the Atlantic menhaden Brevoortia tyrannus, whereas flounders Paralichthys spp. were the least affected. In fish that survived thermal shock during entrainment, respiration apparently returned to normal within a few hours at ambient temperature.

(b) Discharge Area Effects

Biological effects of waste heat from power stations are not necessarily confined to the period of entrainment. Heated water, on being discharged from the cooling system, mixes with water at ambient temperature (p. 1789) and may affect organisms in the receiving water body. This constitutes a considerably milder thermal shock than that experienced by entrained organisms (SAVAGE, 1969).

Two main types of thermal effect occur where a power station's cooling system discharges into the environment. Firstly, planktonic organisms present in the vicinity of the heated effluent may become entrained in the warm-water plume, where they experience raised but decreasing temperatures, less severe and probably of shorter duration than those experienced by organisms that had been entrained within the power station's cooling system. Some organisms, including fish, on entering the plume of heated water, may choose to remain within it (p. 1810). Secondly, benthic organisms occurring in the immediate vicinity of the plume may be heated continuously, the effect decreasing with increasing distance from the discharge

Phytoplankton

Thermal effects in the neighbourhood of power station discharges have been the subject of a number of investigations. For phytoplankton, no consistent reduction in primary production (ISAAC, 1979) or no significant differences in seasonal changes, variety or abundance of the phytoplankton standing crop (HEFFNER and co-authors, 1973) was recorded at power stations in the north-eastern USA. SAVAGE (1969) and BIENFANG and JOHNSON (1980) found little or no effect of temperature increases on phytoplankton in the discharge areas of power stations in the UK and Hawaii, respectively. BIENFANG and JOHNSON detected no thermal effects in simulated laboratory studies with temperature increases of 1·5 and 3·5 C°. In addition, the results of laboratory experiments by HIRAYAMA and HIRANO (1970b) indicated that elevated temperatures in power station discharges into the open sea may not cause great damage to the marine phytoplankton, even within the immediate area of the effluent discharge.

Neither HAMILTON and co-authors (1970) nor CARPENTER and co-authors (1974a) were able to detect any consistent reduction in primary production in thermal discharge areas. However, CARPENTER and co-authors considered that their sampling methods were not adequate to detect small changes that may occur in phytoplankton populations in coastal discharge areas. Some studies in more estuarine conditions have indicated that heated discharges may affect the phytoplankton. For example, in brackish-water phytoplankton (Pamlico river estuary, North Carolina, USA), increased temperatures in a discharge area apparently accelerated successions in late winter/early spring populations, resulting in a more diverse population with larger cells and an increase in numbers of dinoflagellates (CARPENTER, 1973). Increases in phytoplankton primary production, caused by heated discharges, have been reported from estuarine areas in southern England (PANNELL and co-authors, 1962; HOCKLEY, 1963) and from brackish water coastal areas of southern Finland (ILUS and KESKITALO, 1980).

Heated effluents sometimes give rise to algal blooms in the vicinity of the outfall. This may cause severe mortalities among filter feeding benthic animals (NAYLOR, 1965b). A dominance of certain species, e.g. *Gonyaulax polyedra*, in these blooms may initiate 'red tides' such as that observed by BRIAND (1975) near a thermal discharge. 'Red tides' in turn are frequently lethal to fish. However, the massive blooms, mainly of blue–green algae, observed near a power station effluent on the Potomac river, USA (SIMMONS and co-authors, 1974), were apparently not caused by heated power station discharges but rather by urban nutrient effluents.

Zooplankton

Increased densities of zooplankton have been recorded in the vicinity of warm-water discharges located in temperate estuaries. For example, in a series of studies in Southampton Water, England (PANNELL and co-authors, 1962; RAYMONT and CARRIE, 1964; RAYMONT, 1972), greatly increased summer densities of zooplankton were recorded in the area affected by heated discharges from the Marchwood Power Station. These increases were apparently caused mainly by an increased number of larvae of *Elminius nodestus*, a warm-water barnacle, and to a lesser extent by greater densities of larval *Mytilus edulis*, and two warm–temperate copepods, *Acartia tonsa* and *Euterpina acutifrons*. There was also a tendency for *E. acutifrons* to reach its peak density rather earlier in the summer at Marchwood, after the station began discharging heated water, than previously. The above-mentioned authors suggested that these changes may have been caused by a general warming of Southampton Water (JARMAN and DE TURVILLE, 1974), especially in the region of the discharge from Marchwood Power Station. In addition, a number of adult and copepodid stages of the mediterranean *A. grani* have been collected on several occasions in the vicinity of Marchwood.

Acartia tonsa, apparently more tolerant of increased temperatures than other cold-water copepods, replaced these in thermally enriched ecosystems (MARKOWSKI, 1962; RAYMONT and CARRIE, 1964; NAYLOR, 1965b; CLARK and BROWNELL, 1973). In the artificially warmed waters of Cavendish docks, Barrow-in-Furness, UK, MARKOWSKI recorded the disappearance of *Calanus finmarchicus* and *A. clausi* and the appearance of large numbers of *A. tonsa*. Laboratory experiments with *A. tonsa*, collected in temperate waters (GONZÁLEZ and YEVICH, 1977), showed this species to survive and breed at 27 °C but to die at 37 °C and above if exposed for prolonged periods. However, a tropical genotype of *A. tonsa* may survive and breed in a thermally enriched area of a semi-enclosed cove with average temperatures of 37 °C (YOUNGBLUTH, 1976).

In tropical estuaries, warm-water discharges apparently have a more severe effect on zooplankton populations than in temperate estuaries. For example, studies by ALDEN (1979) in the Crystal river estuary, Florida, have shown that heated effluents cause high copepod mortalities in summer, especially among non-estuarine, coastal species entering the estuary, comparable to mortalities experienced by entrained populations. Apparently, subtropical estuarine copepods are naturally adapted to higher temperature ranges with upper levels close to their lethal limits, in contrast to some of the coastal copepods which are more vulnerable to temperature increases.

In more open sites, however, the effects of heated discharges may be significantly reduced owing to effluent water mixing. At a warm-water discharge area in Long Island Sound, no significant changes were observed in zooplankton populations (CARPENTER and co-authors, 1974a). However, CARPENTER and co-authors suggested that their sampling methods were inadequate for detecting small population changes at this site. Laboratory simulation experiments (HIRAYAMA and HIRANO, 1970a) indicated that temperature increases in the immediate vicinity of warm water discharges only rarely exert adverse effects on the planktonic larvae of marine benthic organisms.

Fish

Compared with zooplankton, fish populations in the vicinity of heated discharges are probably affected to a greater extent by raised temperatures, especially as fish are frequently attracted by the warm water (see below). In addition, fish larvae and juvenile fish, exposed to temperature increases during entrainment (p. 1806), tended to be more severely damaged by thermal stress than planktonic invertebrates. In view of this, and the economic importance of fish, possible effects of warm-water effluents on fish have been the subject of a large number of investigations. Descriptions of methods of study and evaluation of thermal effects on fish are given in a series of papers edited by VAN WINKLE (1977). Studies have been carried out in the laboratory and at heated discharges, coastal sites, brackish-water estuaries, as well as in freshwater rivers and lakes. There is an extensive literature on fish in heated freshwater locations (e.g. TALMAGE and COUTANT, 1978).

With respect to warm-water discharges in US marine and brackish-water areas, McERLEAN and co-authors (1973) in the Patuxent estuary and HILLMAN and co-authors (1977) in Long Island Sound were unable to detect any effects on fish populations compared with a background of seasonal and yearly fluctuations.

Attraction, avoidance and diversity

Shoals of fish are frequently attracted to warm-water plumes or effluent canals where they may remain for extended periods (ILES, 1963; ALLEN and co-authors, 1970; GRIMÅS, 1970; BRUNGS, 1973; CLARK and BROWNELL, 1973; YOUNG and GIBSON, 1973; FALKE and SMITH, 1974; BILLI and co-authors, 1975; SYLVESTER, 1975; CAIRNS, 1977; BURTON and co-authors, 1979; ISAAC, 1979). The reviewers have also observed grey mullet (*Mugil* sp.) shoals in the warm-water plumes from Hunterston Power Station on the Firth of Clyde, Scotland (see LANGFORD, 1975, for observations on the attraction of fish to warm water areas in freshwater locations).

The attraction of fish to thermal discharges, however, is apparently a seasonal effect. Whereas fish may be attracted to the warm-water plumes throughout most of the year, and especially during winter, they generally avoid the effluent during the hottest months of the year (CAIRNS, 1956; 1977; ELSER, 1965; MIHURSKY, 1969; GRIMES, 1971; GRIMES and MOUNTAIN, 1971; BADER and co-authors, 1972; MOUNTAIN, 1972; LANDRY and STRAWN, 1973; GALLAWAY and STRAWN, 1974, 1975; SHAPOT, 1978). Most of these records refer to work carried out at estuarine and coastal power stations in Florida and Texas, although some were also derived from studies in the cooler latitudes of the UK and New England (USA). The studies were carried out mainly by trawl surveys, revealing species diversity in warm-water discharge areas to increase in winter and to decrease in summer—in contrast to control areas outside the influence of the heated effluent (GRIMES, 1971; GRIMES and MOUNTAIN, 1971; MOUNTAIN, 1972; GALLAWAY and STRAWN, 1974, 1975).

Fish stocks in warm-water areas decrease in density during the warmest months of the year and undergo changes in population structure, some species being more sensitive to high temperatures than others. In a study of fish stocks in a warm water discharge, GALLAWAY and STRAWN (1974) recorded a reduction in diversity of fish when the effluent temperature exceeded 35 °C but found that different species began avoiding the effluent over a range of temperatures. For instance, Gulf menhaden *Brevoortia patronus* tended to avoid heated effluents at temperatures as low as 30 °C whereas for some other species, such as the sea catfish *Arius felis*, avoidance began at 37 °C, and striped mullet *Mugil cephalus* were even observed in the effluent at 40 °C. These temperature tolerance levels, however, varied according to salinity (see also Volume I: KINNE, 1970, 1971; HOLLIDAY, 1971; ALDERDICE, 1972).

Since the warm-water avoidance pattern only affects a very localized area around a thermal discharge, it has been suggested (GALLAWAY and STRAWN, 1974; SHAPOT, 1978) that this behaviour probably has a negligible effect on the fish populations. However, other investigators (GRIMÅS, 1970; CLARK and BROWNELL, 1973; YOUNG, 1974; LANGFORD, 1975) thought that power station effluents may sometimes disrupt the normal migration patterns of fish by acting as barriers to migrating shoals. Any alteration to the normal migrations of commercially important fish such as salmon, sea trout and menhaden could have serious economic consequences (CLARK and BROWNELL, 1973).

The availability of food in a thermal discharge area is apparently an important factor in attracting some species of fish to the effluent, although raised temperatures are probably the major attraction, especially in winter. The presence of prey organisms in the effluent has been suggested (GALLAWAY and STRAWN, 1974) as the main cause of concentrations of sea catfish *Arius felis* in a power station discharge in Galveston Bay, Texas; entrained prey organisms—injured by intake screens, or killed or disorientated by heat during entrainment—may be food for this fish. NEILL (1969; quoted in INTERNATIONAL ATOMIC ENERGY AGENCY, 1974) also reported intense feeding by freshwater fish on entrained zooplankton in the outfall of a Wisconsin Power Station (USA), the zooplankton having been killed or debilitated sufficiently to be unable to avoid predators. Food supplies were apparently also partly responsible for attracting striped mullet *Mugil cephalus* into the power station effluent in Galveston Bay. This power station produced masses of surface foam at the point of discharge and observations by GALLAWAY and STRAWN (1974) suggested that *M. cephalus* grazed on this foam. They did not enlarge on this statement but similar foaming occurs at power station effluents on the Firth of Clyde, Scotland, studied by the reviewers. Here, the foam appears to collect an algal/organic detritus coating on the air–water interfaces of the bubbles and, assuming a similar phenomenon occurs on the coast of Florida, this may provide an attraction and a food for *M. cephalus*.

A complex of temperature effects and other entrainment factors, therefore, appears to provide food for some of the species of fish attracted to the warm water discharge area. However, temperature effects on food organisms apparently do not necessarily provide fish with an increased food supply. Sometimes fish may be deprived of their normal food organisms by temperature-induced mortalities. For example, work at the Chesapeake Biological Laboratory (MIHURSKY and KENNEDY, 1967) showed that the opossum shrimp *Neomysis americana*, a major food item of juvenile striped bass *Roccus* (= *Morone*) *saxatilis*, was extremely sensitive to raised temperatures. They suggested that high mortalities of this shrimp, caused by heated discharges, may have a detrimental effect on the local striped bass population as the nursery area of these fish lies just upstream from a power station.

Lethal effects: heat shock

Avoidance of extremely high temperatures makes it unlikely that fish will be killed by warm water (ALABASTER, 1963; COUTANT, 1970b; GRIMES and MOUNTAIN, 1971; CLARK and BROWNELL, 1973; LANGFORD, 1975), although caged fish have been killed by warm-water effluents in summer (ALABASTER, 1963; MARCY and co-authors, 1972). Nevertheless, fish kills caused by heat shock do occur (e.g. HOAK, 1961a; MIHURSKY and CRONIN, 1967; MIHURSKY, 1969; FAIRBANKS and co-authors, 1971; YOUNG and

GIBSON, 1973; GALLAWAY and STRAWN, 1974; YOUNG, 1974; ISAAC, 1979; see also the detailed discussion in CLARK and BROWNELL, 1973).

Fish using estuaries as nursery grounds are especially vulnerable to heated discharges from estuarine power stations (CLARK and BROWNELL, 1973). This is especially the case with menhaden *Brevoortia tyrannus* which migrate through estuaries at different stages of their life cycle (YOUNG, 1974) and regularly suffer heavy mortalities as a result of power station entrainment or heated discharges. Salmon are more easily killed by heat shock when acclimating in estuaries on passage from seawater to freshwater than when living in either sea water or freshwater (ALABASTER, 1967; quoted by LANGFORD, 1975).

Fish kills by heat shock occur as a result of fish becoming trapped in the effluent area, possibly by an ebbing tide or by fish swimming accidentally into a warm-water plume at a temperature above their lethal limit. Temperatures above the lethal limits (about 39 °C) of Gulf menhaden *Brevoortia patronus* and sea catfish *Arius felis* apparently occurred in summer in the heated effluent from a power station in Galveston Bay, Texas (USA) and resulted in a few individuals of these two species being killed (GALLAWAY and STRAWN, 1974). These authors, however, point out that these exceptions to the general pattern of high temperature avoidance were rare.

Much more spectacular mass mortalities have been recorded from other power stations. YOUNG and GIBSON (1973) observed that a shoal of juvenile Atlantic menhaden *Brevoortia tyrannus*, migrating down Long Island Sound (USA), suffered a lethal heat shock when they suddenly swam into a warm plume with a temperature of 37 to 38 °C (ΔT about 15 C°) from Northport Power Station; at least 200 000 fish were killed. Massive kills of adult menhaden have been recorded by ISAAC (1979) who noted occasional congregations of this species in a warm-water plume at Plymouth, Massachusetts (USA). When the tide changed the fish sometimes became trapped in much warmer water. The presence of some dissolved chlorine (p. 1835) in the effluent water may have increased the stress by acting synergistically with the thermal stress.

A large proportion of the recorded heat-shock fish kills involved juveniles and adults of Atlantic menhaden *Brevoortia tyrannus*. This species is one of the more sensitive with respect to raised temperatures. Experiments with menhaden suggest that thermal shock above 33 °C may cause rapid death (LEWIS and HETTLER, 1968).

The lethal temperature may vary according to a number of factors. GALLAWAY and STRAWN (1974), for instance, report that suboptimal salinities may lower the temperature at which Gulf menhaden are seriously affected by high temperatures. Acclimation temperatures and the period of time the fish remain at a particular high temperature are also important (MIHURSKY and KENNEDY, 1967; PEARCE, 1969). PEARCE found that winter flounder *Pseudopleuronectes americanus*, acclimated at ambient sea temperature, died when exposed to water at 25 °C for less than 48 h, and yet fish of the same species swam for short periods in water at 31·1 °C in order to feed. MIHURSKY and KENNEDY, referring to fish studies at the Chesapeake Biological Laboratory (USA), reported that one species, when acclimated to 35 °C, could only withstand a temperature rise of 2·2 to 3·9 C° for 24 h before 50% mortality occurred.

Fish kills at power station discharges may sometimes be the result of indirect temperature, non-thermal or synergistic effects (p. 1831). One major fish kill observed by ISAAC (1979) was found, as a result of histological examination of dead fish, to have been caused by thermally induced gas embolism. Dissolved gases in the tissues had apparently come out of solution when the fish swam from relatively cool sea water into a warm-water plume. Gas embolism may also result from the fish swimming into super-saturated warm water at the effluent (CLARK and BROWNELL, 1973; see also p. 1894).

Sudden temperature increases experienced in the warm-water discharge area, there-fore, may have serious effects on fish, depending on acclimation temperature, tempera-ture rise, plume temperature, lethal temperature of the species concerned, duration of the temperature increase, physiological state of the fish, salinity and probably also the concentration of various chemicals and gases. Long-term physiological effects are dis-cussed on p. 1807.

Lethal effects: cold shock

This type of lethal effect can occur if a power station ceases operation (p. 1836). As a result of this, fish living in the heated effluent and acclimated to a relatively high temperature will suffer a physiological shock that may be sufficient to kill large numbers. Cold-shock fish deaths may occur (MCERLEAN and co-authors, 1973; YOUNG, 1974; CAIRNS, 1977; COUTANT, 1977). If migratory fish are induced by a heated effluent to remain in an area they would normally vacate during the colder months of the year and if, during this period, the power station ceases operation (p. 1871).

Cold-shock fish kills have been reported by a number of investigators (CAIRNS, 1972; 1977; CLARK and BROWNELL, 1973; ASH and co-authors, 1974; YOUNG, 1974; COUT-ANT, 1977; BURTON and co-authors, 1979). COUTANT (1977) reviewed the literature on cold-shock fish deaths, but mainly with respect to freshwater power stations. YOUNG (1974) reported several fish kills apparently caused by cold shock. On one occasion at least 500 000 menhaden, wintering in a thermal plume in Barnegat Bay, New Jersey (USA), were killed when the temperature at the power station discharge suddenly decreased from 15 to 3 °C as a result of an unscheduled shutdown of the station. On another two occasions (both in January), several thousand dead fish (mainly adult and juvenile Atlantic menhaden) were found frozen in ice along the edge of a power station discharge canal, again coinciding with the sudden cessation of power station activity. Atlantic menhaden appear to be especially vulnerable to cold shock.

The effects of cooling in a power station discharge were simulated in experiments carried out by HETTLER and CLEMENTS (1978) and BURTON and co-authors (1979). These experiments showed that the rate of temperature decrease was important: the more pronounced the rate of decrease, the greater was the mortality caused by cold shock. BURTON and co-authors suggested, therefore, that winter menhaden kills at power stations could be minimized by regulating the rate at which power stations reduce the temperature of their warm-water discharge when closing down.

Cold effluents

Reductions in temperature also occur in areas receiving industrial effluents at below ambient temperatures. Cold-water effluents are much rarer than heated effluents but occur, for example, in discharges from liquid natural gas conversion plants (p. 1771). The biological effects of cold effluents, apparently, have not been studied in any detail. However, a number of simulation experiments were carried out for the cold effluent from a proposed liquid gas conversion plant in California (SOULE and OGURI, 1974). The results of one of these experiments, on the northern anchovy *Engraulis mordax*, suggested that temperatures below 9·5 °C may seriously affect larval development in this species (BREWER, 1974). Also, it appeared that a significant increase in anchovy mortalities might occur in winter, if the temperature of an appreciable area of the receiving body of water was reduced to 7 °C or below. In addition, even if chilled individuals are not killed by low temperatures, they may become torpid (SOULE, 1974) and, therefore, less able to avoid environmental hazards such as predators and various stress factors (Volume I: KINNE, 1970b, p. 413).

Sublethal effects

Thermal effluents may affect fish populations also in ways that are not immediately obvious. For instance, increased metabolic rates are to be expected in organisms residing in a thermal discharge area compared with those not affected by increased temperatures (e.g. DE SYLVA, 1969a; INTERNATIONAL ATOMIC ENERGY AGENCY, 1974). Changes in growth rates and times of reproduction, in particular, may occur in organisms living in warm-water areas (NAYLOR, 1965a,b; TARZWELL, 1970).

With respect to fish, GRIMES (1971) found no significant difference in either reproduction or growth between effluent and control sites at a coastal power station in Florida, whereas TREMBLEY (1960) and HOAK (1961a) reported early breeding of fish in freshwater locations affected by thermal discharges. In addition, the early occurrence and liberation of young of the sea catfish *Arius felis* in a thermal effluent in Galveston Bay, Texas (USA), may have been the result of elevated temperatures (GALLAWAY and STRAWN, 1974).

Early or prolonged spawning, resulting from artificial but very localized increases in temperature, may not necessarily be beneficial to fish stocks (DE SYLVA, 1969a). Where suitable and sufficient food for larvae is not available outside the normal breeding season of fish, larval mortalities through starvation will increase dramatically (e.g. TARZWELL, 1970; LANGFORD, 1975). This appeared to be the case with unseasonal larvae of some invertebrate species (BARNETT and HARDY, 1969; HARDY, 1977; see also p. 1823). Even if elevated temperatures to stimulate an increase in larval food and result in an enlarged fish stock, they may also cause predator populations to increase. Augmented predator pressure may nullify the increased production of fish (DE SYLVA, 1969b).

The growth of fish in thermal effluents has been studied and, although GRIMES (1971) found no difference between effluent and control sites, other workers (MARKOWSKI, 1962, 1966; ILES, 1963; LANGFORD, 1975) recorded increased growth rates and extended growing seasons in fish from cooling water receiving basins of power stations. In one of these studies, in artificially heated docks (Barrow-in-Furness, UK) MARKOWSKI (1966) found otoliths of flounders *Platichthys flesus* from warm-water areas to be very difficult to read. He attributed this to the raised temperature and abnormally intensive feeding of flounders throughout the winter, which presumably allowed the fish to grow continuously at this time of year; as a result, no clear winter ring was laid down in the otolith. Higher initial growth rates at elevated temperatures have also been reported by KINNE (1963) for *Cyprinodon macularius*.

High growth rates of fish in heated effluents have been studied experimentally with a view to exploiting them commercially. NASH (1968, 1969b, 1970b) cultured plaice *Pleuronectes platessa* and sole *Solea solea* in warm water from Hunterston Nuclear Power Station, Scotland; he found that the fish grew to marketable size 12 months earlier than fish in natural populations. This cultured population, however, was fed artificially on chopped mussel and scallop. The culture of fish in thermal effluents is discussed more fully on pp. 1873–1902.

In natural populations, increased metabolic rates at elevated temperatures will lead o faster growth rates only if there is sufficient food available to meet the increased demand. If the food supply is insufficient to meet the demand then starvation may result (KENNEDY and MIHURSKY, 1972; WOOD, 1973; COUTANT and SUFFERN, 1979). This is especially liable to happen in winter (KENNEDY and MIHURSKY, 1972; CLARK and BROWNELL, 1973) and has been suggested as the reason for the 'skinny fish' reported by MERRIMAN (1972; quoted in COUTANT, 1970b and INTERNATIONAL ATOMIC ENERGY AGENCY, 1974) in the discharge canal of Yankee Power Station, Connecticut, USA, in winter. These fish were found to be losing weight at a much faster rate than fish at control sites.

Combinations of raised temperature and low food supply may weaken fish residing in heated effluents and increase their susceptibility to predation, pathogenic bacteria and parasitic infections (DE SYLVA, 1969a; MIHURSKY and co-authors, 1970; TARZWELL, 1970; INTERNATIONAL ATOMIC ENERGY AGENCY, 1974). In addition, the increased temperatures may stimulate the growth of parasitic organisms. In INTERNATIONAL ATOMIC ENERGY AGENCY (1974), reference is made to the work of MUSSELIUS (1963) on the parasitic flat-worm *Bothriocephalus gowkongenis;* this worm reached sexual maturity in the gut of freshwater carp twice as fast at 22 to 25 °C than at 16 to 19 °C. Although this study was carried out with freshwater fish, similar parasites presumably develop in fish residing in warm-water effluents in estuarine and marine localities. However, MÖLLER (1978b) reports that the number of fish infected with helminth and bacterial parasites in a heated discharge in Kiel Fjord (Federal Republic of Germany) was no higher than that at control sites. Fish occurring in heated docks at Barrow-in-Furness (UK) were shown by MARKOWSKI (1966) to have lower infections of parasitic worms than elsewhere; he suggested that this may have been caused by a combination of raised temperature, reduced salinity and the absence of intermediate host copepods in the plankton of these docks. Bacterial growth may be supressed by the presence of chlorine residues in continuous, low concentrations in heated effluents. For example, NASH (1968, 1969b) found that plaice and sole being cultured in chlorinated cooling water at Hunterston Power Station, Scotland, remained free from disease (p. 1897).

Heated discharges, nevertheless, may have a direct or indirect effect on fish which, even if not immediately lethal, may weaken the fish (DE SYLVA, 1969a; HOSS and co-authors, 1974) or induce changes in behaviour (COUTANT, 1977) and increase vulnerability to predation. For instance, in experimental studies COUTANT (1973) found that young chinook salmon *Oncorhynchus tshawytscha* and rainbow trout *Salmo gairdneri*, which had received a sublethal heat shock, were taken more easily by predators from a mixed shoal of shocked and control fish. Similar increased predation effects have also been observed as a result of cold shock, caused by the sudden cessation of a warm-water discharge (COUTANT and co-authors, 1974; COUTANT, 1977). Increased predation on disorientated entrained organisms is discussed on p. 1807 (zooplankton) and p. 1805 (fish).

In conclusion, heated effluents may have considerable effects on fish in the vicinity of the discharge, especially if the receiving area is of limited extent, as in an estuary. Such localized fish populations appear to be the most seriously affected, whilst phytoplankton populations are least affected. The deleterious effects of temperature on one species may become a beneficial effect for another species if the latter's food supply is thereby increased.

Phytobenthos

With benthic flora the effects of discharges are more pronounced. For instance, a community of benthic diatoms in the discharge canal of a power station in New Jersey, USA, were found by HEIN and KOPPEN (1979) to contain significantly fewer species compared with the intake canal, indicating that the elevated temperature had an effect on the population structure of the diatoms. Laboratory studies showed that the various species of diatoms occurring at the power station had different levels of temperature tolerance. Similar observations were made by SAKS and co-authors (1974), reporting on laboratory simulation studies in relation to a thermal effluent on Long Island, USA. Their results indicated that temperature increases from 10 to 25 °C enhanced competition among epiphytic salt-marsh algae and it was concluded that blooms of particular species of algae would become conspicuous at the temperature elevations ($\Delta T \approx 17$ C°) recorded at the power station. In addition, many diatom species apparently also respond to temperature increases by forming resting stages or by ceasing to divide (CAIRNS 1956).

Shifts in algal community structure towards thermophilic green algae have been recorded by ADAMS (1968) and HECHTEL and co-authors (1970). However, the most frequently quoted example is that of a succession of simple algae from diatoms, with a preferred growth temperature of 18 to 30 °C, to green algae (30 to 35 °C) to blue–green algae (35 to 40 °C) as the temperature increased (CAIRNS, 1956, 1971; CLARK and BROWNELL, 1973).

Blue–green algae also replaced green, red and brown algae which had disappeared from thermal discharge areas in Biscayne Bay, Florida (WOOD and ZIEMAN, 1969; ROESSLER, 1971; THORHAUG and co-authors, 1973) and developed into large mats in areas, also in Biscayne Bay, which had become thermally denuded of the turtle grass *Thalassia testudinum* (BADER and co-authors, 1972; SMITH and TEAS, 1979; THORHAUG and SCHROEDER, 1979: see p. 1818). Large mats of blue–green algae also replaced turtle grass in Guayanilla Bay, Puerto Rico, where a power station discharge temperature exceeded 33·5 °C (THORHAUG and SCHROEDER, 1979). A dominance of blue–green algae in periphyton communities of power station discharge canals (INTERNATIONAL ATOMIC ENERGY AGENCY, 1974) appeared to be associated with temperatures in excess of 30 °C.

Higher temperatures, however, do not necessarily result in complete destruction of temperature-sensitive algae. These forms re-establish themselves if cooler conditions return (CAIRNS, 1956). Seasonal recovery may also occur. For example, in a thermal discharge area in Florida macroalgae underwent a seasonal cycle of summer decrease–winter recovery in the 3 to 4 C° above-ambient zone around the discharge (THORHAUG and co-authors, 1973). In the 5 C° above-ambient zone, however, macroalgae disappeared completely.

In cooler latitudes, thermal effluents may keep estuarine areas ice free in winter and warmer than normal in summer. In one location of this type—Narragansett Bay, Rhode Island, USA—GONZÁLEZ and YEVICH (1977) found extensive growth of the green algae *Ulva* sp. and *Enteromorpha* sp. 2 months earlier than in neighbouring areas not affected by the thermal effluent. Significant increases in *Enteromorpha* were also recorded from the vicinity of a power station in southern Finland (ILUS and KESKITALO, 1980), again associated with a decrease in winter ice cover. In addition, a prolific growth of *Enteromorpha* occurred on experimental slabs placed at a thermal outfall in docks at Barrow-in-Furness, UK, in contrast to intake slabs where there was no algal growth (MARKOWSKI, 1960).

Growth of larger species of brown algae has also been investigated in thermal discharge areas. Several studies were carried out on the giant kelp *Macrocystis pyrifera* in California (USA). A number of these studies were reviewed by ANDREWS (1976). The general impression apparently was that heated effluents may kill most, if not all, of the macroalgae in the actual discharge and drastically change species composition in the vicinity of the discharge.

ANDREWS (1976) found that warm-water effluents may cause discoloration, brittleness and irregular growths in the holdfasts of giant kelp. These discharges also caused a deterioration in the surface canopies of *Macrocystis pyrifera* (NORTH and ADAMS, 1969; NORTH, 1977), apparently as a result of the water temperature at the discharge being raised above the ambient summer temperatures of southern California. An increase of 1·25 C° above summer ambient for several weeks caused canopy deterioration, whereas an increase of 6·25 C° above summer ambient killed nearly all the Californian species of kelp. Canopy deterioration does occur naturally in California during the hottest months, but is more pronounced and more prolonged in the vicinity of thermal discharges (NORTH, 1977).

NORTH (1969) suggested that the absence of macroalgae in thermal effluents may have resulted from algal spores being killed by the higher temperature or prevented from settling by the velocity of the effluent current. He also suggested that high temperatures may have killed the algal gametophytes or sporophytes in the discharge canal. There was no obvious natural replacement of thermally killed cold-water algae by warm-water forms; the entire algal flora in heated discharges on the Californian coast was simply depleted (NORTH, 1969). A very low algal density was also recorded in the neighbourhood of the heated discharge from Hunterson Power Station (CLOKIE and BONEY, 1980); this reduction may also have been a result of thermal stress.

A successful re-introduction of *Macrocystis* was carried out by NORTH (1977) in a depleted area at the discharge of San Onofre Power Station, California, using heat-tolerant varieties of this species from Mexico. NORTH found that the two forms of *Macrocystis* differed in the temperature ranges at which the maximum rates of photosynthesis occurred. With the colder water forms in California this range was 20 to 25 °C, deterioration of surface fronds occurring at temperatures of 25 °C and above. The optimum temperature range for photosynthesis in the Mexican variety was 25 to 30 °C. The Mexican plants were therefore more tolerant of conditions at thermal discharge areas in California.

Forests of macroalgae are typical of inshore areas in the temperate zone whereas in subtropical sea areas the dominant inshore flora consists of rooted vascular plants, such as the turtle grass *Thalassia testudinum*, which form extensive beds on shallow sandy bottoms and help to stabilise the sand (BADER and co-authors, 1972; SMITH and TEAS, 1979). The effect of heated effluents on *T. testudinum* beds have been studied by a number of investigators, mainly in Biscayne Bay and Card Sound, Florida (WOOD and ZIEMAN, 1969; ROESSLER, 1971, 1977; BADER and co-authors, 1972; THORHAUG and co-authors, 1973, 1978; WOOD, 1973; ZIEMAN and WOOD, 1975; SMITH and TEAS, 1979; THORHAUG and SCHROEDER, 1979) and to lesser extents in Guayanilla Bay, Puerto Rico

(SCHROEDER, 1975; THORHAUG and co-authors, 1978; THORHAUG and SCHROEDER 1979) and Trinity Bay, Texas (see discussion in CLARK and BROWNELL, 1973).

These investigations have shown that heated effluents have a severe effect on *Thalassia testudinum* beds. In regions closest to the effluents the beds were destroyed, leaving an area of bare sand, often with extensive patches of blue–green algae. The area of complete destruction of turtle grass ranged from 9 to 35 ha for power station effluents discharging into Biscayne Bay (BADER and co-authors, 1972; THORHAUG and co-authors, 1973; SMITH and TEAS, 1979). This area coincided with temperature increase of 5 C° or more above ambient (THORHAUG, 1974; THORHAUG and co-authors, 1978 THORHAUG and SCHROEDER, 1979). The lethal temperature for *T. testudinum*, as given by various workers, ranged from 'a sustained temperature of 33 °C' (BADER and co-authors, 1972) to 35·5 to 36 °C (SMITH and TEAS, 1979).

Where heating was continuous and prolonged, the destruction of *Thalassia testudinum* leaves was followed by the destruction of roots and rhizomes (WOOD and ZIEMAN 1969). The rhizomes were found to be dead by the end of the first summer of elevated temperatures (THORHAUG and SCHROEDER, 1979). Outside the thermally denuded area, WOOD and ZIEMAN (1969) found that the percentage of plants killed by above-normal temperatures decreased with increasing distance from the effluent, the reduction in biomass being inversely related to temperature (THORHAUG and SCHROEDER, 1979).

Sublethal temperature increases affected an area of about 120 ha around the Turkey point effluent in Biscayne Bay (BADER and co-authors, 1972). Within this area turtle grass productivity was reduced (ROESSLER, 1971; THORHAUG and SCHROEDER, 1979) Slower growth rates were also reported by THORHAUG and SCHROEDER as well as a decrease in the sexual structures of turtle grass, and a pronounced increase in asexual structures, such as short stalks, in response to non-lethal temperature stresses. In addition, increased temperatures near the lethal level had a significant, deleterious effect on the photosynthetic pigments of turtle grass (HEDGPETH, 1977). For instance, an increase in flaccid, mottled and dark coloured leaves in place of the bright green of healthy *Thalassia testudinum* leaves was reported by THORHAUG and SCHROEDER (1979).

Turtle grass beds affected by sublethal temperature increases may recover on a seasonal basis. Studies by ROESSLER (1971, 1977), BADER and co-authors (1972), THORHAUG and co-authors (1973) and THORHAUG and SCHROEDER (1979) revealed that turtle grass productivity in the outer area decreased when the summer temperature exceeded 31 °C (3 to 4 C° above ambient), but improved in winter to become even more productive than in the control areas. Where seasonal recovery does not occur, *Thalassia testudinum* rhizomes may survive for some time under temperature conditions that destroy leaf growth and produce leaves again when more favourable conditions return (WOOD and ZIEMAN, 1969; SMITH and TEAS, 1979). This occurred in an area of Biscayne Bay as a result of diluting the thermal discharge with ambient water to reduce the effluent temperature (THORHAUG and SCHROEDER, 1979).

In areas closer to the thermal discharge, where the turtle grass rhizomes had been completely destroyed, recovery may take a long time, even if thermal conditions returned to normal (WOOD and ZIEMAN, 1969). SMITH and TEAS (1979) suggested that recovery may occur as a result of re-seeding of turtle grass or by the invasion of plants from the surrounding area by vegetative growth from the root system.

The recovery of damaged or denuded *Thalassia testudinum* beds, seasonally or as a result of dilutions of the thermal effluent, appeared to confirm, therefore, that the original damage was primarily the result of increases in temperature (ROESSLER, 1971). In addition, discharge temperatures were apparently above the lethal temperature of *T. testudinum* (SMITH and TEAS, 1979).

Other factors, such as the turbidity of the effluent or the strength of current flow, may sometimes contribute to the effects of thermal damage (THORHAUG and SCHROEDER, 1979). Turbidity reduces illumination and photosynthesis in turtle grass, while current flow may wash away organic material and possibly affect the biomass of *Thalassia testudinum*. In addition, current flow over areas of sand thermally denuded of turtle grass may raise sand into suspension and thus increase turbidity. Increased turbidity, through its effect on illumination and photosynthesis may kill turtle grass growing on the borders of the bare area and thus enlarge the size of the denuded area (BADER and co-authors, 1972; THORHAUG, 1974; SMITH and TEAS, 1979). The size of the thermally denuded area may also increase as a result of erosion around its edge destroying parts of the turtle grass bed (WOOD and ZIEMAN, 1969).

Turtle grass is not the only rooted vascular plant occurring in warm water discharge areas, although it has received the most attention. Other species were investigated by ANDERSON (1969) in a relatively low salinity region of the Patuxent estuary, Maryland (USA). He studied *Potamogeton perfoliatus* and *Ruppia maritima* in detail and found that *P. perfoliatus* tended to replace *R. maritima* in a thermally stressed area. In this area *R. maritima* ceased producing vegetative growth after the first summer of thermal discharge; apparently the effluent temperature had increased above the critical temperature for vegetative growth. Flowering and seed germination in *R. maritima* were also very sensitive to increased temperatures (SETCHELL, 1924, 1964; quoted in ANDERSON, 1969).

Potamogeton perfoliatus was more tolerant of increased temperatures than *Ruppia maritima*, the former apparently being capable of physiological adjustment to temperature elevations. For instance, ANDERSON (1969) showed that, although respiration by young leaves increased with temperature, the more mature *P. perfoliatus* leaves compensated for higher temperatures by a reduction in respiration.

The extent to which warm-water effluents affect plant life depends on a number of factors. For instance, the effect of sublethal elevated temperatures on *Thalassia testudinum* beds depends on the amount and duration of heating, the time of year and the salinity (WOOD and ZIEMAN, 1969). Location also appears to have an effect on estuarine species in areas where large natural physical fluctuations occur possibly having a greater tolerance to thermal increases than those occurring at more open coastal sites (HEDGPETH, 1957; KINNE, 1963; NAYLOR, 1965b, CLARK and BROWNELL, 1973; see also Volume I).

Survival of plants in warm-water effluents is also affected to a large extent by latitude. MAYER (1914), NAYLOR (1965b), WOOD and ZIEMAN (1969) and THORHAUG and SCHROEDER (1979) have pointed out that organisms in tropical and subtropical environments live close to their lethal temperatures, especially in summer, whereas those in cooler latitudes can survive appreciably larger temperature increases. This appears to be confirmed by studies of benthic plants. For example, a temperature increase of only 5 C° in Florida was apparently sufficient to destroy beds of the turtle grass *Thalassia testudinum* (THORHAUG and SCHROEDER, 1979), whereas prolific growths of the green alga *Enteromorpha* sp. developed at thermal discharges with similar temperature elevations in the cooler latitudes of the UK (MARKOWSKI, 1960) and Rhode Island, USA (GONZÁLEZ and YEVICH, 1977).

Increased temperatures, therefore, can have a pronounced effect on benthic plant communities occurring in the vicinity of thermal discharges. In warmer latitudes, especially, the plants may be destroyed, summer growth inhibited, growth rates reduced or local plants may be replaced by other species more tolerant of higher temperatures. Replacement, damage or destruction of benthic plant communities may, in turn, exert severe effects on associated animal communities that depend on them for food and shelter (p. 1839).

Zoobenthos

Heated effluents may have a direct effect on benthic animal communities in the discharge area. Benthic communities are more static than plankton and fish; those in the vicinity of heated discharges are likely to be exposed to elevated temperatures throughout much of their life and may be good indicators of the biological effects of these discharges. For this reason, benthic organisms have been the subject of many investigations into warm water effects. Various investigators have reported temperature effects on breeding times, settlement, growth, metabolism and survival with resulting changes in species composition and fouling by marine organisms (see also Volume I).

Reproduction

Among the most frequently reported effects have been early and prolonged breeding in areas affected by heated discharges. For example, increased larval densities of the immigrant barnacle *Elminius modestus* and the mussel *Mytilus edulis* in the plankton of heated areas in Southampton Water (UK)—noted by PANNELL and co-authors (1962), RAYMONT and CARRIE (1964) and RAYMONT (1972) (see also p. 1809)—were probably caused by increased production by these two species in areas affected by heated discharges. *E. modestus* apparently also continued to breed for a longer time in early autumn compared with its pre-effluent breeding pattern; it is suggested (PANNELL and co-authors, 1962) that the extended breeding of this warm water species might be associated with the increased temperatures in the heated areas.

Another immigrant species, the American bivalve *Mercenaria mercenaria*, has also been studied extensively in Southampton Water and elsewhere in southern England (ANSELL and co-authors, 1964; ANSELL and LANDER, 1967). The English populations of this bivalve had a breeding pattern with a single autumn spawning period, comparable to that at the northern limit of the natural geographical range in North America. However, in clam populations exposed to heated effluents two spawning periods occurred, one in spring and the other in autumn—comparable to the spawning pattern in southern parts of the geographic range. The populations in the immediate vicinity of these effluents were shown to come into breeding condition and to reach a higher spawning potentiality earlier than elsewhere in southern England.

Another bivalve, the oyster *Ostrea lurida*, was studied by ADAMS (1969) in a power station discharge canal in California. He found that increased water temperatures apparently induced spawning. In the eastern USA, severe localized shipworm infestations were reported by MASNIK (1979) from the immediate vicinity of a warm-water plume in Barnegat Bay, New Jersey. In this area, an extended breeding and growing season by a dominant shipworm species (*Bankia gouldi*) was observed in test panel studies by TURNER (1974).

A number of early studies on the activities of wood borers in heated effluents were

eviewed by NAYLOR (1965b). He concluded that elevated temperatures resulted in
more protracted periods of breeding and colonization of new timber by the isopod
Limnoria tripunctata and the shipworm *Teredo navalis,* and that there was evidence for
increased depredations by *T. navalis* in the heated area of Swansea Docks (UK). RAY-
MONT (1976) referred to sudden, very damaging outbreaks of shipworms, such as *Teredo
navalis,* in Swansea Docks and of *Lyrodus pedicellatus* in Shoreham Harbour; these he
considered to be almost certainly due to artificially raised temperatures. Increased
reproduction and tunnelling rates by *Lyrodus pedicellatus* in heated areas were reported by
BOARD, 1971 (quoted in LANGFORD, 1975).

Two species of wood-boring isopod exhibited extended breeding seasons in areas receiving warm-water effluents: *Limnoria tripunctata* in Swansea Docks (NAYLOR, 1965a,b)
and *Limnoria quadripunctata* in the Marchwood area of Southampton Water (HOCKLEY,
1963). Unusually prolonged migrations of *L. quadripunctata* occurred at Marchwood
(ELTRINGHAM and HOCKLEY, 1961; PANNELL and co-authors, 1962), presumably
induced by increased temperatures. The main migration, according to PANNELL and
co-authors, lasted 2 months longer than at a control site, and migrations continued
on a much reduced scale throughout the winter. There was an increase in timber attacks
by *L. quadripunctata*, both at Marchwood and at a control site, but the increase was
maintained longer at Marchwood (HOCKLEY, 1963). With these species and the various
shipworm species, the effect of increased temperatures in extending the breeding season
was to prolong the period when timber structures were exposed to settlement and attack
by fouling organisms. Increased breeding by shipworms in thermally affected areas may
result in a considerable number of larvae being introduced into areas outside the
warm-water plume (MASNIK, 1979) where they may also attack timber structures.
Proliferation of wood-boring organisms, therefore, constitutes a detrimental aspect of
heated effluents when timber structures such as boats and piers occur.

Advanced or prolonged breeding has been recorded for a wide range of benthic
organisms in the vicinity of power station effluents in the UK. In a warm-water area in
Swansea Docks, the amphipod *Corophium acherusicum* bred earlier than at Plymouth,
while the ascidians *Botryllus schlosseri, Diplosoma listerianum, Ciona intestinalis* and *Ascidiella
aspersa* had longer breeding seasons and developed new colonies over a longer period of
the year than elsewhere in the UK (see discussion in NAYLOR, 1965b). Thermally
induced early breeding and settlement was also reported (Ryland, 1960) for two species
of Polyzoa, *Bugula neritina* and *B. stolonifera*.

Early and increased proliferation of benthic animals, such as ascidians, tube worms,
barnacles and mussels, in the vicinity of heated effluents may cause extensive fouling of
marine structures, especially in semi-enclosed docks, bays and estuaries. For example,
barnacles and mussels apparently increased in density in Southampton Water as a result
of small increases in temperature (PANNELL and co-authors, 1962), whilst greater proliferation of ascidians apparently occurred in the warm water of Swansea Docks (NAYLOR,
1965b). Periodic and extensive fouling by marine organisms, especially the tubeworm
Hydroides norvegica have been reported from an area in Sydney Harbour (Australia)
affected by warm-water discharges (WOOD, 1973). In another study, an early proliferation of ascidians, barnacles and hydrozoans occurred extensively within a localized
area of Narragansett Bay, Rhode Island, USA, where heated effluents kept the area
ice-free in winter and warmer than neighbouring areas in summer (GONZÁLEZ and
YEVICH, 1977).

With non-fouling benthic organisms, thermally induced advancements in the breed
ing seasons were frequently regarded as having an overall beneficial effect on the organ
isms concerned (WOOD and ZIEMAN, 1969), although this may not always have been th
case. An example of detrimental effects was shown in the studies of *Mercenaria mercenari*
in Southampton Water (p. 1820), where the heated discharge from Marchwood Powe
Station induced spring spawning. Eggs were released when the normal spring sea tem
peratures were insufficient to promote larval development (ANSELL and LANDER, 1967)
This spawning, therefore, did not contribute to the recruitment of young clams (ANSEL
and LANDER, 1967; RAYMONT, 1976).

Various investigators have produced contradictory results and opinions on whethe
thermally induced early breeding is beneficial or detrimental to benthic populations c
invertebrates. Differing results may have depended on the type and location of th
investigation. However, even where the same investigators studied a variety of organ
isms at the same location, the results varied relative to species. Apparently the effects c
early breeding were dependent on the breeding biology of the species in question. Fo
instance, the spawning time of many species may be altered relative to the seasonal foo
supplies of the larval stages (CLARK and BROWNELL, 1973).

Examples of variability in the effects of early breeding and of the dependence of som
larvae on seasonal food supplies were shown in the results of studies carried out in th
immediate vicinity of the heated discharge from Hunterston 'A' Power Station in Sco
land (BARNETT and HARDY, 1969; BARNETT, 1971, 1972; HARDY, 1977). At this site, a
investigation of the breeding biology of a number of sand-dwelling animals showed tha
certain spring-spawning species had started breeding earlier than those of control popu
lations outside the influence of the warm water effluent. A 2 to 3 month advancement i
the breeding season was recorded for the harpacticoid copepod *Asellopsis intermedia*, th
amphipod *Urothöe brevicornis* and the gastropod *Nassarius reticulatus*.

Breeding *Nassarius reticulatus* attached their egg capsules mainly to certain species c
red algae (BARNETT and co-authors, 1980). Therefore, eggs deposited within a heate
area would have been bathed in warm water throughout their period of developmen
Egg development studies in the laboratory (BARNETT and HARDY, 1969; BARNETT
1972) showed that these capsules hatched in a shorter time at higher temperature
within the range 8 to 14 °C. However, the veliger larvae released from these capsule
were planktonic and those produced precociously would have been released, as wit
Mercenaria mercenaria eggs (see above), into a cooling water mass at a time of year whe
there was very little phytoplankton available as food (BARNETT, 1972). Large larva
mortalities would presumably have occurred as a result of starvation.

As with *Nassarius reticulatus*, the eggs produced earlier than normal by the sand
burrowing copepod *Asellopsis intermedia* remained within the area of influence of th
heated effluent, the eggs developing within an egg sac attached to the female. Egg
development experiments in the laboratory (HARDY, 1977) showed that the eggs of thi
species also developed faster with increasing temperature at least within the range 5 t
14 °C. However, unlike *N. reticulatus*, the nauplius larvae that hatched from these egg
were benthic instead of planktonic and remained within the warm-water area, as did th
subsequent naupliar and copepodid larval stages. Developmental studies with naupli
and copepodids in laboratory conditions (HARDY, 1977) showed that the rate c
development of the larval stages increased with increasing temperature within the 5 t
14 °C experimental range, as did the rate of egg development, suggesting that warm

water effects might be beneficial. However, field studies indicated that very large mortalities occurred in the precociously produced larvae (BARNETT and HARDY, 1969; HARDY, 1977). Apparently these larvae were produced before an adequate food supply of sand-encrusting micro-organisms had developed.

With both *Asellopsis intermedia* and *Nassarius reticulatus*, therefore, heat-induced early spawning was apparently detrimental, even though development rates were accelerated. The mass mortality of precocious planktonic larvae may not have any serious effect on the density of local warm-water populations, such as those of *N. reticulatus*, as recruitment may take place from populations outwith the area affected by the heated effluent and spawning at normal times. With benthic larvae such as those of *A. intermedia*, however, recruitment from outwith the warm water area will be very much slower and the destruction of precocious larvae may have a more serious effect on population densities.

The sand-burrowing amphipod *Urothöe brevicornis*, like *Asellopsis intermedia*, also produces benthic young (BARNETT, 1971). However, unlike *A. intermedia*, this amphipod fed on detrital material and, therefore, its food supply was not so dependent on the time of year. The amphipods produced from the early breeding at Hunterston had a longer growing period and reached a larger final size than those from a control site not affected by the heated effluent (BARNETT, 1971).

The effects of early breeding, therefore, depend firstly on whether the young stages of various benthic organisms are planktonic or benthic and secondly on whether or not the larvae depend on a seasonal food supply. In some cases the effects may be beneficial while in others they may be detrimental, resulting in large larval mortalities.

The three early breeding species studied at Hunterston were spring-spawning animals. However, winter spawners also occurred near the Hunterston effluent and the breeding ecology of one of these, the lugworm *Arenicola marina*, has been investigated (SCOTTISH MARINE BIOLOGICAL ASSOCIATION 1974, 1979, 1980). The heated effluent apparently delayed the spawning of *A. marina* by 2 to 3 wk compared with control populations. Further, a proportion of the population of ripe males failed to spawn and retained mature sperm, densely packed in the coelomic cavity, throughout the following year. In these unspent males, sperm for the next winter's spawning also developed in the coelomic cavity throughout the following year.

Spawning in the bivalves *Mya arenaria* and *Macoma balthica* may also be inhibited or delayed if thermal effluents are discharged into Chesapeake Bay (USA) at high but sublethal temperatures (KENNEDY and MIHURSKY, 1971). In addition, NAYLOR (1965a,b) found that the common shore crab *Carcinus maenas* did not breed in heated areas of Swansea Docks (UK) during the period of highest effluent temperatures (1957–60). Some organisms were especially susceptible to sharp changes in temperature that occurred during the spawning period. For the clam *Mercenaria mercenaria* in New Jersey, USA, sudden changes in temperature may have caused a physiological shock that precluded the clams from spawning (KENNISH and OLSSON, 1979).

Heated effluents may also have more complex effects on reproductive development and general physiology. Simulation studies on the American oyster *Crassostrea virginica* revealed that continuous exposure to 35 °C at first greatly accelerated the rate of gametogenesis (QUICK, 1971a). This was followed by a regression of gametogenesis and a breakdown of gonadal and connective tissue with marked cytolysis. Increased loss of body fluids was recorded, especially in experiments that combined high temperatures with high salinities and 'secondary lips' formed on the oyster shells. Prolonged exposure to elevated temperatures had a marked effect on the condition of oysters, those in good condition with large glycogen reserves generally suffering the greatest decline in condition and exhibiting the heaviest mortalities, especially at higher salinities. In contrast, field studies at the Crystal River Power Station, Florida, USA (QUICK, 1971a) showed that this species could survive internal temperatures as high as 49·5 °C for brief periods.

The growth rate of *Crassostrea virginica* increased when cultured in heated effluent water (HESS and co-authors, 1979); however, its parasitic worms *Polydora ligni* and *P. websteri* also proliferated. These worms apparently burrowed into the oyster shells, making the oyster's appearance less attractive and reducing their market value. Studies with other organisms (p. 1815) also indicated that heated effluents may cause an increase in parasitic attacks. In contrast, QUICK (1971b) found that a continuous elevated temperature of 35 °C was a primary factor inhibiting the parasitic fungus *Labyrinthomyxa marina* present in many of the oysters that he studied.

Benthic animals, in general, are frequently reported to be more prolific in the immediate vicinity of heated effluents than at intakes. For instance, MARKOWSKI (1959, 1960, 1962), in studies at Cavendish Docks, Barrow-in-Furness, UK, recorded earlier settlements and a greater proliferation of epifauna on experimental slabs located at the effluent, compared with similar slabs at the intake. Similarly, in the Patuxent estuary, USA, early settlements of benthic organisms occurred in the vicinity of a heated effluent, and the total dry weight production of epifauna (mainly barnacles and hydroids) was almost 3 times that at the intake (CORY and NAUMAN, 1969; NAUMAN and CORY, 1969), even though the number of species at the effluent decreased in late summer compared with the intake (p. 1826).

Growth

In the Patuxent estuary the monthly growth rate of the barnacle *Balanus improvisus* was 7 mm, nearly 2 mm more than the previously recorded maximum growth of this species in the estuary. The above authors attributed this increased growth mainly to increased temperatures, which were 6 C° higher at the effluent than at the intake. However, it was thought that more highly productive water may have contributed to the results.

Increased growth rates or extended growing periods have been noted in the harpacticoid copepod *Asellopsis intermedia* (p. 1822) and the amphipod *Urothöe brevicornis* (p. 1823) in the vicinity of the heated effluent from Hunterston Power Station in western Scotland. Increased shell growth was observed in the bivalve *Tellina tenuis* inhabiting in large numbers a sandy beach near the Hunterston outfall (BARNETT and HARDY, 1969; BARNETT, 1971). The greatest increase occurred in young *T. tenuis* which grew to a mean length of 3 to 4 mm by their second winter in the pre-effluent years 1960–64, compared with a mean length of about 5·5 mm when affected by the effluent, as in the years 1964–68.

Shell-weight differences were observed in *Nassarius reticulatus* and *N. obsoletus*. Shells from the vicinity of heated discharges at Hunterston, Scotland, and Northport, Long Island, USA, were thinner and lighter than control shells of the same wet flesh weight (BARNETT, 1972), presumably because of increased growth rates.

Other investigators, such as PRICE and co-authors (1976), reported thermally accelerated gametogenesis, growth rates and extended growing seasons for marine organisms, including many commercially valuable benthic species. For instance in Narragansett Bay, Rhode Island, USA, *Mytilus edulis*, which settled in a warm-water discharge canal in autumn, grew at a fast rate and spawned as early as June in the following year (GONZÁLEZ and YEVICH, 1977). Increased growth of *Crassostrea virginica* in heated effluents has already been referred to (p. 1824). In California (USA) elevated temperatures in power station discharge canals provided favourable conditions for settlement and growth of *Ostrea lurida*, *Cardium corbis* and a number of other bivalve species (ADAMS, 1969). ADAMS also recorded an increased growth rate in oysters introduced into the warm waters of a discharge canal in Humboldt Bay. Similarly, in southern England, *O. edulis* colonized an area in the vicinity of the heated effluent from Fawley Power Station (HAWES and co-authors, 1975; RAYMONT, 1976) where it grew at a faster rate than normally outside the heated area.

In southern England *Mercenaria mercenaria* revealed an extended growth period, both at the beginning and end of the normal growing season, in the immediate vicinity of heated discharges (ANSELL, 1963; ANSELL and co-authors, 1964; RAYMONT, 1976). Clams at the heated discharge also grew faster, showing larger annual increments in both shell length and total weight than clams outwith the effects of the heated effluent. In Barnegat Bay, New Jersey (USA), growth of *M. mercenaria* living within 1·6 km of a thermal effluent increased with increasing temperature up to about 25 °C (KENNISH and OLSSON, 1975). Summer effluent temperatures in the heated area apparently exceeded the critical threshold temperature for optimum growth. This was indicated by a significantly lower summer growth rate and a greater number of growth checks in clams living in the vicinity of the effluent compared with those living at a greater distance from the effluent. KENNISH and OLSSON further established that a majority of the growth checks, recorded in the microstructure of clam shells, were correlated with thermal shocks caused by the rapidly fluctuating temperatures of the effluent.

Seasonal effects on growth of benthic organisms were demonstrated in a number of other investigations. Heated discharges from Danish power stations (MØLLER and DAHL-MADSEN, 1979) simulated the growth of *Mytilus edulis* in winter but inhibited it in summer. Maximum mortalities occurred in summer in the immediate vicinity of the heated discharges and resulted in a low biomass of *M. edulis* at the discharges. High biomasses recorded a few hundred metres away were attributed to an optimum combination of water flow and raised temperatures (p. 1836). Low summer growth rates of *Mercenaria mercenaria* were apparently a result of higher than optimum temperatures rather than food supply (KENNISH and OLSSON, 1975); however, ANSELL (1969) suggested that for most of the year growth of *M. mercenaria* was limited by factors other than temperature, the most important being the availability of suitable planktonic food.

Metabolism

The availability of food organisms is especially critical in winter. In temperate regions most marine animals have a reduced food requirement at that time of year and cease growing (KINNE, 1963, 1970b; NAYLOR, 1965b). Unseasonal artifical warming, therefore, may stimulate increased metabolic activity at a time when the nutritional demand is normally low (KENNEDY and MIHURSKY, 1972; HEDGPETH, 1977) resulting in increased feeding and growth rates (NAYLOR, 1965b). Increased metabolic activity during winter may allow certain molluscs, such as *Mercenaria mercenaria* and *Crassostrea virginica*, to utilize early peaks of phytoplankton biomass (ANSELL, 1969; PRICE and co-authors, 1976; HESS and co-authors, 1979). However, if food supplies do not satisfy metabolic needs, loss of weight can result, especially in winter (ANSELL, 1969; ANSELL and co-authors, 1964; MIHURSKY, 1967, KENNEDY and MIHURSKY, 1972). Increased metabolic activity combined with insufficient food may be responsible for the 'thin' oysters (*Ostrea edulis* and *Crassostrea angulata*) in 'obviously poor condition', observed in the immediate vicinity of a warm-water effluent from Bradwell Power Station (UK) (HAWES and co-authors, 1975; LANGFORD, 1975). Similar effects were observed in fish populations near heated effluents (p. 1815).

Warm-water effluents, therefore, by increasing the metabolic activity during periods of food scarcity, may cause loss of weight and increased mortalities. Weakened animals, in turn, may become more vulnerable to disease and predation. In addition, low food supplies in winter/early spring probably also caused the high mortalities of precocious larvae noted previously (p. 1823).

Heat shock

Mortalities near heated effluents also occur in summer, apparently as a result of lethal temperatures. The area around a warm-water outfall in Narragansett Bay, Rhode Island (USA), was characterized by a reduction in the number of species in late summer (GONZÁLEZ and YEVICH, 1977). Temperatures of 27 °C and above caused total mortalities in *Mytilus edulis* after a few days. Similarly, populations of *Mya arenaria*, which had settled in the discharge canal, were destroyed by prolonged exposure to 35 °C and above.

Reductions in population density and/or disappearance of flatworms, amphipods and colonial hydroids in summer occurred at a warm-water effluent in the Patuxent estuary, USA (NAUMAN and CORY, 1969). Large heat-shock crab kills were observed at this outfall, the largest recorded being *ca.* 40 000 *Callinectes sapidus* (MIHURSKY and CRONIN 1967; quoted in GRIMES, 1971). However, excess chlorine in the effluent water may have contributed to the mortalities (see p. 1835). Heated effluents also had a severe effect on coral communities at Tanguisson Power Station, Guam (NEUDECKER, 1976, 1977; BIRKELAND and co-authors, 1979), corals being killed along 2000 m² of reef margin (NEUDECKER, 1977).

Artificially raised temperatures, therefore, may cause mortalities among, or total destruction of, temperature-sensitive benthos at power station outfalls, especially in summer, in warmer latitudes where animals live close to their thermal death points.

Sublethal effects

Where heated effluents are not directly lethal, mortalities may occur from indirect temperature effects. Non-lethal temperature stress may have a detrimental effect on growth, reproduction and other metabolic activities of an animal, lowering its resistance to parasites and disease and decreasing its ability to avoid predators (TARZWELL, 1970; KINNE, 1980; LAUCKNER, 1980). A reduction in the feeding activity of *Mytilus edulis* at near-lethal temperatures was reported by GONZÁLEZ and YEVICH (1977). Sublethal temperatures also had a serious effect on byssus threads. Weakened byssus threads may increase the possibility of predation (PEARCE, 1969), and non-renewal of threads may lead to the sloughing-off of individual mussels or even of portions of a mussel bed (GONZÁLEZ and YEVICH, 1977). The burrowing activities of bivalves can be affected considerably at near-lethal temperatures (ANSELL and co-authors, 1980a,b), possibly increasing vulnerability to predation. Heated effluents had an indirect as well as a direct effect on corals at Kahe Power Station, Hawaii. Here, HEDGPETH (1977) established hat a 2 to 4 C° rise above ambient summer temperatures of 27 to 29 °C caused a umber of coral deaths; many corals survived the heat stress only to die later, the levated temperatures apparently having destroyed the coral's symbiotic algae.

Winter survival

Artificially raised temperatures may help marine organisms survive cold periods during winter and ameliorate the effects of severe winters (NAYLOR, 1965b). It has already been noted (p. 1816) that a warm-water discharge kept an area of Narragansett Bay, Rhode Island, USA, ice-free in winter and that early proliferation of certain marine organisms occurred as a result. In addition, COUGHLAN (1970; quoted in LANGFORD, 975) has shown that *Ostrea edulis* and *Crepidula fornicata* survived the severe winter of 962–63 in the vicinity of a heated outfall. Outside the heated area, the ambient sea emperature dropped as low as 0 to −2 °C, accompanied by extremely large mortalities of *O. edulis* and *C. fornicata*. Likewise, in a study of benthic organisms in Chesapeake Bay, USA, LOI and WILSON (1979) noted an increased abundance of the 8 commonest macro-infaunal species at a heated effluent, compared with control areas. They suggested that a temperature increase of about 3 C° had made the warm-water area relatively more habitable during the unusually cold winter of 1977.

Changes in species composition

Warm-water effluents, therefore, may have a profound effect on the species composition of benthic animals in thermally affected areas as a result of selective mortalities either directly or indirectly) or by depressing or stimulating the breeding and/or the growth of various species, especially during periods of minimum or maximum temperature. Significant changes in species composition have been reported by a number of investigators. For example, NORTH and ADAMS (1969) reported changes in the species composition of the benthic fauna at a power station in Morro Bay, California, USA, both in the discharge canal and on about 0.6 ha of the bottom in the vicinity of the discharge. At this site, a significant increase in benthic, warm water species relative to cold water species was reported from the immediate vicinity of the heated discharge ADAMS, 1969; NORTH, 1969).

Changes in species compositions have also been shown for a number of coastal and estuarine locations in the eastern USA. At Oyster Creek Generating Station, New Jersey, MASNIK (1979) reported an altered shipworm community in the immediate vicinity of the discharge plume (p. 1828). GONZÁLEZ and YEVICH (1977), in Narragansett Bay, Rhode Island, found that the species composition of benthic animals along the edges of a warm-water discharge canal and in the vicinity of the warm water plume was markedly different from that at the intake. A reduction in the number of species at the outfall occurred during the warmest months of the year, when species diversity reached its lowest level. Seasonal changes in the diversity of benthic animals at warm water discharges were also studied by WARINNER and BREHMER (1966) in the York river estuary, Virginia, USA. The lowest diversity of species was again found in summer in the immediate vicinity of a warm-water discharge, whereas in winter the diversity of species was much higher.

LYONS and co-authors (1971) studied the immediate vicinity of a discharge canal outfall in an estuarine area near Crystal river, Florida. Here the heated water from the outfall had little, if any, influence on the composition of the benthic invertebrate fauna.

Apparently, most Crystal river species were wide-ranging forms capable of withstanding large variations in environmental conditions.

In the Patuxent estuary, Maryland, USA, NAUMAN and CORY (1969) noted that both the intake and effluent canals of a power station had generally similar species composi-tions. However, there were differences in species diversity caused by differences in settling times. In addition, some organisms—such as flatworms, amphipods and col onial hydroids—declined in number and/or disappeared from the effluent canal during the warmest months of the year (p. 1826). Similar changes at a warm-water discharge on the south coast of Finland were noted by ILUS and KESKITALO (1980). At this low salinity location, the number of *Chironomus plumosus* larvae on the muddy bottom increased in the immediate vicinity of the discharge. These larvae apparently replaced the tubificid *Potamothrix hammoniensis* as the dominant organism, while the amphipod *Pontoporeia affinis* disappeared from the discharge fauna. Absence of benthic animals in the immediate vicinity of a heated effluent in Biscayne Bay, Florida (ROESSLER, 1977) however, may have been caused by the destruction of turtle grass in the thermally elevated region rather than as a direct result of elevated temperatures (p. 1839).

Immigrant species

Changes in species composition may also result from warm-water immigrant species settling in the vicinity of a heated discharge. Occurrence of warm-water immigrants has been reported mainly where raised temperatures occur in docks or estuaries which receive shipping from warmer latitudes. Presumably adults or larvae of these species are carried amongst the fouling on the ships' bottoms or in ballast water (NAYLOR, 1965b RAYMONT, 1976; BULLIMORE and co-authors, 1978) and, if the temperatures in the receiving area are suitable, may settle and develop breeding populations. The occur rence of immigrant species in British waters has been studied by a number of inves tigators (see reviews by NAYLOR, 1965a,b and RAYMONT, 1976). NAYLOR listed 10 such species recorded in Swansea Docks, South Wales, UK, where a heated effluent entered a fully saline, semi-enclosed dock and raised the dock temperature by 7 to 10 C above ambient. Most of the species reported apparently originated in warmer latitudes, they included the crabs *Brachynotus sexdentatus* and *Neopanope texana sayi*, the isopod *Lim noria tripunctata*, the barnacle *Balanus amphitrite* var. *denticulata*, the kamptozoan *Lox osomella kefersteinii* and various polychaetes and polyzoans. Some of these were also recorded in a warm-water location in Shoreham Harbour (RAYMONT, 1976). *Limnoria tripunctata* also occurred in an artificially heated area in Southampton Water as did the immigrant bivalve *Mercenaria mercenaria,* while *Balanus amphitrite* was recorded from a number of warm-water locations in southern Britain (CRISP and MOLESWORTH, 1951) The crab *Pilumnus hirtellus,* normally confined to the south and west coasts of Britain, has been recorded from the vicinity of the heated effluent of Bradwell Power Station on the cooler east coast of Britain (HAWES and co-authors, 1975).

Warm-water immigrant species have also been reported from artificially heated localities in North America. At a heated discharge in Barnegat Bay, New Jersey, MAS NIK (1979) found four species of shipworm: two native (*Bankia gouldi* and *Teredo navalis*) and two immigrant subtropical (*T. bartschi* and *T. furcifera*). The immigrant species which had not been reported previously from this locality, had developed breeding populations in the discharge creek of the power station.

Most of the warm-water immigrant species apparently remained confined to artifi cially heated areas, although a few have been recorded also from neighbouring areas unaffected by warm water plumes. For example, JONES (1963) found that the immigrant isopod *Limnoria tripunctata* was replacing the related native species *L. quadripunctata* in areas of southern Britain. He suggested that the spread of *L. tripunctata* may have been assisted by the presence of heated effluents which acted as spreading centres for the warmer water species. In addition, ANSELL (1963) produced evidence that the bivalve *Mercenaria mercenaria* had become acclimatized in Southampton Water. Although this bivalve probably originally settled in Southampton Water without the aid of warm-water areas (RAYMONT, 1976), it bred and grew more prolifically in the vicinity of heated discharges than in other areas of southern England (ANSELL, 1963; ANSELL and co-authors, 1964; ANSELL and LANDER, 1967; RAYMONT, 1976; the larvae released from warm-water areas apparently augmented settlements elsewhere. Similarly, populations of shipworm in water at ambient temperatures were apparently increased by the addition of larvae released from warm-water populations (MASNIK, 1979; see also p. 1821). Occasional settlements of *Balanus amphitrite* var. *denticulata* have also been found outwith their warm-water colonies. For example, CRISP and MOLESWORTH (1951) recorded the formation of temporary colonies of this barnacle outside the heated area of Shoreham Harbour, where persistent colonies occurred, during the warmest months of the year.

Return to cooler conditions

Balanus amphitrite was the only species of barnacle to be found near the heated effluent in Swansea Docks, South Wales, UK, during the period of maximum warm-water discharge (1956–60). However, with the reduction in the warm-water output after 1960 and the termination of the outflow in 1975 (BULLIMORE and co-authors, 1978), cooler conditions returned to the docks; two other species, the native *Balanus crenatus* and the temperate water immigrant *Elminius modestus*, extended their ranges closer to the outfall and were then found living in company with *Balanus amphitrite* (NAYLOR, 1965a,b; RAYMONT, 1976; BULLIMORE and co-authors, 1978). The reasons for the continued existence of *B. amphitrite* in the cooler conditions remain unknown, although BULLIMORE and co-authors put forward a number of suggestions. For instance, the barnacles may have become acclimatized to the cooler conditions, or a series of exceptionally warm summers during the study may have allowed breeding to continue or the continued occurrence may reflect the longevity of individuals.

Studies carried out in Swansea Docks during the cooler post-1960 period were reviewed by BULLIMORE and co-authors (1978). The immigrant crab *Neopanope texana sayi* also continued to live in the cooler conditions. This species apparently encounters similar low temperatures at the northern end of its natural American distribution—in the Gulf of St. Lawrence. Some of the other warmer water immigrants, such as the polyzoan *Bugula neritina*, were not recorded after the discharge of heated water ceased. However, a number of native British species, not observed in Swansea Docks during the warmer 1956–60 period, occurred during the subsequent cooler period. These included *Balanus crenatus*, the portunid crab *Macropipus puber*, the bryozoan *Bowerbankia gracillima*, the bivalve *Cerastoderma glaucum* and *Gammarus locusta* (see discussions in NAYLOR, 1965a and BULLIMORE and co-authors, 1978). In addition, *Carcinus maenas* certainly bred in the docks after 1960 although it apparently did not breed during the warmer 1956–60 period (NAYLOR, 1965a,b). The number of generations of the ascidian *Ciona intestinalis* in any one year decreased after 1960.

Reductions in temperature resulting from the cessation of warm-water discharges, therefore, may have a pronounced effect on the biology and distribution of marine invertebrates.

Cold effluents

Temperature reductions may also occur as a result of industrial discharges at below-ambient temperatures (p. 1771). A number of cold effluent simulation experiments with marine invertebrates were carried out for a proposed liquid natural gas conversion plant in California (SOULE and OGURI, 1974). In one of these experiments, NORSE (1974) acclimated four organisms at different temperatures before transferring them to lower temperatures for 24 h. The results suggested that cold effluents might produce a number of biological effects, both lethal and sublethal. For example, byssus-thread formation in *Mytilus edulis* was impaired when temperatures were lowered by 8 C° below acclimation and none developed when temperatures were reduced by 12 or 16 C°. Decreases of this magnitude, therefore, would prevent mussels from moving by the breaking and reattachment of byssus threads.

NORSE (1974) also reported that the epibenthic harpacticoid *Tisbe* sp. could not swim, clean its appendages or feed when transferred to 5 °C from an acclimation temperature of 17 °C; thus delayed mortalities would probably occur. Significant mortalities were recorded when *Tisbe* sp. was acclimated at 21 °C before transfer to 5 °C. Reductions in temperature also lowered reproductive rate, and reproduction ceased at 5 °C. Reductions in mobility and swimming ability were recorded for zoeae larvae of *Pachygrapsus crassipes* and the semi-benthic anthomedusa *Cladonema* sp. when the temperature was decreased by 4 C°. With an 8 C° drop, NORSE found that *Cladonema* was hardly able to swim while *P. crassipes* was unable to swim or hold normally on to the substratum. When transferred to water at 5 °C, mortalities in *Cladonema* sp. reached 100%.

No immediate mortalities were recorded in *Littorina planaxis* and *L. scutulata* when these were transferred to 6 °C. However, the results of cold-effluent simulation experiments by HADLEY and STRAUGHAN (1974) indicated that a reduction in temperature in summer would result in a substantial but temporary reduction in the attachment rate of these gastropods. They suggest, therefore, that wave action may result in a population reduction through the detachment and loss of chilled individuals.

Cold-water discharges, therefore, may affect marine organisms, depending on the species, the amount and rate of temperature decrease and the season when the temperature reduction occurs.

The extent to which temperature affects marine organisms varies according to location and environmental factors. For example, a number of thermally induced biological changes apparently occurred in Swansea Docks (p. 1821), whereas relatively few changes were recorded at the Cavendish Docks in Barrow-in-Furness (MARKOWSKI, 1959, 1960, 1962). The difference between these two docks may be related to the different salinity regimes; Swansea Docks is fully saline whereas much lower salinities occurred in the Cavendish Docks. Wide variations in temperature and salinity normally occur in brackish-water estuarine areas with the result that the typical fauna of estuarine habitats are more tolerant of variations in temperature and salinity than purely marine organisms (p. 1809). The various ways in which salinity modifies the effects of temperature on organisms is discussed fully in Volume 1: KINNE (1970b), ALDERDICE (1972); see also KINNE (1964) and the following section.

(c) Synergistic and Non-thermal Effects

Synergistic Effects of Temperature and Salinity

Temperature and salinity are primary factors in the physical environment of marine and brackish-water organisms (Volume I). Variations in both these factors can have pronounced effects on the biology of organisms and severely limit their distributions. These factors also act synergistically: the biological effects of one may be modified by variations in the other. For comprehensive reviews of the biological effects of temperature and salinity combinations see KINNE (1964, 1970b) and ALDERDICE (1972).

Salinity modifications of temperature effects have been demonstrated in thermal effluent studies with simulation experiments. For example, in experiments with *Crassostrea virginica* it was found that continuous exposure to a temperature of 35 °C had a serious effect and that mortalities increased with increasing salinity (QUICK, 1971a). The manner in which salinity acts with temperature appears to be very complicated, depending on various factors such as species, food, previous salinity experience and habitat (see discussion in GAUDY and co-authors, 1982). In their experimental studies, GAUDY and co-authors showed that within the temperature range 14 to 24 °C, the shortest generation times of the harpacticoid copepod *Tisbe holothuriae* occurred within the normal salinity range of the original biotope of the species (28 to 38‰ S) but the generation times became prolonged at both higher and lower salinities.

Desalination Effluents

Synergistic temperature and salinity interactions are probably especially important where warm-water saline discharges are released into the environment from desalination plants (p. 1780). Although warm saline biotopes occur naturally (BASSON and co-authors, 1977), they are found in discrete, relatively stable, mainly tropical and subtropical environments such as coastal lagoons, and contain limited communities of highly specialized organisms. Desalination effluents, however, tend to be intermittent, with discharges occurring mainly in summer and autumn. Under these circumstances, there will be little time for specialized warm-water saline communities to develop.

Theoretical considerations of the possible effects of desalination effluents on marine organisms, with examples of salinity effects, are discussed in a DOE publication (DEPARTMENT OF THE ENVIRONMENT, 1977) and in publications of the US Office of Saline Water (THOMSON and co-authors, 1969; ZEITOUN and co-authors, 1969a,b). The salinities and temperatures of desalination effluents are normally considerably above natural extremes, salinities ranging from $1 \cdot 5$ to $2 \cdot 0$ times that of the feed water supplying the plant, while the ΔT may sometimes be as great as 15 C° and the actual temperature as high as 43 °C (LEIGHTON and co-authors, 1967). The warm saline effluents of desalination plants tend to sink in the receiving body of water (p. 1796), in contrast to power station effluents where the plumes are normally buoyant (p. 1788). Benthic organisms, therefore, are more likely to be affected by desalination effluents than by the warm-water discharges from power stations.

Very few biological studies appear to have been carried out on the effects of desalination plant effluents. Among the more detailed investigations were those at Point Loma in California (LEIGHTON and co-authors, 1967), at Key West in Florida (CLARKE and co-authors, 1970; CHESHER, 1975) and at La Rosiere on Jersey in the Channel Islands (DEPARTMENT OF THE ENVIRONMENT, 1977). Effluents at La Rosiere and Point Loma were discharged across the intertidal zone whereas those at Key West were released directly into the receiving water body.

At La Rosiere, no living organisms were found in the path of the undiluted effluent, except at extreme low tide level. *Enteromorpha* sp. colonized the area in early spring when the desalination plant was not functioning. However, when the plant began operations practically all algae in the path of the effluent were bleached and killed, except for those growing on the tops of rocks protruding above the effluent. Presumably, dilution by sea water at high tide considerably reduced the effect of the effluent, allowing these 'caps' of algae to continue growing.

Similar effects were recorded at Point Loma in California (USA), where LEIGHTON and co-authors (1967) investigated an intertidal rocky area affected by the desalination plant effluent. Typical algae were found to be absent from the effluent drainage channels in the rocky intertidal zone, the path of the discharge being marked by a growth of filamentous green and red algae covering the rock surface. Animal life in the discharge zone was limited to an occasional shore crab *Pachygrapsus* sp., and two species of anemone that were apparently resistant both to high salinity and high temperature. Of the more typical animals of the area, mussels, barnacles and limpets were confined to the tops of rocks that protruded above the discharge flow channels. The effects of the effluent were therefore very localized. Simulation experiments revealed that full-strength effluent water at 16 to 18 °C was lethal within 24 h to most species examined. The sea urchin *Strongylocentrotus purpuratus* was most seriously affected. Its immediate response was to withdraw its tube feet. In the surf zone this increased the likelihood of detachment and vulnerability to predation.

Extensive studies were carried out at a large desalination plant in Safe Harbor, Key West, Florida, USA (CLARKE and co-authors, 1970; CHESHER, 1975) using various methods such as benthic quadrat counts, transect surveys, settlement panels and discs, the transplanting of selected species and simulation studies in the laboratory. The desalination effluent had a deleterious, but localized, effect on a large proportion of the marine organisms. Even organisms that were numerous in Safe Harbor were adversely affected in the immediate vicinity of the desalination plant discharge. Among the organisms excluded from the area were the ascidian *Ascidia nigra,* various gastropods, echinoids and a number of algal species. Bryozoan colonies were smaller and much less numerous in the immediate vicinity of the discharge than in control areas. Gorgonians were damaged or killed when transplanted to a site close to the effluent but not when transplanted to control areas (CLARKE and co-authors, 1970). In addition, it was shown by laboratory studies that photosynthetic activity in the turtle grass *Thalassia testudinum* underwent a 50% reduction when exposed to 12% effluent for 24 h (CHESHER, 1975).

The desalination effluent appeared to have a beneficial effect on some other Safe Harbor organisms. For example, CLARKE and co-authors (1970) observed that certain fish and the stone crab *Menippe mercenaria* were attracted to the effluent and that reproduction was apparently stimulated in *M. mercenaria* and in the aplysiid sea slug *Bursatella* sp. In addition, barnacles and filter-feeding tube worms were more abundant in effluent-affected areas, presumably because the current flow provided more planktonic food organisms.

While the effects observed in these studies were apparently related to the properties of the effluent, CLARKE and co-authors (1970) were unable to ascertain whether these were mainly temperature and/or salinity related and to what extent metal ions and/or descaling chemicals contributed. However, they suggested that the damage to gorgonian colonies may have been caused mainly by acids in the effluent or by scouring by sediment particles at the discharge.

Metal Ion and Synergistic Effects

CHESHER (1975) cautioned that the effluent from Key West desalination plant may not necessarily have a beneficial effect on fish. Fish attracted by one physical parameter, such as temperature or food, may be damaged by toxic substances. For instance, CHESHER observed epidermal lesions in fish from the effluent and found, by histological examination, that the livers of some of the fish had been adversely affected by copper concentrations in the effluent.

Biological damage caused by the Key West plant was especially serious when the plant started discharging after being shut down for a period (CHESHER, 1975). An increase in copper in the effluent was noted at these times and the extreme toxicity observed, especially with respect to sabellid worms and *Ascidia nigra*, was confirmed by laboratory studies. The increase in copper ions apparently originated from corrosion of metallic components in the plant (DEPARTMENT OF THE ENVIRONMENT, 1977). Laboratory studies by CHESHER also showed that even relatively low concentrations of copper, compared with copper concentrations in desalination effluents, considerably lowered cell division rates. (For a detailed review on pollution due to heavy metals and their compounds, consult Chapter 3.)

ZEITOUN and co-authors (1969b) reported that copper concentrations of the order of 0·5 ppm at 20 °C, when supplied continuously, were toxic to diatoms and dinoflagellates, two of the most important primary producers in the marine environment. At sublethal concentrations copper may be transferred along the food chain to higher trophic levels and, ZEITOUN and co-authors suggest, may be the cause of greening in oysters (p. 1892).

Toxic effects of copper may be altered synergistically by both temperature and salinity. For example, ZEITOUN and co-authors (1969a) reported that the copper uptake of certain benthic organisms may be doubled by a temperature increase of 10 C°. On the other hand, increased salinities apparently reduced the toxicity of copper ions (ZEITOUN and co-authors, 1969a,b).

With power station effluents, SAVAGE (1969) suggested that temperature increases may produce synergistic effects by increasing the toxicities of trace substances in the heated effluents. In fish, synergistic effects between elevated temperatures and certain trace metals have been reported by HAWES and co-authors (1975) and HOSS and co-authors (1975). The latter investigators conducted simulation experiments, in which fish were subjected to thermal shock following an increase in the copper content of the water (presumably originating from corrosion of condenser tubes); fish mortalities turned out to be much higher than those caused by either stress acting alone. Substantial increases in copper (MIHURSKY and CRONIN, 1967; GRIMES, 1971; MOUNTAIN, 1972) and zinc (GRIMES, 1971) have also been reported in the flesh of oysters grown in discharge canals or in the immediate vicinity of power station discharges. These increases may be partly a result of increased metal ion content of the effluent water and partly increased rate of metabolism—and, therefore, an increased ion uptake—of oysters living in the heated water (see also p. 1892).

Heated effluents may also alter the toxicity of metal ions and pesticides that may be present in a receiving body of water. The long-term toxicity of mercury has been studied by HUISMAN and co-authors (1980) using cell cultures of *Scenedesmus acutus*, to which mercuric chloride was added. Experimental results showed that the toxicity of mercury increased with increasing temperature and that a decline in cell growth was accompanied by a decrease in the rate of photosynthesis.

Chlorine and Synergistic Effects

Chlorine ions may also affect aquatic organisms both directly and/or synergistically with some other stress. Chlorine is one of a complex of stress factors (Volume III: KINNE, 1977, pp. 1077–1081); (see also p. 1796); it can seriously affect marine organisms, especially during entrainment in power station cooling systems. Usually as chlorine gas, it is frequently added to cooling water systems as an antifouling agent (p. 1797). In cooling systems, chlorine reacts with sea water, resulting in the formation of a number of additional chemicals, including bromine and chloramine (p. 1888) all of which may produce biological effects. For convenience, these are generally referred to as chlorination effects. Desalination plant effluents may also contain chlorine as a biocide (ZEI-TOUN and co-authors, 1969a; DEPARTMENT OF THE ENVIRONMENT, 1977).

Chlorination can have an extremely deleterious effect on a wide range of entrained organisms. High mortalities have been reported for estuarine bacteria (HAMILTON and co-authors, 1970; INTERNATIONAL ATOMIC ENERGY AGENCY, 1974). Chlorination caused a very substantial reduction in primary production (HAMILTON and co-authors, 1970; CARPENTER and co-authors, 1972), a reduction in ^{14}C uptake (LAUER and co-authors, 1974) and a significant decrease in chlorophyll *a* and, therefore, photosynthesis (HEINLE, 1969; MORGAN and STROSS, 1969; BROOK and BAKER, 1972). Zooplankton productivity was severely damaged by chlorination (CLARK and BROWNELL, 1973), very high mortalities occurred (HEINLE, 1969, 1976; LAUER and co-authors, 1974; GONZÁLEZ and YEVICH, 1977; ISAAC, 1979) and the growth of copepods that survived entrainment in the cooling system during chlorination periods was retarded (HEINLE, 1969). However, some organisms were much more sensitive to chlorination than others. For example, larvae of *Ostrea edulis* were unharmed by chlorine concentrations up to 10 ppm, and a substantial proportion survived and still grew at concentrations up to 20 ppm (WAUGH, 1964). In contrast, the nauplii of *Elminius modestus* were far more sensitive, the majority dying at concentrations as low as 2 ppm.

With respect to fish, MARKOWSKI (1962) suggested that entrained individuals were not affected by chlorine in cooling water. However, later studies (ALDERSON, 1972, 1974; BRUNGS, 1973; LAUER and co-authors, 1974; HOSS and co-authors, 1975) indicated that chlorine had a very severe effect on fish eggs and larvae, sometimes even at very low concentrations. As with zooplankton, some fish were apparently more susceptible than others (BRUNGS, 1973). Sensitivity also varied with the stage of development (ALDERSON, 1972, 1974; LAUER and co-authors, 1974), eggs being relatively more tolerant than larvae. Sublethal effects of chlorination were reported by CAPUZZO and co-authors (1977). They observed reductions in metabolic activity and suggested that growth and maturation might be affected as well as an increased vulnerability to disease and predation.

Additional accounts of effects of chlorination during entrainment are given by HIRAYAMA and HIRANO (1970a,b), ALDERSON (1972), BRUNGS (1973), SCHUBEL (1974), CAIRNS (1977), CAPUZZO and co-authors (1977), MORGAN and CARPENTER (1978), CAPUZZO (1979) and others. Their results indicate that chlorination is likely to have far more severe effects on organisms than raised temperatures and other entrainment stresses, except at very low concentrations.

Chlorine stress may also affect organisms in the warm-water discharge area of a power station, where the effects are likely to decrease with increasing distance from the effluent. Various studies, however, disagree as to the severity of these effects. For example, the results of simulation experiments in Japan (HIRAYAMA and HIRANO, 1970b) indicated that most of the residual chlorine is dissipated by the time the power station discharges into the open sea and may not cause great damage to the marine phytoplankton, even within the immediate vicinity of the discharge. In contrast, CLARK and BROWNELL (1973) concluded that residual chlorine levels at the effluent may sometimes be sufficient to cause fish deaths. BRUNGS (1973) warned of the danger of highly chlorine-sensitive fish such as trout and salmon being attracted to the warm-water effluents where they may encounter residual chlorine or chlorine derivatives in lethal concentrations. Several kills of juvenile menhaden have been reported by FAIRBANKS and co-authors (1971), who attributed the deaths to chlorine residuals of 0·8 to 1·5 ppm at the discharge. Each kill apparently involved between several hundred and several thousand fish. On the other hand, continuous, relatively low levels of chlorine (0·02 to 0·1 ppm) in the cooling water from Hunterston Power Station (Scotland) did not have any detrimental effects on plaice *Pleuronectes platessa* and sole *Solea solea*, reared successfully in the heated effluent water from this station (NASH, 1968, 1969b, 1970b; p. 1815).

The presence of chlorine in warm-water discharges also apparently affected benthic organisms. In the Patuxent estuary (USA), the complete destruction of epifaunal organisms and a high mortality of blue crabs *Callinectes sapidus* in a power station effluent canal may have been caused by an overdose of chlorine (NAUMAN and CORY, 1969) in addition to the high temperatures (p. 1826). In addition, the presence of chlorine has been reported (ANSELL, 1969) to depress the activity and to considerably reduce the growth rate of the immigrant bivalve *Mercenaria mercenaria* which occurred in the vicinity of thermal discharges in southern England. However, molluscs in general tend to be less susceptible to chlorine effects than fish (LANGFORD, 1977).

For a more detailed account of chlorine effects, especially with respect to the culturing of marine organisms in power station effluents, consult p. 1888. Extensive accounts of the effects of chlorine on aquatic organisms are presented in reviews by BRUNGS (1973), WHITEHOUSE (1975), BLOCK and co-authors (1977) and COUGHLAN and WHITEHOUSE (1977).

A reduced viability caused by one stress, however, is apparently liable to increase an organism's vulnerability to other stresses. Increases in heat, for instance, can act synergistically with most other stresses (KINNE, 1970b; CLARK and BROWNELL, 1973; SOULE, 1974) and, where a species was already subject to one stress, the addition of heat might result in death at a lower temperature than normal or within a shorter period at a given temperature (PEARCE, 1969). Synergistic effects of temperature and chlorine have been reported by several authors (STOBER and HANSON, 1974; HOSS and co-authors, 1975; CAIRNS, 1977; CAPUZZO, 1979; ISAAC, 1979). According to ISAAC, fish kills in thermal discharges in Massachusetts (USA) were often caused by a combination of heat and chlorine. HOSS and co-authors have shown by entrainment simulation studies with flounder *Paralichthys* sp., striped mullet *Mugil cephalus* and menhaden *Brevoortia tyrannus* that increased temperatures reduced the numbers of fish surviving near-lethal chlorine concentrations. The effects of temperature increases on the toxicity of a number of other chemicals, in addition to chlorine, was reviewed by CAIRNS and co-authors (1975).

Organisms may be affected, therefore, both during entrainment and in the discharge area by temperature changes and also by the various chemicals present in the cooling water system which may act synergistically with temperature.

Current Flow and Impingement Effects

The strength of flow of the cooling water system may also affect organisms before, during or after entrainment. The current flow before and after entrainment apparently provides food for bottom-living animals and fish and affects fish distributions. For example, the habitat of bay anchovy *Anchoa mitchilli* in a discharge area in Galveston Bay, USA, was reduced by the current flow (GALLAWAY and STRAWN, 1974). Current flow was important at a power station in Kiel Fjord, Federal Republic of Germany: according to MÖLLER (1978b), the current brought in large quantities of zooplankton which, in turn, attracted cod *Gadus morhua* and eel *Anguilla anguilla* to feed at both intake and discharge. MÖLLER also reported that the plankton brought in by the current flow apparently provided food for benthic animals living in the power station intake and discharge areas and that these animals starved when the station closed down. Crab and eel then moved in to scavenge on the dead and dying benthic animals. An increased biomass of benthic organisms, because of the water flow and raised temperatures near to, but not in the immediate vicinity of, a thermal discharge in Denmark was reported by MØLLER and DAHL-MADSEN (1979). The current from a power station's thermal discharge may cause changes in the populations of marine organisms in the immediate vicinity of the discharge by removing the fine material from the sediment (ADAMS and co-authors, 1970). The adverse effect of this on beds of turtle grass *Thalassia testudinum* was discussed earlier (p. 1819): increased currents and temperatures may aggravate each other's effects by producing a larger denuded area within the seagrass bed than would either stress acting alone.

Current flow during entrainment may also have an effect on organisms. In this case, however, the effects may be more serious, especially with respect to larger organisms such as fish. Entrained organisms may be injured mechanically by being buffeted against the sides of the cooling system and condenser tubes or by impingement on the intake screens.

Coarse and fine screens are installed at the intakes to the cooling water systems of power stations to prevent larger fish and debris from entering the system (p. 1777 and p. 1848). Although MÖLLER (1978b) claimed that the number of fish killed by the screens was low, other workers (ADAMS and co-authors, 1970; GALLAWAY and STRAWN 1974; YOUNG, 1974; ISAAC, 1979) have recorded high fish mortalities. In fact, screen kills may sometimes be very large, blocking the screens with dead fish and forcing the power station to shut down (CLARK and BROWNELL, 1973). This, in turn, may cause cold-shock deaths among fish at the discharge area (p. 1813).

Planktonic organisms, including larval and juvenile fish, pass through these intake screens to become entrained in the cooling water system. During entrainment, some of the larger juvenile fish and zooplankton may suffer mechanical injury when passing through pumps or by being buffeted against projections in the cooling system (MAR-KOWSKI, 1959; CLARK and BROWNELL, 1973; MARCY, 1973; CARPENTER and co-authors, 1974b; HAWES and co-authors, 1975; BEINFANG and JOHNSON, 1980). How-ever, although CARPENTER and co-authors reported a 70% copepod mortality at a power station in Long Island Sound, USA, caused by mechanical or hydraulic stresses during entrainment, MARKOWSKI was of the opinion that the mechanically damaged organ-isms, which he observed at a power station in Barrow-in-Furness, UK, formed a neglig-ible percentage of the total zooplankton. Apparently, water-flow velocities and sizes of planktonic organisms are important in determining the amount of mechanical damage sustained during entrainment (see MARCY, 1975, for a review of the literature on mechanical damage during entrainment).

Pressure Effects

In addition to mechanical stresses, entrained organisms are exposed to sudden changes in hydrostatic pressure (Volume I: FLÜGEL, 1972; KINNE, 1972) as they pass through the cooling water system. Entrained fish are most likely to be affected (CLARK and BROWNELL, 1973) because of the presence in most fish of a gas-filled swimbladder which may not be able to compensate for sudden pressure changes and so ruptures. This possibility was studied experimentally in larval and juvenile fish of a size that could pass through the intake screens of power stations (LAUER and co-authors, 1974; COUTANT and KEDL, 1975; BLAXTER and HOSS, 1979; HOSS and BLAXTER, 1979); abrupt pres-sure changes normally had a negligible effect.

Early entrainable stages (<40 mm) of Atlantic herring *Clupea harengus* were generally unaffected by pressure; mortalities, however, did occur when the larvae were about 25 to 30 mm in length. HOSS and BLAXTER (1979) attributed this period of pressure sensitiv-ity to the development of paired gas-filled bullae behind the brain, as part of the auditory system. These bullae were unable to compensate for pressure changes and ruptured as a result. Older larvae were less affected because of the development of the swim bladder, connected to the bullae by means of gas ducts, and capable of adapting to pressure changes occurring during entrainment.

Studies by HEFFNER and co-authors (1973) and LAUER and co-authors (1974) in the Hudson river estuary and BIENFANG and JOHNSON (1980) in Hawaii suggest that physi-cal factors such as pressure, turbulence and mechanical stress do not have significant effects on the mortalities of entrained phytoplankton. A more detailed account of various physical factors (pressure, acceleration and shear forces, abrasion and collision) that may affect entrained organisms is given in a review by MARCY and co-authors (1978).

(d) Ecosystem Effects

The effects on the ecosystem of mortalities due to entrainment are difficult to predict, some authorities suggesting that entrainment losses would have a serious effect on local populations whereas others claim that the effects would be negligible. For example, at a power station in Florida (USA), the primary production of phytoplankton, reduced during entrainment, had recovered by the time the phytoplankton had reached the end of the effluent canal (FOX and MOYER, 1973), therefore the number of phytoplankters killed by this subtropical power station had a negligible effect on the environment.

Likewise, in field studies by HAMILTON and co-authors (1970), whatever the magnitude of loss during entrainment, no consistent reduction in primary production was detected in the vicinity of the outfall from an effluent canal. In addition, laboratory studies (CAIRNS, 1977) suggested that the effects would not be particularly severe even if all the entrained organisms were killed.

Californian phytoplankton populations returned to their original density within a day following entrainment in winter; however, no such recovery occurred following entrainment at temperatures above 31 °C, as in September (p. 1801). Thus, during the warmest months, phytoplankton biomass in the effluent areas was depressed rather than elevated and this, in turn, would have had a significant effect on the standing stock of local phytoplankton populations (BRIAND, 1975). Similarly, CLARK and BROWNELL (1973) concluded from previous investigations that power station-induced plankton kills exerted a significant adverse effect in the vicinity of the effluents and that immediate and pronounced damage to productivity was caused by temperature increases during entrainment in the warm season.

Even population changes during entrainment may exert an impact. For instance, thermally induced shifts from diatoms to blue–green algae (p. 1801) may affect the stocks of certain fishes (BRIAND, 1975). Some fish depend on diatoms as food and are seriously affected when diatoms are replaced by blooms of blue–green algae, which are apparently unsuitable as food (BADER and co-authors, 1972; CLARK and BROWNELL, 1973; CAIRNS, 1977). Population changes may also cause 'red tides' (p. 1808), which frequently have a lethal effect on fish (see also Volume V: PERES, 1982c).

Loss of fish eggs and larvae through physical, chemical and thermal damage during entrainment may ultimately affect the populations of the species involved (ISAAC, 1979). Estuarine and coastal power stations are apparently responsible for the destruction of large numbers of entrained planktonic larvae of inshore fish (CLARK and BROWNELL, 1973; HOSS and co-authors 1974; YOUNG, 1974; ENRIGHT, 1977). In fact, power stations have been described by some authors (COUTANT, 1971a; SCHUBEL, 1974; ENRIGHT, 1977) as acting like large artificial predators of zooplankton, including fish eggs and larvae. The destruction of fish larvae in particular may have serious economic consequences, especially for populations of commercially important fish such as menhaden, striped bass and salmon, which spend part of their early life cycles in estuaries (e.g. WARINNER and BREHMER, 1966; CLARK and BROWNELL, 1973; YOUNG, 1974; CHADWICK and co-authors, 1977). For these reasons CLARK and BROWNELL (1973) and YOUNG (1974) cautioned against the siting of power stations on estuaries.

Estuarine power stations may cause greater damage to local aquatic organisms than open-coast stations (CLARK and BROWNELL, 1973). CARPENTER and co-authors (1974b) calculated that about 70% of the zooplankton entrained in Millstone Power Station were not returned to Long Island Sound (USA) and that this loss in biomass would lead to a loss in production of 0·1 to 0·3% of the zooplankton in the eastern part of the sound. Power stations also have a significant influence on the food web of an estuary as a result of entraining zooplankton in the cooling system (DAWSON, 1979).

The presence in warm-water discharges of planktonic organisms (including fish larvae), killed or disorientated during entrainment, was apparently at least partly responsible for attracting shoals of fish into the warm-water plumes where they fed on these organisms (p. 1811). Other fish may be attracted by concentrations of algae in foam at the effluent (p. 1811) or because of a preference for above-ambient temperatures, especially if the species in question is near the northern limit of its geographical range. Power stations, therefore, may have an affect on distribution, behaviour and food chains of fish in a receiving body of water, especially if the receiving body is of limited size, as in an estuary.

Fish that remain in the heated effluent may become acclimated to the high temperatures and are liable to suffer mass mortalities owing to cold shock if the power station suddenly ceases operation (p. 1813). Nevertheless, occasional cold kills may have little general ecological impact even though spectacular localised fish kills may occur (COUTANT, 1977).

Fish kills in power station discharges—although apparently very large—are small compared with the total stock, and sufficiently infrequent (HOAK, 1961a,b) not to affect the stocks as a whole. Thus losses of menhaden in Long Island Sound caused by heat shock had little impact on the fishery, but there may be serious consequences of projected increases in warm water discharges into the Sound (YOUNG and GIBSON, 1973). Power station effluents probably have a greater effect on relatively small localized fish populations, such as those of the winter flounder *Pseudopleuronectes americanus*, than on more widely dispersed populations (HORST, 1977).

Heated effluents have apparently helped certain species to extend their distributional range. For example, an unspecified species of fish extended the northern limits of its range in the USA as a result of overwintering in warm water discharges (COUTANT, 1977). In Southampton Water, thermal effluents apparently assisted *Mercenaria mercenaria* to establish itself in southern England (p. 1828). In addition, a number of tropical and subtropical organisms became established in the warm water of Swansea Docks (p. 1829).

In the vicinity of discharges warm waters may also exert indirect effects on the ecosystem if they destroy, damage or cause population changes in benthic plant communities. Such effects are most severe in warm-temperate, subtropical and tropical environments (p. 1819), especially where the receiving areas are of limited extent. Plant communities provide shelter and food for associated animal communities. Specific animal communities are associated with beds of the turtle grass *Thalassia testudinum* and forests of the giant kelp *Macrocystis pyrifera* (NORTH and ADAMS, 1969; BADER and co-authors, 1972; CLARK and BROWNELL, 1973; NORTH, 1977). Therefore, thermally induced changes in these plant communities may have secondary effects on associated communities. Deterioration or destruction of turtle grass beds in Florida, for example, resulted in a decrease in the numbers of individuals and species of animals, including some of commercial importance (ROESSLER, 1971; BADER and co-authors, 1972; CLARK and BROWNELL, 1973; THORHAUG and co-authors, 1973). The same applied to forests of *Macrocystis pyrifera* in California, where increased temperatures caused a deterioration of the canopies or destruction of the kelp. Animals that depended on *M. pyrifera* for food and shelter were forced to move out of the area (NORTH and ADAMS, 1969; NORTH, 1977).

Even the replacement of a dominant plant species by a more temperature tolerant one may influence animal communities. For example, the shift in algal community structure from diatoms to green algae to blue–green alga (p. 1801) probably had an adverse effect on algae-eating fish (p. 1838). Similarly, mats of blue–green algae in areas thermally denuded to *Thalassia testudinum* (p. 1816) provide neither shelter nor food for animals normally associated with turtle grass beds.

Animal numbers, however, do not invariably decrease in thermally damaged *T. testudinum* beds. Dead turtle grass provides food for detritus-feeding animals that move briefly into the area to feed and then move out leaving the area denuded of vegetation (THORHAUG and co-authors, 1973; THORHAUG and SCHROEDER, 1979).

Effects of heated effluents may be magnified because of the interdependence of species. Even if not directly lethal, temperature increases can disrupt reproduction and growth or reduce the food supply, thus causing imbalances in the ecosystem (QUICK, 1971a). In warmer latitudes, even a relatively small temperature rise may have this effect (see discussion in QUICK, 1971a).

In summary, organisms entrained in cooling water systems or in areas receiving thermally altered effluents may be variously affected by temperature changes. The raised or lowered temperature of the water system may exert a direct lethal effect, depending on the magnitude and duration of the temperature change, species or growth stage involved, season and latitude. Temperature changes may also act indirectly by reducing the organism's chance of survival through changes in metabolism and/or behaviour which increase its vulnerability to predation or to parasitic infection (e.g. KINNE, 1980; LAUCKNER, 1980). Increased temperatures sometimes stimulate increased metabolic and growth rates or early reproduction at a time of food scarcity and mass mortalities may result, in both adult populations and their offspring. As a consequence, the relative importance of certain species may be significantly altered in the immediate vicinity of an effluent or the species may cease to occur. Thermally altered effluents, therefore, sometimes cause permanent or seasonal changes in the population structure of a receiving area (INTERNATIONAL ATOMIC ENERGY AGENCY, 1974).

Temperature effects may further act synergistically with other stress factors in a cooling water system or at an effluent, producing combined effects more serious than those caused by a single factor. Factors such as chlorine, pressure or mechanical damage exert serious effects on their own. Chlorination apparently may cause more damage to organisms than temperature alone.

Biological damage caused by the effluent may have repercussions at higher levels of the food chain, especially if the affected organism is an important link in the food web or provides shelter for other organisms. Consequently, various methods have been tried, both voluntarily and as a result of legal action, to reduce potentially negative effects of heated effluents.

(5) Management of Waste-heat Impacts

In view of the effects reviewed above, management of waste-heat impacts is ecologically desirable and in many cases necessary. In principle, there are three main categories for managing thermal discharges into coastal waters and estuaries: (i) limitation of temperature deformations at source; (ii) reduction of unavoidable temperature changes before or as they are imposed on the environment; (iii) constructive utilization of warm or cold effluents prior to discharge.

All three categories are subject to considerable research efforts throughout the world. In recent years several conferences have dealt with various aspects in marine and freshwaters (see particularly INTERNATIONAL ATOMIC ENERGY AGENCY, 1974, 1975; LEE and SENGUPTA, 1979a; TIEWS, 1981).

Category (i) is particularly important with regard to waste heat. By far the main ecological problem, waste-heat involves wastage of about two thirds of the energy of conventional fuels, much of which is discharged to lakes, rivers, estuaries and oceans. The reviewers consider it important to examine critically all causes of heat release with a view to managing detrimental effects and reducing the production of waste heat.

Category (ii) concerns the management of technically unavoidable temperature changes, aiming at a reduction of detrimental effects. We consider here categories (i) and (ii). Constructive utilization is dealt with in the next section.

(a) Changing Emphasis from Environmental to Energy Conservation

There are two reasons for managing and regulating waste-heat release from power stations and other industries: to alleviate the effects on the surrounding environment and to conserve energy. In the past, when fuels were cheap, there were fewer incentives for their economical use. Waste-heat regulation was achieved mainly by legislation designed to protect the environment, but it was considered more expensive to install devices such as cooling towers to dispose heat to the atmosphere than to discard it directly to aquatic environments like lakes, rivers and estuaries.

As many modern power stations became much larger and required quantities of cooling water greater than the cooling capacities of most rivers, estuarine and coastal sites became increasingly attractive. It was assumed that this created little need to construct devices such as cooling towers since plentiful supplies of sea water provided a convenient way of discarding waste heat.

GORDON (1975) described the 1960s as the period when the energy supply industries were faced with numerous environmental problems such as the siting of power stations and oil refineries and concern about radioactive waste and other pollution resulting from the production, conversion and use of energy. Yet despite these environmental pressures, GORDON emphasized that the availability of cheap energy was taken for granted. The Department of Economic and Social Affairs of the UNITED NATIONS SECRETARIAT (1974) reported that no physical shortages of fuels were forecast on a global basis for the medium-term future, although it recognized that national and institutional controls of oil supplies were beginning to have an effect on prices and were creating serious foreign exchange problems for many developing countries. In 1973, however, certain oil-producing countries reduced their supplies and the Organization of Petroleum Exporting Countries (OPEC) introduced large oil price increases which have been repeated at intervals in subsequent years.

These events had a profound effect on the world's economy and were to highlight the great international importance of energy as a resource influencing all levels and costs of industrial production, including employment, living standards and environmental conditions (GORDON, 1975). Different attitudes about energy production and utilization are now prevalent among industrialized nations; energy conservation is the main incentive and reason for regulating waste-heat. According to CLARK (1975), this emphasizes how political situations concerning energy supplies can be beneficial by preventing the indiscriminate use of world energy resources. Thus, some countries have been forced to reconsider their programmes of electrical energy generation, particularly the types of fuel. Nuclear energy has become even more economically attractive, with the added attraction of being free from significant air pollution (GORDON, 1975), an aspect of particular importance close to larger cities (MIKOLA and co-authors, 1975). However the emotive issues of possible radiation hazards can provoke considerable public antagonism to the construction of nuclear power stations.

Now that the need for greater energy conservation has become more important for economic reasons, it is to be hoped that this will be for the lasting benefit of the environment. GORDON (1975) predicted that an increased transition in generating methods would result in changes in attitudes towards environmental protection, which could be limited to avoiding hazards to human health and assets of national importance, whilst protection on the grounds of amenity or aesthetics would be deferred.

(b) Improved Industrial Methods and Energy Use

Power Station Efficiencies

One way of using some of the fuel energy lost in cooling waters is to increase the efficiencies of power stations so that more of the fuel energy is converted into electricity and less is wasted to the environment. Small increases in thermal efficiencies can yield large financial savings and economic incentives for not wasting increasingly expensive energy are very considerable (p. 1774).

As older and less efficient generating plant is replaced by modern equipment, less fuel energy should be discarded with cooling waters. Increasing the unit capacity of electrical generating installations and reducing the share supplied by small-sized units are two of the principal ways of improving energy economy at present (LYBERG and co-authors, 1975).

Although increased efficiency results in better use of fuel and in a relative decrease in overall thermal pollution for a given generating capacity, the large size of plant means that more waste heat is discharged by individual power stations, with greater thermal pollution—possibly at fewer individual sites. This could provide better opportunities for the management and exploitation of waste heat.

Even with modern plants, about 60% of energy is still discarded, mainly to the aquatic environment, and with conventional steam turbines driving turbo-alternators efficiencies of about 40% are close to the maximum (FRAAS, 1975). There is now a great deal of research into new thermodynamic cycles and working fluids. Various alternatives are being investigated to increase the efficiency of fuel energy conversion. For example, magnetohydrodynamic (MHD) generation operates thermodynamically with a cycle similar to the gas turbine (SHEINDLIN and JACKSON, 1975); it has prospects of considerable improvements in efficiency over conventional steam plant. An electrically conducting working fluid is made to expand through a magnetic field producing an electrical output directly, due to motional electromagnetic induction. Reviewing MHD generation, SHEINDLIN and JACKSON (1975) described cooperative work between the USA and USSR, and reported the first pilot-scale installation on the outskirts of Moscow supplying electricity to the grid. They thought that MHD generation should be commercially available by the mid-1980s.

Both FUSHIMI (1975) and SHEINDLIN and JACKSON (1975) referred to MHD generation thermal efficiencies of 50 to 60% and these should result in impressive reductions in the heat discharged in cooling waters. The latter authors emphasized that this would produce smaller discharges to lakes and rivers or would require smaller cooling towers. Alternatively, once the heat absorbing capacity of a particular site had been determined, an MHD generating station could have a larger capacity than a conventional plant. There appear to be considerable prospects for reduced thermal pollution by future MHD generation. For example, VEERARAGHAVACHARY (1975) described a potential 500 MW (e) system in which the first 300 MW(e) was provided by direct MHD conversion and the remaining 200 MW(e) by conventional steam plant using 'waste heat' from the first stage. This would approximately halve the cooling water requirement and would require only 70% of the fuel compared with conventional 500 MW(e) plant. There appear to be possibilities for raising the conversion efficiencies of existing power stations from 35–40% to about 60% by using a MHD generation topping cycle (GEORGE, 1975). VEERARAGHAVACHARY (1975) suggested that an alternative to such binary MHD systems would be the MHD generator with gas (air) instead of steam turbines and this would eliminate the use of cooling water for steam condensation. Further, the hot combustion gases and hot air rejected from these air turbines would be at a higher temperature than steam turbine condenser cooling water and could be readily used for the generation of low-pressure process steam for heating purposes, desalination, etc. HENRY and FAZZOLARE (1979) described the use of a bottoming cycle to recover heat from a high-temperature gas-cooled reactor/gas turbine power system. Again, the potential advantages would be reduced thermal pollution and conservation of fuel for a given electrical output. The prospect of carrying this out with existing power stations might be very attractive.

DIETRICH (1975) forecasts that rapid introduction of new and more efficient systems of generating electrical power will be reduced by the development of fast breeder reactors which will burn the plutonium produced by present-day reactors, the useful period of which will be limited only by uranium resources. According to DIETRICH, generation by both light-water and fast-breeder reactors promises to solve the nuclear fuel supply problem permanently without exceeding the world's resources of uranium fuel. If thermonuclear energy is developed sufficiently it will be, by far, the most promising source of useful energy (BRAAMS, 1975). Most of the energy from fusion is carried by neutrons and must be recovered by means of a thermal cycle. Thus, the long-term prospects for the introduction of alternative and more efficient power generation with much lower thermal pollution may not be particularly good. It would appear that conventional steam cycles will continue to be used to generate electricity for a long time, and this will continue to require large quantities of cooling water.

Combined Heat and Power Generation

Possibilities for reducing waste-heat release to the environment that are arousing much interest nowadays involve combining district heating by steam, or industrial steam requirements, with the generation of electricity. This principle is known as 'combined heat and power generation' (CHP) (MARSHALL, 1979) or 'co-generation' (CONSERVATION COMMISSION OF THE WORLD ENERGY CONFERENCE, 1978). Considering both the world-wide energy crisis and environmental deterioration caused by waste heat, the best solution would be combined industrial plants that supply both electrical power and thermal energy (DZUNG and MÜHLHAÜSER, 1975). In district heating, thermal energy is supplied directly to households, eliminating the many small individual installations that are not only less efficient but create more atmospheric pollution because of their use of fossil fuels.

As already pointed out, even with modern generating stations about 60%, or more, of the fuel energy is discarded, mainly to the aquatic environment, with the condenser cooling waters. A conventional steam turbine has a steam outlet, at the condensers, with an absolute pressure of only about 0·05 atm (KARKHECK and POWELL, 1979). This produces maximum pressure differences between the steam inlet and outlet, resulting in more efficient operation of the turbine. The temperature of this outlet is normally about 38 °C, too low for district heating which would require enormous pipes and heat exchangers. However, it is possible to design steam turbines so that the exhaust steam provides a more useful temperature for domestic and industrial heating. According to MARSHALL (1979), there are two methods of achieving this: firstly, by using back-pressure turbines in which the steam outlet pressure is raised to about 1 atm and the steam passing through the turbine condenses at higher temperatures of 85 to 110 °C (Fig. 8-10), compared with about 38 °C for conventional plants. The second method is by the use of 'pass-out' turbines in which some steam is taken from the turbine at an intermediate stage, between the high- and low-pressure turbine stages, for use in district heating, whilst the remaining steam continues to expand through the low-pressure stage and is condensed in the same way as in a conventional turbine (Fig. 8-10).

Raising the condenser temperatures in these turbines reduces the efficiency of electricity generation alone from about 33% to about 25% (KARKHECK and POWELL, 1979), but the total energy utilization of the plant is greater (MARSHALL, 1979) than that of a plant producing only electricity because heat energy is being supplied as well as elec-

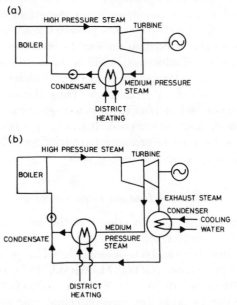

Fig. 8-10: Methods of obtaining heat and electrical power from a combined heat and power station. (a) Back pressure turbine; (b) pass-out turbine. (After MARSHALL, 1979; reproduced by permission of the Controller of Her Majesty's Stationery Office, London.)

Table 8-3

Useful energy (GJ) obtained in 3 different ways from 1 ton of coal with a calorific value of 26 GJ (After MARSHALL, 1979; reproduced by permission of the Controller of Her Majesty's Stationery Office, London)

	Conventional power station	Conventional industrial boiler	Combined heat and power
Electricity	9·1	—	6·6
Heat	—	20·8	13·8
Total	9·1	20·8	20·4
Fuel conversion efficiency	35%	80%	78·5%

ricity (Table 8-3). The loss in electrical efficiency is compensated by the gains in heating and the overall improvements in energy conversion using combined heat and power systems with back-pressure turbines are very considerable. For example, KARKHECK and POWELL (1979) quote overall efficiencies as high as 85% for fossil-fuelled power stations and 95% for light-water nuclear reactors in the USA. These estimates may be optimistically high in practical terms although similar estimates by MARSHALL (1979) suggest efficiencies as high as 80% (Table 8-3). CHP generation can result in very large fuel savings in the region of 28% (MARSHALL, 1979) compared with the fuel required to generate the same quantities of heat and power in separate conventional power stations. In an age of escalating energy costs this represents enormous economic advantages for the system and largely accounts for the present great interest. It also reduces waste heat.

The distribution of useful heat from CHP for district heating means the installation of expensive networks of pipes but most authors are of the opinion that the method is particularly worthwhile for areas of high-density housing (e.g. GERSDORF and SOMMER, 1975). The method is not new: the second largest system for district heating has been in operation in Philadelphia, USA, for more than 90 yr (CLYMER, 1979). In Europe, there is extensive use of CHP generation for district heating, and MARSHALL (1979) reported that it contributed just over 8% of Denmark's total heating load, 12% in Sweden, about 5% in the Federal Republic of Germany and about 9% in Finland.

Whilst CHP district heating appears to offer very considerable advantages in reduced thermal pollution and fuel savings, there can be problems in its use resulting in increased thermal pollution. CHP generation is efficient as long as the heating phase is required but if electricity alone is required then the efficiency is less than that of conventional plants. In estuaries and rivers with limited cooling capacities the greatest limitations for power generation are during summer when ambient temperatures are at their highest (GERSDORF and SOMMER, 1975). This is the time when district heating is not required in temperate regions and CHP stations would operate at low efficiency during the electricity-only phase in summer, with even greater heat loads being discharged to the environment than from a conventional power station. GERSDORF and SOMMER suggest that in such cases power stations could be provided with cooling towers so that the larger amounts of waste heat in summer could be discharged directly to the atmosphere. In winter the smaller amounts of waste heat, due to CHP operation, could be discharged directly to the ocean, estuary or river at a time when ambient temperatures were lower and when the use of wet cooling towers might be precluded if they caused precipitation with the colder atmosphere (p. 1852). Presumably, this arrangement would only be successful as long as greater efficiency gains from dual generation in winter more than offset reduced summer efficiencies caused by electricity-only operation. This would probably be the case in countries with prolonged heating seasons and where plant could be shut down for servicing during part of the summer.

The dual facility of CHP is best planned at the design stage of the power station since specially designed plants give greater flexibility for the extraction of larger amounts of steam compared with conventional power stations. Nevertheless, MARSHALL (1979) referred to one manufacturer of a large steam turbine plant who claimed that it is feasible to extract suitable quantities of steam from existing turbines, so that district heating is a possible way of reducing thermal pollution from existing conventional plant. This would mean less efficient electricity generation during the periods when steam was being extracted for heating, but in periods when this was no longer required, the power station could resort to full electrical capacity again. TIMMERMAN (1979) proposed a new system that would use the turbine condenser cooling water instead of the steam passing through the turbines. This may help by making it possible to use existing power station plant without much modification. The turbine condenser cooling water temperature is raised to the maximum possible with existing designs, about 38 to 49° C. This is still not high enough for adequate space heating, and TIMMERMAN (1979) suggested using heat pumps in the cooling water to allow district heating systems to operate at higher temperatures. This would also have the advantage of isolating the heating system from salt cooling waters of coastal and estuarine power stations, with which there could be corrosion and fouling problems.

Clearly, CHP district heating systems need to be located in areas where full advantage can be taken of the dual role. In many cases, they could be utilized in towns and cities in coastal regions. In estuarine localities, their use could do much to reduce heated discharges, perhaps in combination with cooling ponds or towers for periods when district heating is not required. Their efficient application could possibly result in the siting of more power stations in inland urban areas to take advantage of the heating, with reductions in numbers of proposed power stations on coastal sites where they might otherwise be placed to take advantage of plentiful supplies of cooling water.

Systems such as CHP with district heating would be most useful in countries of higher latitudes where there is a demand for space heating. In these regions, thermal pollution is less of a problem than in lower latitudes, because of lower ambient temperatures, although CHP instead of conventional plants on rivers, lakes and enclosed estuaries would do a great deal to alleviate thermal pollution there. However, the need for reduced thermal pollution is greatest in subtropical and tropical countries (p. 1803) and these are countries in which there is little requirement for CHP district heating for buildings.

Another aspect of CHP generation is producing steam for use in industrial processes rather than district heating (MARSHALL, 1979), and major process industries, especially paper, chemical and associated industries are the main users. In the USA, combined industrial power stations generating both steam and electricity did so with overall efficiencies of 60 to 75% compared with about 35% for conventional plant and using only half to two-thirds of the fuel (LEE, 1979).

There are many economic and technical problems to be overcome with CHP generation and the widespread application of the method will clearly require great cooperation between governments and the various sections of industry and communities, particularly in the planning stages of generating stations. In the interests of both environmental pollution and fuel conservation this must be encouraged as far as possible.

Energy Use

Official agencies in the USA are committed to a reduction in the rate of energy growth by improving utilization, reducing waste and reducing the use of less essential energy YAROSH (1977). This could have significant effects on the amount of generating capacity required to produce electricity, and on the amounts of waste heat discharged to the environment. There has been considerable progress both industrially and domestically, e.g. improvements in the standards of insulation of industrial and domestic buildings to cut down heat losses in temperate and colder climates and increased efficiency in industrial processes. A great deal of waste is incorporated into economic systems of industrialized and affluent societies, e.g. built-in obsolescence in industrial products, advertising to stimulate the urge to buy goods, widespread use of unnecessary packaging and inefficient use of the motor car (MURTHY, 1975). Changing attitudes towards energy use could help a great deal; it is a profligate society which, in winter, insists that living accommodation temperatures should not fall below 24 °C whilst in the summer they should not rise above 16 °C (CLARK, 1975). Fortunately, although the reasons for increased efficiencies in energy use are mostly economic incentives, the benefits for the environment, including reduced thermal pollution, could be considerable, especially in industrialised and developing countries where energy demands will increase in the future.

Desalination Techniques

With most desalination plants there appears to be little attempt to reduce the temperatures of the brine effluent before discharge to the environment. Although volumes are considerably less than those discharged by power stations (p. 1780, Table 8-2), temperature elevations can be higher and the salinity well in excess of ambient levels.

A recently developed method of desalination not only reduces the temperature and salinity of discharges, but also utilizes heated cooling sea water from industrial plants such as power stations and factories as the source of feed water (MOBERG, 1977). It employs the principle of the Torricellian vacuum in which sea water under vacuum at normal temperature and pressure can be supported as a column about 10 m high with a water vapour phase in the vacuum above the upper water surface. The principle is incorporated into an inverted U-tube acting as a siphon with the vapour phase at the top functioning as the evaporator. Heated sea water flows through the siphon maintaining the supply in the evaporator from which vapour passes to adjacent condensers which are cooled by sea water of ambient temperature. The system is satisfactory as long as the temperature elevation of the heated sea water is $7\,C°$, or more, above that of the ambient cooling sea water. An installation producing $1000\ m^3\ d^{-1}$ of freshwater will discharge $6000\ m^3\ h^{-1}$ of evaporator effluent with a ΔT of $2·8\ C°$ and $7000\ m^3\ h^{-1}$ of condenser cooling water with a ΔT of $3·6\ C°$. Mixing of the two discharges would result in a ΔT of less than half that of the original power station discharge, although the volume discharged would be more than twice that taken in from the power station. Lower operating temperatures compared with most conventional plants should mean greatly reduced corrosion and less metal contamination of the effluent. The salinity increase in the effluent should be very small, probably rising from 35 to about 35·24‰ S, a very considerable improvement on conventional desalination plants. The method may have great potential for use in conjunction with conventional power stations and could help to reduce ΔT values of power station cooling waters as well as reducing increased salinities and temperatures from desalination plants.

(c) Reducing Impingement and Entrainment

Most of the environmental problems associated with heated effluents have been with open-circuit or once-through cooling systems of power stations because of the very large volumes of water used. In the cooling system, before any effects due to thermal deformations occur, there can be problems with mortalities, particularly of fish, as a result of impingement on screens specially sited at the intake to prevent organisms being entrained. In some cases, very high mortalities of fish have been reported as a result of impingement (p. 1836). These problems are discussed particularly by CLARK and BROWNELL (1973) and in HOCUTT and co-authors (1980). Recent conferences have specialized in the problems (JENSEN, 1974, 1976, 1978).

HOCUTT (1980) has reviewed the literature on various behavioural barriers and guidance systems designed to repel fish from intake structures to prevent or reduce impingement and entrainment. The various methods include electric screens, air-bubble curtains, illumination, acoustics and changes in current direction and velocity. Their effectiveness has been shown to be very variable and careful siting of the power station may be one of the most effective methods of reducing impingement and entrainment (p. 1858). Impingement is one of the most important reasons why CLARK and BROWNELL (1973) recommended strongly that open-circuit cooling systems should not be used on estuaries where some organisms are frequently concentrated and which are often vital nursery areas for commercially important fish species (p. 1812).

Organisms that pass through intake screens are entrained in cooling water and this can result in damage due to chlorination, temperature shock, pressure shock and mechanical damage (pp. 1834–1837). Chlorination as a biocidal treatment is discussed on pp. 1888–1890 with respect to thermal aquaculture. It is usually carried out intermittently or continuously at low concentrations. These various entrainment effects are a further reason for not siting power stations with open-circuit cooling systems on estuaries because, in vulnerable areas, larval stages of important species can be drawn through the screens at the intake and may be killed by any of the above effects or by various combinations (CLARK and BROWNELL, 1973).

Thus, careful siting of a power station can do much to reduce impingement and entrainment effects. Non-thermal entrainment effects, such as pressure changes and mechanical damage, can be eased to some extent by careful design of the cooling water system to reduce these effects to the minimum. However, a great deal can depend on the rate at which cooling water is pumped because slower rates can reduce pressure changes around the pumps and mechanical damage due to impingement and shear effects caused by turbulence. This can cause a conflict: slower pumping rates causing smaller mechanical and pressure effects produce greater temperature effects due to slower flow rates through the condensers (COMMITTEE ON ENTRAINMENT, 1978b; SCHUBEL and co-authors, 1978; see also pp. 1866 and 1867).

(d) Power Station Condenser Temperatures

There are engineering and economic reasons for having the condenser cooling water as cold as possible to begin with, as this produces low steam outlet pressures in the turbines which then operate with maximum efficiency (WIUFF, 1977; KARKHECK and POWELL, 1979; p. 1774). Where cold ambient water cannot be obtained, low condenser temperatures are achieved mainly by pumping large quantities of cooling water, and this is costly (KERR, 1976a,b). For economic reasons such costs must be assessed against the benefits to be derived from greater turbine efficiency due to lower condenser cooling temperatures. Pumping costs can be important; CLARK and BROWNELL (1973) referred to a proposed power station in which cooling water was to be pumped a distance of 4 km. In order to reduce sizes and costs of pumps and pipework as well as the running costs of the pumps, the proposed temperature elevation was to be as high as 25 C°. Pumping large volumes of water can cause other problems, e.g. high fish mortalities on fish screens at the intake headworks. According to CLARK and BROWNELL (1973), fish mortalities are directly proportional to the volume of cooling water taken in and are just as likely to be caused by flow volumes as by flow velocities. Hence, the penalty for pumping larger cooling water volumes to maintain low condenser temperatures could be higher fish mortalities in addition to higher pumping costs (COMMITTEE ON ENTRAIN-MENT, 1978b; SCHUBEL and co-authors, 1978) (pp. 1866 and 1867). There are limi-tations, therefore, to the extent to which temperature elevations of cooling waters can be minimized as they leave the turbine condensers. Unless this heat can be reclaimed and utilized constructively it must be disposed to either the aquatic or terrestrial (atmosphere) environment, or both.

(e) Reducing Heated Effluent Temperatures Before Discharge

Various methods may be used to reduce temperatures of cooling waters before dis-charge to aquatic environments (Fig. 8-2): (i) dilution of cooling water with water of ambient temperature; (ii) cooling in special ponds, lakes, reservoirs or canals; (iii) cooling in wet cooling towers; (iv) cooling in dry cooling towers. Cooling devices such as reservoirs, lakes and towers can be used with both open and closed-circuit systems.

The majority of coastal and estuarine power stations, particularly in temperate re-gions, do not have water cooling before discharge but simply rely on dispersal of heat to the aquatic environment using open-circuit cooling systems discharging directly to the estuary or sea.

Dilution of Cooling Water

When there is fear of discharged cooling water damaging adjacent marine com-munities and where statutory limits may have been imposed on discharge temperatures, the water may be deliberately mixed with water of ambient temperature before discharge (Fig. 8-2b). HIDU and co-authors (1974) referred to this as 'tempering'. In the USA, mixing of cooling water with dilution water has been practised, for example, at power stations in Florida (SHAPOT, 1978; JOHNSON, W. J., 1979). Further north, in New Jersey, special dilution pumps are used to dilute cooling water when its temperature reaches 35 °C in summer (KENNISH and OLSSON, 1975). This still causes organisms, entrained in the cooling water and passing through condensers, to be subjected to temperature shock, but subsequent dilution and mixing with ambient water causes a rapid return to near-ambient temperatures. The technique is primarily a method of satisfying statutory obligations to protect the aquatic environment outside the confines of the power station. Dilution pumping can introduce problems of impingement and cause mechanical damage to entrained organisms (p. 1869).

Cooling in Special Ponds, Lakes, Reservoirs or Canals

These techniques are used where it is necessary to reduce temperatures of cooling waters before discharge (Fig. 8-2c,d). They are employed mainly with freshwater inland sites and also with power stations on estuaries and shallow bays. The methods have the disadvantage of requiring large areas of land which may be both expensive and in short supply, particularly in industrialized, urban and high-amenity areas. One estimate is of about 607 ha 1000 MW(e)$^{-1}$ (CLARK and BROWNELL, 1973). In some cases, open-circuit systems are used with these methods so that some of the waste heat passes to the atmosphere by radiation and evaporation before the cooling water is discharged to the aquatic environment at a lower temperature. An example is at Cedar Bayou generating station on Galveston Bay, Texas, USA; here cooling water of variable salinity is drawn along a 9·8 km long canal to the station, thence along a 9 km canal to a cooling lake with an area of 1053 ha, from which water then flows back to the sea (HOLT and STRAWN, 1977). Both outfall canal and cooling lake provide cooling to the atmosphere. A disadvantage is that entrained organisms will be subjected to elevated temperatures for longer periods than in an open-circuit system without canal or lake, discharging directly to the sea.

Ponds, lakes, reservoirs or canals are mostly incorporated in closed-circuit systems in which as much heat as possible is passed directly from the cooling water to the atmosphere before the water is recycled for further use. During operation, closed-circuits take in only about 2 to 4% of the water used by open-circuit systems (CLARK and BROWNELL, 1973). Most of this is used to make up water loss by evaporation and to adjust the quantity of cooling water. Closed circuits also discharge small quantities of heated water, called 'blow-down'. The smaller quantities of entrained organisms compared to an open circuit are most likely to be killed (CLARK and BROWNELL, 1973; REYNOLDS, 1980). In localities where cooling ponds are required it is likely that an alternative open-circuit system would probably kill most, if not all, of the entrained organisms; a closed-circuit system would kill a much smaller proportion.

Various cooling alternatives have been used during early stages of power station operations. For example, the Turkey Point power stations on the subtropical lagoon of Biscayne Bay, Florida, USA, originally had an open-circuit system discharging at 5 C° above ambient into the lagoon, which had an annual ambient temperature range of 9 to 35 °C (ROESSLER, 1977). Extensive damage was caused to the adjacent biota (p. 1818).

Hence the system was modified to provide dilution with equal volumes of ambient water to lower the temperature before discharge through a relocated outfall about 15 km further south. It was later replaced by a closed-circuit system in which cooling is achieved by a unique network of specially constructed inland canals (ROESSLER, 1977; for details consult HENDERSON, 1979). Condenser cooling water is discharged into 32 adjacent, parallel, shallow cooling canals, each about 8 km long, 61 m wide and 0·6 to 1·2 m deep. A collecting canal at one end returns water to the power station via another group of parallel canals, the complete cycle taking about 2 d. To control salinity, restricted amounts of make-up sea water are taken in and cooling water discharged, but apart from these, the system is essentially closed. The total cooling water surface is about 1560 ha for the 2320 MW(e) power station, equivalent to about 672 ha 1000 MW(e)$^{-1}$. This is similar to CLARK and BROWNELL's (1973) estimate of 607 ha 1000 MW(e)$^{-1}$.

One particular advantage of large cooling ponds and lakes is that in certain localities they may have considerable potential for development as sites for thermal aquaculture and recreation. A good example is the 648 ha freshwater lake specially built to provide cooling for an 833 MW(e) power station near Colorado City, Texas, USA (PETERSON and SEO, 1977). Recreational facilities are provided (swimming, boating, fishing), municipal water and rearing facilities for channel catfish *Ictalurus punctatus*. A 1093 ha lake near Rochester, Illinois, was one of the first specially constructed cooling lakes to be developed for recreation. The water flows about 8 km around the lake in about 3 d, providing cooling water for a 1200 MW(e) power station. Eight species of freshwater fish were reported to thrive there whilst part of the shore provided a winter haven for wildfowl. There appear to be few similar recreational developments with power stations using sea water although there are examples of water being used for thermal aquaculture (p. 1899). Pressures for development of specially constructed cooling lakes are greatest at inland sites because of cooling water shortages, but at vulnerable estuarine localities and open coast sites there may be considerable potential for similar facilities, although land shortages in coastal areas might create difficulties.

Cooling efficiencies of ponds, canals and reservoirs used with closed-circuit systems can be improved by the use of water sprays to increase evaporation. This method is employed by a number of estuarine power stations, particularly in the USA (CLARK and BROWNELL, 1973). Water is pumped through sprays above the pond surface, the droplets cooling rapidly during their flight through the air. The cooling rate depends particularly on the duration of droplet flight, wet-bulb thermometer temperature and wind strength. The last two factors are uncontrolled but spray ponds can be a very effective and economical means of dispersing waste heat. In large spray ponds the temperature of the cooling water can be as close as 8 to 11 C° to the wet-bulb temperature (BAIRD and MYERS, 1979).

An alternative to the spray pond is the spray canal. A good example is the Sacramento–San Joaquin delta in California, USA (CLARK and BROWNELL, 1973). Performances of spray canals have been discussed by PORTER and CHATURVEDI (1979). Disadvantages of sprays are the additional capital and running costs, and salt spray drift on to the surrounding countryside which can create terrestrial pollution. Although wide areas are sprayed, according to WEINSTEIN and co-authors (1979) the spray drift is probably fairly small at the source and less than that caused by the next type of cooling device.

Wet Cooling Towers

In closed-circuit cooling systems (Fig. 8-2d), wet cooling towers appear to be the most popular devices. Good descriptions are given in CLARK and BROWNELL (1973). There are two types: (i) natural draught and (ii) mechanical draught.

In a natural draught cooling tower, cooling water from the turbine condensers is sprayed on to rows of wooden slats near the base of the tower, over which the water flows and is cooled by evaporation. It then collects in ponds at the base of the tower and is recirculated for use in the condensers again. The upward natural draught of air in the tower draws air through inlets around the tower base and this greatly improves evaporative cooling. To achieve good draught, towers can be as high as 120 m (CLARK and BROWNELL, 1973). They are usually broad structures so that for considerations of visual amenity they form obtrusive features on a landscape, especially in low-lying coastal areas and they are expensive to build. They have the further disadvantage of creating salt spray when used with saline water, and in certain meteorological conditions of low temperatures and high humidities they create cloud plumes (CHAMPION and co-authors, 1979; PARK and co-authors, 1979). Wet towers consume large amounts of water owing to evaporative losses; this is particularly serious with limited freshwater supplies (DINSMORE, 1979). Visual obtrusiveness of wet towers can be reduced by using smaller mechanical draught towers with a height of ca. 21 m (CLARK and BROWNELL, 1973), in which airflow is enforced by large electrically powered fans. However, problems still exist with cloud, precipitation and salt spray. According to CLARK and BROWNELL (1973), natural draught towers produce the widest spread of salt spray but the salt fallout at some distance is very small. LASKOWSKI and WOODARD (1979) compared the two types, used with brackish and salt water; they concluded that natural draught towers had the least effect on the terrestrial environment. In some inland localities wet cooling towers are not suitable because of high freshwater consumption and shortage of water (DINSMORE, 1979). In such places a water reservoir is necessary for use during water shortages and DINSMORE suggests that the area required could be twice that of a cooling pond used to provide the only cooling source.

As with cooling ponds and lakes, cooling towers can be used in association with open-circuit systems to lower temperature before cooling water discharge. However, this does not reduce the volume of water drawn in from the cooling water source, unlike closed-circuit systems which take in only about 2 to 4% of the water volumes pumped by open-circuit systems. Thus, CLARK and BROWNELL (1973) recommend that towers and ponds should be combined with closed rather than open systems because of the much smaller volumes of water used. They also suggest that spray canals and ponds are better choices than wet cooling towers for brackish and salt water use because of greater salt spray localization, smaller visual impact and lower construction costs.

Dry Cooling Towers

Although these have been used in the chemical processing industry for many years (BELTER, 1975), their significance for the electricity supply industry has been recognized only recently. They can be used to avoid problems of local fog, rain or visible plumes. Heat from cooling water is transferred to the atmosphere via heat exchangers without direct contact between water and air. In dry towers the cooling effect of the latent heat of evaporation is lost. Hence they must be much larger than wet towers or the forced-draught fans must be more powerful so that construction costs exceed those for wet towers (MOORE, 1979). Whereas wet cooling towers increase generating costs of electricity by 5 to 6%, dry cooling towers increase it by *ca*. 20 to 25% (MESAROVIĆ, 1975). Dry towers consume virtually no water and produce neither fog nor heated effluent, but since they are larger they are visually more obtrusive than wet towers. Because of greater buoyancy of warm air plumes, dry cooling towers could generate local storms, providing there is a considerable concentration of generating capacity, in the region of 10 000 to 50 000 MW(e) as in a power park (HANNA and GIFFORD, 1975). Dry towers lose cooling capacity during periods of high ambient air temperatures. This is a problem particularly in warmer parts of the world (e.g. southern USA). When maximum electrical energy is required during warm seasons to provide for air conditioning, dry cooling towers are most inefficient. ENGLESSON and co-authors (1979) suggested a combination of wet and dry cooling for these situations: dry towers could be used most of the time, more efficient supplementary wet towers during warm periods of high electricity demand. In this way, the disadvantages of greater water consumption and undesirable plumes associated with wet towers would be restricted to relatively short periods. Indeed, during warm periods visible plumes from wet cooling towers would be less likely.

For freedom of choice, WIDMER and co-authors (1975) suggest direct ammonia cooling of condensers, with ammonia cooling in dry, atmospheric cooling towers. Where water is used to deluge heat exchangers containing ammonia during periods of high ambient temperature, ALLEMAN and co-authors (1979) propose ammonia as a heat exchange liquid between condensers and dry/wet cooling towers. Wet cooling is the most economic method in most circumstances, provided sufficient water is available to make up water loss. Only if this is limited and other environmental problems become significant can wet/dry cooling tower combinations become economic alternatives. In vulnerable aquatic environments such as some estuaries, dry cooling alone or wet/dry cooling would be the best methods, virtually avoiding or at least reducing both the amount of make-up water required and the discharge of blow-down water, compared with wet tower cooling in closed circuits. However, these facilities are more expensive to construct.

In recent years there has been considerable discussion in the literature about the relative merits of different cooling systems and whether open-circuit cooling systems are as bad as they are often said to be. Closed-circuit systems can present problems for marine and estuarine environments. Where wet cooling towers are used and make-up water is taken from salt-water sources, the salinity of the cooling water will gradually increase due to evaporative losses. The blow-down water from these systems can be discharged to the environment with salinities as high as 51‰ S and contain accumulations of chemicals including biocides (MINER and WARRICK, 1975). Furthermore, organisms taken in with make-up water will be almost certainly killed (CLARK and BROWNELL, 1973; REYNOLDS, 1980).

Some authors have suggested that when these aspects are considered with other problems—such as salt spray in the atmosphere, damage to surrounding vegetation, creation of fog conditions, consumption of freshwater at inland sites, decreased electrical efficiency and increased construction costs—open-circuit systems may have considerable advantages. According to MINER and WARRICK (1975), open-circuit systems at coastal sites in general cause less damage than closed-circuits to aquatic and terrestrial environments as a whole. In considering the risks of different systems, REYNOLDS (1980) sees great prejudice against the balanced development of cooling-system alternatives in USA. He suggests the imbalance could be corrected by changes of policy favouring open-circuit cooling and cooling reservoirs where conditions are suitable. However, such general statements need to be considered very carefully for areas where impingement and entrainment problems are likely to be serious and in warm regions where there is less scope for thermal deformations of the marine environment.

(f) Reducing the Impact of Thermal Discharges

Types of Discharge

Aspects of fate and dispersal of heated effluents are discussed on pp. 1788–1796. Here, we review various methods of reducing the impact on discharge.

The simplest type of discharge is from a large-diameter (*ca.* 3 m) pipe or from a discharge canal from which heated water flows to form a surface plume. In some cases, culvert pipes open vertically on the seabed from which heated water flows vertically to spread out on the sea surface.

Plumes of heated water from these low-velocity discharges may spread out over considerable areas of sea surface. They usually have little dilution near the discharge and can cause excessively high temperatures over large areas (STAUFFER and EDINGER, 1980). Regulations concerning the extent and magnitude of such plumes vary considerably. Some of the most stringent have been imposed by regulatory agencies in the USA. For example, the State of New York specified that outside an area of 1·5 ha (diameter 138 m) plume-temperature elevation should not exceed 0·8 C° in summer and 2·2 C° in autumn, winter and spring (BRODFIELD, 1977). Such restrictions are difficult to achieve with the simple type of outfall, and alternative designs have been built. Surface or submerged jets at the end of the discharge produce high-velocity currents that entrain some of the surrounding ambient water. This dilutes waste heat, producing lower temperatures over smaller areas. STAUFFER and EDINGER (1980) formulate time-excess temperature exposure (EDINGER and co-authors, 1974). The current velocity produced by the jet decreases with distance:

$$\frac{U}{U_o} = \left(\frac{X}{X_o}\right)^n \tag{1}$$

where U_o = velocity at the jet, X_o = discharge distance, X = distance along the jet centre line, X_o = 6 to 10 times the jet diameter, and $n = -0·5$ for a two-dimensional jet intersecting the bottom and $= -1$ for a jet that can expand both vertically and laterally. Distribution of excess temperature along the jet centre-line follows the equation

$$\frac{\theta}{\theta_o} = \left(\frac{X}{X_o}\right)^n \tag{2}$$

where θ_o = temperature elevation at the jet. Combining Equations 1 and 2 gives the following time–temperature relation:

$$t = \frac{X_o}{U_o(1-n)} \left[1 - \frac{(1-n)}{n} \ln\left(\frac{\theta}{\theta_o}\right)\right]$$

where $t = O$ when $\theta = \theta_o$ at the discharge. According to STAUFFER and EDINGER (1980) this relationship applies as far as the decaying jet velocity (Equation 1) becomes identical with that of the ambient current.

The only alternative that provides adequate rapid mixing with the surrounding water is the multiport diffuser (e.g. BRODFIELD, 1977; ADAMS and STOLZENBACH, 1979). Instead of a simple opening at the end of the outfall culvert pipe, the multiple diffuser consists of an extended pipe on the seabed, frequently several hundred metres long, provided along its length with a series of smaller jets that discharge heated water at high velocity to produce rapid entrainment and mixing with the cooler ambient water (Fig. 8-11). ISAACSON and co-authors (1979) give an example of two discharge pipes in California, USA, each with diffuser pipes 768 m long and 63 nozzles (nozzle diameter: 5·2 cm). Jets are designed to discharge at various angles to the horizontal. There are several designs of diffusers (e.g. ADAMS and STOLZENBACH, 1979) depending on aspects such as the orientation of individual jet nozzles, orientation of the diffuser pipe carrying the nozzles, diffuser pipe division into branches (e.g. T-shaped) and direction of natural water currents. Various mixing patterns can be achieved (Fig. 8-12). FISCHER and co-authors (1979) describe the design of ocean waste-water discharge systems, including multiport diffusers and thermal effluents, and PEASE and SKELLY (1975) discuss the design basis for submerged diffusers. BRODFIELD (1977) exemplifies diffuser effectiveness for rapidly cooling heated water upon discharge and compares it with the same discharge horizontally on the shore without diffusers (Fig. 8-13).

Restriction of these thermal mixing zones to the smallest possible areas is particularly important close to the shore where warm surface water can spread over shallow areas, possibly causing damage to the benthos. Multiport diffusers sited some distance from the shore can, to a great extent, avoid this particular problem (MILLER and BECK, 1975).

The rapid dilution of heated effluent is particularly important in temperate areas where large populations of fish may be attracted to an artificially warmed area and held there during periods when they would normally migrate to warmer waters during the autumn and winter months (pp. 1810 and 1871). While it is important to avoid power stations ceasing discharge during these periods, there are occasions when mechanical

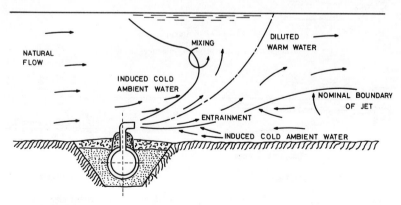

Fig. 8-11: Dilution of heated effluent by jets. (After BRODFIELD, 1977; reproduced by permission of the Institut de la Vie, Paris.)

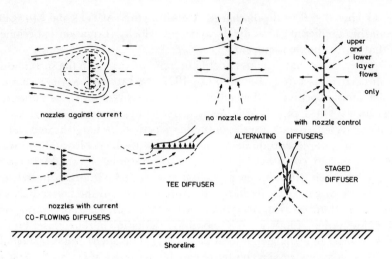

Fig. 8-12: Mixing patterns of heated discharges using multiport diffusers of various designs. (After ADAMS and STOLZENBACH, 1979; reproduced by permission of the Hemisphere Publishing Corporation, Washington, DC.)

problems may dictate otherwise. It is better to avoid attracting the fish in the first place, hence the importance of rapid dilution which will also help to avoid building up attractively high temperatures close to the preferred levels of fish (PILATI, 1976). The lower the temperature elevation the less chance there is of attracting a large fish population

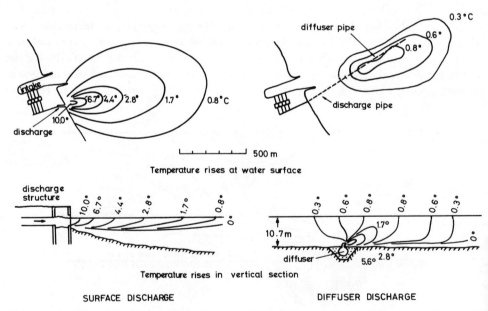

Fig. 8-13: Mixing patterns of heated effluents produced by surface discharge and diffuser discharge on the seabed. (After BRODFIELD, 1977; reproduced by permission of the Institut de la Vie, Paris.)

and the less chance of mortalities from cold shock in the event of a power station shut-down.

Where heated plumes sink, problems can arise owing to lack of atmospheric cooling at the sea surface. Heated water may become trapped between an upper layer of less dense water and either a lower layer of denser water or the sea bed; here it can spread to form a thin warm layer over an extensive area.

Negatively buoyant heated plumes from power stations (e.g. PANNELL and co-authors, 1962; CLARK and BROWNELL, 1973; GALLAWAY and STRAWN, 1974) are discussed on p. 1795. Although distillation desalination plants discharge much smaller water volumes than power stations, they produce the worst types of negatively buoyant heated plumes: these are warmer than most power station discharges, and have a very high salinity. The effluents can spread over extensive areas of the sea bed and exert direct detrimental effects on the benthos (p. 1832), both in intertidal areas (LEIGHTON and co-authors, 1967) and subtidal areas (THOMSON and co-authors, 1969; ZEITOUN and co-authors, 1969a,b). GOTO and HAKUTA (1977) emphasize the importance of diffusing brine when discharging into the sea. WADA and co-authors (1977) used simulation analysis with numerical models to predict the range of diffusion for both temperature and salinity increases and confirmed results obtained with hydraulic models and field surveys. Compared with a buoyant discharge the important difference is that in water of sufficient depth, the vertical jet of a dense effluent will rise, entraining and mixing with some of the surrounding water, but its upward momentum will eventually slow down and stop at a level some distance below the water surface owing to the neutralizing effect of gravity. The effluent then starts to fall back to the sea bed, entraining some of the ambient water from these shallower depths. With a vertical jet in a static ambient water body the effluent falls back towards the sea bed around the rising jet. Thus, the rising jet will only entrain descending effluent discharged slightly earlier, without entraining ambient water until it starts to descend. This effect will largely be avoided if there is an ambient current-flow past the vertical discharge or if the jet is set at an angle to the vertical.

Once the effluent returns to the sea bed it spreads out, still entraining some of the overlying water as long as there is sufficient flow to cause some turbulent mixing. A multiport diffuser with small nozzles produces much better mixing than a single vertical or horizontal discharge. ISO (1977) described such a system for desalination plants in Japan in which three separate types of nozzle were used. PINCINE and LIST (1973) designed a line diffuser for releasing a hypersaline waste to an estuary and suggested that this would dilute the effluent sufficiently, permitting safe disposal. Since it may be necessary to dilute desalination effluents by 4 or 5 times to reduce copper contaminants to supposedly safe levels of about $0 \cdot 02$ mg l^{-1}, this may provide sufficient dilution for other potentially harmful effluent properties, such as elevated temperatures and salinities (US OFFICE OF SALINE WATER, 1968). Mixing desalination effluents with sewage effluents has been suggested as one means of reducing temperature and salinity elevations (DEPARTMENT OF THE ENVIRONMENT, 1977). Desalination plants are sometimes installed close to power stations that provide power for desalting. In these cases effluent from both units can be mixed before discharge, usually to produce a buoyant plume, despite the increased salinity and because of the higher temperature, which spreads out on the sea surface, the higher salinity resulting in a layer of greater thickness than one from the power station alone (WADA and co-authors, 1977).

The other type of negatively buoyant plume is that with temperature levels below the ambient sea water temperature, produced as a result of using ambient water as a source of heat to vaporize the gas in liquid natural gas conversion plants (PARKER and UHL, 1975; p. 1771). In the USA, in the Los Angeles and Long Beach Harbor areas, unloading of liquid natural gas from Alaska was planned and the cold sea water discharge from heating the gas was to be 4 C° below ambient (REISH, 1977). HEINLE (1977) described a similar situation in Chesapeake Bay, Maryland, USA. Both authors stress that the best way of using such cold effluent is for cooling power station condensers; although according to HEINLE one power station could supply the necessary heat for all US liquid natural gas plants. Power station cooling would be a good method of avoiding the problem. Otherwise, multiple diffusers could provide adequate mixing with ambient sea water to avoid or reduce the spreading of cold sea water on the sea bed and thus creating unseasonal temperatures at certain times of the year.

Siting Considerations

It is important to determine the heat-receiving capacity of a power station site before deciding on location, particularly in enclosed areas. For minimizing detrimental effects of waste-heat discharges, particularly in estuaries, pertinent physical characteristics must be determined, such as patterns of current-flow, temperature, salinity and dissolved oxygen (JEFFERS, 1972). The distribution and magnitude of excess temperatures imposed by the power station must be predicted. Finally, criteria must be designed for the protection of the biota at the site.

For the marine environment, particularly estuaries, the best cooling systems are closed circuits, using towers, ponds or canals, which facilitate adequate treatment of blow-down or purge water. These permit siting of power stations almost anywhere on sea coasts and estuaries, provided other aspects like amenity and atmospheric pollution will not interfere (pp. 1850–1854). The purpose of the close-circuit system is to transfer heat directly to the atmosphere using only a relatively small quantity of water which is almost self-contained in the plant.

There are significant advantages for choosing open-circuit cooling: lower costs, lower water consumption compared with wet evaporative cooling towers and ponds (pp. 1850 and 1852) and visually less obtrusive. Open-circuit cooling is, therefore, likely to remain a first choice by engineers as long as they are permitted to use it.

There are considerable economic incentives for power stations to be located on estuaries because of the closeness to industrial areas around port facilities, which also provide access for bulk supplies of fuels such as coal, oil and liquid natural gas. Estuaries are national resources for a variety of reasons: they are important for port facilities and maritime access to the hinterland and for fisheries. Unfortunately for the environment, a large majority of the voting public and governments generally do not appreciate the biological importance of estuaries and the need to protect them (POLLARD, 1976). The reclamation of vast areas of biologically highly productive intertidal sandflats, mudflats, saltmarsh and mangroves is testimony to this attitude. All too often these areas are regarded by planners as being of supreme importance for industrial development, especially where land is scarce and expensive. There is no doubt, however, that for many ecological reasons critical pollution of estuaries must be avoided, including their uses as cooling water sources (e.g. CLARK and BROWNELL, 1973; CLARK, 1977). As a general principle, if ecologically vulnerable estuaries cannot be avoided as power station sites, closed-circuit cooling should be used.

Differing opinions have been expressed about the siting of desalination discharges in estuaries. One suggestion is that because these areas are subjected to considerable natural salinity fluctuations, they might provide good disposal areas (DEPARTMENT OF THE ENVIRONMENT, 1977). However, it is unlikely that many desalination plants would be sited in estuaries because of the proximity of freshwater.

Open coasts are generally the best sites for power stations with open-circuit cooling. Here, shallow bays should be avoided and areas chosen with reasonably deep water close inshore so that intakes may be located as deep as possible to take in cold water. In some localities advantage may be taken of nutrient-rich cold water from depth to supply aquaculture facilities after it has been warmed in power stations, e.g. Hawaii (GUNDERSON and BIENFANG, 1972), Virgin Islands (ROELS and co-authors, 1976), a technique that offers enormous scope for future development (p. 1883).

Prevailing wind direction can be an important factor because offshore winds carry heated water away from intertidal areas whereas onshore winds cause it to build up there for fairly prolonged periods (BARNETT, 1971). SUNDARAM and WU (1973) and YIH (1978) have studied effects of winds on buoyant surface plumes (p. 1791), and the former authors showed that wind-induced effects are among the most important factors affecting these plumes. With negatively buoyant plumes, particularly from desalination plants, the wind effect will be negligible, stressing the great importance of mixing the saline water with, say, power station effluent (WADA and co-authors, 1977) or sewage effluent (DEPARTMENT OF THE ENVIRONMENT, 1977) to make it buoyant or to provide good diffusers on the outfall to ensure good mixing with the surrounding water.

It is important to choose areas with good tidal mixing to help dissipate the heat by mixing it with ambient water. PINGREE and MADDOCK (1977) have stressed the value of headlands for this purpose for most effluents. Headland and offshore discharges are both methods of removing thermal mixing zones from more vulnerable shallow water areas.

There are very important geographical considerations in siting power stations because of the large amounts of heat discharged. It is clear from Section 4 that tropical and subtropical areas of the world are most vulnerable to thermal pollution. In general, marine organisms here live at temperatures which are much closer to their thermal death points than in temperate and colder regions of the world and there is far less scope for temperature increases. In these warm regions siting of power stations must be carried out with the greatest care. Even on open coastal sites it may be necessary to provide special cooling arrangements to reduce thermal discharges to a minimum and avoid entrainment.

The arguments against temperature elevations in warmer regions also apply to desalination plant discharges. In certain parts of the world, particularly the arid countries of the Middle East, the increasing prosperity from oil exploitation in recent years has seen very large increases in the numbers of desalination plants situated especially on the coasts of the Arabian Peninsula. Recently, about a third of the world's desalination plants were estimated to be in the Middle East, most of them in Kuwait (HARASHINA, 1977). Numbers of these plants will probably increase considerably in the future. For example, Saudi Arabia has about 20 desalination plants at present but could have about 30 more by the end of the century, some of them nuclear powered (SARDAR, 1981).

There is little information on the effects of the negatively buoyant plumes on the benthic fauna of these regions, which must be very vulnerable to the temperature and salinity elevations usually associated with most of the installations. However, concern has been expressed recently by the Arab League Educational, Scientific and Cultural Organisation (ALESCO) that future wastes from nuclear desalination plants will be dumped into the Red Sea and that the combination of cooling waters from desalination plants and mining operations could raise ambient Red Sea temperatures by 5 C° (SADAR, 1981). Perhaps there is scope in these regions for the desalination technique described by MOBERG (1977) which produces only small temperature and salinity elevations (p. 1847).

More efficient generating of electricity will do much to reduce the total quantities of waste heat but energy parks, using combined heat and power generation, will concentrate the impact locally, e.g. to approximately 10 to 50 GW(e) (HANNAH and GIFFORD, 1975) instead of the 2 to 3 GW(e) of large present-day power stations. Although generation efficiency should be much higher, about 60 to 70% instead of 30 to 40%, the concentration of waste heat will be considerable, requiring very careful siting.

A recent development is the proposed construction of floating nuclear power stations which could be moored off coasts and would use open waters as a heat sink (HANSON, 1974; ZECHELLA and HAMMOND, 1975). Presumably such structures would need to be moored behind protective breakwaters. Similarly, artificial multipurpose industrial port islands have been proposed on which power stations and desalination plants could be sited, in addition to other industrial projects (BIGGS, 1977). However, dense, heated effluents from distillation desalination plants would require careful mixing and dilution before disposal to prevent damage to benthic communities in particular.

(g) Criteria for Mitigating the Effects of Temperature Changes on Organisms

Temperature changes imposed on the marine environment should be considered and regulated very carefully in relation to the temperature tolerances of organisms, particularly of ecologically important species. Temperature changes fall into three categories: (i) elevated temperatures in cooling water inflicted on organisms by entrainment or by discharge to shallow coastal seas and estuaries; (ii) elevated temperature conditions caused by (i) to which organisms and communities become acclimated and which may suddenly disappear following, say, the breakdown of a power station, inflicting subsequent cold shocks; and (iii) lower temperatures, less than ambient, caused by the discharge of cold effluent.

Temperature Elevations

In an important paper, COUTANT (1972) considered the scientific basis for defining water temperature standards at power stations. He emphasized the difference between temperature criteria and temperature standards. Criteria are the ecological thermal requirements or limitations of aquatic organisms and these can be different from thermal standards, which are the legal restrictions applied to discharge temperatures to ensure that ecological requirements (criteria) are satisfied. Agreement between these two aspects can be extremely complicated. Organisms usually have different thermal requirements, depending on the species in question and the stages of development. It is well known that poikilothermic organisms have upper and lower temperature limits with optimum levels for growth, development, metabolism, reproduction, locomotion, migration, etc. (e.g. BRETT, 1960, 1970; KINNE, 1964, 1970b; COUTANT, 1972). These limits depend on the previous temperature history or acclimation of the organism.

In natural conditions thermal changes may produce temperature regimes either above or below optimum levels, perhaps fluctuating from one to the other, and thermal deformations must be judged against these natural fluctuations. In some cases, added heat may enhance biological processes, a situation exploited in thermal aquaculture (p. 1873). However, in other cases heat may be to the organism's disadvantage. Further, when heat has been discharged to an area for a considerable period the cold shock following a cessation of discharge may also have considerable adverse effects on the biota. COUTANT (1972) emphasized the problem of knowing how far artificial temperature changes can be superimposed on natural temperature fluctuations without having adverse effects. Because of the wide variations in natural temperatures for geographical, locational, seasonal, diurnal, tidal and other reasons it is important to appreciate that no single temperature limit can be applied over wide regions. The requirements must be closely related to the conditions in each body of water and to the ecological requirements of communities living in that water. For management of thermally deformed discharges, COUTANT (1972) further considers it important to ensure the maintenance of any natural seasonal thermal cycle; changes during spring and autumn should take place gradually.

Since, within the communities affected by thermal effluents, there is a range of species with different thermal requirements, a serious problem is knowing which species should provide the basis for management criteria. COUTANT (1972) suggests that important species could be distinguished as follows: (i) those of value to commercial or sport fisheries; (ii) those with a high biomass in the community; (iii) those significant as links in food chains of other important species; and (iv) endangered or otherwise unique species.

COUTANT (1972) has further made the following points. The thermal limits set should not be simply elevations above ambient since ambient temperatures may be above or below the optimum for a given species. It is better to set limits in relation to particular temperature tolerances of a species. Temperature criteria may need to be defined separately for the different thermal requirements of a particular species: (i) maximum sustained temperatures should be consistent with maintaining desirable productivity levels; (ii) maximum levels of metabolic acclimation to higher temperatures should not be so high as to prevent a return to the ambient winter temperatures if the artificial heating ceases, without causing damage; (iii) temperature–time limitations for survival following brief exposures to excessively high or low temperatures should not be exceeded; (iv) there should be restricted temperature ranges for various stages of reproduction; (v) there should be thermal limits for diverse species community compositions, particularly where important food chains might be altered or species with a nuisance value introduced. In addition, COUTANT (1972) suggests that thermal requirements of downstream organisms in rivers should be carefully taken into account. All these aspects must be considered in relation to the effects of non-thermal stress due to impingement and entrainment. Temperature changes resulting from thermal discharges may be imposed by (i) entrainment of organisms in the cooling water system; (ii) turbulent entrainment in the heated discharges flowing from power stations; and (iii) heated discharge flowing over shallow areas and affecting benthic organisms. It is, of course, very important to distinguish between these different effects when assessing the possible impact of thermal discharges because temperature changes can affect organisms in different ways. During entrainment, organisms are exposed to higher elevated temperatures for shorter periods than those living in shallow water areas over which heated effluent flows. COUTANT (1972) discusses two types of exposure in relation to the establishment of temperature criteria.

Short-term exposure

COUTANT (1971b, 1972) elaborated what has now become a well established framework for predicting thermal effects on entrained organisms. More recently, he developed this framework further in association with other workers (SCHUBEL and co-authors, 1978). The framework is based on the results of many laboratory experiments with a variety of different organisms, usually adults, and on a much smaller number of field studies at working power stations. COUTANT's framework relies on appropriate data for species for which predictions about temperature effects are required. There is a very large literature on the effects of temperature on marine and freshwater organisms (Volume I: KINNE, 1964, 1970b; KENNEDY and MIHURSKY, 1967; COUTANT, 1968, 1969a, 1970a, 1971a; BRETT, 1970; COUTANT and GOODYEAR, 1972; COUTANT and PFUDERER, 1973, 1974; RANEY and co-authors, 1973; BELTZ and co-authors, 1974; COUTANT and TALMAGE, 1975, 1976; TALMAGE and COUTANT, 1978). However, there is considerable variation in the experimental strategies adopted by different workers and often data are not appropriate for entrainment studies because most experiments are not designed to detect short-term effects. Further, with field observations it is difficult to separate thermal from other factors acting synergistically (p. 1831).

Based on FRY (1947, 1957a,b), different temperature levels have been used to assess the effects of increased temperatures on marine organisms. These levels have been discussed for entrainment by COUTANT (1970b, 1972) and SCHUBEL and co-authors (1978). The ultimate incipient lethal temperature is the highest temperature to which organisms can be exposed continuously for an indefinite period without increasing the mortality rate. The incipient lethal temperature can usually be increased according to the previous acclimation history of the organism. This was first shown in the elegant experiments with salmonids by BRETT (1952) (Fig. 8-14) and since documented for many invertebrates.

As an alternative to lethal temperatures, some authors have used the critical thermal maximum temperature, the level at which an organism's loss of equilibrium results in it being unable to escape from conditions leading to death. According to SCHUBEL and co-authors (1978), it cannot be used for predicting entrainment mortalities because there are difficulties in the determination of the temperature end points. Both temperature and time are involved and different heating rates have been used with organisms of different body sizes to ensure that the rate of temperature change is as uniform as possible within all the tissues of the organism. This can result in different heating rates for organisms of different sizes.

SCHUBEL and co-authors (1978) particularly warned against the use of the 24, 48 and 96 h exposure experiments for power station entrainment studies. They quote, for example, BURTON and co-authors' (1976) entrainment experiments with the mysid *Neomysis awatchensis* that showed no mortalities after 10 °C-acclimated individuals were exposed to 35 °C for 6 min, although HAIR (1971) reported that the 48 h LT_{50} for 22 °C-

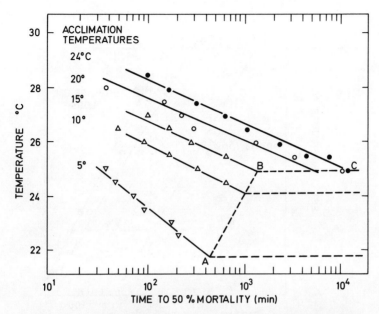

Fig. 8-14: Time to 50% mortality at elevated temperatures of organisms acclimated to different temperatures. For each acclimation temperature there is an incipient lethal temperature (broken lines) which is the highest temperature to which the organism can be continuously exposed without increasing its mortality rate. Line A–B represents the lethal temperature threshold which rises with acclimation temperature until it ceases at B–C, the ultimate incipient lethal temperature. (After BRETT, 1952, as modified by SCHUBEL and co-authors, 1978; reproduced by permission of Academic Press, New York.)

acclimated individuals was 25 °C. In another case, LAUER and co-authors (1974) concluded that the use of 24 to 96 h times in studies on the Hudson river would have indicated 100% mortalities whereas their experimental times showed 100% survival with nearly all their test animals.

Most of the data on thermal resistances are for fish (North American species listed by COUTANT, 1972), and COUTANT has emphasized that these usually follow dose–response curves of the type used in toxicity experiments. According to SCHUBEL and co-authors (1978), there are sufficient data in the literature for the suitability of the dose–response concept to be established for small invertebrates and larval stages; however, the ability to predict the survival of plankton following entrainment must depend on knowledge of the time–temperature survival data for many species and there is a great need for much more information on this aspect.

Since organisms are affected by increased temperatures before lethal levels are reached (p. 1814), debilitation can have considerable effects on survival. For example, pre-death debilitation in young rainbow trout *Salmo gairdneri* can result in increased vulnerability to predation at temperature levels below those causing death (COUTANT, 1973; Fig. 8-15). At higher sublethal levels there may be a loss of equilibrium (COUTANT and DEAN, 1972). Benthic animals in heated areas close to discharges could be similarly affected; some burrowing bivalves fail to burrow at elevated temperatures (ANSELL and

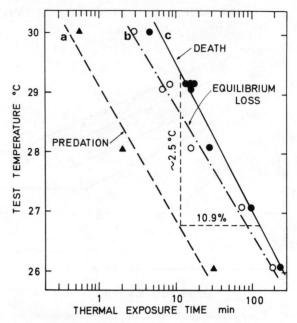

Fig. 8-15: *Salmo gairdneri*. Effects of acute thermal shock on a sibling group of juveniles acclimated to 15 °C. (a) Time to initial vulnerability to predation; (b) median time to loss of equilibrium; (c) median time to death. (After COUTANT, 1973; reproduced by permission of the Minister of Supply and Services, Canada.)

co-authors, 1980a,b). SCHUBEL and co-authors (1978) quote and discuss various results, and point out that the different levels for both increase in vulnerability to predation and loss of equilibrium follow dose responses similar to, but at lower levels than, those for lethal temperatures. Effects could be defined by temperature and time. FRY and co-authors (1941) and BLACK (1953) have shown that in some freshwater fish reducing temperatures by 2 C° below lethal levels, for a given exposure period, can eliminate mortalities. If COUTANT's (1973) results for salmonids are typical, the difference between the temperature level causing increased vulnerability to predation and the lethal temperature for the same exposure period is about 2·5 C°. This may provide a useful correction factor for lethal temperature levels when applied to power station impact predictions (COUTANT, 1972; NATIONAL ACADEMY OF SCIENCES, 1973; SCHUBEL and co-authors, 1978). MIHURSKY (1969) suggested the use of the 24 h median tolerance limit (LT_{50}) minus 2·2 C° as a method of estimating the upper temperature limit to protect a species in the field. Application of a correction factor to lethal temperature data may be more convenient because lethal temperature experiments are usually easier to perform than experiments to detect increased vulnerability to predation (SCHUBEL and co-authors, 1978).

Debilitation and disorientation are two aspects of the sublethal effects of temperature elevations but there are various others. SCHUBEL and co-authors (1978) have discussed these in relation to entrainment, emphasizing the possible importance of sublethal temperature effects for eggs and larvae by affecting cell and tissue differentation.

COUTANT (1972) and SCHUBEL and co-authors (1978) considered various ways in which temperature–time mortality data can be used to predict the impact of temperature elevations on the mortality of entrained organisms. A simple method (NATIONAL ACADEMY OF SCIENCES, 1973; SCHUBEL and co-authors, 1978) uses the temperature–time graph. For example, in Fig. 8-16 an intake temperature of 20 °C, a temperature elevation of 10 C° and an exposure (transit time) to 30 °C for 5 min results in a mortality considerably less than the 50% level that would be expected after 3 h exposure. Presentation of different levels of mortality on the graph would give an indication of the mortality, if any, that might be expected from given temperature elevations. At any particular mortality level, survival time at temperature extremes may be calculated from the following relationship (COUTANT, 1972; SCHUBEL and co-authors, 1978):

$$\log Time_{min} = a + b\,(Temp_{°C})$$

where a = intercept and b = slope. By rearrangement and setting the right-hand side of the equation to be equal to or less than unity we obtain:

$$1 \geq \frac{Time_{min}}{10^{[a + b(Temp°C)]}}$$

and the equation can be used to define the conditions for survival. The 2 C° or other (x C°) correction factor for sub-lethal effects may also be incorporated:

$$1 \geq \frac{Time_{min}}{10^{[a + b(Temp°C + xC°)]}}$$

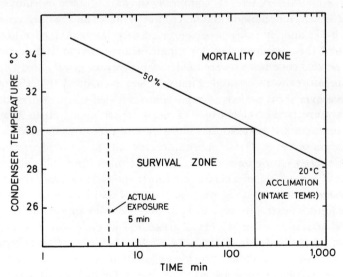

Fig. 8-16: Use of a thermal mortality line to predict whether or not mortalities of entrained orgnisms are to be expected. In this example acclimation and intake temperatures are 20 °C, temperature elevation 10 C°, and transit time 5 min followed by instantaneous cooling. (After SCHUBEL and co-authors, 1978; reproduced by permission of Academic Press, New York.)

Where cooling occurs in the discharge plume from a power station, the varying temperature conditions may be treated as a series of discrete temperature exposures (COUTANT, 1972; SCHUBEL and co-authors, 1978) for which the effects are additive:

$$1 \geq \frac{Time_1}{10^{[a + b(Temp_1)]}} + \frac{Time_2}{10^{[a + b(Temp_2)]}} + \ldots + \frac{Time_n}{10^{[a + b(Temp_n)]}}$$

The authors mentioned explain how these methods can be used by engineers to select safe combinations of temperatures and exposure periods in the design of power station cooling systems by the appropriate choice of pumps, discharge structures and mixing zones. COUTANT (1972) gives some very useful examples, with discussions, of how the criteria for temperature–time exposures may be applied to existing and proposed power stations, and how the various designs of cooling water system satisfy these criteria. For example, a long discharge canal or offshore pipe will prolong the period of exposure to higher temperatures by entrained organisms. Inadequate mixing of the discharged plume with the surrounding ambient water or the use of a large cooling pond prior to discharge will have the same effect of prolonging exposure for entrained organisms although permitting some cooling to the atmosphere before discharge to the nearby environment.

SCHUBEL and co-authors (1978) developed the application of temperature–mortality data to entrainment predictions by selecting data in the literature that satisfied three criteria: (i) ΔTs were applied instantaneously; (ii) mortalities were noted according to both temperature and time; and (iii) time periods extended from only a few minutes to about 2 h. They then accumulated the selected data into the four different environmental categories of salmonid rivers, lacustrine, estuarine and open coastal, of which only the last two will be considered here. By combining the data for 50% mortalities for fish eggs, larvae, juvenile fish, zooplankton and macroinvertebrates for the open coast into a single graph (Fig. 8-17) and by using separate graphs for the combined estuarine fish data (Fig. 8-18) and the combined data for estuarine zooplankton and macroinvertebrates (Fig. 8-19), several conclusions were made from this very generalized approach. Acclimation temperatures were ignored. Firstly, it was concluded that thermal death is a dose–response over short periods of exposure to elevated temperatures, although maximum temperature is the main cause of death. After about 20 to 30 min, mortality depends on temperature alone so that in the case of exposures in power station cooling systems in excess of about 20 min, mortalities due only to elevated temperatures should be predictable solely on the basis of the exposure temperature. This does not, of course, take account of synergistic effects of temperature interacting with other factors such as chlorination and various physical stresses. SCHUBEL and co-authors concluded that the temperature limit for exposure periods up to about 20 min appears to be about 27 °C for estuarine and coastal marine fish eggs, larvae and juvenile fish. For zooplankton and macrobenthic invertebrates it seems to be slightly higher so that criteria designed to protect various fish stages should provide adequate protection for invertebrates.

With heated effluents there are considerable difficulties in attempting to establish entrainment temperature criteria because of interactions with non-thermal effects within the cooling system due to biocidal chlorination and physical stresses imposed by pumping, turbulence and pressure changes. These aspects have been considered by the COMMITTEE ON ENTRAINMENT (1978a–c). The problem is that in order to maintain

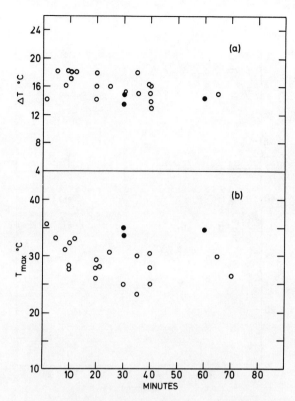

Fig. 8-17: Time to 50% mortality for combined data for open-coast fish eggs, larvae, juveniles, zooplankton and macrobenthos abruptly exposed to (a) temperature elevations, ΔT C°, being the thermal increment above the base (acclimation) temperature, and (b) the actual temperature experienced by the organisms, T_{max} °C. Data derived from experiments using different base (acclimation) temperatures. (After SCHUBEL and co-authors, 1978; reproduced by permission of Academic Press, New York.)

lower temperatures in the condensers to reduce thermal mortalities, larger volumes of cooling water must be pumped. However, with open-circuit cooling this means entraining more organisms, possibly causing greater mortalities due to the physical and chemical stresses of the cooling system. If smaller volumes of water are pumped, causing higher temperature elevations, experience shows that mortalities from chemical stresses decrease because fewer organisms are entrained.

In a detailed discussion, the COMMITTEE ON ENTRAINMENT (1978b) developed a model that takes account of thermal, physical and chemical stresses. The most commonly used measure of damage is f, the entrainment mortality fraction, which is the proportion of the number of entrained organisms killed. However, a better measure is the entrainment mortality rate R, the number of organisms killed per unit time (COMMITTEE ON ENTRAINMENT, 1978b). Accordingly, entrainment mortality rate may be

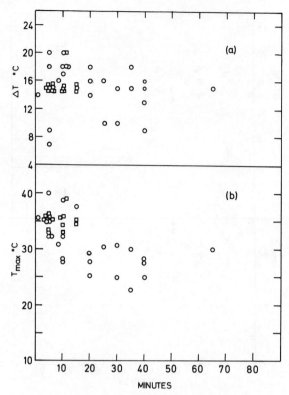

Fig. 8-18: Time to 50% mortality for combined data for estuarine ichthyoplankton. Details as for Fig. 8-17. (After SCHUBEL and co-authors, 1978; reproduced by permission of Academic Press, New York.)

expressed as a function of excess temperatures as follows:

$$R(\Delta T) = K(\Delta T)^{-1}\left[f_t(\Delta T) + f_p(\Delta T) + f_c(\Delta T)\right]$$

where $f_t(\Delta T), f_p(\Delta T)$ and $f_c(\Delta T)$ are fractional mortalities for temperature, physical and chemical causes, respectively, caused either directly, as in $f_t(\Delta T)$, or indirectly, as in $f_p(\Delta T)$ and $f_c(\Delta T)$. $K(\Delta T)^{-1}$ characterizes the particular power station and the population density. K is identically equal to $\eta H/PC_p$, where η = organism density, H = heat to be disposed, P = coolant density and C_p = coolant heat capacity. The best operating excess temperature is the one that is least damaging to the organisms and this is determined by the minimum R value. The COMMITTEE ON ENTRAINMENT (1978b,c) show that to cause the least harm to entrained organisms, power plants should not be operated at small ΔT values—lower than maximum biologically permissible levels—because this would expose the organisms to other entrainment stresses. Physical stresses caused by greater flow usually cause more mortalities than thermal stresses, so that ΔT values for particular power stations need to be chosen with care and may require to be changed according to season. The Committee's conclusions (COMMITTEE ON ENTRAINMENT, 1978b) are that at many plants the ΔT is well below the minimum desirable temperature to minimize R and should be raised by decreasing flow rates to

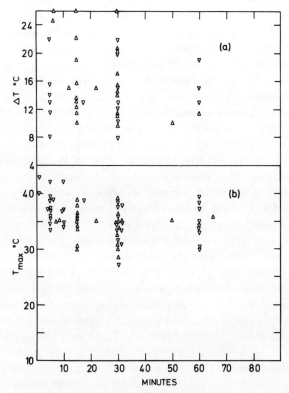

Fig. 8-19: Time to 50% mortality for combined data
for estuarine invertebrates. Details as for Fig. 8-17.
(After SCHUBEL and co-authors, 1978; reproduced
by permission of Academic Press, New York.)

reduce the numbers of entrained organisms. Care may be required during winter to
prevent damage from cold shock effects in the event of an unscheduled shut-down. The
Committee suggest that much can be done to reduce entrainment losses in power
stations. For example, timing maintenance shut-downs to coincide with periods of
abundance of important organisms would help. However, this would require a great
deal of cooperation between biologists and engineers who may be dictated to by more
direct economic considerations and pressures. Intakes could be designed to draw water
from various depths, chosen to avoid high densities of vulnerable organisms. Pumping
rates might be varied according to organism densities, perhaps diurnally, tidally or
seasonally, although this would require continuous and careful monitoring of populations by biologists. Dilution pumping (p. 1849) is not recommended because of the
increased physical stresses, despite being a useful technique to reduce heat stresses,
especially on the surrounding environment.

Long-term exposure

There are various sublethal effects of long-term temperature elevations imposed on
the environment adjacent to a thermal discharge (pp. 1807, 1814 and 1839). The most
important question is what are the effects of changes in the biology of individual organisms

on populations. At this level, the consequences of thermal changes become very difficult to analyse. For example, if organisms are induced to spawn out of phase with the natural reproductive cycle, what are the consequences for feeding, survival and growth of larvae on recruitment to the population? Some species, with benthic larvae that remain in a heated area, may have an adequate food supply, particularly if they are detritus feeders. In contrast, planktonic larvae, prematurely produced before a natural spring phytoplankton bloom, may face starvation in the plankton (p. 1822) and if they are carried out of an artificially warmed area will be subjected to low ambient temperatures, which, if not lethal, will delay development and prolong exposure to predation.

Criteria for temperature elevations may be applied in different ways. COUTANT (1972) lists the following three general aspects which should be satisfied in considering prolonged exposures: (i) a criterion for warmer seasons, (ii) another for winter and (iii) additional criteria for gradual changes required during reproductive periods. Most organisms have a limited range of geographical distribution according to the temperature tolerances for their various physiological functions. These functions usually have optimum temperature levels although this does not mean that organisms tend to live only at these levels; they may also thrive at higher or lower temperatures. COUTANT suggested that there must be some temperature between the optimum and upper incipient lethal levels that is limiting during prolonged exposure. Net growth rate is the most sensitive function; there is a temperature at which growth ceases, despite the presence of food. This is the temperature of zero net growth. There are examples where the temperature level for the sustained natural occurrence of animals is very close to the mean of the optimum temperature and the temperature of zero net growth. COUTANT suggests this as a useful method of estimating a weekly limiting mean temperature, on condition that the maximum temperature does not exceed the limit for short-term exposure. If it is applied to the most sensitive stages of the most sensitive species of ecological or commercial importance it should provide the temperature criterion for a discharge to a particular locality. Levels of zero net growth are not known for many species and the ultimate incipient lethal temperature might be taken as a provisional alternative in calculating any criteria, although in such cases it would be prudent to introduce a suitable safety factor.

For applying criteria to reproductive effects of long-term exposures to elevated temperatures (p. 1820) COUTANT (1972) lists the following: (i) periods of gamete maturation should be preserved; (ii) there should be no temperature barriers to spawning migrations, an aspect of particular importance in estuaries and rivers; (iii) temperatures should not be so high as to prevent spawning of winter breeding species; (iv) abrupt temperature changes should not be introduced into spawning areas; (v) where spawning is synchronous with a seasonal food supply, temperature changes should not alter the period of reproduction so that synchrony is broken. In some localities, these temperature criteria for reproduction may need to take priority over other long-term criteria.

In some localities it may be necessary to establish criteria to prevent the species diversity being altered and the introduction of species that may have a particularly adverse nuisance value (COUTANT, 1972).

Although there are considerable data available in the literature on temperature effects, some of them appear to be of limited value in establishing temperature criteria for the marine and estuarine environment and much more work is required. Particularly important and vulnerable areas appear to be tropical and subtropical regions of the world where heated discharges will undoubtedly increase in number as industrial development occurs.

Another type of heated discharge for which criteria are difficult to establish is the warm brine from desalination plants, particularly with respect to benthic organisms. The interactions of temperature and salinity make it so much more difficult to establish criteria. Although salinity has received the greatest attention as an environmental factor that modifies thermal tolerances (Volume I: KINNE, 1970b, 1971), the data available in the literature are more limited than those for temperature alone (Volume I: MACLEOD, 1971; GESSNER and SCHRAMM, 1971; KINNE, 1971; HOLLIDAY, 1971). The number of species for which criteria can be established is, therefore, smaller. Further, the areas of the world where most desalination plants operate (i.e. Middle East) are areas in which the fauna and flora are less well known. There is great scope for research in this field before satisfactory criteria can be established and applied.

Cold Shock Following Heated Discharges

This aspect is especially important in areas where fish may be attracted to warm water in large numbers (p. 1810). On occasions the ΔTs of heated discharges have been raised because power generation engineers have taken advantage of the colder ambient water in winter to pump less cooling water (COUTANT, 1977). If a power station has to shut down during winter, rates of temperature decrease can be very rapid indeed, especially if the cooling water pumps continue to operate; COUTANT gives an example of a discharge temperature of 26·5 °C dropping to only about 10·5 °C in less than 10 min.

Long-term exposures to elevated temperatures in winter should be such that the lower limit of the range of thermal tolerance represents the level at which there would be no significant mortalities of important species if the elevated temperatures were suddenly lowered to ambient temperature (COUTANT, 1972). The elevated temperature limit in such cases should be the acclimation temperature corresponding to the lower lethal threshold temperature equal to the ambient water temperature. This means that the temperature elevations during the winter may need to be limited unless the continuous discharge of heated water could be guaranteed, which would be virtually impossible because of the possibility of power station breakdowns. According to COUTANT, a possible way of ensuring this would be to build more than one power station on the site. However, even if the discharge of heated water could be guaranteed during winter, there have been cases where winter storms have broken up a heated plume so that fish that had been attracted to a heated discharge, and had remained there past the normal migration time, were killed in the ambient cold water (SILVERMAN, 1972). Most of the problems have been restricted to colder latitudes but the patterns of responses have been similar to natural cold kills that have occurred in waters of warmer latitudes. These are usually associated with the onset of cold weather conditions after fish have been attracted to abnormally warm water.

Three criteria should be applied to limit cold-shock effects (COUTANT, 1977): (i) lethal temperature limits; (ii) limiting loss of equilibrium; (iii) limiting increased vulnerability to predation. Most data are available for lethal temperature tolerances although the other two criteria are to be preferred for limiting cold-shock effects.

An important consideration in mitigating cold-shock effects is to avoid creating large areas of heated water in which fish can congregate (p. 1810). Careful siting of a power station discharge so that it can discharge into open coastal areas, preferably with strong tidal currents, should do much to ease the problem. Rapid mixing with ambient water by using jet diffusers will help greatly (COUTANT, 1977). This is because initial discharge velocities will be too great for fish to reside in that part of the plume, whilst further away rapid mixing with ambient water should create less warm conditions proportionately less attractive to fish. According to COUTANT, passing discharge water along a canal may lessen cold shock damage following a shut-down by allowing slower cooling of water to the atmosphere.

Cold Effluents

Cold effluent discharges superimposed on naturally occurring extreme low temperatures can have significant local effects on marine organisms (p. 1813) since exceptionally low natural winter temperatures have been shown to cause considerable and extensive damage to populations (e.g. KINNE, 1970b). This may be particularly true of benthic populations since cold effluent, being less dense than water at ambient temperature, will sink and spread out on the seabed. There is little information on criteria for cold effluents. COUTANT's (1977) criteria for cold shock following exposure to warm water should be applicable to cold effluent effects, but without the preceding acclimation to temperatures warmer than ambient. Lower temperature thresholds for loss of equilibrium, zero net growth and mortality could be used in a manner similar to upper thresholds for heated effluents (COUTANT, 1972; SCHUBEL and co-authors, 1978). Entrainment losses in cold effluents may be fairly small since relatively small volumes of water are involved compared with power stations (p. 1771). However, long-term effects on the adjacent biota may be more serious, particularly if little mixing occurs with the overlying ambient water, so that large reservoirs of cold water may become established on the bottom with significant effects for the benthos. Low temperatures could have particularly important effects on gametogenesis and spawning, perhaps even preventing reproductive development in spring and summer spawners, and criteria for cold effluents should take account of these aspects.

Careful siting and the use of high velocity jet discharges to provide rapid mixing should do much to avoid problems with cold effluents.

(6) Management of Waste-Heat Uses

It is unsatisfactory to take a natural resource, such as oil or coal, to use only about one third for making useful energy available, wastefully disposing of the other two thirds in a way that deforms living systems, whilst, at the same time, waste treatment systems are using other energy inputs (e.g. sewage) to produce virtually no usable products (DIAZ and TREZEK, 1979). Such poor management may have been more acceptable in an age of cheap energy but integration of different processes is desirable and essential for the future. Hopefully, the whole question of energy use will be gradually rationalized.

Engineers have regarded waste heat from power stations as a lost resource which should have other uses following discharge (NASH, 1969a; HAWES and co-authors, 1975). The justification for doing so on economic grounds was not particularly good in the past and there was slow development of satisfactory systems. More efficient utilization may result from escalating costs of energy production.

It is very important that there should be a great deal more cooperation between power station engineers and waste-heat users (KINNE and ROSENTHAL, 1977), particularly in the design and planning of power stations, in order to develop new methods of using waste heat. This may be difficult to achieve since the development of constructive uses is still in its early stages. Until effective demonstrations of sound and cost-effective uses can be made by developers there may be some reluctance by the power generation industry to become involved, especially if it involves modification of power station operating schedules. In addition, development of effective schemes is usually costly but investment of funds and provision of facilities by public bodies such as the White Fish Authority, Central Electricity Board and South of Scotland Electricity Board in the UK, Electricité de France in France and the Oak Ridge National Laboratory and Tennessee Valley Authority in the USA have been made to allow investigations of commercial possibilities for waste-heat utilization through construction of pilot plants for aquaculture and for horticultural and agricultural heating (see below). A particular problem is the dependence of such schemes on operating schedules and reliability of associated power stations, control of which may not necessarily be influenced by the requirements of the secondary user. Projects utilizing waste heat may require alternative methods of heating and circulating water in the event of a power station shut-down. Modern nuclear stations may have certain advantages in this area because they are usually designed to run continuously as 'base-load' stations in order to achieve maximum efficiency and this ensures a more continuous and reliable supply of heated water.

Methods of waste-heat utilization can be divided into four main types: (i) thermal aquaculture, (ii) sport fishing, (iii) agricultural and horticultural uses and (iv) combined heat and power generation to use a much higher proportion of the primary fuel. At present, thermal aquaculture is the largest user of waste heat and appears to offer the greatest potential for future development. Sport fishing is almost an incidental user, benefiting from the attraction to some thermal effluents of particular species of fish. Agricultural and horticultural uses are in the early stages of development and are mostly associated with power stations using freshwater cooling, although there may be considerable scope for power stations on marine sites, despite problems associated with salt water cooling. Combined heat and power generation providing district heating and industrial process heat, as steam, has been discussed in the previous section since it is a good method of regulating waste heat disposal to the environment by making more efficient use of primary fuel. It is a good constructive use for energy that would otherwise be lost as low-grade heat, using it at higher temperatures that are more useful for industrial and domestic heating purposes. Low-grade heat discarded to the environment has only very limited constructive uses, principally in the area of enhancing and controlling biological processes.

Various conferences have considered the question of waste-heat management and utilization (DEVIK, 1975; INTERNATIONAL ATOMIC ENERGY AGENCY, 1975; MAROIS, 1977; LEE and SENGUPTA, 1979a; TIEWS, 1981).

(a) Thermal Aquaculture

The best and most highly developed use of waste heat in cooling waters from industry is in thermal aquaculture, where it is used to promote the growth of commercially important species. Basically, there are two types of aquaculture: extensive and intensive

(Volume III: KINNE and ROSENTHAL, 1977). Extensive aquaculture is carried out in natural waters with little or no control of the environment and appears to have little use for heated effluents. Intensive aquaculture facilitates optimum utilization of heated effluents. As KINNE and ROSENTHAL point out, in highly industrialized countries the economic feasibility of extensive mariculture is bound to decrease in the future with increasing costs of labour and coastal land. Instead, the significance of intensive operations is more likely to increase in such countries. For temperate regions, KINNE and ROSENTHAL (1977) predict progressive application of greenhouse methods, comparable to controlled-environment agriculture. These are particularly appropriate for utilizing heated effluents from power stations.

Thermal aquaculture is a combination of three industries: aquaculture, geothermal heat utilization and waste-heat utilization (PETERSON and SEO, 1977).

In certain areas (e.g. Iceland, Italy, Japan, USA) geothermal energy is used as a heat source for thermal aquaculture (KOMAGATA and co-authors, 1970; YAROSH and co-authors, 1972; MILNE, 1975; JOHNSON, W. C., 1979; GIORGETTI and co-authors, 1981), solar heating has been used in Canada (AYLES and co-authors, 1981; VAN TOEVER and MACKAY, 1981), while oil-fired boilers (ANONYMOUS, 1980a) or electric heating (MØLLER and BJERK, 1981) have been used to provide heat for salmon rearing in Scotland and Norway, respectively. MEANEY (1973) discusses cases where direct heating with steam boilers has been used to advantage in aquaculture despite high costs. In northern Norway, KJØLSETH (1981) used sea water to raise the temperature of freshwater employed to rear Atlantic salmon smolts in 2 yr instead of the natural 4 to 5 yr, with considerable savings in fuel costs. The warm fresh water from turbine and transformer cooling systems of hydro-electric generating plants has been used in Norway for Atlantic salmon smolt production (KITTELSEN and GJEDREM, 1981). The large potential for this source of heating in Newfoundland and Labrador (eastern Canada) has been investigated by SUTTERLIN (1981).

While there are advantages in using heat for commercial aquaculture, oil price increases since 1973 have made some heating methods very costly, in some cases prohibitive, so that the use of waste heat from power stations is now even more attractive than previously (e.g. MILNE, 1976).

Use of Heated Effluent in Aquaculture

Heated effluents discharged by different industries have been utilized for aquaculture to a limited extent. In Japan, effluents from paper making, silk reeling, pharmaceutical and cement manufacturing industries supplied profitable commercial aquaculture enterprises selling highly priced fish such as eel and *Tilapia* species (CHIBA, 1981b). Eels have been farmed in heated effluent from a cement factory in England (ANONYMOUS, 1980b); in Scotland, heated effluents from whisky distilleries have been proposed (KERR and THAIN, 1977). Most of these heated effluents are freshwater; they are usually discharged at higher temperatures and with much smaller volumes than those from electricity generating stations (CHIBA, 1981b).

However, the power-generating industry has large numbers of plants with sea water effluents and these discharge very large volumes of water with relatively small temperature elevations, too low for space heating but excellent for aquaculture.

In several countries, a considerable amount of research has been carried out in recent years into the application of waste heat for aquaculture; e.g. France, FOULQUIER and DESCAMPS (1978); Federal Republic of Germany, KUHLMANN (1979a), KOOPS and KUHLMANN (1981a,b); Japan, TANAKA (1979), CHIBA (1981a); UK, NASH (1969a,b, 1970a,b), KERR (1976b), KERR and KINGWELL (1977); USA, MUENCH (1976), GODFRIAUX and co-authors (1977), HOLT and STRAWN (1977), PETERSON and SEO (1977), FARMANFARMAIAN and MOORE (1978), CHAMBERLAIN and STRAWN (1977), GUERRA and co-authors (1979a), HESS and co-authors (1979).

At least five advantages have been suggested, and in some cases proved, for using heated effluents in aquaculture: (i) reduction of detrimental effects of waste heat on the environment and some otherwise wasted energy can be used constructively; (ii) costs of providing seafoods can be significantly reduced; (iii) increased productivity can supply alternative seafoods as major fisheries become over-exploited; (iv) growth enhancement, including raising and rearing outside normal growth periods; (v) species from warmer regions can be cultivated in cooler areas.

Reducing detrimental effects of waste heat by constructive use

It has been suggested that thermal aquaculture may help to reduce the effects of waste heat on the environment. However, BELTER (1975) concludes from engineering analyses that it is unlikely beneficial uses of waste heat will reduce potential thermal pollution problems from electricity generating stations, and will be of little help in resolving siting problems. While reductions of waste-heat stress are ecologically desirable, in many cases it seems doubtful whether thermal aquaculture can make significant contributions, despite optimistic forecasts (e.g. MUENCH, 1976). Volumes of heated effluent discharged by modern generating stations with open-circuit cooling are much larger than those required by even the largest fish farms and aquaculture facilities are unlikely to provide much cooling. Instead, lakes specially constructed for cooling could provide facilities for aquaculture (e.g. HOLT and STRAWN, 1977; PETERSON and SEO, 1977; pp. 1850 and 1899).

Despite the small proportions of waste heat likely to be used for aquaculture, at least some of the energy otherwise wastefully discharged to the environment can be recycled and used constructively, thus making more efficient use of original fuel (MUENCH, 1976). In an age of steadily increasing shortages of most of the more commonly used sources of energy, such as fossil fuels, any method of utilizing constructively even a small proportion of waste heat is very desirable. Thermal aquaculture is becoming one of the best developed and most widely used of these constructive uses.

Reduction of seafood-production costs

Waste heat will probably make considerable contributions to reducing energy expenditures for providing seafoods in general and reducing the operational costs of aquaculture in the future. Collection of marine fish is costly in fuel (KINNE and ROSENTHAL, 1977). The fuel requirements of various marine fishing activities have been calculated (LEACH, 1976; EDWARDSON, 1977). However, it is difficult to compare these with fish obtained from mariculture operations because mariculture research has largely neglected considerations of the amount of energy required to produce food for human consumption (KINNE and ROSENTHAL, 1977).

While EDWARDSON (1976) made comparisons of gross energy requirements for various fish farming activities throughout the world, these were all related to freshwater

Table 8-4

Gross energy requirements (GER) per unit of production for various selected fish farms and fishing activities (After EDWARDSON, 1976; reproduced by permission of Fish Farming International)

Production categories	GER (GJ t^{-1} protein)	GER (GJ t^{-1} whole fish)
Carp		
Philippines, ponds (fertilization and feeding)	18	2
Germany, ponds (fertilization and feeding)	250	25
Catfish		
Thailand, ponds (feeding only)	523	56
USA, ponds (feeding only)	891	95
Milkfish		
Taiwan, ponds (fertilization and feeding)	52	7
Philippines, pens (feeding only)	9	1
Tilapia		
Africa, ponds (fertilization only)	5	0·4
Thailand, ponds (fertilization and feeding)	160	11
Trout		
UK, ponds (feeding only)	389	56
Herring		
Inshore sea fishing	26	3
Cod		
Deep sea fishing	309	30

(Table 8-4). A few of the extensive systems required less than 1 GJ ton^{-1} of fish, which is less than that estimated for inshore fishing. Some intensive types of aquaculture required more energy than that for distant-water deep trawling. This resulted mainly from supplementary feeding involving a high input of fish meal. According to EDWARDSON, reducing this input could achieve large savings and the lowest energy costs were recorded in cases without supplementary feeding. VAN TOEVER and MACKAY (1981) emphasized the energy intensiveness of aquaculture in North America; rainbow trout production required an energy subsidy of about 80 MJ kg^{-1} fish (80 GJ ton^{-1}).

While there appear to be no equivalent data for mariculture the evidence at hand from freshwater systems suggests that mariculture, particularly with carnivorous species requiring supplementary feeding, could be more energy expensive than inshore fishing and, in some cases, more so than deep-sea fishing. However, mariculture would appear to have considerable scope as a commercial means of providing greater quantities of fish for human consumption and thermal mariculture could play a significant role in this development.

The economics of marine fish farming have been discussed by PALFREMAN (1973), who gives the equation

$$C = a_1 xw + \frac{a_2 yt}{dh} + B$$

where C = annual cost per unit of output, a_1 = price of fish food, x = food conversion ratio, w = one unit of output, a_2 = price of variable capital, y = number of fish per unit of output, t = time period in which organisms are retained, d = stocking density, h = holding facility and B = other costs such as fixed capital and maintenance per unit of output. Some of these factors can be modified significantly by improvements in temperature conditions, particularly food conversion ratios and culture periods. Since energy costs of intensive aquaculture are high, especially with artificial heating, feeding, waste treatment and disposal, as well as pumping (KERR, 1976a; TANAKA, 1979), waste heat provides an alternative to costly fuel heating systems.

Despite the attractiveness of thermal aquaculture for reducing production costs, economic limitations must also be judged against other forms of culture, such as less costly extensive aquaculture and, in some cases, thermally cultured species might have to compete with fish obtained from ocean ranching (GUERRA and co-authors, 1976). Another important cost aspect is that as the technology develops waste heat will be regarded more as a chargeable resource by producers. It would be a mistake to assume that waste heat from power stations or other industries would in the long run be available without charges to the aquaculturist. At least two reports (ANONYMOUS, 1974, 1978) mention electricity-generating organizations charging for waste heat. However, a pollution-conscious industry should welcome and encourage constructive uses of its waste heat, and keep those charges at reasonably low levels. Certainly, providing waste heat free of charge or at only low cost will continue to do a great deal to encourage research and development in the commercial exploitation of this otherwise wasted resource.

Augmentation of food supplies

In Volume III, KINNE and ROSENTHAL (1977) discuss present and future contributions of mariculture to supplies of seafoods for human consumption (see also KINNE, 1982 and in press). In 1973, aquaculture provided about 7·5% of the total world fish catch. For the period 1974–84, increases in fish catches would probably increase supplies of fish protein by about 5% yr^{-1} whilst aquaculture protein supplies would increase by about 10% yr^{-1}, making increasingly significant contributions to supplies of seafoods. Thermal aquaculture could have an important role in this development. In 1975, the White Fish Authority in the UK assessed the sea farming potential of the country at about 50 000 tons yr^{-1}, plus about 8000 tons yr^{-1} using industrial waste heat (KERR and KINGWELL, 1977). Altogether this potential production represented about 5% of the total UK fish consumption and thermal aquaculture could clearly make a significant contribution.

KERR (1976a) considered the production potential of power stations for flatfish. He gives the production intensity (tons ha^{-1} of farm area) as

$$P = \frac{\rho i}{10d \, w_{f\,max}} = \frac{H}{A}$$

where ρ = tank stocking density (kg m^{-3}), i = intensity of stocking (%), d = tank depth (m), $w_{f\,max}$ = maximum live weight held to produce 1 kg at harvest (kg), H = annual harvest weight (tons) and A = area of farm (ha). The farm area consists of tank area plus the area occupied by buildings and services. Tank surface area would be as low as 30% of the total on small farms but as high as 50% on larger farms. While much depends on flow rates through the tanks, KERR calculated that the practical maximum production intensity, ρ, could lie between 59 and 98 tons ha^{-1}, although he suggested that a power station discharging about 100 000 m^3 h^{-1} could theoretically provide enough warm seawater to support the production of about 2000 tons of fish annually.

It is clear that increased productivity as a result of using waste heat with suitable species could make important contributions to supplies of alternative seafoods as major fisheries become over-exploited (e.g. MUENCH, 1976; Volume III: KINNE and ROSENTHAL, 1977). There appears to be considerable scope for the development of luxury seafood markets in more wealthy nations using heated effluents to culture more exotic, warm-water species in cooler, more temperate regions of the world.

Growth enhancement

In mariculture, organisms are often cultivated at ambient temperatures below those producing maximum growth. By raising and controlling temperatures through the use of heated effluents the growth of cultured organisms can often be increased dramatically, particularly with artificial feeding. For the USA, STICKNEY (1979) lists present and potential aquaculture species according to their temperature requirements for optimum growth. Cold-water species have temperature optima for growth at or below 15 °C, warm-water species at 25 °C or more and mid-range species between 15 and 25 °C. Many of these species are suitable for thermal aquaculture, including some of the so-called cold-water species such as the rainbow trout *Salmo gairdneri* (SAUNDERS, 1976). OPPENHEIMER and BROGDEN (1976) list temperature and salinity limits and optima for various species of fish, crab and shrimp in the Gulf of Mexico.

In the UK, power station heated waters were used to enhance growth and extend the growing seasons in plaice *Pleuronectes platessa*, sole *Solea solea* (NASH, 1968, 1969b, 1970b; KERR, 1976a,b) and turbot *Scophthalmus maximus* (KERR, 1976a,b; JONES and co-authors, 1981). Seasonal temperature variations in heated water supplies were adjusted to optimum growth temperatures by mixing effluent with suitable quantities of cooler ambient water (Fig. 8-20b). *S. maximus* body weights increased by about 1·3% d^{-1} at 10·5 °C but by 3·5% d^{-1} at 18 °C, a rate 2·7 times greater (Fig. 8-21; JONES and co-authors, 1981).

Temperature can affect growth rates in various ways; for example, feeding rates, food conversion ratios and growth efficiencies (KINNE, 1960; BOTSFORD and GOSSARD, 1978). However, optimum temperatures for these different functions may differ for any one species, e.g. *Scophthalmus maximus* (Fig. 8-21; JONES and co-authors 1981), *Oncorhynchus nerka* (BRETT and co-authors, 1969). In contrat, in channel catfish *Ictalurus punctatus* (ANDREWS and co-authors, 1972) and desert pupfish *Cyprinodon macularius* (KINNE, 1960) these temperatures were close together. It is important to know how these optimum temperatures differ for a particular species being considered for thermal aquaculture. Clearly, it is most convenient when temperatures coincide or are close together. When differences do occur, then choice must be made between the advantages of a rapid turnover of fish produced by using optimum temperatures for growth and the advantages of reduced food costs by using optimum temperatures for food conversion.

Food conversion ratios may be improved not only by choosing correct culture temperatures but also by the general development and improvement of feeds (WEATHERLY, 1976; STICKNEY, 1979). Many other aspects of fish biology affecting growth rates in heated systems receive treatment in works on general aquaculture (e.g. Volume III: KINNE and ROSENTHAL, 1977; STICKNEY, 1979).

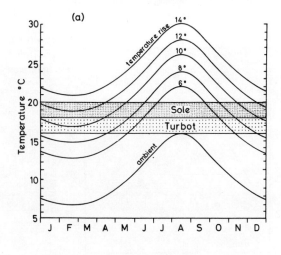

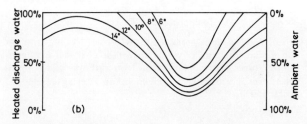

Fig. 8-20: (a) Seasonal temperatures in cooling water
produced by various power station temperature elevations, and their relation to the desired temperature
ranges for sole *Solea solea* and turbot *Scophthalmus maximus*. (b) Seasonal variations in the relative proportions
of heated and ambient water flows to achieve a tank
temperature of 18 °C. (After Kerr, 1976a; modified;
reproduced by permission of the Institute of Energy,
London.)

In many cases fish species with wide salinity tolerances are being cultured, including
some freshwater species being grown to advantage in heated brackish waters. Many are
listed in Volume III (KINNE and ROSENTHAL, 1977). European and Japanese eels
Anguilla anguilla and *A. japonica* are being used extensively because they are tolerant of
changing conditions in culture tanks and have high market values. With an optimum
growth temperature of 26·5 °C (KUHLMANN, 1979b) *A. anguilla* is well suited for
thermal aquaculture and is the most important fish species for this purpose in European
waters. It has been grown very successfully in pilot plants in heated fresh, brackish and
sea water effluents in a number of European countries, e.g. France (freshwater)
BALLIGAND and co-authors (1981), DESCAMPS and co-authors (1981); Germany
(brackish water) KOOPS and KUHLMANN (1981a,b), KUHLMANN and KOOPS (1981);
UK (sea water) JONES (1976, 1980). *A. japonica* is also reared successfully in heated sea
water effluents in Japan (CHIBA, 1981a,b).

Salmonids are receiving increasing attention for thermal aquaculture. Although

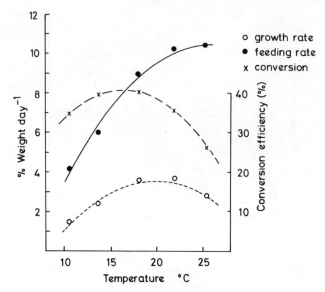

Fig. 8-21: *Scophthalmus maximus*: growth rate, feeding rate and conversion efficiency of young grown for 21 d at five different temperatures. (After JONES and co-authors, 1981; reproduced by permission of Heenemann Verlagsgesellschaft mbH, Berlin.)

rainbow trout *Salmo gairdneri* are suitable for farming in sea water (SEDGWICK, 1970) their use in thermal aquaculture appears to have been restricted to freshwater effluents (GUERRA and co-authors, 1976, 1979a; GODFRIAUX and co-authors, 1977; FARMANFARMAIAN and MOORE, 1978; ROGERS and CANE, 1981). Salmon, however, have been reared in heated seawater, e.g. *Oncorhynchus kisutch* grown to 'pan-sized' yearlings (PETERSON and SEO, 1977); development of fry and smolts shortened (e.g. SAUNDERS, 1976).

Some commercially important species may be reared outside their normal seasons of growth and reproduction. In the UK, the growth season for *Scophthalmus maximus* was extended throughout the year (KERR, 1976a, Fig. 8-20a). JONES and co-authors (1981) applied this on a commercial basis at a power station in Wales, although interestingly their work describes the anomaly of having to revert to ambient water entirely between May and December to avoid the risk of mortalities from chlorine-dosing of the cooling water system, which takes place at this time, thus losing the advantages of higher temperatures for growth during spring and summer. However, turbot were raised successfully in about 2 yr attaining marketable sizes of 1·5 to 2·5 kg primarily as a result of higher temperatures producing an extended period of growth. In Japan TANAKA (1979) compared the growth of yellowtail *Seriola quinqueradiata* in ponds of heated sea water with that in the sea (Fig. 8-22). When ambient sea water temperatures fell below 20 °C, heated effluent was mixed with water supplying culture tanks and growth in ponds was maintained during winter. Fish of 0·5 kg in early October grew to 2 kg by April compared with only about 1·25 kg at ambient temperatures. CHIBA (1981b) describes some culture ponds in Japan being supplied with heated effluent only during

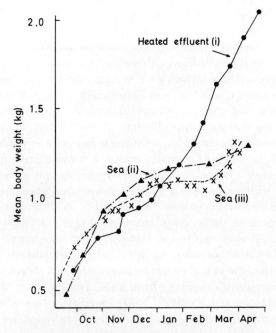

Fig. 8-22: *Seriola quinqueradiata*: growth of yellow-tail (i) in a pond warmed with heated effluent, (ii) in the sea (Japanese Prefectural Experimental Station) and (iii) in the sea (commercial culturist). (After TANAKA, 1979; modified; reproduced by permission of the Food and Agriculture Organization of the United Nations, Rome.)

the winter months. This method is used in thermal aquaculture particularly in cases where summer temperatures would be too high for a particular species with additions of thermal effluent.

In most cases of thermal aquaculture, the combination of enhanced growth on the one hand and raising and rearing outside normal seasons on the other is used to advantage, and there are many examples of this in different parts of the world. In Japan, where some of the earliest research on thermal aquaculture was started in the early 1960s, most power stations are sited on the coast and many commercial mariculture enterprises utilize the plentiful supplies of heated sea water having average temperatures of about 7 C° above ambient. Various species are reared in ponds, tanks and cages in the sea (TANAKA, 1979; CHIBA, 1981b). CHIBA lists seven examples of fish farms in various parts of Japan rearing at least six different species of fish, two of crustaceans and one mollusc. The species included *Seriola quinqueradiata*, *Mylio macrocephalus*, *Rhabdosargus sarba*, *Chrysophrys major*, *Anguilla japonica*, *A. anguilla*, *Penaeus japonicus* and the abalone *Nordotis discus hannai*. In the USA, there has been considerable research to exploit waste heat in thermal aquaculture ranging from *Mugil cephalus*, *Pogonais cromis* and *Micropogon undulatus* in sea water in Texas (BRANCH and STRAWN, 1978) to freshwater culture of *Salmo gairdneri*, *Ictalurus punctatus* and tropical freshwater prawn *Macrobrachium rosenbergii* in temperate waters of New Jersey (GUERRA and co-authors, 1976, 1979a; GODFRIAUX and co-authors, 1977; FARMANFARMAIAN and MOORE, 1978).

Mollusc culture in heated effluents is discussed by RYTHER and TENORE (1976). Bivalves have been reared successfully, although feeding remains a problem. With their low trophic levels they transform plant protein economically and can be grown at high population densities under controlled conditions (Volume III: KINNE and ROSENTHAL, 1977). Greater exploitation of spring phytoplankton blooms by *Mercenaria mercenaria* and *Crassostrea virginica* was due to heated effluents (ANSELL, 1969: HESS and co-authors, 1979). No winter growth occurred despite suitable temperatures. However, TINSMAN and MAURER (1974) found that *C. virginica* grown in warm effluent in Delaware, USA, grew faster during winter, developed higher flesh weights and glycogen contents than controls, with lower winter mortalities. Summer temperatures became too high, causing high mortalities and production of lower flesh weights. With bivalves, excessively high temperatures can result in depletion of stored food reserves (ANSELL and SIVADAS, 1973). LOUGH (1975) showed that temperatures at which maximum growth could occur may stress bivalve larvae abnormally, resulting in high mortalities. The importance of adequate feeding with algae at higher temperatures has been shown for various bivalve larvae, e.g. *Crassostrea gigas*, *C. virginica*, *Ostrea edulis*, *Tapes japonica*, *Mercenaria mercenaria* and *Mytilus edulis* (MANN and RYTHER, 1977). Successful commercial developments have taken place in the USA, especially with *C. virginica*, by feeding with algae at the larval stages and by taking advantage of the high productivity of coastal waters combined with heated effluents. Notable examples are at a power station at North Port, Long Island (MUENCH, 1976; PETERSON and SEO, 1977) and Moss Landing, California (RUTHERFORD, 1975), where growth periods before marketing are reduced considerably by warmer water. Mature oysters of marketable size were being produced in about 15 mo instead of 3 to 4 yr in California.

For organisms with short life cycles the advantages of enhanced growth rates can be used to shorten the growth period to marketing sufficiently to allow a second successive crop to be grown in the same facility but this time using elevated temperatures to produce growth outside the normal season. Some of the crustaceans appear to be particularly suitable for this technique. The production cycle of *Penaeus kerathurus* at a power station in Italy, for example, started with post-larvae in September developing to commercial size by April, followed by another production cycle during the summer yielding marketable sizes by September (PALMEGIANO and SAROGLIA, 1981).

In some species of fish the spawning season is altered. For example, red sea bream *Chrysophrys major* could be induced to spawn 2 mo earlier than normal, in February instead of April–May. In some cases, it was possible to repeat seedling production in the same ponds during the year because of the higher temperatures and shorter periods of growth. Seedling of larger size could be obtained and these produced better survival rates when planted out in natural waters compared with normal-sized seedlings (CHIBA, 1981b).

The use of elevated temperatures is often restricted to earlier stages of an organism's life cycle. Examples are *Solea solea* (PERSON-LE RUYET and co-authors, 1981), *Chrysophrys major*, *Mylio macrocephalus*, *Plecoglossus alteivelis*, *Penaeus japonicus*, *Nordotis discus*, and *N. discus hannai* (CHIBA, 1981b). Juvenile salmonids are suitable for rearing in heated waters. For example, in the Baltic *Salmo salar* fingerlings reared in heated brackish water reached the smolt stage as yearlings instead of the normal 2 to 3 yr using conventional methods in freshwater (TUUNAINEN and co-authors, 1981; VIRTANEN and co-authors, 1981). In northern Norway heated freshwaters from hydro-electric power stations were mixed with sea water to rear salmon smolts in warm brackish water (INGEBRIGTSEN and TORRISSEN, 1981). In the USA heated effluents accelerate the growth of five species of *Oncorhynchus* (CARROLL and co-authors, 1981) and of salmon fry for release (SAUNDERS, 1976). Ocean ranching uses heated water to reduce, from 18 to 6 mo, the freshwater stages of salmon smolts released to the ocean or pen-reared in the sea.

There appears to be considerable scope for using heated effluents for nursery culture of bivalves. CLAUS and co-authors (1981) described research with various species in Belgium, stressing the importance of applying sufficient food to spat grown at elevated temperatures. There are good examples of heated effluent being used in the early stages of oyster growth, e.g. *Crassostrea virginica* (RUTHERFORD, 1975; MUENCH, 1976; PETERSON and SEO, 1977; CARROLL and co-authors, 1981). A commercial facility in California uses heated effluent to produce seed of various bivalve species for planting out and takes advantage of local upwelling of nutrient-rich water stimulating good phytoplankton growth in a shallow tidal embayment (RUTHERFORD, 1975).

The deliberate use of deep nutrient-rich water for thermal mariculture may have considerable potential. Such water is cold and must be heated for mariculture use (see discussion in RAYMOND and co-authors, 1974). Its use with only solar heating has been demonstrated by the pilot project of ROELS and co-authors (1976) on the Caribbean Virgin Island of St. Croix. Intermediate Antarctic water was pumped from a depth of 870 m into shore-based ponds, heated by the sun, and diatoms grown as a food source for seven species of bivalves which rapidly grew to market size. Effluent from these culture tanks was used to grow spiny lobsters *Panulirus argus*, queen conch *Strombus gigas* and carrageenin-producing seaweeds. The main advantages of deep water, apart from high nutrient levels, is the absence of disease and pollution, making it ideal for mariculture. Artificial upwelling in a pipe would also avoid any dilution of nutrient-rich water with nutrient-deficient surface water. ROELS and co-authors (1976) anticipated that with large-scale systems, deep cold water at only 5 to 7 °C, compared with surface water of 26 to 29 °C, would provide sufficient temperature differential to permit desalination or generation of electricity by the CLAUDE (1930) process (HICKLING and BROWN, 1973), or by the planned modern OTEC (ocean thermal energy conversion) systems in a land-based form (p. 1922). After generating electricity, the deep water would then be used for mariculture. Another possibility is that deep cold water could be used for cooling purposes in conventional power stations where it would be heated in the turbine condensers before being used for mariculture (GUNDERSEN and BIENFANG, 1972). Where this could be carried out in tropical regions it could mean virtually no thermal pollution of surface sea water. Not many coastal states are situated conveniently close to deep water, but where they are the technique could have great potential. In Hawaii, for instance, surface waters are practically devoid of nutrients (GUNDERSEN and BIENFANG, 1972). Another suggestion has been to discharge the heated cooling water from power plants at depths where cold nutrient-rich water occurs and to use its buoyancy to entrain cold water in a vertical pipe and bring it to the surface by natural convection (MARTINO and MARCHELLO, 1968). The idea was to increase surface water productivity and hence fish production but the economics seemed to be doubtful and the method has not been tested. RYTHER and TENORE (1976) discussed integrated systems for mollusc culture using deep water, heated effluents and sewage effluents, and ROELS and co-authors (1978) tested a pilot-scale sewage-effluent/aquaculture/tertiary-treatment plant for shellfish and seaweed production. Thermal effluents were not used and it was thought that the method might be best used in warmer latitudes where the ranges of seasonal fluctuation in temperature and solar radiation are small. KEUP (1981) suggested mixing heated discharges with waste waters, including sewage, to maintain higher temperatures in treatment/aquaculture systems during the winter in temperate and northern climates. The integration of intensive aquaculture with waste-heat use and agricultural systems has some potential (CARROLL and co-authors, 1981), although the agricultural aspect may be more relevant to freshwater thermal aquaculture. For example, PETERSON and SEO (1977) described a geothermally heated water system being built in California which would raise catfish *Ictalurus punctatus* in raceways from which the water would flow to algal-rearing ponds, then to culture ponds containing tilapia, mullet, crayfish, freshwater clams and other filter feeders before being used to irrigate fields of the forage crop alfalfa.

Culture of warm water species in cooler regions

If tropical and sub-tropical species are being considered for thermal aquaculture in temperate regions, growth periods would probably have to be restricted to about 6 mo because, even with the use of heated effluents, winter temperatures would probably be too low for growth and survival. Supplementary heating during these periods would probably be prohibitively expensive. However, during the cooler season another, cooler water species could be cultured. The two advantages of being able to use heated effluents to grow organisms outside their normal season of growth, during the cooler months, and of being able to rear animals from warmer regions of the world in more temperate areas, during summer, can be used for what is now termed di-seasonal thermal aquaculture. Some of the greatest temperature extremes occur in freshwater and one of the best demonstrations of the technique was at the Mercer power station on the Delaware river, New Jersey, USA. Rainbow trout *Salmo gairdneri* were reared during winter and tropical freshwater shrimp *Macrobrachium rosenbergii* during summer (GODFRIAUX and co-authors, 1977; FARMANFARMAIAN and MOORE, 1978). A commercially more profitable summer crop was later shown to be the channel catfish *Ictalurus punctatus* (GUERRA and co-authors, 1979b). Although seasonal temperature extremes are greater in freshwaters than in the sea, the range can still be considerable in estuaries and shallow coastal waters in temperate regions and there may be considerable scope for di-seasonal thermal mariculture in the future.

The results of only a few demonstration projects have been published so far. Of particular concern is that adequate temperatures be maintained for warm-water species during periods of power station shut-down and for maintaining brood stock during the cold season. For the USA, STICKNEY (1979) lists present and potential species suitable for warm-water aquaculture. Further examples are penaeid prawns (CARROLL and co-authors, 1981) and *Penaeus japonicus* (TANAKA, 1979; CUZON and co-authors, 1981).

There is little scope for using heated effluents to culture warm-water species in the localities in which they occur naturally and thermal aquaculture has little value for tropical and sub-tropical regions (SYLVESTER, 1975).

Problems in the Use of Heated Effluents in Aquaculture

A number of important conditions must be satisfied if thermal aquaculture is to be successfully established at a site: (i) power station operating schedules must be suitable and heated water available over a long period; (ii) sufficient land should be available; (iii) ambient water conditions must be suitable; (iv) heated effluent should be of suitable quality; (v) aquaculture waste-treatment and disposal must be satisfactory; (vi) disease and stress should be at low levels; (vii) correct choice of culture organisms; and (viii) choice of suitable holding methods.

Power station operating schedules and water availability

Various authors have pointed out the importance of this aspect in thermal aquaculture (e.g. GAUCHER, 1968; ASTON and co-authors, 1976; HARVEY, 1979; ASTON, 1981; CHIBA, 1981b). The long-term availability of the heated water must be assured. Since modern power stations have life expectancies of about 25 to 30 yr (nuclear) and about 40 yr (fossil-fuelled) (ASTON, 1981), their heated effluents should be suitable for long-term aquaculture.

For economic reasons most large modern stations generate electricity as continuously as possible. These are called base-load stations (ASTON, 1981; NASH and PAULSEN, 1981) and they produce continuous, heated discharges most suitable for thermal aquaculture (ASTON, 1981). They are usually taken out of service for maintenance work for minimal and generally predictable periods. Other types of power stations may be used intermittently, either on a daily and/or seasonal basis, to provide electricity during periods of high demand. These are usually older and less efficient plants although occasionally modern stations using highly priced oil-fuel have been designed specifically for intermittent operation to satisfy peak demands (SOUTH OF SCOTLAND ELECTRICITY BOARD, 1978). These have irregular thermal discharges and are, therefore, of less interest for thermal aquaculture. Power stations in warmer regions and used only during summer—mainly to provide electricity for air conditioning—might have limited uses for thermal aquaculture due to already high ambient temperatures (STICKNEY, 1979). They would not be operational during winter when they would have greatest use for aquaculture. In contrast, in cooler regions both base-load plants and power stations used only during winter months of high electrical demand can be used for culturing some species. As long as the power station is operational throughout winter ambient temperatures for the rest of the year may be adequate for the continued culture of fish like sole and turbot (KERR, 1976a; JONES and co-authors, 1981). A number of factors can influence schedules such as maintenance, breakdowns and economic factors such as demand for electricity and cost of fuel. Seasons when maintenance shutdowns occur can vary; in temperate regions these occur mostly during summer when electricity demand is at its lowest, whilst in warmer areas maintenance periods are often in cool seasons because the high electrical demand for air-conditioning is during the summer months (STICKNEY, 1979). Seasonal and diurnal variations in demand for electricity can affect operating schedules and amounts and temperatures of thermal effluents being discharged. Changes in operations of power stations can cause short and long-term fluctuations in discharge temperatures (FORD and co-authors, 1979). Some power stations produce daily temperature cycles with changes of 6–9 C° (ALABASTER, 1963) and in many power station discharges constant temperatures are maintained for only a few hours at a time (ALABASTER, 1963; ANSELL, 1969; SPEAKMAN and KRENKEL, 1971; YEE, 1971). Problems might arise when, with little warning, economic or meteorological conditions reduce the demand for electricity resulting in periodic shut-downs of power stations, or engineering faults might require a shut-down at short notice for indeterminate periods. In such cases, without alternative sources of heating, there would be an interruption to growth and intolerant species may die (STICKNEY, 1979). Alternative heating might have to be installed to allow for these occasions unless the cultured organism has a wide temperature tolerance.

The implications of these temperature changes for aquaculturists have been discussed by various authors (e.g. McNEIL, 1970; SNYDER and BLAHM, 1971; YEE, 1972; HUGUENIN and RYTHER, 1974; FELIX, 1978).

Another factor of importance to a continuous supply of warm water is the number of generating units present in the power station (ASTON, 1981). For example, stations with four generating units are less likely to have them all out of use at any one time for repair than stations with only two units. This factor could prove to be a disadvantage with some of the large modern power stations which, for reasons of efficiency, are designed with much larger but fewer generating units than previously. Some new stations have only two units, each of 660 MW (SOUTH OF SCOTLAND ELECTRICITY BOARD, 1977).

Land availability

The question of land availability in mariculture has been discussed in Volume III (KINNE and ROSENTHAL, 1977). There must be sufficient land available for the establishment of a commercially viable mariculture farm. Where heated effluents occur on industrialized estuaries or open coasts of high amenity value, land can be both scarce and expensive and in such cases aquaculture facilities would need to be very profitable to justify the cost of highly priced land, probably requiring very intensive culturing of highly valued commercial species. The land would also need to be available for suitably long periods, preferably with scope for expansion if a profitable market developed (MUENCH, 1976). These aspects may impose considerable constraints on mariculture developments in such areas since fish farms occupy fairly large areas of land. KERR (1976a) analysed production potentials of marine flatfish at power stations and gives a method of calculating farm areas (p. 1877) depending on the production intensity required.

Ambient water

The ambient water at the site must be of suitable quality for cultivation (Volume III: KINNE, 1976). Diurnal, seasonal and sporadic changes in physical parameters must be determined. Large natural fluctuations in temperature and salinity could cause problems with culture facilities, possibly limiting choice of species. In extreme cases, cooling waters may be fresh during heavy rainfall and fully saline during dry periods (HOLT and STRAWN, 1977), thus restricting the choice of culture organisms to those with wide salinity tolerances. Fluctuating physical conditions, however, might have considerable advantages in keeping marine fouling to a minimum and reducing the necessity for certain antifouling measures in power station operating schedules.

Ambient water must be free of pollution and potential pollution (MUENCH, 1976). Environmental pollution in relation to general mariculture is discussed in Volume III

(KINNE and ROSENTHAL, 1977). Many power stations are unsuitable because they are situated on polluted estuaries. However, power stations requiring large quantities of cooling water and sited on open coasts provide plentiful supplies of pollution-free heated sea water very suitable for aquaculture.

An adequate supply of good quality water at ambient temperature might be necessary to dilute the heated effluent and provide greater temperature control (KEUP, 1981). To maintain certain culture tank temperatures varying proportions of ambient sea water need to be mixed with effluent depending on season (Fig. 8-20b; KERR, 1976a,b).

There may be occasions when the quality of the power station cooling water deteriorates to such an extent that the supply to the aquaculture facility has to be stopped. If the supply cannot be restored within a reasonable period then an ambient supply may be needed to maintain the stock in good condition, albeit at a lower temperature. However, costs to aquaculturists of pumping ambient water would be greater than for heated effluent (KERR, 1976a; see also p. 1900).

Water quality of heated effluent

Water quality is a vital factor in the location of an aquaculture facility (NASH and PAULSEN, 1981). Assuming that the ambient water is suitably free of pollution, treatment of that water in the power station, when used for cooling, can cause problems for thermal aquaculturists. A comprehensive review of the chemicals likely to be found, and of the toxicities, is by BECKER and THATCHER (1973), whilst RYTHER and TENORE (1976), ASTON (1981) and NASH and PAULSEN (1981) briefly reviewed water quality particularly for thermal aquaculture. ASTON advised any potential fish farmer contemplating the use of heated effluents to consult BECKER and THATCHER's review, which suggested that there are chemicals associated with thermal effluents that have a potentially harmful effect (SAROGLIA and co-authors, 1981). However, MUENCH (1976) emphasized that the water must be low in concentrations of heavy metals and chlorinated hydrocarbons. There is no universal chemical composition to power station cooling waters, which is usually unique to each situation (BECKER and THATCHER, 1973). Various chemicals are added for neutralizing water, for pre-operational cleaning, for preserving the strength of the wooden slats in the cooling waters of closed-circuit cooling systems as well as various chemicals to control scaling and corrosion of condenser tubes and biological fouling of the cooling system. Most thermal aquaculture is carried out with open-circuit cooling water systems, although closed circuits have been used on occasions (ASTON and co-authors, 1976), mostly in freshwaters.

There are two main sources of contamination of heated effluents from power stations: (i) chemical treatment of the cooling water to reduce corrosion, scale formation and fouling and to maintain the efficiency of the system; (ii) various discarded wastes from the power station that, for convenience, may be passed into the cooling system for discharge.

Chemical and mechanical treatments. Chemicals are often added to cooling water to control corrosion of turbine condenser tubes. Ferrous sulphate added to the cooling water of power stations deposits a thin film of protective iron oxide on the tube surface which inhibits corrosion (EFFERTZ and FICHTE, 1976). The short-term doses at concentrations of about 1 ppm, about 2 or 3 times per day, should mean that this need not be a serious problem for aquaculture because of the high dilution factors. Phosphonates and polyphosphates may be added to inhibit formation of scale on condenser tubes

(McCoy, 1974) and polyol esters to keep precipitates of calcium and magnesium salts in suspension (MATSON, 1977), particularly with closed-circuit cooling systems. With open-circuit systems, more likely to be used for thermal aquaculture, dilution factors usually prevent formation of scale and there is less need for chemical treatment than with closed systems.

Biological fouling consists principally of (i) bacterial slimes and (ii) sessile animals such as mussels and barnacles. Bacterial slimes form particularly in condenser tubes, seriously affecting power station efficiencies (ROENSCH and co-authors, 1978). The planktonic larvae of mussels and barnacles settle on the surfaces of cooling systems where they attach and grow, causing serious problems by reducing culvert pipe diameters, restricting flows of cooling sea water and causing metal erosion and corrosion of condenser tubes (HOLMES, 1970a,b; COUGHLAN and WHITEHOUSE, 1977).

A variety of biocides are used to control biological fouling but chlorine is the most common (BECKER and THATCHER, 1973: MORGAN and CARPENTER, 1978; NASH and PAULSEN, 1981). It is usually injected as gas but sometimes as hypochlorite (MARTIN, 1937; ESTES, 1938). Since high chlorine levels cause fish mortalities (ROBERTS and co-authors, 1975) care is needed when using chlorinated heated effluents for thermal aquaculture.

Reviews of chlorination of power station cooling waters, both marine and freshwater, and its biological effects were published by BECKER and THATCHER (1973), WHITEHOUSE (1975) and MATTICE and ZITTEL (1976). ZEITOUN and REYNOLDS (1978) give a recent general account of chlorination in power stations. MATTICE and ZITTEL (1976) document the toxicities of chlorine to various species of marine organisms, repeated, with modifications, by SAROGLIA and SCARANO (1979). Various aspects of the environmental effects and fate of chlorine in coastal waters, some of which are of interest to aquaculturists, have been discussed at recent conferences (JOLLEY, 1976; BLOCK and co-authors, 1977; JOLLEY and co-authors, 1978). NASH (1974a), ASTON (1981), NASH and PAULSEN (1981) and SAROGLIA and co-authors (1981) discuss chlorination and thermal aquaculture.

The fate of chlorine in sea water is extremely complicated. When injected it dissolves rapidly, hydrolyzing to form hypochlorous and hydrochloric acids. NASH (1974a) defines this as the 'free available chlorine residual'. Some combines with organic and inorganic compounds and certain dissolved gases, termed the 'combined available chlorine residual' (NASH, 1974a). It combines with ammonia to form chloramines (BECKER and THATCHER, 1973), with its bromine counterparts and with organic compounds containing nitrogen (CAPUZZO and co-authors, 1977; JOHNSON, 1977). However, there is still considerable uncertainty about how chlorine reacts with sea water chemicals, often making it very difficult to interpret which toxicants are being assayed (BLOCK and co-authors, 1977; JOHNSON, 1977). According to NASH and PAULSEN (1981), the amount of free chlorine as hypochlorous acid determines the effectiveness of the treatment. Measured residuals may not be chlorine alone but a variety of different compounds which, for convenience, are frequently referred to simply as chlorine (NASH, 1974a).

Chlorine is injected into cooling waters either periodically or in continuous doses (BEAUCHAMP, 1969; CAPUZZO and co-authors, 1977; ASTON, 1981; NASH and PAULSEN, 1981). The frequency, concentration and duration of dosing can vary considerably and is of great importance for thermal aquaculture. Intermittent or 'slug' doses are used at most power stations (NASH and PAULSEN, 1981), particularly at inland sites and at stations with closed-circuit cooling systems, to control bacterial slimes. At inland stations in the UK intermittent doses with residuals of about 0.5 ppm are given for periods of 10 to 20 min every hour (ASTON, 1981). In the USA intermittent chlorination for 2 to 3 h d^{-1} is commonly used (BECKER and THATCHER, 1973) and chlorine residuals ranging from 0·05 to 5·00 ppm have been reported (MARSHALL, 1971; BRUNGS, 1973) in efforts to remove mussels and barnacles from condenser tubes. Mussel fouling is a particularly serious problem with sea water cooling systems (BEAUCHAMP, 1969; HOLMES, 1970a,b; BECKER and THATCHER, 1973; ZEITOUN and REYNOLDS, 1978) and even high levels of intermittent dosing are ineffective because mussels close their shells until the dosing period has passed. To control mussel settlement, coastal and estuarine power stations in the UK commonly use continuous low-level chlorination from March to November (JAMES, 1967; BEAUCHAMP, 1969; HOLMES, 1970a; JONES and co-authors, 1981). Chlorine is usually injected to give an initial concentration of 0·5 ppm and residual of about 0·2 ppm (ASTON, 1981), although residuals ranging from 0·02 to 0·3 ppm (NASH, 1974a) and 0·02 to 0·05 ppm (JAMES, 1967) have been reported. The continuous low levels are not intended to be lethal but to deter the settlement of larvae and to irritate any newly settled mussels so that they move along the culverts and condenser tubes (HOLMES, 1970a,b). It is possible for even these low chlorine concentrations to stress and kill organisms used in thermal aquaculture. Adequate chlorine control methods or alternatives to chlorine must be found if coastal power station cooling waters are to be used for aquaculture (LANGFORD, 1977). Without these, cultured species will be restricted to those tolerant of chlorine and even non-lethal concentrations might have adverse effects on growth or on market values of produce by changing taste, colour or texture (SYLVESTER, 1975).

The chlorine tolerances of organisms have been shown to vary considerably (MATTICE and ZITTEL, 1976); *Mugil cephalus* and *Anguilla anguilla* are both tolerant (HEATH, 1977; ASTON, 1981) and are being used increasingly in thermal aquaculture, whilst *Dicentrarchus labrax* and *Penaeus kerathurus* are much less tolerant (SAROGLIA and co-authors, 1981). There may be differences in the chlorine tolerances of different developmental stages of the same species; *Pleuronectes platessa* eggs were more tolerant of chlorine than newly hatched larvae, and during development to metamorphosis larvae of this species and *Solea solea* showed increasing chlorine tolerance (ALDERSON, 1974). Although LANGFORD (1977) suggested that, in general, fish are less tolerant of chlorine than molluscs, bivalves are particularly sensitive since even low chlorine levels depress their activity and reduce the growth rates (ANSELL, 1969). Crustaceans such as *P. kerathurus* are less tolerant of chlorine than fish (SAROGLIA and co-authors, 1981). In fact, this species was the most sensitive tested by these authors who suggested that an acceptable level for culturing should be just below 0·05 ppm total residual chlorine. Flatfish in the UK tolerated this concentration continuously (KERR, 1976a). However, this suggested limit is considerably lower than concentrations arising from power station chlorination (BEAUCHAMP, 1969; MARSHALL, 1971; BRUNGS, 1973; ASTON, 1981; JONES and co-authors, 1981). In practice, by the time water reaches culture tanks, chlorine levels may be low enough to be safe for fish farming, although this cannot be guaranteed. Chlorination usually takes place at the cooling-water intake whilst heated water for aquaculture is tapped at some point between the condensers and outfall (NASH, 1974a), by which time there may have been a considerable reduction in chlorine.

ALDERSON (1974) reported that where continuous chlorination takes place to maintain residuals of about 0·5 ppm at the condensers, this concentration may decrease to about 0·1 ppm at the outfall.

Various methods may be used to reduce chlorine concentrations. Before the development of ultraviolet sterilization, water for intensive aquaculture was usually treated with chlorine gas to kill bacteria and some viruses (LIAO, 1981) and subsequently dechlorinated by bubbling air through the water or allowing it to stand for an extended period. Considerable aeration occurs when water is pumped through culverts and over spillways, etc., of power stations. KERR (1976a) noted that tank-inlet cascades and vigorous aeration helped to reduce chlorine concentrations to 0·05 ppm in tanks despite the water supply containing up to 0·3 ppm. Allowing chlorinated water to stand in tanks exposed to sunlight reduces chlorine concentrations; other meteorological factors may contribute to this effect (NASH, 1974a). Water flow through fish tanks is considerably slower than in the cooling system, allowing more time for chlorine uptake by the higher levels of dissolved organic matter present in tanks. According to ALDERSON (1974), chlorine in fish tanks receiving 4 to 6 water changes daily seldom rose above 0·02 ppm despite main effluent concentrations of 0·35 ppm. In some cases it might be necessary to allow heated effluent to cool in a pond before pumping into fish tanks (FLATOW, 1981) or to mix it with ambient sea water. Such methods would also reduce chlorine concentrations.

There may be occasions when faults with chlorine gas equipment results in large accidental doses and warning devices may have to be installed (PAGE-JONES, 1971).

Various strategies have been adopted to avoid the risk of mortalities from chlorine in thermal aquaculture. JONES and co-authors (1981) reported that cooling water used to rear turbot in a UK fish farm sometimes contained chlorine values higher than the 0·04 to 0·08 ppm regarded as safe for turbot and sole (ALDERSON, 1974). JONES and co-authors supplied their tanks with heated sea water only during autumn and winter when no chlorination occurred, whilst during spring and summer ambient sea water was used. At power stations employing intermittent chlorination low chlorine values can be maintained if exchange times for water in fish tanks are considerably longer than chlorine injection periods (ASTON, 1981). Partial or complete recirculation of water in fish tanks has been used in the USA during chlorination periods (GUERRA and co-authors, 1979b). Heat exchangers to isolate culture water from chlorinated cooling water are another possibility but may be large and expensive (p. 1902).

Water for intensive aquaculture used to be sterilized with chlorine and then dechlorinated before use in fish tanks (LIAO, 1981). This suggests that, with good management, chlorinated cooling waters could be beneficial to fish farmers since there may be some protection from disease. Freshwater aquaculture installations, using unheated municipal water with lethal concentrations of chlorine, pass the water through beds of activated charcoal or treat with sodium sulphite (STICKNEY, 1979), an expensive technique in large or open-circuit systems.

There are various biocidal alternatives to chlorine, e.g. bromamines and copper sulphate, 1,2-benzisothiazolin-3-one and ozone (COUGHLAN and WHITEHOUSE, 1977), and solutions of biodegradable dodecylguanidine salts (BOIES and co-authors, 1973). Mainly used in freshwater closed-circuit cooling systems, these alternatives are easier to handle than chlorine and some have smaller environmental effects but they are more expensive.

Non-chemical alternatives to chlorine are unlikely to cause any problems for thermal aquaculturists. Ultrasonic (FISCHENKO and co-authors, 1973) and mechanical, abrasive techniques such as 'Taprogge', 'Amertap' or 'Mann' systems (ASTON and co-authors, 1976; ASTON, 1981) of squeezing sponge-rubber balls through the condenser tubes may be effective for condensers but not for the main culverts. However, the ball system has been used at a Long Island Power Station, USA, supplying heated water to a large and successful oyster farm (ASTON and co-authors, 1976; MARGRAF, 1977). Unusual fouling control occurs at Kahe Power Station, Hawaii, where the abrasive action of sand passing through the cooling system avoids the need for chemical antifouling treatments (BIENFANG and JOHNSON, 1980).

In conclusion, chlorine is one of the most persistent serious problems likely to face fish farmers at coastal and estuarine power stations. Nevertheless, successful culture appears to be possible as long as chlorine levels in culture tanks are kept to tolerably low levels, and this requires good management and cooperation between aquaculturists and power station engineers. If sufficiently low levels cannot be achieved, then alternative chemical or mechanical methods or heat exchangers would be other possibilities.

Disposal of other power station wastes. These are frequently discharged into open cooling systems of coastal and estuarine power stations for convenience. Enormous mixing and dilution takes place when these wastes enter the cooling system and again when the heated effluent is finally discharged to the environment (NASH, 1970b; REYNOLDS, 1980). BECKER and THATCHER (1973) give a tentative list of the chemicals used, or likely to be used, and provide information on the complex effects that can occur when toxicants are introduced to the environment. The listed compounds are many and varied—each situation is usually unique depending on local conditions.

Boiler cleaning agents are frequently discharged with the cooling water and may contain such toxic compounds as zinc corrosion inhibitors and hexavalent chrome (MATSON, 1977). Metal-cleaning compounds are usually alkaline or acid cleaners, alkaline chelating rinses or organic solvents. They frequently contain copper, iron, nickel, zinc and chromates (HART and DELANEY, 1978).

In power stations with cooling towers the tower purge water or blow-down usually forms the largest component of the effluent and comprises make-up water, anti-fouling and anti-corrosion chemicals (MATSON, 1977). HART and DELANEY (1978) list the numerous compounds likely to be found in this type of effluent.

Wastes arise from stack scrubbers in fossil-fuelled stations when large quantities of ash may enter the cooling system through carelessness; this ash will contain carbon soot, organometallic compounds and oxides and salts of vanadium, nickel and iron (HART and DELANEY, 1978).

Drainage water from coal store yards at coal-fired stations can enter the cooling systems. This can result in metal sulphides and sulphuric acid forming which, in turn, can dissolve other compounds to release heavy metals into the drainage water (HART and DELANEY, 1978). NASH and PAULSEN (1981) identified certain treatment and waste chemicals of concern, particularly the highly toxic and persistent polychlorinated biphenyl (PCB) compounds and hexavalent chrome and zinc corrosion inhibitors. However, GUERRA and co-authors (1979a) reported permissible levels of PCBs in the freshwater shrimp *Macrobrachium rosenbergii* and rainbow trout *Salmo gairdneri* and of heavy metals in rainbow trout reared in a freshwater aquaculture facility at a coal-fired station on the Delaware river, USA. European eels *Anguilla anguilla* reared in a freshwater thermal aquaculture facility in England had lower concentrations of heavy metals than either controls from unheated water or wild fish (ROMERIL and DAVIS, 1976). This was thought to be due to metal accumulating in lower concentrations in faster growing fish. MÉLARD and PHILIPPART (1981) reported that the freshwater tilapia *Sarotherodon niloticus*, grown in the heated effluent of a Belgian nuclear power station, had heavy metal concentrations below the levels permitted by law. SAROGLIA and co-authors (1981) measured concentrations of various heavy metals (Hg, Cd, Pb, Ni, Zn, Co, Mn and Fe) in the tissues of European eel *Anguilla anguilla*, bass *Dicentrarchus labrax*, sea bream *Sparius aurata* and shrimp *Penaeus kerathurus* grown in the discharge water from a power station in Italy. Comparisons with specimens from other places, away from the power station, showed that the organisms cultured in elevated temperatures did not accumulate the metals and in some cases contained concentrations below those of wild animals.

However, there have been cases of organisms near power stations accumulating fairly high concentrations of copper in their tissues. The phenomenon of 'greening' in oysters is well known and has been attributed to the operation of power stations (ROOSENBURG, 1969; MARGRAF, 1977) although the reasons are not clear. ROOSENBURG suggested the possibility that effluent stressed the oysters in a way that disrupted the physiological mechanisms controlling metal uptake although the possibility remained that the source of copper was leaching from the condenser cooling tubes. MARGRAF (1977) suggested replacing tubes containing copper with titanium tubes, which have a much longer operational life. The copper concentrations in oysters were non-lethal but animals were unfit for human consumption (ROOSENBURG, 1969). However, there are a number of cases of oysters being reared in heated effluents from power stations apparently without any such problems (e.g. TINSMAN and MAURER, 1974; PRICE and co-authors, 1976) whilst the highly successful rearing of the oyster *Crassostrea virginica* in Long Island Sound (PETERSON and SEO, 1977) relies on growth in a heated effluent for several months during the early stages.

In industrial countries there are increasing numbers of nuclear power stations. Although these have no flue gases resulting in treatment wastes, they use chemical treatments that produce other wastes similar to conventional fossil-fuelled steam stations. They also produce small amounts of radioactive wastes (MITCHELL, 1967–78; GUERRA, 1975; HETHERINGTON, 1976; AMERICAN NUCLEAR SOCIETY, 1977; HUNT, 1979; MAUCHLINE, 1980). Irradiated fuel rods are placed in special cooling ponds and slight leakage of the fuel-containing cans may occur with small quantities of neutron-induced radioisotopes and some fission products being leached into the cooling pond water. This water may be periodically discharged into the cooling water discharge, but the radioisotope quantities are normally so low that they cannot be distinguished from background levels or those originating from other sources (MITCHELL, 1967–78; HETHERINGTON, 1976; HUNT, 1979). Some of the recent nuclear power stations use closed-circuit cooling systems and much of the blow-down containing radioactive substances can be recovered for disposal by other means (NASH and PAULSEN, 1981). Radioactive wastes can be retained and concentrated by organisms in the food web (BAPTIST, 1966; SALO, 1969; BAPTIST and co-authors, 1970; BAPTIST and RICE, 1977), particularly through filter feeding and sedimentation. This was clearly shown in the oyster *Crassostrea virginica* at the Yankee Power Station in Maine, USA (PRICE and co-authors, 1976). Levels of radionuclides in oysters grown near the outfall were 5 times

hose of oysters grown near the intake. These workers suggested a model to predict levels
n shellfish and the release schedules required to minimize uptake so that the best sites
could be chosen for future aquaculture (HESS and co-authors, 1975, 1977, 1979; PRICE
and co-authors, 1976). Oysters grown in a thermal aquaculture unit at the Hunterston
Power Station, UK, had levels well below the limits set by the International Commission for Radiobiological Protection (MITCHELL, 1975). Radionuclide levels in fish
grown in heated effluents have been shown to be very low. At the Hunterston installation, levels for plaice *Pleuronectes platessa* were also well below the ICRP limits (MITCH-
ELL, 1975), whilst at a Belgian freshwater facility the mouthbrooding tilapia *Sarotherodon
niloticus* contained radionuclide concentrations below legal levels (MÉLARD and PHILIP-
PART, 1981).

Although radionuclide levels in cultivated organisms do not appear to offer many
problems, it is clear that acceptable levels will have to be assured otherwise this could
cause marketing problems (SYLVESTER, 1975). In addition to the safety aspect, radioactive wastes are a highly emotive subject and there could be considerable public
prejudice against produce from nuclear power stations.

With all the wastes discharged by power stations, most releases, with the exception of
continuous chlorination, occur for short periods and these could be programmed in
collaboration with aquaculturists so that water supplies to fish tanks could be
temporarily stopped. Another alternative would be for the aquaculture water supply to
be tapped before the wastes discharge point (ASTON, 1981), although this would not
prevent chemicals intended to treat the cooling system from being taken in. A
combination of the two methods would appear to be best with, in the event of continuous
chlorination, adequate safeguards to ensure that chlorine concentrations in fish tanks do
not become too high.

Dissolved gases: oxygen. The importance of dissolved gases to marine organisms has
been reviewed in Volume I: KALLE (1972), RHEINHEIMER (1972), VERNBERG (1972),
VIDAVER (1972); see also DAVIS (1975). Oxygen levels in cooling water can be affected
by passage through the cooling system. The concentration at the outfall is generally
higher than at the intake due to pumping and aeration. However, if the intake water is
saturated then the outfall concentration may be slightly lower owing to the 8 to 10 C°
temperature elevation at the condensers (ROSS and WHITEHOUSE, 1973), despite the
higher rate of oxygen exchange from gas to solution at higher temperatures (BEWTRA,
1970). There may be occasions when gas supersaturation occurs (see p. 1894).

In a modern intensive culture system oxygen requirements of fish are usually higher
than the oxygen available in the water, hence aeration or oxygenation is necessary
(ASTON, 1981). In addition, artificial feeding causes increased oxygen demands and so
do higher temperatures. For example, WIENBECK (1981) showed that the oxygen
consumption of fed European eels *Anguilla anguilla* was twice that of unfed eels. Where
fish are fed optimum amounts of food, growth may depend on oxygen levels close to
saturation levels (ALABASTER, 1973). In contrast, if the level of feeding is below optimum
values, growth may depend on oxygen levels considerably below saturation (ANDREWS
and co-authors, 1973). Different oxygen demands play a significant role in the choice of
species suitable for thermal aquaculture. For example, *Scophthalmus maximus* is far less
active in intensive rearing tanks than salmonids, and low oxygen consumption rates are
reported (JONES and co-authors, 1981). This could be a particularly important aspect
where adequate oxygenation is a problem in thermal aquaculture.

Since the main purpose of thermal aquaculture is to take advantage of elevated temperatures for obtaining increased rates of food conversion, there is little point in providing large amounts of food below optimum levels of dissolved oxygen. Maintaining adequate oxygen supplies is, therefore, very important. Aeration has been used to maintain oxygen levels in intensive fish tanks by means of spillways, bubblers, diffusers, surface aerators, aspirators, cascades, tray aerators, airlift pumps, submerged pumps and venturi and U-tubes (KINNE, 1976; LIAO, 1981; SOWERBUTTS and FORSTER, 1981). However, it is important to avoid excessive water movement (WIENBECK, 1981) and nitrogen supersaturation (see below). In culture tanks, heavy bubbling might cause damage to delicate organisms (KINNE, 1976). To maintain concentrations of 5 mg O_2 l^{-1}—a minimum level desirable for many marine organisms—the rate of aeration at 25 °C has to be twice that at 15 °C (ASTON, 1981).

Where aeration is insufficient, high stocking densities can only be maintained by oxygenation (JONES and co-authors, 1981; WIENBECK, 1981). One method used with eels is to start oxygenating when they are fed and to continue for a certain period afterwards, when the oxygen consumption of the fish is still high (WIENBECK, 1981). Proper oxygenation can reduce to one tenth the volume of water required in intensive culture systems (MEADE, 1974; FORSTER and co-authors, 1977). Where desirable oxygenation permits supersaturation (see below) to be maintained (SOWERBUTTS and FORSTER, 1981). Improved fish growth has been reported at oxygen levels up to 250% air saturation (SHABI and HIBBERD, 1977, quoted by SOWERBUTTS and FORSTER, 1981). However, oxygen is expensive and cultivators need to ensure that maximum dissolution occurs in the water to reduce wastage (SOWERBUTTS and FORSTER, 1981). These authors review gas exchange, aeration and re-oxygenation in fish farming and WICKINS (1981) gives a brief review of the effects of varying oxygen levels. For an important account of aeration, its dynamics and theory consult Volume III: KINNE (1976).

Dissolved gases: supersaturation. Gas supersaturation is a potential problem in thermal aquaculture since it can give rise to 'gas-bubble disease' in fish (Volume III: KINNE, 1976, 1977; see also MARCELLO and STRAWN, 1973; SYLVESTER, 1975; SOWERBUTTS and FORSTER, 1981). RUCKER (1976) and NEBEKER and co-authors (1976) showed that gas-bubble disease is caused by total gas pressure although supersaturation by nitrogen is most harmful to fish (KITTELSEN and GJEDREM, 1981). Both oxygen and nitrogen supersaturation can occur in cooling water during passage through a power station for a variety of reasons: raising saturated water from depth, entraining air when the water passes over spillways or through pumps, pressurizing the water and heating saturated water at condensers. The conditions for supersaturation can be found in many power station cooling systems. Although this has been reported not to cause many problems for aquaculture (HUGUENIN and RYTHER, 1974), there have been adverse reports. According to TUUNAINEN and co-authors (1981) gas-bubble disease was the main source of mortalities in Baltic salmon *Salmo salar* fingerlings raised in heated brackish water, and problems have been experienced with salmonids reared in heated freshwaters (INGEBRIGTSEN and TORRISSEN, 1981; KITTELSEN and GJEDREM, 1981). Salmonid tolerance levels for nitrogen saturation are 100 to 105% (LIAO and MAYO, 1972) whilst saturation values for river water could be as high as 110 to 120% (INGEBRIGTSEN and TORRISSEN, 1981). Similar levels can be encountered under certain conditions in sea water cooling systems. JONES and co-authors (1981) report mortalities of turbot

Scophthalmus maximus in an intensive thermal acquaculture facility (North Wales, UK) at 118% nitrogen supersaturation of the incoming warm water.

A simple and most effective means of reducing gas supersaturation is by bubbling air through the water, which can reduce levels of 110 to 120% down to 100 to 102% saturation in heated river water (INGEBRIGTSEN and TORRISSEN, 1981). Aeration has also been used to reduce supersaturation in heated brackish water (TUUNAINEN and co-authors, 1981). KITTELSEN and GJEDREM (1981) passed supersaturated freshwater through a series of perforated aluminium plates; this method could be used with sea water provided the plates were made of a non-corroding material such as plastic sheet.

Gas supersaturation should not present difficulties for thermal aquaculture as long as it is detected in time and adequate degassing facilities are available, e.g. in pond and tank culture. Where fish are reared in cages placed directly in the cooling-water discharges of power stations, degassing is difficult or impossible and alternative methods are required. CHAMBERLAIN and STRAWN (1977) reared 7 species of estuarine fish in cages in a power station discharge canal in Galveston Bay (Texas, USA). Mortalities of 99% in 14 d due to gas-bubble disease occurred in floating cages whilst fish in cages submerged at a depth of 3 m remained unaffected. In submerged cages winter growth rates of unfed fish were higher than those of regularly fed fish kept at ambient temperature.

Waste treatment and disposal

Intensive cultivation at elevated temperatures results in large amounts of faecal and metabolic wastes, such as carbon dioxide and nitrogenous wastes. These can build up in aquaculture ponds and tanks, causing environmental pollution. They require careful management and application of appropriate treatment techniques (see important reviews and discussions by KINNE, 1976; STICKNEY, 1979; LIAO, 1981; WICKINS, 1981).

Disease, parasites and stress

Culturing organisms at elevated temperatures may induce higher incidences of disease (ASTON and co-authors, 1976; KINNE, 1980). Disease control in intensive systems is necessary not only to prevent mortalities but for obtaining larger, stronger and healthier products (FLATOW, 1981). However, published evidence suggests few serious problems so far in thermal aquaculture. Salmon smolts reared in heated freshwaters showed no disease over a 9-yr period (KITTELSEN and GJEDREM, 1981) and Baltic salmon reared in heated brackish water had neither infections nor parasites (TUUNAINEN and co-authors, 1981). In different intensive aquaculture systems in the Federal Republic of Germany, bacterial and viral infections turned out to be less deleterious with improved environmental conditions (SCHLOTFELDT, 1981).

In Japan bacterial and parasitic diseases, fairly common in commercial fish net pens, were seldom found in flow-through culture systems employing thermal effluents from power stations (EGUSA, 1981). Apparently, in open flow-through systems disease agents were more readily washed away; both pathogens and parasites were seldom found in open-coast waters where modern power stations are often built. Cultured species are often introduced from other geographical areas. Hence the risks of introducing foreign disease agents from different geographical areas may be considerable (ROSENTHAL, 1981; MUNRO and FIJAN, 1981). An epizootic causing extensive damage to the

Japanese eel (*Anguilla japonica*) industry was apparently due to a virus introduced with elvers of European eel *A. anguilla* (EGUSA, 1979). In contrast, the European eel was found to be particularly vulnerable to various parasites in Japan. Hence EGUSA warned against the translocation of these parasites. Disease may also be introduced with cultivated animals from wild indigenous stock. JONES and co-authors (1981) found a *Vibrio* sp. infection in wild, locally caught turbot reared at a power station in North Wales, UK, causing daily mortalities as high as 10% of the culture population. Application of oxytetracycline was effective. Similarly, MCVICAR (1975), reported a protozoan parasite *Glugea stephani* in the intestines of plaice *Pleuronectes platessa* from a southern stock, reared in the heated effluent of a Scottish power station. This infection would probably have been avoided by using imported eggs or an isolated broodstock (MUNRO and FIJAN, 1981).

Temperature often exerts a clear and selective influence on the types of diseases in culture systems (SNIESZKO, 1974; ROBERTS, 1975). This is particularly important in places where wide fluctuations in daily and seasonal temperatures prevail (MEYER, 1970). Changes in water temperature can alter the incidence of both bacterial and viral diseases as well as the responses of fish to pathogen invasions. This subject has been reviewed in detail by MEYER (1970), ROBERTS (1975) and KINNE (1980). It is important, therefore, that temperature changes involved at power stations should be considered in relation to the endemic diseases of the locality considered for aquaculture (COUTANT, 1971c). Some disease agents can become increasingly virulent at higher temperatures whilst others may become less active (SNIESZKO, 1974). Disease agents may have optimum temperatures somewhat different from those of their hosts. This can affect disease manifestation. Thus *Aeromonas* infections of salmonids were highest at 17·8 to 20·5 °C and lowest at 3·9 to 6·7 °C (GROBERG and co-authors, 1978). In contrast, bacterial kidney disease (agent: *Corynebacterium* sp.) caused maximum mortalities at 6·7 to 12·2 °C but was maximally suppressed at 20·5 °C (SANDERS and co-authors, 1978). Bacterial disease agents reproduce within certain temperature limits. REICHENBACH-KLINKE (1981) lists these limits for numerous species, as well as temperatures at which no infections occur. He also points out that some viral diseases of fishes disappear at higher temperatures and that infections are almost absent from fish kept at tropical or subtropical temperatures. This is partly due to maximum activation of the immune system at higher temperatures. BISSET (1947) and CUSHING (1971) reviewed the effects of temperature on bacterial agents and immune responses of fish. While at low temperatures cellular and humoral defences are not active and the bacteria are unable to multiply (SNIESZKO, 1974), at high temperatures the fish may become diseased and die or active bacterial stages disappear from the tissues.

The majority of bacterial diseases in thermal aquaculture appear to be caused by *Vibrio* spp. (e.g. EGUSA 1981; JONES and co-authors, 1981). EGUSA (1981) reported low mortalities from these diseases, which could be effectively treated, and suggested that bacterial agents were flushed away in open culture systems. This could be one of the advantages of using power station cooling waters because there would be a plentiful supply of heated water for this purpose. Most of the diseases affecting thermal aquaculture are common to intensive aquaculture. There are several recent accounts of importance to thermal aquaculture: EGUSA (1981) reviews general aspects of disease and its control in Japanese heated effluents. MUNRO and FIJAN (1981) discuss similar aspects for European waters and provide advice for minimising disease effects by monitoring fish health employing prophylactic and chemotherapeutic treatments. REICHENBACH-KLINKE (1981) describes the effects of temperature on outbreak and intensity of fish diseases and SCHLOTFELDT (1981) surveys experiences with diseases in intensive freshwater systems over many years. ROBERTS (1976) gives an account of bacterial diseases of farmed fish, and STICKNEY (1979) deals with disease and its treatment in cultured organisms used, particularly for warm-water aquaculture. Fish pathology is reviewed in ROBERTS (1978).

Fish stressed and debilitated by winter conditions succumbed more readily to disease (LIEBMANN and co-authors, 1960). The significance of stress in outbreaks of infectious diseases of fish has been emphasized by MEYER (1970) and WEDEMEYER (1970). Pollution can augment the level of stress, thus making fish more vulnerable to disease (SNIESZKO, 1974). An important paper on quantifying stress response in fish kept in intensive aquaculture systems, particularly those using recirculated heated water, has been presented by WEDEMEYER (1981). He offers recommendations for hatchery management to reduce stress intensities. Stress effects have been investigated in *Ictalurus punctatus* grown in heated effluents (HILGE and co-authors, 1981) and in *Anguilla anguilla* grown in intensive aquaculture systems (PETERS and co-authors, 1981). Both papers discuss stress in terms of SELYE's (1950, 1956, 1973) 'general adaptation syndrome' for higher vertebrates, the stress response being similar in fish (GRONOW, 1974). Stress control is important for managing thermal aquaculture facilities by maintaining satisfactory water quality in the cooling water.

It is usually easier to prevent a disease than to cure it (e.g. KINNE, 1980). MUNRO and FIJAN (1981) therefore advocate the establishment of large-scale pilot aquaculture units on proposed sites in order to determine the effects of water quality before installing large-scale commercial enterprises. The introduction of disease-free stock is a first requirement, and MUNRO and FIJAN highlight the need for more research to find suitable chemicals for disinfecting eggs and to increase the numbers of available vaccines. Thermal aquaculture using recirculated water involves greater health risks to fish than once-through systems. Hence careful monitoring of disease incidence is required throughout cultivation. While treatment of cooling water with chlorine, to prevent slime and fouling growths in power station condensers, may suppress disease agents (NASH, 1968, 1969b; KEUP, 1981), chlorine levels must be reduced before the water is pumped into the cultures (p. 1890). Additional methods of controlling disease agents include the use of elevated water temperatures to incapacitate or kill cold-water pathogens, ozonation (e.g. KINNE, 1976; OAKES and co-authors, 1979; STICKNEY, 1979; ROSENTHAL, 1981; LIAO, 1981), ultraviolet irradiation (e.g. KINNE, 1976; FLATOW, 1981). Ozone is very toxic and must be given adequate time to convert back to molecular oxygen before passing to the culture system (e.g. STICKNEY, 1979). Measurable amounts of ozone should be avoided in culture systems (WHEATON, 1977). SAUNDER and ROSENTHAL (1975) recommend activated charcoal filters for reducing critically high ozone levels. Some ultraviolet lamps include a wavelength of 1849 Å, thus producing ozone; hence care must be taken to ensure that toxic concentrations are avoided (FLATOW, 1981). STEWART (1973) used 5 ppm potassium permanganate at unspecified intervals in fish pools supplied with heated effluent and presumed that this treatment was effective because no disease problems were encountered. In the event of a major disease outbreak, MUNRO and FIJAN (1981) suggest killing the whole stock and carefully disinfecting the facilities and equipment before restocking.

In summary, heated effluents appear to have four advantages for disease control: firstly, many modern power stations are sited on open coasts, well away from some sources of contamination. Secondly, chlorination of cooling water tends to suppress virulent organisms. Thirdly, open-circuit cooling systems provide large amounts of heated water that support once-through culture facilities; these increase the likelihood of flushing pathogens and parasites out of the culture facilities. Since power stations with open-circuit cooling are designed to avoid re-circulation of cooling water as far as possible, there should be little chance for recycling of pathogens assuming that the aquaculture discharge is either into, or close to, the power station outfall and not near the intake. Fourthly, elevated temperatures tend to reduce the chances of cold-water pathogen development. However, at the same time warm-water pathogens may be stimulated into active growth.

Choice of organisms for culture

The careful choice of organisms with physiological tolerances suitable for rearing in artificially heated water is a very important aspect of thermal aquaculture (MUENCH, 1976). Suitable animals could be tropical, or species with wide temperature tolerances or temperate species. Some temperate region species show faster growth rates than tropical ones when grown in constant warm conditions. Growth characteristics sometimes have adverse effects in thermal aquaculture. For example, striped mullet *Mugil cephalus* in coastal waters of the Gulf of Mexico show depressed growth during summer and winter with most rapid growth in spring and autumn (CECH and WOHLSCHLAG, 1975). Similarly, mullet grown in the discharge canal of a power station showed a depression of growth because of the simulated summer conditions (CHAMBERLAIN and STRAWN, 1977). Suitability of a species for thermal aquaculture depends on factors such as growth rate, food conversion ratio and growth efficiency (BOTSFORD and GOSSARD, 1978), all of which are affected by temperature. Whilst nutritionists develop feeds which produce better growth rates or conversion ratios (WEATHERLY, 1976) and geneticists select for higher growth rates (IVERSON, 1968), temperature still plays an important role in improving production. In some localities, such as estuaries, cultured organisms must be tolerant of wide fluctuations of salinity (HOLT and STRAWN, 1977). These authors also thought it important to choose species that could be harvested in a single growing season, although some facilities rear fish successfully over more than one season. Larvae should be easily cultured and suitable for brood stock management. MUENCH (1976) suggested that high economic value was one of the first requirements; many aquaculture developments grow species for high-value gourmet markets. NASH (1974b) discusses various aspects of species selection for open-sea mariculture of interest to thermal aquaculturists.

Holding Methods in Thermal Aquaculture

There are three different ways in which heated effluents may be used for aquaculture: (i) direct culture in heated discharges; (ii) culture in ponds, tanks or raceways; (iii) indirect use of heated effluents. The way adopted affects the choice of culture technique. For example, cages are best for fish cultivation directly in discharge canals or in cooling ponds, lakes or reservoirs, whilst culture ponds, tanks or raceways are used where heated water is pumped or diverted from the cooling circuit. The reviewers are not aware of any cases where fenced enclosures and extensive cultures have been used with thermal discharges. However, in freshwater CARROLL and co-authors (1981) mention the intensive culture of channel catfish *Ictalurus punctatus* in raceways at a geothermal aquaculture unit in Idaho, USA, and the subsequent use of this raceway effluent for the extensive culture of tilapia fingerlings in ponds. Similar systems might eventually be developed for mariculture. There appear to be only small economic differences between cage, raceway and fenced enclosure culture methods (COLLINS and DELMENDO, 1979) and details concerning the various designs, methods of use and economics are in the literature. For reviews consult Volume III: KINNE (1976), KINNE and ROSENTHAL (1977); see also MILNE (1972, 1976) and STICKNEY (1979). KINNE and ROSENTHAL list important pertinent books and papers. The 1976 FAO Technical Conference on Aquaculture in Kyoto, Japan, dealt with different culture systems with various species in different parts of the world (PILLAY and DILL, 1979), and the FAO/EIFAC Symposium in Bergen (TIEWS, 1981) considered the use of heated effluents and recirculation systems for intensive aquaculture, including holding methods.

Direct culture in heated discharges
    Fish can be cultivated in floating cages placed in the discharge canal or at the mouth of the heated discharge into the sea, or in cooling ponds or lakes into which the power station discharges. Growing fish in cages in the sea near a heated discharge might be the most vulnerable of these methods. Unless cages could be sited in sheltered parts of the coast, rough weather conditions might preclude their use, although there have been recent efforts to design floating breakwaters for the benefit of aquaculturists (KATO and co-authors, 1979). HUGUENIN and ROTHWELL (1979) discussed the design of cage systems for exposed areas on tropical coastal sites (e.g. Hawaii). Since these involve bottom-mounted cages to avoid surface wave action, the application is limited for thermal aquaculture. Biofouling of cages is a serious problem in mariculture and it may be an even greater problem for cages in thermal effluents. ANSUINI and HUGUENIN (1978) suggest that cages with metal meshes made from copper–nickel alloys may reduce the difficulties and provide greater strength under exposed conditions.
    Organisms cultivated in heated effluents must have suitable temperature tolerances where fluctuations of water currents result in cages being intermittently exposed to water at ambient temperatures. Thermal stratification, with heated effluent forming a surface layer, could cause problems with floating cages below the heated layer. Despite these problems some fish, not suitable for pond culture on land, could be grown in floating cages moored in the sea where power plant effluents discharge; in Japan successful trials have been carried out with the yellowtail *Seriola quinqueradiata* (CHIBA, 1981b). Particular problems were that summer temperatures became too high and cages either had to be towed or sunk into cooler water.
    Cultivation in discharge canals or cooling ponds is more reliable than in open waters because of more even temperatures. However, in discharge canals, cultured organisms would be exposed directly to any deterioration in water quality as a result of power station operations and there would be little opportunity of rectifying this by stopping or diverting the water flow, or of adjusting water quality. Culture cages suffered from heavy fouling in discharge canals (MARCELLO and STRAWN, 1973; CHAMBERLAIN and STRAWN, 1977) and excessive flow rates caused cages to break open. In addition, it is inadvisable to use culture cages in canals that empty whenever the power station ceases pumping sea water.
    Cooling ponds and lakes can provide good facilities for mariculture with large

sheltered areas available for cage culture (HOLT and STRAWN, 1977), as long as environmental conditions such as excessive salinity fluctuations are not prohibitive. While there appear to be a few cases where cooling ponds have been used for mariculture, limniculture cooling lakes have been used very successfully to raise rainbow trout *Salmo gairdneri* in floating cages (REID, 1981; ROGERS and CANE, 1981). Freshwater cooling lakes are not only used for rearing fish in floating cages, but also for recreational facilities such as angling, boating and swimming (p. 1851; PETERSON and SEO, 1977).

Culture in ponds, tanks or raceways

In most cases of thermal aquaculture heated water is pumped to ponds or tanks in a shore-based mariculture unit. The heated discharge is tapped at some point between condensers and discharge outfall and part of the heated effluent is either diverted or pumped to the aquaculture facility. This provides maximum control: the water flow can be stopped during periods of chemical waste discharge or of high chlorine dosing, or whenever water temperatures become too high for cultivated organisms. In the last event, ambient water can be pumped and mixed with the heated water to maintain or re-establish the required temperatures or the heated water can be cooled by spraying onto the surfaces of ponds and tanks. It can also be pumped through a special cooling pond prior to use (PARKER and STRAWN, 1976). Furthermore, pumped water supplies can be easily aerated or oxygenated to raise the oxygen concentration and they can be aerated or passed over cascades or through sprays to remove excessive chlorine or reduce gas supersaturation.

However, pumping water is costly (KERR, 1976a; TANAKA, 1979; CARROLL and co-authors, 1981; CHIBA, 1981b). In Japan, CHIBA (1981b) recommends avoiding the use of pumps where possible and to control water flow into culture ponds by gravity in order to increase the profitability of commercial thermal aquaculture. TANAKA (1979) points out that because the discharge water flows at a high rate its energy can be used to raise the water to culture pond levels by narrowing the water channel. While TANAKA maintains that this keeps pumping costs to a minimum, this presumably refers to the costs for the aquaculturist, with the greater cost being borne by the power station operators. However, taking advantage of the energy supplied by the very large pumps used in power stations could be the most efficient overall method of pumping for thermal aquaculture.

CARROLL and co-authors (1981) give an example of a commercial company in the USA that found pumping costs to maintain sufficiently high water temperatures too high to permit commercial exploitation, although the company was continuing a research culture unit. KERR (1976a) gives a detailed analysis of the pumping costs of thermal aquaculture. He emphasizes particularly the importance of allowing for the costs of pumping ambient sea water as well as heated water. This is an important aspect because during summer, effluent temperatures may be too high for the cultured organisms and it might be necessary to mix the heated water with water at ambient temperatures. The problem with an ambient sea water supply is that the pumping head to the fish farm is higher than that from the heated water supply and distances will be almost invariably greater, resulting in higher frictional losses in the pipe work. A possible source of cooling sea water at a more convenient distance from a mariculture unit might be the cooling water supply prior to entering the steam turbine condensers. However, if this water is chlorinated it is likely to contain a higher chlorine concentration than in the discharge and hence may not be suitable. Thus, pumping ambient water from the sea is likely to be more expensive than pumping the same volume of heated water. KERR (1976a) thought that the cost of pumping water could dictate the amount of warmed water that was utilized for fish farming, even in the event of the power generating industry providing the water free of charge. Much will depend on the value of the marketed products.

Despite high pumping costs, thermal aquaculture using ponds and tanks has the advantage of providing opportunities for water quality control, and commercial enterprises are becoming increasingly established in various parts of the world.

Alternatives for holding cultured stock are tanks, ponds or raceways. Tanks and ponds appear to be commonest and can be constructed from a variety of materials (e.g. MILNE, 1972, 1976; Volume III: KINNE, 1976). Earthen ponds of approximately 1000 m² surface area were used by BRANCH and STRAWN (1978) in polyculture experiments with a thermal effluent in Texas. In Japan, where much of the pioneering work on thermal as well as general aquaculture has been carried out, ponds and tanks for thermal aquaculture were built of concrete, glass-fibre and iron frames with linings of vinyl chloride (MILNE, 1972; CHIBA, 1981b). MILNE suggested that the lighter types of construction provided greater flexibility because these tanks could be more easily rearranged at a later date than those constructed of concrete. Areas ranged from 15 to 100 m² and sometimes as large as 300 m². Pond depths were usually 1 to 1·5 m with water depths of about 0·6 to 1·2 m (CHIBA, 1981b). Shapes were square, rectangular and circular, the last type usually having a central drainage.

Some thermal aquaculture farms have used raceways (e.g. GUERRA and co-authors, 1976, 1979a; GODFRIAUX and co-authors, 1977; CARROLL and co-authors, 1981). The use of raceways for thermal aquaculture was advocated by GAUCHER (1968) and YEE (1972). MILLER (1966) considered various aspects of their design (see also Volume III: KINNE, 1976; STICKNEY, 1979). In industrial localities and other places where land for aquaculture is either scarce or expensive, raceways can facilitate the high intensity of culture necessary for a good economic return (GODFRIAUX and co-authors, 1977).

In some cases, particularly in temperate regions, culture tanks have been housed in greenhouses to provide protection and conserve heat. KINNE (1976; Volume III), points to the development of this theme as a parallel to 'controlled-environment agriculture'. In Japan, eels are now cultivated in ponds of heated freshwater in greenhouses (CHIBA, 1981a). The technique turned out to be so successful that it is being adopted by most of the commercial eel culturists of Japan. FORREST (1976) describes the technique, including polythene and glass-covered ponds in which elvers are raised. Although heat is provided by special boilers that circulate steam or hot water through pipes in the ponds, the possibilities for thermal aquaculture using waste heat are promising.

The construction of shore-based holding facilities such as ponds, tanks and raceways is expensive, especially if they must be provided with shading roofs to control seaweed growth (TANAKA, 1979) or with greenhouses to conserve heat. Added to the costs of pumping water the expenses may be so high as to require high-density culture of high-value species. Because of this TANAKA suggests that for large-scale cultures, floating sea cages in the outfalls of power stations have greater potential.

Indirect use of heated effluents

Water from power stations with closed-circuit cooling systems might present problems because of the build up of chemical wastes in cooling water following evaporative losses in cooling towers. Open-circuit systems are usually much less of a problem unless the power station employs excessively high concentrations of chemicals. Clearly, much depends on the tolerances of cultured organisms. Bivalves are particularly sensitive and even low chlorine levels depress their activity and reduce the growth rates (ANSELL, 1969).

Where effluent water quality is poor, it may be necessary to avoid direct contact between cultured organisms and heated effluent. One way of achieving this is to use heat exchangers between the contaminated cooling water and the aquaculture water supply. CARROLL and co-authors (1981), for example, employed heat exchangers in two prawn culture systems, using condenser cooling water, to avoid contamination by radionuclides and toxic chemicals. CLAUS and co-authors (1981) found the indirect use of heated water particularly useful for bivalve culture. However, since most cooling waters contain only low-grade heat the exchangers must be large and, consequently, they are expensive. HUGUENIN (1975) used interlocking steel piling as a cheap form of heat exchanger. In northern Norway long plastic piping was submerged in the sea to raise the temperature of water supplied for salmon, but fouling became a problem in a short time (KJØLSETH, 1981; WANDSVIK and WALLACE, 1981). Temperature control in hatcheries is conventionally achieved with a boiler and chiller connected to a heat exchanger (LIAO, 1981) for which the best materials are stainless steel (AYLES and co-authors, 1981) or titanium (KINGWELL and co-authors, 1977; GUERRA and co-authors, 1979a). BRETT (in MACCRIMMON and co-authors, 1974) refers to the availability of plate heat exchangers with an efficiency of 80% although the material was unspecified. With any kind of heat exchanger, it is most important to use materials that will neither corrode nor contaminate, and a design that allows easy cleaning to remove fouling.

As thermal mariculture becomes more established and commercially viable, electricity-generating engineers might design special heat exchangers for installation into, or next to, turbine condensers, which would supply heat to mariculture heat exchangers without using the low-grade heat of cooling water. The longer distance from the mariculture unit would require greater lengths of pipework. A similar arrangement has been used successfully in Romania for heating greenhouses (VAN DER HORST, 1972) in which the heat exchanger was placed between the steam turbines and condensers of a power station (p. 1907). Such a system might prove to be particularly useful for cultivation of bivalves. With present technologies the transfer of heat from cooling waters to aquaculture facilities via heat exchangers may not be economically viable (CARROLL and co-authors, 1981) and the development of steam-heated exchangers, in or near the steam turbine condensers, or the use of heat pumps in the discharge water similar to the system proposed by TIMMERMAN (1979) for heating domestic buildings (p. 1846) might eventually produce more viable systems.

(b) Sport Fishing

Some species of fish are attracted to the outfalls of heated cooling waters; the popular angling press in the UK frequently refers to the enhanced angling success near power station outfalls. References in the scientific literature, whilst less common, confirm the potential attractiveness of thermal effluents to certain species of fish (p. 1810; see reviews in COUTANT, 1979b; LANGFORD, 1972) both in freshwater (e.g. ELSER, 1965; GIBBONS and co-authors, 1972; MARCY and GALVIN, 1973) and marine environments (e.g. ALLEN and co-authors, 1970; GALLAWAY and STRAWN, 1974, 1975).

In many localities—from open sea coasts and estuaries to freshwater rivers and lakes—the attraction of fish to heated effluents has been shown to be largely seasonal (ALABASTER, 1963; DE SYLVA, 1963; ELSER, 1965; NAYLOR, 1965b; TREMBLEY, 1965; MIHURSKY, 1969; ALLEN and co-authors, 1970; FAIRBANKS and co-authors, 1971; BARKLEY and PERRIN, 1972; LANDRY and STRAWN, 1973). While in warm seasons fish are less attracted, they occur in high abundance during the colder months of the winter. Effluent discharge temperatures may become too high to attract fish during summer. For example, LANDRY and STRAWN (1973) refer to fish moving out of temperatures >35 °C into ambient waters. Attraction to warm water in winter can support a seasonal sport fishery, concentrated near the discharge area (ALABASTER, 1963; TREMBLEY, 1965; LANDRY and STRAWN, 1973; MARCY and GALVIN, 1973).

Waste heat is not always beneficial to sport fishing. Discussing the implications of waste heat for the sport fisheries in the USA, DOUGLAS (1968) suggested that a small amount of heated waste water may sometimes be of benefit to a fishery, but only in limited situations. Again, ALLEN and co-authors (1970) emphasized that the changes brought about by heated discharges depend on local conditions; only in some instances have they recorded beneficial effects. On the coast of northern California, at Humboldt Bay, they found enhanced sport angling in the vicinity of the discharge from the Buhne Point nuclear generating plant. Sampling by anglers equipped with line, hooks and bait (catch per angler hour) revealed that 3 out of 12 fish species analyzed preferred the warmer waters discharged by the generating plant. These were the walleye surfperch *Hyperprosopon argenteum*, the pile perch *Rhacochilus vacca* and the jacksmelt *Atherinopsis californiensis*. Another 3 species showed some attraction to the outfall. The walleye surfperch was attracted most commonly and many of these fish had been in the heated water long enough to acquire body temperatures equal to those of the effluent.

In the Patuxent estuary of Maryland, USA, over 50% of all angling visits in the area were made to the heated effluent canal of the Chalk Point Power Station (MOORE and co-authors, 1973). Over 90% of these visits occurred during the cooler season from October to April at a time when few visits were being made to other parts of the estuary. Quantitative assessments of the angling pressures and harvests in the area are given by MOORE and co-authors (1973). In Galveston Bay, Texas, LANDRY and STRAWN (1973) described an annual cycle of sport fishing activity in a warm effluent discharging to an outfall area of 9·9 ha.

In many industrialized countries with a tendency towards shorter working weeks and higher standards of living, angling is becoming increasingly important. In the USA for example, there is a large and expanding demand for aquatic recreational activities and sport fishing is an important part of that demand (COUTANT, 1977). In freshwaters, management and restocking takes place with a view to improving angling facilities. According to COUTANT, some reservoir fisheries have been so successful that the construction of new USA reservoirs are sometimes partly justified on the basis of the sport fisheries they are expected to provide. In the UK, some fish farms are organized to provide facilities for anglers. This aspect could be developed in marine farms. Facilities might include sport fisheries encouraged and developed in the outfalls of power stations and accessible to anglers. Most power stations are generally sited reasonably close to urban areas of dense population where easy access to such fishing facilities could be enjoyed and appreciated by large numbers of sea anglers. The construction of suitable cooling ponds might be a good method of providing these facilities in some localities. In freshwater, for example, one of the first purpose-built cooling lakes, in Illinois (USA), was converted into a recreational lake in which many species of fish thrive (PETERSON and SEO, 1977).

Whilst constructive uses of waste heat may be found in sport fisheries attracted to the heated plumes from power stations, this form of activity will be dependent on the continued operation of the power station involved. When such a station shuts down, temporarily, loss of that fishery might occur with large mortalities due to cold shock (p. 1813). Restoration of the fishery would undoubtedly take much longer than the duration of the shut down. Thus, such periodically *ad hoc* uses might have considerable disadvantages, although for most of the time they would provide useful recreational facilities. COUTANT (1977) suggests that one way of avoiding the cold-shock problem would be to have more than one power station in operation at a given site or, alternatively, to restrict the temperature elevations to levels that permit the fish to return to ambient temperatures without significant mortalities. The best way of utilizing waste heat would be in circumstances where man could have complete control over the system, as in suitably designed installations used for thermal aquaculture (HEDGPETH and GONOR, 1969).

(c) Agricultural and Horticultural Uses

Three main agricultural and horticultural methods are being developed for using waste heat from industry, particularly power stations: (i) soil warming in open fields, (ii) greenhouse heating and (iii) animal shelters. In addition, attempts have been made to use such heat for drying fruit and vegetables after harvesting. Most developments have been in association with inland power stations employing freshwater cooling. This is probably because of the greater environmental pressures for alleviating waste-heat problems on freshwaters and the larger number of inland compared with coastal power stations. There are problems in using sea water for these purposes because of greater fouling and corrosion and because accidental leakages of salt water into soil and greenhouses would be intolerable. YAROSH (1977) suggested that with all methods of using waste heat the biggest problems are availability of land and the constraints on its use. This is particularly true of agricultural and horticultural uses where large areas of land are usually required. Despite these difficulties it is likely that agricultural and horticultural uses could be adapted to coastal and estuarine sites.

Soil Warming

LUCKOW and REINKEN (1979) were very enthusiastic about the prospects for using waste heat for soil warming in fields next to power stations, as an alternative to dry cooling towers, to reduce temperatures before cooling water was returned to rivers. This is a particularly pressing problem in countries such as the Federal Republic of Germany with limited coastlines and, therefore, with rivers that are already heavily used for industrial purposes and where overheating becomes a problem. It is less of a problem with estuarine and open coast sites, at least in temperate regions, but there could be economic advantages in using the waste heat from coastal stations in this manner. The technique is to bury a network of pipes, through which warmed water is pumped to heat the surrounding soil. In the Federal Republic of Germany, LUCKOW and REINKEN (1979) buried pipes in fields on an experimental site using an oil-fired boiler to supply warm water, and on two sites next to a 2100 MW(e) lignite power station and a 237 MW(e) nuclear station, respectively. The following significant increases in crop yields occurred in the heated fields compared with adjacent unheated fields: corn, 57%, sugar from sugar beet, 70%; winter wheat, 40%; and pastureland 48%. Spring potatoes not only increased yields by 60% but also ripened 4 weeks earlier whilst soya beans, which do not normally grow in Germany, yielded 5·5 tons ha^{-1}.

GRAUBY (1977) described similar work in France. Much of the work on soil warming in the USA was reviewed by YAROSH (1977) and by DEWALLE (1979), who also described a study in Pennsylvania of the economic and technical feasibility of soil warming as an alternative to other methods of dissipating waste heat. In general, earlier harvests and increased yields for crops grown in heated open fields have been reported. Increases as high as 300% were quoted by DEWALLE for cabbages and snap beans in North Carolina and 85% and 100% for bush beans and broccoli, respectively, in Oregon. Here, it proved possible to grow two crops of bush beans in succession where only one had been grown previously (BOERSMA, 1970). In Minnesota, early varieties of potatoes matured 2 to 3 wk earlier owing to earlier planting in otherwise frozen soil (ALLRED, 1975). RYKBOST and co-authors (1975) found earlier maturation of nearly all the 13 crops they investigated in Oregon. However, some problems were experienced by these workers and in Alabama and North Carolina, summer-planted crops did not benefit from the heated soil. Further north, in Minnesota, freezing of the upper parts of potato plants occurred with warmed soil in the early spring (ALLRED, 1975) although studies in Oregon (PRICE, 1972) examined the value of warm water sprays, using warm water from a nearby mill, for providing frost protection to fruit crops. This allowed both earlier and later harvests and extended growing seasons. Direct spraying with cooling water, of course, would only be suitable where freshwater was used for cooling and the technique would not be feasible at estuarine and coastal power stations discharging brackish or fully saline cooling water without considerable modification. Even with power stations using freshwater cooling there may be limitations for crop irrigation since that water usually contains chemical contaminants added for various reasons.

A problem in using saline cooling waters to heat networks of pipes would be the considerable fouling, unless adequate and probably expensive antifouling precautions were taken. However, most modern power stations use some form of controlled chlorination to prevent fouling and by using modern plastic piping to prevent corrosion, the technique might eventually be feasible for coastal and estuarine stations. Another problem would be ensuring that no leakage of saline water took place into the surrounding soil with disastrous effects on the crops being grown. A remedy would be to have a closed-circuit freshwater-filled system for soil warming, heated with suitable heat exchangers installed between the power station turbines and condensers (p. 1907; VAN DER HORST, 1972). If such a system could be used for coastal and estuarine power stations it would avoid problems of fouling and leakage.

In general, results suggest that the value of a scheme would depend on season and geographical location. In warmer regions, open-field soil warming appears to be useful for boosting crops during cooler seasons of autumn, winter and spring, but has no value during warm seasons. In cooler, temperate regions, the crop response is favourable in spring, summer and autumn, as long as there is adequate moisture and nutrients, but in winter and early spring, freezing of the aerial parts of crops can be a problem (DEWALLE, 1979).

The feasibility of using waste heat from coastal and estuarine power stations for soil warming has apparently never been tested. OLSZEWSKI and co-authors (1979) assessed the relative merits of different methods of utilizing waste heat in an attempt to find the best technologies. In their list of 6 priorities, soil warming was ranked as low as fifth, presumably based on assessments carried out with freshwater cooling systems. DEWALLE (1979) quotes the conclusions of JOHNS and co-authors (1971) and RYKBOST and co-authors (1974) that the investment in soil warming was probably too great to be offset by agricultural advantages alone unless crops of high value were grown. They thought that soil warming over large areas would also need to be justified by the need for heat dissipation. If this is the case, then soil warming would have a greater environmental value in restricted estuaries than at good open-coast sites where, in temperate regions at least, there is probably little environmental requirement for pre-cooling before effluent is discharged to the sea. However, there have been very large increases in agricultural fuel costs in recent years, and the method might yet prove to be economically useful in the coastal zone. Many estuarine and coastal power stations, if they are not in urban and industrialized areas, are situated on or near good agricultural land in which soil warming might be attempted. This is particularly true of nuclear power stations where, for safety considerations, they are usually located away from densely populated areas.

Although soil warming might be a useful way of using some waste heat, it is unlikely to be a very feasible way of cooling water to ambient temperatures before returning the water to the environment, unless it were carried out on a very large scale. Various estimates have been made of the areas of land needed to provide adequate cooling. According to PLUMMER and RACHFORD, cited by DEWALLE (1979), 1820 ha of land would be required to dissipate the heat from a 1500 MW(e) nuclear power station. The cost at that time was estimated to be 54% above the cost of conventional natural-draught wet cooling towers but was some 40% less than dry cooling towers. Presumably heat dissipation in soil would be less likely to create cloud and fog conditions than wet cooling towers (p. 1852) and would be comparable to dry cooling towers. One of the problems is that cooling water being returned to the power station from soil warming would not be as cold as engineers would like and additional fuel would be required to generate electricity. Because of this, PLUMMER and RACHFORD estimated that heat dissipation in soil was at its optimum when the cost of additional fuel was balanced against the cost of using a larger area for soil warming to achieve greater cooling. As a result, their estimates of the required land area were considerably less than those of other workers who had divided the total heat load of the power station by the heat loss per hectare.

Greenhouse Heating

Although OLSZEWSKI and co-authors (1979) placed greenhouse heating below soil warming at the bottom of their list ordering the merits of various methods of utilizing waste heat, it does have the important advantage over open-field soil warming of better control of aerial environmental temperatures. As with soil warming in open fields, most schemes tried out so far used fresh cooling water from inland power stations, but their use at coastal stations using saline cooling water may be less feasible because of fouling and leakage possibilities. One of the earliest schemes appears to have been built in Romania and copied in the USSR (VAN DER HORST, 1972). At Ploesti, the Romanians have operated a complex of greenhouses covering 80 ha for many years, utilizing heat from a local power station. Cucumbers, tomatoes, eggplants and lettuces were being produced in commercial quantities and exported to other east European countries. The scheme had evidently been successful since, by 1972, another 50 ha of greenhouses had been added to the complex. In Japan the electricity generating industry has, for many years, used heated discharges from power stations to heat greenhouses by pumping water over the roofs and along the walls to maintain inside temperatures of 15 to 25 °C in which vegetables have been successfully grown (INTERNATIONAL ATOMIC ENERGY AGENCY, 1974). Schemes have been built and tested in France, (GRAUBY, 1977; FOURCY and co-authors, 1979), Arabia (JENSEN, 1972) and the USA (reviewed by YAROSH, 1977; DEWALLE, 1979; OLSZEWSKI and co-authors, 1979).

The greenhouse heating scheme devised by the Oak Ridge National Laboratory in the USA (OLSZEWSKI and co-authors, 1979) was tested on a large scale by the Tennessee Valley Authority in Alabama (BURNS and co-authors, 1979). Although it did not use waste heat from a power station, an oil-fired boiler simulated station cooling water, and the greenhouses produced crops of tomatoes and cucumbers with considerable success. Further north, in Minnesota, a greenhouse system utilizing heat from a 1360 MW(e) power station (ASHLEY and HIETALA, 1979) produced a variety of successful crops of flowers, tree seedlings, tomatoes, lettuce and green peppers at various times of the year and in marketable quantities of excellent quality. However, yields were poorer than those from commercial greenhouses. In Oregon, lettuce, tomato and cucumber crops were successfully grown in greenhouse experiments with soil heating (RYKBOST and BOERSMA, 1973), at the University of Pennsylvania.

The French greenhouses near a nuclear power station at Grenoble, described by FOURCY and co-authors (1979), cover an area of 9 ha. These have produced a wide variety of crops such as strawberries, tomatoes, cucumbers, aubergines, Jamaica peppers, cuttings of ornamental plants, chrysanthemums and rose trees.

In hot climates, such as the Arabian Gulf, natural sea water has been used to cool plastic greenhouses during the day and heat them at night to produce crops of cucumbers, tomatoes, lettuce and other vegetables (JENSEN, 1972).

In greenhouse systems using heat from power stations, two methods have been employed to tap waste-heat sources. In Minnesota (ASHLEY and HIETALA, 1979) condenser cooling water from one of the wet cooling towers was used with a minimum design temperature of 29·4 °C. In Romania, however, a heat exchanger was installed between the steam turbines and turbine condensers (VAN DER HOST, 1972). This resulted in temperatures of 90 to 150 °C, considerably higher than in the Minnesota experiments. This method used a closed-circuit, pressurized system of pumped water to distribute heat to the greenhouses. Of these two methods, the former may be less likely to be applied to marine and estuarine power stations using salt or brackish water for cooling purposes because of corrosion, fouling and leakage problems. The second method could be more feasible for these stations since a closed-circuit system would be filled with freshwater.

For heat losses, the distance heated cooling water has to be pumped before it can be used for greenhouse heating may not be a serious problem. For example, in Minnesota, (ASHLEY and HIETALA, 1979) the cooling water was pumped more than 0·8 km to the greenhouses, apparently with very little loss of temperature, whilst in Romania water of

90 to 150 °C was pumped under pressure through insulated pipes over a distance of 6 km with only a 1 C° temperature drop (VAN DER HORST, 1972). However, pumping costs could be prohibitive (p. 1900).

Various methods have been used to transfer heat into greenhouses. In the Romanian scheme, water was pumped through steel tubes, presumably heating the air in the greenhouses, and for this there were 60 km of tube ha^{-1}. In Alabama (BURNS and co-authors, 1979), water was pumped through evaporative pads in the greenhouses, which acted as primary heat exchangers providing both heat and humidity to the greenhouse atmosphere. In addition, finned-tube heat exchangers provided dry heat and a lower relative humidity when required, for example, to help control, together with chemical spraying, fungal attacks (*Botrytis* spp.) on tomato crops. A similar combination of dry and evaporative heat exchangers was being planned in Czechoslovakia (HATLE and LAMPAR, 1975). FREEMYERS and INCROPERA (1979) compared evaporative and dry heat exchangers in model studies and concluded that wet evaporative systems were better because they had the additional advantage of latent heat exchange through evaporation and the system could also be used for cooling purposes during the summer. If such a system is ever developed for use with salt water, care may need to be taken to ensure that the air circulation system does not pick up droplets of salt water from the evaporative pads and transfer them to the leaves of salt-intolerant crops. However, JENSEN (1972) mentions air in greenhouses at a University of Arizona facility in Mexico being circulated through a spray of naturally warm sea water to heat or cool it and maintain temperatures of about 27 °C year round.

Some heated greenhouses have also used soil heating in addition to air heating. ASHLEY and HIETALA (1979) described the system in Minnesota that used a combination of forced air with a finned-tube heat exchanger for heating the atmosphere and a buried network of polyethylene pipes to heat the soil and maintain warmth at root levels. Their system demonstrated the effective use of waste heat from a 1360 MW(e) station. Despite outdoor winter air temperatures as low as −40 °C, and the water being pumped a distance of 0·8 km from the power station cooling towers, greenhouse temperatures of 12·8 to 15·6 °C were maintained using cooling water at 32·2 to 37·8 °C.

The French system at Grenoble (FOURCY and co-authors, 1979) used an ingenious method of heat transfer which provided both atmospheric and soil warming from the one method: plastic heat exchangers laid on the soil surface. These were flat, flexible PVC sleeves of 'lay-flat' tubing through which welded eyelets of different diameters and spacing provided holes through which plants grew. Water from the nearby nuclear power station was pumped through the plastic sleeves at 12 to 40 °C, passing heat to the underlying roots and the overlying foliage. This system has the advantage of reducing moisture losses from the soil, because the plastic acts as a mulch, and the sleeves can be easily moved or rolled up out of the way to allow soil cultivation for the next crop. If black PVC were to be used, the sleeves could also act as solar panels to remove excess heat from the greenhouses during warm sunny weather, perhaps to be stored for later use maintaining greenhouse temperatures during the night.

There do not appear to be any cases in the literature where soil warming or greenhouse heating have been used with coastal stations. Historically, this may be because there have been greater pressures to try and use, and therefore reduce, the heat discharged into restricted freshwaters. Certainly the problems of applying the technique to seawater cooling systems may be greater because of the risks of fouling and leakage of sea water into the soil. These problems may prove that the application of the method to the coastal zone will be difficult but the examples of glasshouses in Romania, where freshwater for a closed-circuit system is warmed in a heat exchanger between the turbines and the condenser (VAN DER HORST, 1972) and the Mexican greenhouses in which air is circulated through sea water sprays (JENSEN, 1972) may be ways of solving the problem. The ultimate success of schemes will probably depend on the economic advantages of growing worthwhile crops out of season, especially in temperate zones. In an age when costs of conventional fuels for greenhouse heating are increasing very significantly the economic case for using waste heat for this purpose should be improving continually. Indeed, in the UK many commercial greenhouse businesses have been liquidated recently because of the high costs of oil for heating. MARSHALL (1979) mentioned UK developments for greenhouse heating in North Yorkshire using waste heat from a large power station with freshwater cooling. The authors understand that on this site commercial greenhouses covering about 8 ha are now producing large quantities of tomatoes. In 1981, the crop was expected to exceed 2000 tons.

Animal Shelters

Another use of waste heat in agriculture is heating animal shelters to increase food conversion values and growth rates. OLSZEWSKI and co-authors (1979) refer to early work in the USA by the Oak Ridge National Laboratory (see also BEALL, 1971, 1973; HIRST, 1973). The type of greenhouse described earlier for horticulture (p. 1908), heated by finned-tube dry heat exchangers and wet evaporative pads, was used in studies of the engineering and technical feasibility of the method. BELTER (1975) mentioned various US farm animal studies in which environmental temperature was an important factor in increasing growth rates. YAROSH (1977) referred to work in Connecticut and Maryland on the effects of temperature on food consumption and weight gains in broiler poultry and to similar work with pigs by HEITMAN and co-authors (1958). In colder climates, particularly, improvements in weight gains have been significant. For economic and performance reasons, animal shelters ranked second only to external pond culture in OLSZEWSKI and co-authors' (1979) list of potential uses of waste heat; the prospects for this use appear to be good. Their work was carried out with inland power station effluents but it seems likely that the technique could be extended to coastal and estuarine stations without too much difficulty.

The possible limitations of using sea water for greenhouse and soil warming, due to potential corrosion, fouling and leakage, might not be so great in the case of animal shelters since leakage would be of less consequence. However, OLSZEWSKI and co-authors (1979) also mentioned the use of both open and closed heated water systems for algal and aquaculture work. In the open system, the condenser cooling water was used directly whilst in the closed system a heat exchanger separated the condenser cooling water from that used for aquaculture. If a similar closed system could provide a high enough temperature for heating animal houses then it should avoid any problems of using salt water. Alternatively, heat exchangers placed in the exhaust steam between the turbines and the condensers, as used in Romania (p. 1907) (VAN DER HORST, 1972), could provide heated fresh water in a closed system at higher temperatures.

It would seem, though, that the use of waste heat in animal shelters would have its main use in stimulating farm animal growths rather than in using large amounts of waste heat to the extent of significantly reducing aquatic environmental effects in estuaries and on coastal sites. For example, BELTER (1975) estimated that a 1000 MW(e) nuclear power station could provide sufficient waste heat to raise about 1 billion broiler chickens or 10 million pigs yr^{-1}. A typical poultry farm produces about 50000 birds annually and a large pig farm about 5000 pigs. BELTER suggested that typical existing operations are 10 to 100 times smaller than would be needed to use 1% of the waste heat from a modern power station.

The incentives for the eventual application of waste heat from coastal and estuarine power stations for agricultural and horticultural uses will almost certainly be economic rather than environmental. Since the costs of heating greenhouses are increasing greatly in temperate regions, the potential for using waste heat for this purpose at coastal and estuarine power station sites would appear to be considerable. However, it would utilize only a small proportion of the heat and would be of less value in subtropical and tropical regions where thermal pollution is more serious.

(d) Other Constructive Uses of Thermal Effluents

Heated Effluents

The constructive use of heat from power stations by combined heat and power generation has already been discussed (p. 1843). Various other constructive uses for waste-heat discharges have been summarized by DAUGARD and SUNDARAM (1979). They include integration of power stations with sewage treatment plants where sewage effluent is passed through and warmed in the power station cooling system to enhance the biological degradation process. There may be considerable scope for combining this with the type of effluent/aquaculture/tertiary treatment system tested by ROELS and co-authors (1978) (p. 1884). DAUGARD and SUNDARAM concluded that effluent/ aquaculture systems (not using waste heat) might provide alternatives to conventional tertiary treatment systems in areas with populations of 60 000 to 70 000 producing about $3785\,m^3\,d^{-1}$ of sewage. This would be feasible where there was adequate land but not in densely populated urban areas. These authors suggested that the southern USA would be a good region for these developments because of the smaller fluctuations in seasonal temperatures. Power station waste heat might allow these developments in cooler regions (KEUP, 1981).

Other suggested uses for heated effluents include winter de-icing of roads and airport runways, heating outdoor recreational swimming areas, maintaining ice-free shipping lanes and desalination (DAUGARD and SUNDARAM, 1979). These authors refer to the use in Saudi Arabia and Kuwait of condenser cooling water at fairly high temperatures in sea water evaporation systems to produce freshwater in desalination plants. The disadvantages with distillation desalination plants is that they produce effluent with a high temperature and salinity. MOBERG (1977), however, described a new desalination method that utilizes warm sea water discharged by various industrial plants, such as power stations and factories, as sources of feed water (p. 1847). The original ΔT of the feed water (power station discharge) is approximately halved but the total volume of discharge is more than doubled. The method shows considerable energy savings over conventional desalination plants and MOBERG (1977) claims that the energy used with the new method is less than $1·7\%$ of the energy required to heat water to vaporization in other desalination systems. Economically the method is claimed to be competitive with other systems, even using water with ΔT values of only 7 C°. At higher values it should be even more competitive but this would result in higher discharge temperatures.

A possible use for desalination plant brines may be the generation of electricity. If pools of brine, overlain by an insulating layer of freshwater, are exposed to solar radiation, the brine will collect up to 90% of the incident solar energy and the heat trapped can be used to produce electricity (HUDEC and SONNENFELD, 1974). The method has been used in Israel, particularly in high-salinity ponds and lakes adjacent to the Dead Sea. Further development may eventually result in its widespread use in warm climates, possibly using heated brines from the increasing numbers of desalination plants in regions like the Middle East, although the amounts of brine that could be used in this way would be limited. The technique draws attention to a further problem with waste brines released into shallow sea areas; where waste brines are overlain by waters of lower salinity, the absorption of solar energy by the underlying hypersaline water could result in considerable temperature elevations. This may occur already to some extent in certain warm coastal waters where high salinity occurs naturally. In shallow bays, lagoons and salt lakes along the Gulf coast of Saudi Arabia, salinities of 60 to 200‰ are common owing to evaporation (BASSON and co-authors, 1977).

Cold Effluents

The regulation of cold-water discharges from liquid natural gas installations has been briefly discussed (p. 1872). These discharges result from the use of ambient sea water to gasify the liquid gas as it is pumped into distribution pipelines. The obvious use for such effluent is as a source of cooling water for other industrial processes, e.g. power station cooling (HEINLE, 1977; REISH, 1977). There should be few problems in using all such discharges as long as there is adequate cooperation between the industries concerned at the planning and construction phases, since HEINLE pointed out that one power station could provide nearly all the heat from its cooling waters to gasify all the liquid natural gas likely to be imported into the USA. Whilst these cold effluents may help to increase slightly the efficiency of industrial cooling, they would have only a very small effect in reducing thermal pollution.

(7) Conclusions and the Future

(a) Conclusions

Maximum thermal deformations of estuarine and marine environments result from discharges of power station cooling water. Heated effluents from distillation desalination plants are fewer and of smaller volume than power station effluents. Nevertheless, they could be more serious for the surrounding environment because they are warm, of high salinity, negatively buoyant and spread across the seabed, possibly with drastic effects on the benthos. Careful siting of discharges and rapid mixing with the surrounding ambient water can do a great deal to mitigate the effects of these discharges. In many parts of the world the numbers of desalination plants are increasing. In the event of water shortages this will almost certainly create thermal/salinity problems in the future if distillation continues to be used extensively. There appears to be little information on the environmental effects of these effluents and there is great need for more research.

Cold effluents from liquid gas plants are much less common. There is virtually nothing known about their effects and more research is needed. Care should be taken in their siting to avoid the cold, dense discharges spreading over the sea bed. There should be effective mixing with ambient water to reduce potential thermal deformations. Cold discharges should be useful for power station cooling and the combination invites more attention.

In contrast to negatively buoyant effluents, much is known about the impact of heated water from electricity generating stations. A vast and varied literature was produced during the last 20 yr. Generalization is difficult because the effects of a power station can vary greatly according to geographical location, siting and topographic, hydrographic, meteorological and biological conditions. Power stations with open-circuit cooling exert the greatest impact. They can affect organisms thermally, physically and chemically. Organisms may be entrained in the cooling water, impinged on intake screens, treated with biocides, undergo considerable physical stress by pumping and turbulence and be subjected to heat shock in the condenser tubes before being discharged in a plume of warm water which cools and mixes with the ambient water.

In some localities the more serious effects of power stations are not caused by heating but by mechanical and chemical damage due to entrainment. This is particularly true of biologically rich estuaries where spawnings and migrations of commercially important species take place. In some estuaries, serious damage has occurred when large numbers of young fish have been caught on intake screens and killed through impingement. Smaller stages passing through screens have died as a result of mechanical damage and/or temperature elevations and/or biocidal treatments to reduce fouling. The problem is particularly complicated by the synergistic effects of temperature, chemicals and mechanical damage. Pumping more water to reduce temperature elevations may cause higher total mortalities by exposing more organisms to impingement, mechanical damage and biocides. It is clearly very important to maintain, as far as possible, the optimum balance between these various factors in existing power stations.

Even if organisms are not entrained, the distribution of heated effluents from the open-circuit cooling of power stations may cause considerable effects in the surrounding estuarine or coastal environment, depending on geographical and local conditions. Thermal effects are greatest in tropical and subtropical regions of the world where marine and estuarine organisms live close to their lethal temperature levels (MAYER, 1914). There is far less scope for discharging thermal effluents in the tropics than in temperate waters without causing damage to the fauna (MIHURSKY and PEARCE, 1969). In the tropics and subtropics, highest entrainment mortalities may occur through thermal shock, especially in combination with other chemical and physical factors. Even in more temperate waters of the world considerable care is required since damage has been caused by existing power stations, particularly in enclosed estuarine sites. A particular problem here is that the fauna, especially fish attracted to the vicinity of a heated discharge during colder months of the year, may be subjected to a lethal cold shock if the power station suddenly ceases operations.

There is still a great lack of information on the effects of entrainment on most of the species likely to be pumped through power station cooling systems. Similarly, the wider ecological implications of thermal and synergistic effects are poorly understood. Only in a few studies has a power station been treated as an inanimate predator cropping a certain proportion of the local fauna (COUTANT, 1972; SCHUBEL, 1974; ENRIGHT, 1977). In management terms this could mean using the principles of managed fisheries (Volume V: GULLAND, 1983) to establish acceptable limits for mortalities imposed by power stations.

In regulating the environmental impact of thermal effluents much can be done to reduce adverse effects. Improvements in the efficiencies of power generation can help by reducing the amount of fuel energy discarded as waste heat. Recent increases in oil fuel prices have resulted in considerable economic incentives to reduce the wastage of fuel energy in cooling waters, possibly by introducing new and more efficient methods of generating electricity. However, this may be offset by the increased construction of nuclear power stations and by the future introduction of fast-breeder reactors which will use the plutonium produced by the present generation of fission reactors. For a given generating capacity, nuclear power stations discharge more waste heat to their cooling water than conventional stations and their fuel supplies may be sufficient to reduce the economic incentives for fuel conservation.

Better methods of generating electricity include combined heat and power generation which can do much to utilize otherwise wasted thermal energy. Although the efficiency of electricity generation alone is reduced, the overall efficiencies of the combined heat/electricity generation is so much higher than with conventional power stations that there are significant reductions in the waste heat discharged in cooling waters.

Various methods can be used to reduce the impacts of thermal effluents being discharged to the marine environment. These include careful siting, both geographically and locally, the use of closed-circuit cooling systems, with cooling ponds, lakes or. reservoirs, wet or dry cooling towers or combinations of these, and satisfactory mixing of heated effluents with the ambient water. Wet cooling towers can, however, create other problems due to atmospheric effects.

CLARK and BROWNELL (1973) made the following specific recommendations for the use of these different types of cooling system:

(i) Open-circuit cooling systems should only be used on open coastal sites away from estuaries, where minimal damage is likely to be caused to the biota. At these sites the intakes and outlets should be located offshore if this minimises prospects of entraining organisms like fish. SCHUBEL and co-authors (1978) emphasized that open-circuit cooling systems can be designed to reduce entrainment losses, particularly from thermal effects.

(ii) Open-circuit cooling systems should not be used at vulnerable estuarine sites. Estuaries are usually very important nursery areas for many species and there is a high risk of entrainment. Considerable damage may be inflicted by power station operations.

(iii) In estuaries not vital to the spawning of some species, closed-circuit cooling systems using cooling towers or spray ponds should be acceptable as long as the discharge water is treated adequately for temperature and waste chemicals used for various purposes in the power station.

(iv) In estuaries that are vital spawning areas for some species (e.g. striped bass in the USA), even closed-circuit cooling systems can create problems and should not be used because of the entrainment of organisms in make-up water and the discharge of blow-down water.

Since CLARK and BROWNELL's (1973) recommendations were made, there have been some developments in the design and construction of dry cooling towers and in the design of wet/dry combinations for use with closed circuit systems. These could provide alternative cooling systems for vulnerable estuarine areas, because of the reduction in volumes of make-up and blowdown water. However, the economic costs of construction and reduced generating efficiency may be prohibitively high. SCHUBEL and co-authors (1978) have emphasized that in most cases survival of organisms can be assured without expensive 'over-design' of the cooling system.

Constructive uses of waste heat may help to reduce the amounts discarded to the environment, particularly in the case of combined heat and power generation. However it is questionable whether constructive uses will be able to reduce the total discarded waste heat significantly. Nevertheless, research and development should be continued on an increased scale to take advantage of this otherwise lost resource. Thermal aquaculture is the best developed constructive use with many commercial enterprises being established. Horticultural and agricultural uses have some promising prospects.

(b) The Future

There has been considerable speculation that man's future activities could have pronounced effects on the climate. FLETCHER (1969), HEDGPETH (1977) and LAMB (1977) discussed various examples of ambitious proposals to tamper with water movements in the oceans, some of which could affect climate and sea temperatures. BORISOV's (1962, 1969) proposal to dam the Bering Strait was intended to stop the northward flow of cold Pacific water and increase the inflow of warm Atlantic water into the Arctic Ocean to free it of ice, possibly by pumping 500 km^3 d^{-1} of Arctic water to the Pacific Ocean. However, various authors have criticized the scheme (discussed by LAMB, 1977). FLETCHER (1969) mentioned two proposals to deflect the Gulf Stream with a dam between Cuba and Florida and weirs across the Grand Banks from Newfoundland. FLETCHER (1969) and LAMB (1977) discussed a prosposal to control flows in the narrow Tatarsk Strait in Japan to increase the flow of the warm Kuroshio Current into the Sea of Okhotsk to reduce the winter ice. There have also been various more likely proposals to tamper with the flow of freshwater into the oceans which could give rise to considerable thermal deformations. For example, DAVIDOV's proposal (FLOHN, 1963) to divert the flow of the large rivers of Siberia from the Arctic Ocean to the more arid regions of Central Asia might result in temperature and salinity changes in Arctic waters and increased ice cover with consequent changes to the climate (GABOR and co-authors, 1978; LAMB, 1982). Altering the freshwater supplies of North America, Africa and Siberia by large impoundments and canals could have enormous temperature and salinity effects on coastal and estuarine regions (FLETCHER, 1969) and consequent effects on the distribution of marine organisms.

Grandiose schemes such as these are very much in the future and highly speculative. There are, however, other activities by man which may have much more generalized effects on global and ocean temperatures by affecting the earth's heat budget.

Virtually all the earth's energy income is from short-wave radiation from the sun. Approximately 99% of this is between 0·17 and 4 μm, with about half within the visible spectrum of 0·4 to 0·74 μm (LAMB, 1972). Some of this radiation is reflected back into space and is of no importance to the earth's heat budget (SVERDRUP and co-authors, 1963). However, radiation is absorbed by the atmosphere (particularly by oxygen and ozone), oceans and land masses (LAMB, 1972). About 99% of terrestrial radiation has a much longer wavelength than solar radiation, between 3 and 100 μm (LAMB, 1972). Carbon dioxide, water vapour and various trace gases in the atmosphere absorb within the spectrum of the outgoing terrestrial radiation and so retain some of this heat. From the entire system, long-wave radiation returns heat to space in a manner that balances the radiation received, thus maintaining overall average temperatures in the atmosphere and oceans that remain fairly constant (SVERDRUP and co-authors, 1963). The total mass of the world's oceans is about 270 times that of the atmosphere, and the specific heat of the waters is about 4 times that of dry air at constant pressure (LAMB, 1977). Thus, the oceans act as a heat sink that greatly influences the world's climate.

Various human activities have been cited as being potentially responsible for changes in the earth's heat budget and climate, giving rise to much discussion and controversy in view of natural climatic variability and long-term changes. A recent synthesis by LAMB (1982) and his important reviews (1972, 1977) consider pertinent literature. The WORLD METEOROLOGICAL ORGANIZATION (1972) and the NATIONAL ACADEMY OF SCIENCES (1975) conclude that human effects on both regional and world climate may now be just detectable and will gradually increase in importance. Probably of greatest importance is the growing evidence that increasing quantities of carbon dioxide gas in the atmosphere from fossil-fuel combustion and deforestation is augmenting the 'greenhouse' effect, trapping more of the long-wave radiation passing from the earth's surface into space (e.g. LAMB, 1972, 1977, 1982; NATIONAL ACADEMY OF SCIENCES, 1979; HANSEN and co-authors, 1981). This could have serious long-term implications for ocean temperatures. LAMB (1972) quotes various authors who suggest that long-term variations in the atmospheric carbon dioxide could make a difference of ± 1% in the earth's surface emissions of long-wave radiation absorbed by the atmosphere. According to PLASS (1956), doubling the present atmospheric concentration could raise world temperatures by 3·6 C°. He suggests that the observed 1 C° rise between the beginning of the industrial age and 1950 is due to the increase in atmospheric carbon dioxide caused by burning fossil fuels. While MATTHEWS (1959) assesses the effect to be about a quarter of PLASS's figures, MANABE and WETHERALD (1967) believe it to be about half. However, various other natural causes have also been cited (reviewed by LAMB, 1972, 1977, 1982; CUSHING and DICKSON, 1976).

LAMB (1972) points out that although fossil-fuel combustion was adding 9×10^9 ton yr^{-1} of carbon dioxide, not all was being retained in the atmosphere since the oceans act as an enormous reservoir for this compound, holding about 27 times the amount present in the atmosphere. Furthermore, much of this becomes fixed as calcium carbonate in the deposits of the sea floor. According to LAMB (1977), only 50 to 75% of the carbon dioxide produced remains in the atmosphere, the rest passing to oceans and biosphere. However, since the vertical circulation is slow, LAMB (1972) suggests that it would take 10 000 to 50 000 yr for any changes in atmospheric concentrations to stabilize between the 3 reservoirs of atmosphere, oceans and ocean deposits. LAMB (1977) thought that although the tendency for increased carbon dioxide was to increase the general warming this effect may be much smaller than the commonly accepted estimates. Various aspects of the fate of fossil-fuel carbon dioxide in the atmosphere and oceans are discussed in numerous papers in ANDERSEN and MALAHOFF (1977).

HANSEN and co-authors (1981) add considerably to the view that increasing concentrations of atmospheric carbon dioxide and various trace gases are enhancing the 'greenhouse' effect. Their model studies suggest that in the future this will result in important changes to the world climate because of global warming. If so, some of the most important thermal deformations in the oceans will arise not simply from the discharge of heated effluents but from small temperature increases in the surface waters, over very extensive areas, in response to the general warming (TAKANO, 1981).

According to HANSEN and co-authors (1981), the general effect of increased carbon dioxide is to absorb more of the earth's outgoing radiation so that the general global temperature increases until the outgoing radiation again becomes equal to the solar energy absorbed by the earth. In their model studies, they draw attention to the great heat capacity of the oceans and the fact that the upper 10 m is rapidly mixed. The heat capacity of the continental land masses can be ignored because the ground is a good insulator. However, the rapid exchange of air between land masses and oceans means that there is a horizontal atmospheric heat flux and a rapid exchange of heat between the atmosphere and the ocean surface layers that removes from the atmosphere any net heat derived from the continental land masses. Discussions of relationships between long-term weather and sea surface temperature anomalies were presented by SAWYER (1965), LAMB (1972, 1977) and CUSHING and DICKSON (1976). HANSEN and co-authors (1981) stress that the general lack of knowledge of ocean processes creates uncertainties about the time dependence of global carbon dioxide warming. They considered the case where heat is rapidly mixed with the upper 100 m layer alone and the case where this layer mixes through the thermocline, diffusing to a depth of 1000 m with a diffusion coefficient of $1 \text{ cm}^2 \text{ s}^{-1}$. The global warming between 1880 (293 ppm CO_2) and 1980 (335 ppm CO_2) was estimated to be $0.4 \text{ C}°$ in the first case and $0.25 \text{ C}°$ in the second. Using temperature records from several hundred stations, HANSEN and co-authors extracted 5-yr running mean temperature trends for northern, southern and low latitudes and showed that the global surface air temperature rose about $0.4 \text{ C}°$ in the past century—a rise consistent with the value derived from their model, although this included global cooling in the early 1960s.

Other man-made changes may have resulted from testing nuclear weapons in the atmosphere during the 1950s and early 1960s (KONDRATYEV and NIKOLSKY, 1979). These authors suggest that large emissions of nitrogen oxides from nuclear explosions break down ozone in the stratosphere by photochemical reactions. This absorbs solar energy before it reaches the lower troposphere, resulting in a cooling effect. GRIBBIN (1981) cites KONDRATYEV and NIKOLSKY's unpublished work pointing out that these increases in nitrogen oxides coincided with the weather pattern in the northern hemisphere in the early 1960s, when there was an increased frequency of negative temperature anomalies over the whole globe with maximum drops in 1964, 1965 and 1966. It must be emphasized, however, that this cooling has also been attributed to a number of other natural causes (see discussions in CUSHING and DICKSON, 1976; LAMB, 1977, 1982). The average winter air temperature in 1962–63 was $0.4 \text{ C}°$ below the long-term mean, and GRIBBIN (1981) pointed out that in the UK this was one of the two worst winters in living memory (see also LAMB, 1977, 1982). The effects of this cold winter on the marine environment have been extensively documented and reviewed in Volume I: KINNE (1970b).

Interestingly, this general cold period of 1963–66 also coincided with the beginning of a reversal in the 'Russell cycle' (CUSHING and DICKSON, 1976) in the western English Channel in 1965 (SOUTHWARD, 1980), for which various responsible factors have been suggested (SOUTHWARD, 1980; BUTLER and SOUTHWARD, 1981). The general cooling also agrees with observations for that period from the surface waters of the Rockall Trough in the northeast Atlantic (ELLETT and MARTIN, 1973), with a cooling phase in quasi-cyclical changes in northeast Atlantic waters generally and with a cool period in the North Sea (COLEBROOK and TAYLOR, 1978). Interpretation of these findings is considerably more complex than the suggestion that they may be a result of a general global cooling. For example, northeast Atlantic temperatures appear to show an inverse relationship to Gulf Stream flow (MARTIN, 1972) and to Trade Wind strength (COLEBROOK, 1976), although COLEBROOK and TAYLOR (1978) stressed that anomalous surface exchange, local advection and upwelling must occur as well. The latter authors conclude that for the open North Atlantic Ocean, secular temperature changes appear to be mainly advected and determined by variations in the North Atlantic Current. In contrast, COLEBROOK and TAYLOR hold that the sea surface temperatures of the European shelf appear to be determined mainly by direct heat exchange with the atmosphere.

The extents to which climates are actually influenced by human activities is difficult to determine. Natural fluctuations take place in the world's climate (extensively described by LAMB, 1972, 1977, 1982) and exert considerable effects on the marine environment (e.g. CUSHING and DICKSON, 1976; LAMB, 1977, 1982; CUSHING, 1978; SOUTHWARD, 1980; Volume V, PERES, 1982a,b). Some natural climatic fluctuations may be the result of solar cycles. Apparently, these have been stable features of the sun over many millions of years (RICHTER-BERNBERG and WESTOLL, cited by LAMB, 1972, p. 20; see WILLIAMS, 1981, for a recent discussion of Pre-Cambrian cycles and bibliography). Other changes may result from increased amounts of volcanic aerosols and particulate material in the atmosphere (LAMB, 1972, 1977, 1982; HANSEN and co-authors, 1978, 1981). These may reduce the amounts of solar energy collected by the earth and LAMB (1982) suggests that volcanic dust may be of greater importance than carbon dioxide. However, LAMB emphasizes that the possibility of global warming has to be balanced against the possibility of cooling on a different, longer time-scale as the natural climate develops. Neither side of the balance is yet clearly known or understood.

It is possible that man-made changes interact with natural fluctuations, thus producing either an enhanced or reduced effect. For example, the general global cooling in the early 1960s (HANSEN and co-authors, 1981) coincided with a period of reduced sunspot activity (SOUTHWARD, 1980), increased input of volcanic aerosols to the stratosphere (HANSEN and co-authors, 1978, 1981) and an increase in stratospheric nitrogen oxides from nuclear bomb testing (KONDRATYEV and NIKOLSKY, 1979). However, it is not known what proportion of the cooling may be ascribed to each of these possible factors, or to various factor combinations.

The work of HANSEN and co-authors (1981) suggests that such cooling effects are of a temporary nature and that a general global warming occurs as a result of carbon dioxide emissions, which will emerge from background variations towards the end of the present century with an overall global elevation of about $0 \cdot 2$ C$^{\circ}$ since 1950 (see also LAMB, 1982). It is not clear what temperature elevations will be involved in the oceans and a great deal will depend on the distribution of the additional heat. HANSEN and co-authors cite references stating that, on the basis of palaeoclimatic evidence, the overall warming in the high latitudes will be 2 to 5 times the mean global warming.

Alterations in the regional climatic variations will occur with any general warming and this presumably will have great significance for man because of changes in precipitation, location of deserts and fertile regions. Coastal areas may become wetter and melting of ice sheets might occur with elevations in sea levels. A doubling of atmospheric carbon dioxide could result in a general global warming of about 2 C° some time in the next century (HANSEN and co-authors, 1981) which might cause a warming of about 5 C° in higher latitudes. The West Antarctic ice sheet is evidently vulnerable to rapid melting (MERCER, 1978) which might take place over a century or less and cause a rise in sea level of about 5 to 6 m with very serious flooding implications (HANSEN and co-authors, 1981). These authors point out that much depends on the growth of energy production in the future and show the amounts of carbon dioxide produced by various types of fuel (Table 8-5).

At present, oil is the worst offender, with coal in second position. However, coal is by far the worst potential source of atmospheric carbon dioxide whilst nuclear and some renewable energy sources produce no increase at all. Clearly, much depends on future policies concerning the production of energy but according to HANSEN and co-authors (1981) there is little prospect of political and economic considerations being influenced until there is convincing evidence of the results of global warming, presumably towards the end of the century. They give projections for increased global temperatures to the end of the next century according to various rates of growth and using different types of fuel (Fig. 8-23). Even with no growth at all the projected ΔT would be about 0·5 C° by the year 2020 and about 1 C° by 2080. It seems clear that if these predictions are correct man can no longer afford to discharge carbon dioxide to the atmosphere at the present rate because the climatic, social, political and economic consequences will be too

Table 8-5

Energy supplied and CO_2 released by various fuels (After HANSEN and co-authors, 1981; copyright 1981 by the American Association for the Advancement of Science; reproduced by permission of the American Association for the Advancement of Science and H. J. Hansen)

Fuel	Energy supplied in 1980[a]		CO_2 release per unit energy (oil = 1)	Airborne CO_2 added in 1980[a]		CO_2 added through 1980 (ppm)	Potential airborne CO_2 in virgin reservoirs[b] (ppm)
	$J \times 10^{19}$	%		%	ppm		
Oil	12	40	1	50	0·7	11	70
Coal	7	24	5/4	35	0·5	26	1000
Gas	5	16	3/4	15	0·2	5	50
Oil shale, tar sands, heavy oil	0	0	7/4	0	0	0	100
Nuclear, solar, wood, hydroelectric	6	20	0	0	0	0	0
Total	30	100		100	1·4	42	1220

[a] Based on late 1970s.
[b] Reservoir estimates assume that half the coal above 914 m can be recovered and that oil recovery rates will increase from 25 to 30% to 40%. Estimate for unconventional fossil fuels may be low if techniques are developed for economic extraction of 'synthetic oil' from deposits that are deep or of marginal energy content. It is assumed that the airborne fraction of released CO_2 is fixed.

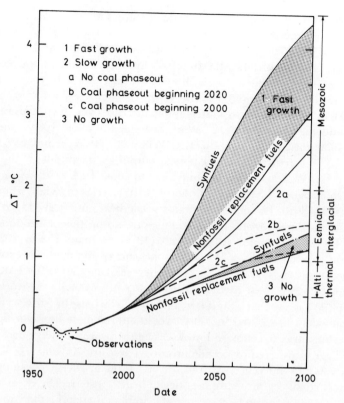

Fig. 8-23: Projections of global temperature increases. Diffusion coefficient beneath the ocean mixed layer is $1\cdot2$ cm^2 s^{-1}, as required for the best fit of the model and observations for the period 1880 to 1978. Estimated global mean warming during earlier warm periods indicated on right. (After HANSEN and co-authors, 1981; copyright, 1981, by the American Association for the Advancement of Science; reproduced by permission of the American Association for the Advancement of Science and H. J. Hansen.)

serious. The case for changing to generating power by some of the alternative energy sources becomes very strong when considered in the context of global carbon dioxide warming and for this reason the generation of electricity by nuclear power has obvious advantages over fossil-fuelled power stations—in addition to the absence of sulphur dioxide emissions which cause acid rain.

The general production of waste heat by man has been causing considerable concern about how much heat can be added to the world before it starts to affect world temperatures and climate. LAMB (1977) emphasized that whilst forward projections of waste-heat output are very uncertain because of the human, political and economic uncertainties, nevertheless there are likely to be climatic problems from increasing waste heat disposal unless the world maintains near-zero growth. Various authors have made predictions about future quantities of waste heat. WILCOX (1977) calculated that if the world's consumption of energy continues to increase at its present rate of 5 to 6% yr^{-1}

then in another 100 yr the input of low-grade heat from man's activities would be about 1% of the solar energy received by earth. In 150 yr it would be 10% and in 200 yr it would equal the solar energy input, which has been estimated as $1 \cdot 8 \times 10^{17}$ W at the earth's surface (TAKANO, 1981). LAMB (1977) cites a United Nations report on world energy supplies that calculated that with a growth rate of about 5% yr^{-1} the output of waste heat would rise from $5 \cdot 5 \times 10^{12}$ W in 1970 to $9 \cdot 6 \times 10^{12}$ W in 1980 and $31 \cdot 8 \times 10^{12}$ W in 2000, an approximately 6-fold increase over 30 yr. This rate of increase would be producing about $1 \cdot 8 \times 10^{15}$ W of waste heat in about 100 yr time, equivalent to 1% of the solar energy, thus agreeing with WILCOX's (1977) estimate. WEINBERG and HAMMOND (1970) predicted that waste-heat output by the middle of the next century would be 3×10^{14} W. This would be equivalent to about $1 \cdot 24 \times 10^{15}$ W in 100 yr, about $0 \cdot 7\%$ of the solar figure. LAMB (1977) quotes BUDYKO (1966), who assumed a 12 to 15 times increase by 2050. By that time, man's waste-heat output, averaged over all land masses, would be about 10% of the present average absorption of incoming radiation at the surface. LAMB also quotes SCHNEIDER's (1974) estimate that by 2020 waste heat would amount to a few tenths of 1% of the solar energy and might be expected to raise global temperatures by 'several tenths of one degree'. Comparison of TAKANO's (1981) figures for solar energy at the earth's surface ($1 \cdot 8 \times 10^{17}$ W) with the world energy consumption (10^{13} W) shows that if the latter continues to increase by 5 to 6% yr^{-1}, in 100 yr it would be equivalent to 1% and in 200 yr would approximate the solar figure.

It is interesting, also, to compare TAKANO's figure for world energy consumption with that for the heat transferred by ocean currents. These are 10^{13} and 10^{15} W respectively, with only a 100-fold difference. At 5 to 6% yr^{-1} growth, energy consumption in 79 to 94 yr time will equal the present quantities of heat transferred by ocean currents.

With present growth rates it seems inevitable, therefore, that waste heat will eventually become a serious problem. LAMB (1977) thought the increasing output of heat is likely to 'gain the upper hand' in the next century, although this aspect of nuclear energy appears to present fewer problems than the carbon dioxide produced by burning fossil fuels (LAMB, 1982). According to WILCOX (1977), the increase in world temperatures associated with waste-heat production would be sufficient to cause melting of the ice sheets and eventual flooding of the world's largest cities and best farmland (see also LAMB, 1972, 1977; HANSEN and co-authors, 1981). HEDGPETH (1977) warned that there are limits to thermal additions or calefaction, emphasizing that we cannot afford a 'try it and see' attitude. It would appear, however, that mankind is doing exactly that. WILCOX (1977) stressed the need to reduce the growth rate of man-made energy production to avoid an eventual overwhelming of natural temperature conditions and changes. Most of this energy production releases heat over very short periods of time compared with the millions of years taken to build up the reserves of fossil fuels. Combined with the possibility of additional warming as a result of increasing atmospheric carbon dioxide the prospects of global warming seem considerable, with a general warming of the oceans and considerable regional variations as a result of climatic changes. LAMB (1977) cites a statement from the World Meteorological Organization in 1977 which places the greatest emphasis on the increasing quantities of atmospheric carbon dioxide and waste-heat production. These were expected to produce a global warming effect that would become dominant over natural fluctuations of climate by the end of the century and would have very serious consequences in the next 50 to 100 yr by changing vegetation belts and melting ice caps. Because of their capacities as heat sinks the oceans are likely to be more important than the atmosphere in maintaining long-term changes in climate (NATIONAL ACADEMY OF SCIENCES, 1975; CUSHING and DICKSON, 1976; LAMB, 1977). Clearly, the need for careful monitoring of climatic and oceanic changes is extremely important if man-made effects are to be assessed against the background of natural long-term fluctuations in climate.

The long-term implications for general temperature increases in the oceans are considerable as a result of small, subtle changes of temperature over extensive areas of earth. There has been a warning that the warming of surface layers will weaken the vertical mixing by stabilizing the vertical stratification of the surface water layer (NATIONAL ACADEMY OF SCIENCES, 1979). Thus, the surface layers may become deficient in nutrient salts, with profound effects on world fisheries. According to SOUTHWARD (1980), there is general agreement that the long-term fluctuations in the fauna of the western English Channel known as the 'Russell cycle' (CUSHING and DICKSON, 1976), are mostly related to changes in climate (RUSSELL and co-authors, 1971; CUSHING and DICKSON, 1976; MADDOCK and SWANN, 1977), possibly associated with changes in sunspot activity (LAMB, 1972; SOUTHWARD, 1980). This cycle passed through a warmer period between the early 1930s and 1960s when a reversal to a cooler phase started. SOUTHWARD (1980) emphasizes that no single climatic factor stands out as the controlling factor although relationships between sea temperature trends and some biological changes up to the 1970s support the hypothesis that temperature was the controlling factor (SOUTHWARD and co-authors, 1975). Mean sea water temperature changes during these fluctuations were only about $0.5\,C°$ but there have been considerable economic implications through changes in commercial fisheries (SOUTHWARD, 1980).

CUSHING and DICKSON (1976) and CUSHING (1978) reviewed biological responses in the seas to climatic changes. The correlations between long-term fluctuations in sea water temperatures and the distribution and abundances of commercially important estuarine and marine species have been demonstrated (e.g. DOW, 1962, 1964, 1972, 1977a,b, 1978). Various papers in PARSONS and co-authors (1978) deal with biological effects of variations in the oceans.

There is now a great deal of interest in the development of alternative energy sources which do not add to the overall heat budget of the world, although they may create considerable warming effects in localized areas (e.g. GABOR and co-authors, 1978; MCGOWN and BOCKRIS, 1980). The alternatives include hydro-electric power, solar and geothermal energy, wind, waves, tides, ocean currents, thermal and salinity gradients. Only some of these involve thermal deformations in the oceans. Most hydro-electric, wind and wave energy schemes are unlikely to cause marine thermal deformations. However, BACK (1978) discussed the possibilities of using existing electricity power stations, during periods of low demand, to pump sea water either for storage in upper reservoirs or to empty disused sea water-flooded mines, so that sea water from the upper reservoirs could be used to generate hydro-electric power during periods of high demand. These systems would probably result in some small, localized thermal deformations. WILCOX (1977) described a scheme off the Californian coast to farm the giant seaweed *Macrocystis pyrifera* over large areas as a means of utilizing solar energy in the seas. The weed would be used for the production of foods, fuels, fertilizers, plastics and other items. Deep nutrient-rich water would be brought to the surface to stimulate growth. Although this would not add to the overall heat budget, there could be

Table 8-6

Estimated amounts of different types of potentially usable energy available in the oceans (After TAKANO, 1981; Table 2 from *Intergovernmental Oceanographic Commission Technical Series*, 21. ©UNESCO 1981. Reproduced by permission of UNESCO)

Source	Energy (W)
Currents	10^{10} to 10^{11}
Tides	10^{10} to 10^{11}
Ocean swell	10^{12}
Thermal gradient	10^{13}
Salinity gradient	10^{12}

considerable redistribution of heat energy in the sea, particularly as a result of mixing cold deep water with the warm surface layers.

TAKANO (1981) estimated potentially usable forms of energy in the oceans (Table 8-6). Ocean currents have already been discussed (p. 1914). Tides, however, are a more feasible possibility for the immediate future by exploiting enclosed areas with a large tidal range. The La Rance tidal power station near St. Malo (N.W. France) is one of the few examples (ANDRE, 1978), with a peak capacity of about 240 MW. This scheme can also be used for limited pumped storage during periods of low electricity demand. This is particularly advantageous when the pumping can take place during periods of small tidal differences between the storage basin and the open sea. However, ANDRE (1978) pointed out that the choice of sites in the world is very limited. Thermal deformations would probably be small and restricted to changes resulting from altered water flows.

The potential from ocean swell or wave energy is the same as TAKANO's (1981) estimate of the world capacity of electricity power stations. Whether, in reality, this quantity of energy would ever be extracted from waves is highly questionable but there are parts of the world where it might eventually be economically feasible. There is considerable interest, research and development in the subject (GLENDENNING, 1978; MIYAZAKI and MASUDA, 1978). Wave-energy utilization is unlikely to result in significant thermal deformations.

Utilizing the salinity gradient would entail the use of osmotic pressure to raise heads of water of about 245 m, using freshwater and sea water of 35‰ S (TAKANO, 1981). The water head would be used to generate electricity. However, the concept would require the development of satisfactory semi-permeable membranes and is only in the early stages of research. In contrast, the thermal gradient in the oceans has been of interest for generating energy since the 19th century (TAKANO, 1981) when the French physicist D'Arsonval suggested the idea as a method of applying the Rankine cycle (FULLER, 1978). An OTEC (ocean thermal energy conversion) unit using the CLAUDE (1930) method was constructed in Cuba in the 1920s but was destroyed by a storm. Since the 1973 oil crisis there has been a revival of interest and considerable research is now taking place into this form of energy production (HANSON, 1974; GOSS and co-authors, 1977; DUNCAN and co-authors, 1978). Modern OTEC uses the surface layers

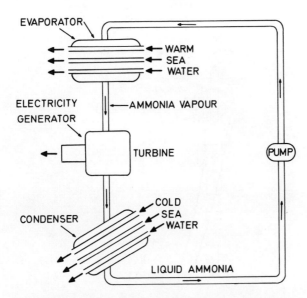

Fig. 8-24; Electricity generating cycle for the power
module of an ocean thermal energy conversion
(OTEC) system. (After FULLER, 1978; modified;
reproduced by permission of Elsevier, Amsterdam.)

of the oceans as a solar collector and energy store, and converts this thermal energy
into electricity using a closed-cycle heat engine. Warm surface water (Fig. 8-24) is
pumped through a heat exchanger which evaporates a working fluid, such as ammonia,
having a low boiling point. The ammonia vapour then drives a turbine generator, which
produces electricity, and the vapour is condensed in a second heat exchanger or
condenser using cold sea water pumped from depth. The pressure difference across the
turbine is about 3 atm (FULLER, 1978). The greater the temperature difference
between the warm surface water and cold deep water, the greater is the efficiency,
so that the method would be confined mainly to tropical and sub-tropical regions of
the world. FULLER (1978) considered that a temperature difference of between 16·5
and 22·2 C° was suitable for a viable system. Designs presently being developed are
based on long slender structures floating vertically in the sea, with a power-generating
module at the surface and a long pipe extending about 500 m into deep water (Fig.
8-25). The Caribbean as a likely potential site is discussed in detail by DUNCAN and
co-authors (1978). A small 50 kW(e) unit was working in Hawaii in early 1980, a
1 MW(e) experimental device was tested there in the same year and a 40 MW(e)
demonstration unit is planned for 1984–85 (VADUS, cited in ANONYMOUS, 1980c).
This author predicted that 400 MW(e) units will displace about 750 000 tons and may
have to be moored in depths of over 1000 m; this will present enormous mooring
problems. VADUS mentions the US National Oceanic and Atmospheric Administration's support for work on ships that will search for areas with the highest
temperature differences and use the energy to produce hydrogen, ammonia and other
raw materials. OTEC is unlikely to achieve an efficiency greater than about 5% but
this was thought to be sufficient to produce electricity at costs comparable to those

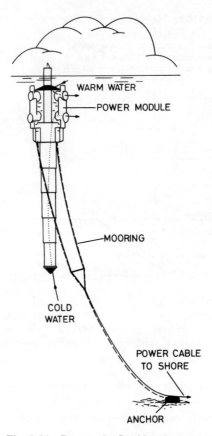

Fig. 8-25: Proposed Lockhead ocean thermal energy conversion platform. (After FULLER, 1978; modified; reproduced by permission of Elsevier, Amsterdam.)

of nuclear power stations (ANONYMOUS, 1980c). Despite these low efficiencies and the problems in development the OTEC system appears to have considerable potential for the future although the CONSERVATION COMMISSION OF THE WORLD ENERGY CONFERENCE (1978) suggested that the difficulties would be too severe to attract much research development.

The environmental effects of OTEC could be significant with large volumes of cold sea water being mixed with the warmer surface waters of the tropics and subtropics. There have been suggestions that this could stimulate primary productivity in the surface waters by increasing the nutrient content, affect surface water temperatures considerably and could affect the carbon dioxide balance between the oceans and atmosphere (FULLER, 1978). With recent increases in atmospheric carbon dioxide from burning fossil fuels, FULLER suggested that the upwelling created by OTEC systems could increase carbon dioxide absorption by the oceans. However, he also states that the opposite might occur as a result of raising cold water rich in carbon dioxide. Any warming effects caused by increased atmospheric carbon dioxide would tend to stabilize the vertical temperature stratification of the oceans. According to TAKANO (1981), large-scale use of thermal gradients might help to reduce harmful effects produced by higher global temperatures, such as changes in climate and increased vertical stability in the oceans, which could decrease the productivity in surface layers through nutrient depletion.

These projects would produce energy without increasing the overall heat budget of the earth, although they may result in some redistribution of heat. The main disadvantage of alternative energy sources is that the technology to exploit them has not been developed sufficiently compared with more conventional methods of generating energy. For one thing, these alternatives are usually low-density energy sources and require very extensive schemes to harness them, whether they be solar panels, windmills or wave energy, etc. Compared with conventional power sources they are far more extensive than, say, a nuclear power station in which the highly concentrated energy source makes a very convenient unit producing vast amounts of relatively easily managed energy. None of the alternative energy sources, except hydro-electric power and geothermal energy, can produce the energy equivalent of a 2000 MW(e) power station on a site covering only a few hectares. No doubt there will be and must be big future developments in the technology to exploit alternative energy sources but they will not be without environmental problems.

In conclusion, the future will see a continued growth in the utilization of energy for man's ends. This will continue to create thermal deformations, not only at the local level, but in the total heat budget of the earth. The issue is extremely complex because of a large variety of political, economic and social implications. In this wider context, the urgent warning of the CONSERVATION COMMISSION OF THE WORLD ENERGY CONFERENCE (1978, p. 245) is of significance to future thermal deformations:

'It is necessary to come to grips with the problem of future energy supply without further delay. We still have the option of providing future generations in all countries with ample supplies of energy in forms environmentally acceptable by today's standards. But if we fail to respond *now* we are likely to lose this opportunity. By doing nothing—or not doing enough—we could in less than ten years in effect have locked ourselves, our children and grandchildren, in an energy-constrained future—a situation which, on a global scale, has never been encountered before and which is likely to cause social unrest and political conflict. Those who can least afford it are likely to be the first to suffer'.

The eventual necessity for strict regulation and control of man's energy budget is obvious, especially since it is uncertain at what level the earth's heat processes and climate will be negatively affected by man-made activities. With present rates of exploitation it is inevitable that the climate will change on a global scale (TERJUNG, 1974).

While thermal deformations have had some serious effects at local levels in the marine and estuarine environments, in the long-term future the effects of global warming, directly through waste-heat production and indirectly through increased atmospheric carbon dioxide could be much more significant for all environments. Effective overall regulation and control will be difficult to achieve; growth and living standards continue to increase in the industrial nations, and if underdeveloped countries are to attain comparable living standards they will require very large additional amounts of energy. Energy requirements will have to be considered, developed, harmonized, controlled and managed very carefully. Even though the extent and effects of climatic and temperature changes are uncertain, LAMB (1982) agrees with the considerable disquiet about the likely distruptions. He warns that there is little time left in which to obtain the knowledge to resolve questions about the outcome and to adopt national and international habits and policies in energy use.

Acknowledgements. We are especially grateful to the Editor, Professor O. KINNE, and to Dr. J. MAUCHLINE for their criticisms and helpful suggestions with the manuscript. We acknowledge with grateful thanks the invaluable help received from MissELEANOR WALTON, Librarian at the Dunstaffnage Marine Research Laboratory, for her patient work in locating many papers in the extensive literature, and Mrs. ELSIE MacDOUGALL for typing the various drafts of the manuscript so patiently. Miss ALISON CHADWICK, Miss RITA GOW, Mr. J. WATSON and Dr. A. MUNRO gave invaluable assistance at various stages. We have received much help and advice in discussions and thank Mr. J. CURRIE, Mr. J GRANT, Mr. J GRAY, Dr. A. PLATT and Mr. J. SMITH of the South of Scotland Electricity Board, Mr. F. JOHNSON and Mr. D. SIMPSON of the North of Scotland Hydro-Electric Board and Dr. T NEEDHAM and Dr. S. SEDGWICK.. Colleagues at Dunstaffnage have also provided help and advice, particularly Dr. A. ANSELL, Dr. MARGARET BARNES, Dr. J. BLAXTER, Mr. A. BULLOCK, Mr. D. ELLETT, Dr. R. GIBSON, Dr. J. GORDON and Dr. T. PEARSON. We express our warmest thanks to all of them. Finally, we are grateful to the Director, Professor R. I. CURRIE, the Scottish Marine Biological Association, the Natural Environment Research Council and the South of Scotland Electricity Board for providing facilities and support for this work.

Literature Cited (Chapter 8)

ABRAHAM, G. (1960). Jet diffusion in liquid of greater density. *J. Hydraul. Div. Am. Soc. civ. Engrs*, **86**, 1–13.

ABRAHAM, G. (1965). Horizontal jets in stagnant fluid of other density. *J. Hydraul. Div. Am. Soc. civ. Engrs*, **91**, 139–154.

ADAMS, E. E. and STOLZENBACH, K. D. (1979). Comparison of alternative diffuser designs for the discharge of heated water into shallow receiving water. In S. S. Lee and S. Sengupta (Eds), *Waste Heat Management and Utilization*, Vol. 2, Proceedings of a Conference, Miami, 1977. Hemisphere, Washington. pp. 771–790.

ADAMS, J. R. (1968). Thermal effects and other considerations at steam electric plants: a survey of studies in the marine environment. *Pacif. Gas Electric Co., Res. Rep.*, No. 6934, 87.

ADAMS, J. R. (1969). Ecological investigations around some thermal power stations in California tidal waters. *Chesapeake Sci.*, **10**, 145–154.

ADAMS, J. R., GORMLY, H. J. and DOYLE, M. J., JR. (1970). Thermal investigations in California. *Mar. Pollut. Bull.*, NS, **1**, 140–142.

ALABASTER, J. S. (1963). The effects of heated effluents on fish. *Int. J. Air Wat. Pollut.*, **7**, 541–563.

ALABASTER, J. S. (1967). The survival of salmon and sea trout in fresh and saline water at high temperature. *Wat. Res.*, **1**, 717–727.

ALABASTER, J. S. (1973). Oxygen in estuaries: requirements for fisheries. In A. Gameson (Ed.), *Mathematical and Hydraulic Modelling of Estuarine Pollution. Tech. Pap. Wat. Pollut. Res. D.S.I.R.*, **13**, 16–23.

ALDEN, R. W., III (1979). Effects of a thermal discharge on the mortality of copepods in a subtropical estuary. *Environ. Pollut.*, **20**, 3–19.

ALDEN, R. W., III, MATURO, F. J. S. JR. and INGRAM, W., III (1976). Interactive effects of temperature, salinity and other factors on coastal copepods. In G. W. Esch and R. W. Mac-farlane (Eds), *Thermal Ecology*, Vol. II, Proceedings of a Symposium, Augusta, Georgia, 1975 (CONF–750425). Technical Information Centre, US Energy Research and Development Administration, Oak Ridge, Tennessee. pp. 336–348.

ALDERDICE, D. F. (1972). Factor combinations: responses of marine poikilotherms to environmental factors acting in concert. In O. Kinne (Ed.), *Marine Ecology, Vol. 1, Environmental Factors,* Part 3. Wiley, London. pp. 1659–1722.

ALDERSON, R. (1972). Effects of low concentrations of free chlorine on eggs and larvae of plaice, *Pleuronectes platessa* L. In M. Ruivo (Ed.), *Marine Pollution and Sea Life.* Fishing News (Books) Ltd., Surrey. pp. 312–315.

ALDERSON, R. (1974). Seawater chlorination and the survival and growth of the early developmental stages of plaice, *Pleuronectes platessa* L. and Dover sole, *Solea solea* (L.). *Aquaculture,* **4,** 41–53.

ALLEMAN, R. T., JOHNSON, B. M. and SMITH, G. C. (1979). Dry/wet cooling towers with ammonia as intermediate heat exchange medium. In S. S. Lee and S. Sengupta (Eds), *Waste Heat Management and Utilization,* Vol. 2, Proceedings of a Conference, Miami, 1977. Hemisphere, Washington. pp. 991–1006.

ALLEN, G. H., BOYDSTUN, L. B. and GARCIA, F. G. (1970). Reaction of marine fishes around warm water discharge from an atomic steam-generating plant. *Prog. Fish Cult.,* **32,** 9–16.

ALLRED, E. R. (1975). Use of waste heat for soil warming and irrigation. In *Proceedings of the American Society of Civil Engineers,* Irrigation and Drainage Division Conference, Logan, Utah, 1975. pp. 287–301.

AMERICAN NUCLEAR SOCIETY (1977). Trace metals, organic and non-organic emissions from nuclear plants. *Trans. Am. nucl. Soc.,* **26,** 106.

ANDERSEN, N. R. and MALAHOFF, A. (Eds) (1977). *The Fate of Fossil Fuel CO_2 in the Oceans,* Plenum, New York.

ANDERSON, R. R. (1969). Temperature and rooted aquatic plants. *Chesapeake Sci.,* **10,** 157–164.

ANDRE, H. (1978). Ten years of experience at the 'La Rance' tidal power plant. *Ocean Mgmt,* **4,** 165–178.

ANDREWS, J. H. (1976). The pathology of marine algae. *Biol. Rev.,* **5,** 211–253.

ANDREWS, J. W., KNIGHT, L. H. and MURAI, T. (1972). Temperature requirements for high density rearing of channel catfish from fingerling to market size. *Prog. Fish Cult.,* **34,** 240–241.

ANDREWS, J. W., MURAI, T. and GIBBONS, G. (1973). The influence of dissolved oxygen on the growth of channel catfish. *Trans. Am. Fish. Soc.,* **102,** 835–838.

ANONYMOUS (1974). Sell waste heat for fish farms. *Fish Farm. Int.,* **2,** 122.

ANONYMOUS (1978). Why not sell warm water? *Fish Farm. Int.,* **5** (2), 46.

ANONYMOUS (1980a). Boilers warm the water for salmon. *Fish Farm. Int.,* **7** (2), 37.

ANONYMOUS (1980b). Cement maker to grow eels. *Fish Farm. Int.,* **7** (1), 8.

ANONYMOUS (1980c). Ocean energy scales new heights. *New Scient.,* **85,** 928.

ANRAKU, M. and KOZASA, E. (1978). The effects of heated effluents on the production of marine plankton (Takahama nuclear power station–1). *Bull. Plankton Soc. Jap.* **25,** 93–110. (Japanese; summary in English).

ANSELL, A. D. (1963). The biology of *Venus mercenaria* in British waters, and in relation to generating station effluents. *Rep. Challenger Soc.,* **3** (15), 38.

ANSELL, A. D. (1969). Thermal releases and shellfish culture: possibilities and limitations. *Chesapeake Sci.,* **10,** 256–257.

ANSELL, A. D. and LANDER, K. F. (1967). Studies on the hard-shell clam, *Venus mercenaria,* in British waters. III. Further observations on the seasonal biochemical cycle and on spawning. *J. appl. Ecol.,* **4,** 425–435.

ANSELL, A. D. and SIVADAS, P. (1973). Some effects of temperature and starvation on the bivalve *Donax vittatus* (da Costa) in experimental laboratory populations. *J. exp. mar. Biol. Ecol.,* **13,** 229–262.

ANSELL, A. D., BARNETT, P. R. O., BODOY, A. and MASSE, H. (1980a). Upper temperature tolerances of some European molluscs. I. *Tellina fabula* and *T. tenuis. Mar. Biol.,* **58,** 33–39.

ANSELL, A. D., BARNETT, P. R. O., BODOY, A. and MASSE, H. (1980b). Upper temperature tolerances of some European molluscs. II. *Donax vittatus, D. semistriatus* and *D. trunculus. Mar. Biol.,* **58,** 41–46.

ANSELL, A. D., LANDER, K. F., COUGHLAN, J. and LOOSMORE, F. A. (1964). Studies on the hard-shell clam, *Venus mercenaria,* in British waters. 1. Growth and reproduction in natural and experimental colonies. *J. appl. Ecol.,* **1,** 63–82.

ANSUINI, F. J. and HUGUENIN, J. E. (1978). The design and development of a fouling resistant marine fish cage system. In J. W. Avault (Ed.), *Proceedings of the 9th Meeting World Mariculture Society*. Louisiana State University, Baton Rouge. pp. 737–745.

ANWAR, H. O. (1977). The flow structure in the front of a moving surface layer. *Ing.-Arch.*, **46**, 143–156.

ASH, G. R., CHYMKO, N. R. and GALLUP, D. N. (1974). Fish kill due to 'cold shock' in Lake Wabamun, Alberta. *J. Fish. Res. Bd Can.*, **31**, 1822–1824.

ASHLEY, G. C. and HIETALA, J. S. (1979). The Sherco Greenhouse: a demonstration of the beneficial use of waste heat. In S. S. Lee and S. Sengupta (Eds), *Waste Heat Management and Utilization*, Vol. 3, Proceedings of a Conference, Miami, 1977. Hemisphere, Washington. pp. 2409–2421.

ASTON, R. J. (1981). The availability and quality of power station cooling water for aquaculture. In K. Tiews (Ed.), *Aquaculture in Heated Effluents and Recirculation Systems*, Vol. 1, Proceedings of a World Symposium, Stavanger, 1980. Heenemann, Berlin, pp. 39–58.

ASTON, R. J., BROWN, D. J. A. and MILNER, A. G. P. (1976). Heated water farms at inland power stations. *Fish Farm. Int.*, **3** (2), 41–44.

AUDUNSON, T., LAND, J. and RYE, H. (1975). Computations of the temperature response of stratified sill fjords to cooling-water discharges. In *Environmental Effects of Cooling Systems at Nuclear Power Plants*, Proceedings of a Symposium, Oslo, 1974. International Atomic Energy Agency, Vienna. pp. 113–139.

AYLES, G. B., SCOTT, K. R., BARICA, J. and LARK, J. G. I. (1981). Combination of a solar collector with water recirculation units in a fish culture operation. In K. Tiews (Ed.), *Aquaculture in Heated Effluents and Recirculation Systems*, Vol. 1, Proceedings of a World Symposium, Stavanger, 1980. Heenemann, Berlin, pp. 309–325.

BACK, P. A. A. (1978). Hydro-electric power generation and pumped storage schemes utilizing the sea. *Ocean Mgmt*, **4**, 179–206.

BADER, R. G., ROESSLER, M. A. and THORHAUG, A. (1972). Thermal pollution of a tropical marine estuary. In M. Ruivo (Ed.), *Marine Pollution and Sea Life*. Fishing News (Books) Ltd., Surrey. pp. 425–428.

BAIRD, R. D. and MYERS, D. M. (1979). The thermal performance characteristics of large spray cooling ponds. In S. S. Lee and S. Sengupta (Eds), *Waste Heat Management and Utilization*, Vol. 2, Proceedings of a Conference, Miami, 1977. Hemisphere, Washington, pp. 929–938.

BALABAN, M. (Ed.) (1977a). *Proceedings of the International Congress on Desalination and Water Reuse, Tokyo, 1977*, Vol. 1. *Desalination*, **22**, 1–524.

BALABAN, M. (Ed.) (1977b). *Proceedings of the International Congress on Desalination and Water Reuse, Tokyo, 1977*, Vol. 2. *Desalination*, **23**, 1–588.

BALLIGAND, P., TRONEL-PEYROZ, J., DESCAMPS, B., GRAUBY, A., FOULQUIER, L. and DUMAS, M. (1981). Realization d'un pilote industriel utilisant des eaux de refrigeration d'une usine pour le grossissement des anguilles. In K. Tiews (Ed.), *Aquaculture in Heated Effluents and Recirculation Systems*, Vol. 2, Proceedings of a World Symposium, Stavanger, 1980. Heenemann, Berlin. pp. 593–600.

BAPTIST, J. P. (1966). Uptake of mixed fission products by marine fishes. *Trans. Am. Fish. Soc.*, **95**, 145–152.

BAPTIST, J. P., HOSS, D. E. and LEWIS, C. W. (1970). Retention of ^{51}Cr, ^{59}Fe, ^{60}Co, ^{65}Zn, ^{85}Sr, ^{95}Nb, ^{141m}In and ^{131}I by the Atlantic croaker. (*Micropogon undulatus*). *Hlth Phys.* **18**, 141–148.

BAPTIST, J. P. and RICE, T. R. (1977). Radioactivity. In J. R. Clark (Ed.), *Coastal Ecosystem Management*. Wiley, New York. pp. 687–691.

BARKER, S. L., and STEWART, J. R. (1978). Mortalities of the larvae of two species of bivalve after acute exposure to elevated temperature. In L. D. Jensen (Ed.), *4th National Workshop on Entrainment and Impingement, Chicago, Illinois, 1977*. Ecological Analysts Inc., New York. pp. 203–210.

BARKLEY, S. W. and PERRIN, C. (1972). The effects of the Lake Catherine Steam Electric Plant effluent on the distribution of fishes in the receiving embayment. *Proc. a. Conf. S.East. Ass. Game Fish Commn*, **25**, 384–392.

BARNETT, P. R. O. (1968). Distribution and ecology of harpacticoid copepods of an intertidal mudflat. *Int. Rev. ges. Hydrobiol.*, **53**, 177–209.

BARNETT, P. R. O. (1971). Some changes in intertidal sand communities due to thermal pollution. *Proc. R. Soc. (Series B)*, **177**, 353–364.

BARNETT, P. R. O. (1972). Effects of warm water effluents from power stations on marine life. *Proc. R. Soc. (Series B)*, **180**, 497–509.

BARNETT, P. R. O. and HARDY, B. L. S. (1969). The effects of temperature on the benthos near the Hunterston Generating Station, Scotland. *Chesapeake Sci.*, **10**, 255–256.

BARNETT, P. R. O., HARDY, B. L. S. and WATSON, J. (1980). Substratum selection and egg-capsule deposition in *Nassarius reticulatus* (L.). *J. exp. mar. Biol. Ecol.*, **45**, 95–103.

BARR, D. I. H. (1958). A hydraulic model study of heat dissipation at Kincardine power station. *Proc. Instn civ. Engrs*, **10**, 305–320.

BASSON, P. W., BURCHARD, J. E., HARDY, J. T. and PRICE, A. R. G. (1977). *Biotopes of the Western Arabian Gulf*, Arabian American Oil Company, Aramco Department of Loss Prevention and Environmental Affairs, Dhahran, Saudi Arabi.

BEALL, S. E. (1971). Waste heat uses cut thermal pollution. *Mech. Engng*, **July**, 15–19.

BEALL, S. E. (1973). Conceptual design of food complex using waste warm water for heating. *J. environ. Qual.*, **2**, 207–215.

BEAUCHAMP, R. S. A. (1969). The use of chlorine in the cooling water system of coastal power station. *Chesapeake Sci.*, **10**, 280.

BECKER, C. D. and THATCHER, T. O. (1973). *Toxicity of Power Plant Chemicals to Aquatic Life*, US Atomic Energy Commission, Washington, DC (Publication WAS–1249 UC–11).

BEER, L. O. and PIPES, W. O. (1969). *The Effects of Discharge of Condenser Water into the Illinois River*, Industrial Bio-Test Laboratories, Inc., Northbrook, Illinois.

BELTER, W. G. (1975). Management of waste heat at nuclear power stations, its possible impact on the environment, and possibilities of its economic use. In *Environmental Effects of Cooling Systems at Nuclear Power Plants*, Proceedings of a Symposium, Oslo, 1974. International Atomic Energy Agency, Vienna. pp. 3–23.

BELTZ, J. R., JOHNSON, J. E., COHEN, D. L. and PRATT, F. B. (1974). An annotated bibliography of the effects of temperature on fish with special reference to the freshwater and anadromous fish species of New England. *Bull. Mass. agric. Exp. Stn*, No. 605.

BENJAMIN, T. B. (1968). Gravity currents and related phenomena. *J. Fluid Mech.*, **31**, 209–248.

BERGAN, P. (1960). On the blocking of mitosis by heat shock applied at different stages in the cleavage divisions of *Trichogaster trichopterus* var. *sumatranus*. Teleostei: Anabantidae). *Nytt Mag. Zool.*, **9**, 37–121.

BEWTRA, J. K. (1970). Effect of temperature on oxygen transfer in water. *Wat. Res.*, **4**, 115–123.

BIENFANG, P. and JOHNSON, W. (1980). Response of subtropical phytoplankton to power plant entrainment. *Environ. Pollut. (Series A)*, **22**, 165–178.

BIGGS, R. B. (1977). Artificial multi-purpose industrial port islands. In P. D. Wilmot and A. Slingerland (Eds), *Technology Assessment and the Ocean*. IPC Science & Technology Press, Guildford pp. 168–176.

BILLI, B., CHIANTORE, G. and GASPARINI, R. (1975). The environment and electric power generation. In R. R. Ferber and R. A. Roxas (Eds), *Transactions 9th World Energy Conference, Detroit, 1974*, Vol. III. US National Committee of the World Energy Conference, New York. pp. 812–831.

BIRKELAND, C., RANDALL, R. H. and GRIMM, G. (1979). Three methods of coral transplantation for the purpose of re-establishing a coral community in the thermal effluent area at the Tanguisson Power Plant. *Tech. Rep. mar. Lab. Univ. Guam*, **60**, 1–24.

BISSET, K. A. (1947). Bacterial infection and immunity in lower vertebrates and invertebrates. *J. Hyg., Camb.*, **45**, 128–135.

BLACK, E. C. (1953). Upper lethal temperatures of some British Columbia freshwater fishes. *J. Fish. Res. Bd Can.*, **10**, 196–210.

BLAND, R. A., HISER, H. W., LEE, S. S. and SENGUPTA, S. (1979). Aerial remote sensing of thermal plumes. In S. S. Lee and S. Sengupta (Eds), *Waste Heat Management and Utilization*, Vol. 2, Proceedings of a Conference, Miami, 1977. Hemisphere, Washington. pp. 1439–1449.

BLAXTER, J. H. S. and HOSS, D. E. (1979). The effect of rapid changes of hydrostatic pressure on the Atlantic herring *Clupea harengus* L. II. The response of the auditory bulla system in larvae and juveniles. *J. exp. mar. Biol. Ecol.*, **41**, 87–100.

BLOCK, R. M., HELZ, G. R. and DAVIS, W. P. (1977). The fate and effects of chlorine in coastal waters: summary and recommendations. *Chesapeake Sci.*, **18**, 97–101.

BOARD, P. A. (1971). *The Effect of a Warm Effluent on the Biology of Shipworms*. Part III. The Effects of Temperature and other Factors on the Tunnelling and Reproduction of *Lyrodus pedicellatus* Quatrefages, Central Electricity Research Laboratories, UK (Report RD/L/N270/71).

BOERSMA, L. (1970). Warm water utilization. *Proceedings of the Conference on the Beneficial Uses of Thermal Discharges, Albany, New York, 1970*, pp. 74–106 (quoted by YAROSH, 1977).

BOIES, D. B., LEVIN, J. E. and BARATZ, D. (1973). Technical and economic evaluations for cooling systems blowdown control techniques. *U.S. Environmental Protection Agency Report*. EPA-660/2-73-026.

BOLUS, R. L., FANG, C. S. and CHIA, S. N. (1973). The design of a thermal monitoring system. *J. mar. Technol. Soc.* **7**, 36–40.

BORISOV, P. M. (1962). The problem of the fundamental amelioration of climate. *Izv. vses. geogr. Obshch.*, **94**, 304–318 (quoted by LAMB, 1977).

BORISOV, P. M. (1969). Can we control the Arctic climate? *Bull. atom. Scient.*, **25**(3), 43–48.

BOTSFORD, L. W. and GOSSARD, T W. (1978). Implications of growth and metabolic rates on costs of aquaculture. In J. W. Avault (Ed.), *Proceedings of the 9th Annual Meeting*, World Mariculture Society. Louisiana State University, Baton Rouge. pp. 413–423.

BRAAMS, C. M. (1975). Thermonuclear energy. In R. R. Ferber and R. A. Roxas (Eds), *Transactions 9th World Energy Conference, Detroit, 1974*, Vol. V. US National Committee of the World Energy Conference, New York. pp. 181–191.

BRANCH, M. R. and STRAWN, K. (1978). Brackish water pond polyculture of estuarine fishes in thermal effluent. In J. W. Avault (Ed.), *Proceedings of the 9th Annual Meeting*, World Mariculture Society, Louisiana State University, Baton Rouge. pp. 345–356.

BRETT, J. R. (1952). Temperature tolerance in young Pacific salmon, genus *Oncorhynchus*. *J. Fish. Res. Bd Can.*, **11**, 265–323.

BRETT, J. R. (1960). Thermal requirements of fish: three decades of study, 1940–1970. In C. M. Tarzwell (Ed.), *Biological Problems in Water Pollution*, Transactions of the 1959 Seminar (Tech. Rep. W 60-3) US Department of Health, Education, and Welfare, R. A. Taft San. Eng. Center, Cincinatti, Ohio. pp. 110–117.

BRETT, J. R. (1970). Temperature. Fishes: introduction and functional responses. In O. KINNE (Ed.), *Marine Ecology, Vol. I Environmental Factors*, Part 1. Wiley, London. pp. 515–560.

BRETT, J. R., SHELBOURN, J. E. and SHOOP, C. T. (1969). Growth rate and body composition of fingerling sockeye salmon, *Oncorhynchus nerka*, in relation to temperature and ration size. *J. Fish. Res. Bd Can.*, **26**, 2363–2394.

BREWER, G. (1974). Preliminary observations on the lower minimal temperature requirements of the northern anchovy. In. D. F. Soule and M. Oguri (Eds), *Marine Studies of San Pedro Bay, California, Part III, Thermal Tolerance and Sediment Toxicity Studies*. Allen Hancock Foundation and the Office of Sea Grant Programs, Los Angeles. pp. 21–43.

BRIAND, F. J.-P. (1975). Effects of power-plant cooling systems on marine phytoplankton. *Mar. Biol.*, **33**, 135–146.

BRODFIELD, B. (1977). Influence of ecological considerations on the engineering and siting of steam electric plants. In M. Marois (Ed.), *Proceedings of the World Conference Towards a Plan of Actions for Mankind, Vol. 3, Biological Balance and Thermal Modifications* (Institut de la Vie, Paris, 1974). Pergamon Press, Oxford. pp. 239–247.

BROOK, A. J. and BAKER, A. L. (1972). Chlorination at power plants: impact on phytoplankton productivity. *Science, N.Y.*, **176**, 1414–1415.

BRUNGS, W. A. (1973). Effects of residual chlorine on aquatic life. *J. Wat. Pollut. Control Fed.*, **45**, 2180–2193.

BUCK, J. D. (1970). Summary of Connecticut River microbiology study. In D. Merriman (Ed.), *The Connecticut River Investigation*, 10th Semiannual Progress Report to Connecticut Water Resources Commission.

BUCK, J. D. and RANKIN, J. S. (1972). Thermal effects on the Connecticut River: bacteria. *J. Wat. Pollut. Control Fed.*, **44**, 47–64.

BUDYKO, M. I. (1966). The possibility of changing the climate by action on the polar ice. In M. I.

Budyko (Ed.), *Sovremennye Problemy Klimatologii* (Contemporary Problems of Climatology) (Russian; English translation by US Joint Pub. Res. Service, JPRS 43, 482, MGA 20.5-182, 1967; quoted by LAMB, 1977). Leningrad. pp. 347–357.

BULLIMORE, B., DYRYNDA, P. E. J. and BOWDEN, N. (1978). The effects of falling temperature on the fauna of Swansea Docks. In J. C. Gamble and R. A. Yorke (Eds), *Progress in Underwater Science*, Vol. 3 (New Series of the Report of the Underwater Association), Proceedings of the 11th Symposium of the Underwater Association, 1977, British Museum (Natural History). Pentech Press, London. pp. 1–279.

BURNETT, J. M., MCMILLAN, W., MACQUEEN, J. F., MOORE, D. J. and SHEPHERD, J. G. (1975). Cooling of power stations. In R. R. Ferber and R. A. Roxas (Eds), *Transactions 9th World Energy Conference, Detroit, 1974*, Vol. III. US National Committee of the World Energy Conference, New York. pp. 333–356.

BURNS, E. R., PILE, R. S. and MADEWELL, C. E. (1979). Waste heat use in a controlled environment greenhouse. In S. S. Lee and S. Sengupta (Eds), *Waste Heat Management and Utilization*, Vol. 3, Proceedings of a Conference, Miami, 1977. Hemisphere, Washington, pp. 2095–2120.

BURTON, B. W. (1974). Water movement patterns from aerial photographs. *Rapp. P.-v. Réun. Cons. int. Explor. Mer*, **167**, 213–221.

BURTON, D. T., ABELL, P. R. and CAPIZZI, T. P. (1979). Cold shock: effect of rate of thermal decrease on Atlantic menhaden. *Mar. Pollut. Bull., N.S.*, **10**, 347–349.

BURTON, D. T., RICHARDSON, L. B., MARGREY, S. L. and ABELL, P. R. (1976). Effects of low ΔT power plant temperatures on estuarine invertebrates. *J. Wat. Pollut. Control Fed.*, **48**, 2259–2272.

BURWITZ, P. and TOBIAS, W. (1977). Das Luftbild als Hilfsmittel bei oekologischen Untersuchungen ueber thermische Belastungen des Unter mains. *Natur Mus., Frankf.*, **107** (3), 65–73.

BUTLER, I. and SOUTHWARD, A. (1981). Plenty more fish in the sea. *New Scient.*, **92**, 110–112.

CAIRNS, J. (1956). Effects of increased temperatures on aquatic organisms. *Ind. Wastes*, **1**, 150–152.

CAIRNS, J. (1971). Thermal pollution: a cause for concern. *J. Wat. Pollut. Control Fed.*, **43**, 55–66.

CAIRNS, J. (1972). Coping with heated waste water discharges from steam electric power plants. *Bioscience*, **22**, 411–420.

CAIRNS, J. (1977). Effects of temperature changes and chlorination upon the community structure of aquatic organisms. In M. Marois (Ed.), *Proceedings of the World Conference Towards a Plan of Actions for Mankind, Vol. 3, Biological Balance and Thermal Modifications* (Institut de la Vie, Paris, 1974). Pergamon Press, Oxford. pp. 129–144.

CAIRNS, J. and LANZA, G. R. (1972). Effects of heated waste waters upon some micro-organisms. *Bull.Va polytech. Inst. State Univ. (Wat. Resour. Res. Center)*, No. 48.

CAIRNS, J., HEATH, A. G. and PARKER, B. D. (1975). The effects of temperature upon the toxicity of chemicals to aquatic organisms. *Hydrobiologia*, **47**, 135–171.

CAPUZZO, J. M. (1979). The effects of temperature on the toxicity of chlorinated cooling waters to marine animals: a preliminary review. *Mar. Pollut. Bull., N.S.*, **10**, 45–47.

CAPUZZO, J. M., GOLDMAN, J. C., DAVIDSON, J. A. and LAWRENCE, S. A. (1977). Chlorinated cooling waters in the marine environment: development of effluent guidelines. *Mar. Pollut. Bull., N.S.*, **8**, 161–163.

CARPENTER, E. J. (1973). Brackish-water phytoplankton response to temperature elevation. *Estuar. coast. mar. Sci.*, **1**, 37–44.

CARPENTER, E. J., ANDERSON, S. J. and PECK, B. B. (1974a) Copepod and chlorophyll *a* concentrations in receiving waters of a nuclear power station and problems associated with their measurement. *Estuar. coast. mar. Sci.*, **2**, 83–88.

CARPENTER, E. J., PECK, B. B. and ANDERSON, S. J. (1972). Cooling water chlorination and productivity of entrained phytoplankton. *Mar. Biol.*, **16**, 37–40.

CARPENTER, E. J., PECK, B. B. and ANDERSON, S. J. (1974b). Survival of copepods passing through a nuclear power station at northeastern Long Island Sound, USA. *Mar. Biol.*, **24**, 49–55.

CARROLL, B. B., HUBERT, W. A. and WARDEN, R. L. (1981). Flow-through/systems northern America. In K. Tiews (Ed.), *Aquaculture in Heated Effluents and Recirculation Systems*, Vol. 2, Proceedings of a World Symposium, Stavanger, 1980. Heenemann, Berlin. pp. 373–382.

CARSTENS, T. (1975). Trapping of heat in sill fjords. In *Environmental Effects of Cooling Systems at Nuclear Power Plants,* Proceedings of a Symposium, Oslo, 1974. International Atomic Energy Agency, Vienna, pp. 99–112.

CARTER, L. (1976). Marine pollution and sea disposal of wastes. The disposal of wastes to tidal waters. *Chemy Ind.,* **19**, 825–829.

CASPERS, H. (1975). BOD experiments on the effect of thermal pollution in rivers and estuaries. *Verh. int. Verein. theor. angew. Limnol.,* **19**, 3192–3198.

CECH, J. J. and WOHLSCHLAG, D. E. (1975). Summer growth depression in the striped mullet. *Mugil cephalus* L. *Contr. mar. Sci.,* **19**, 91–100.

CHADWICK, H. K., STEVENS, D. E. and MILLER, L. (1977). Some factors regulating the striped bass population in the Sacramento–San Joaquin Estuary, California. In W. Van Winkle (Ed.), *Proceedings of the Conference on Assessing the Effects of Power-Plant-Induced Mortality on Fish Populations, Gatlinburg. Tennessee, 1977.* Pergamon Press, New York. pp. 18–35.

CHAHINE, M. T. (1980). Infrared remote sensing of sea surface temperature. In A. Deepak (Ed.), *Remote Sensing of Atmospheres and Oceans.* Academic Press, New York. pp. 411–434.

CHAMBERLAIN, G. and STRAWN, K. (1977). Submerged cage culture of fish in supersaturated thermal effluent. In J. W. Avault (Ed.), *Proceedings of the Eighth Annual Meeting, World Mariculture Society.* Louisiana State University, Baton Rouge. pp. 625–645.

CHAMPION, E. R., GOODMAN, C. H. and SLAWSON, P. R. (1979). Field study of mechanical draft cooling tower plume behavior. In S. S. Lee and S. Sengupta (Eds), *Waste Heat Management and Utilization,* Vol. 2, Proceedings of a Conference, Miami, 1977. Hemisphere, Washington. pp. 939–947.

CHANNABASAPPA, K. C. (1977). Membrane technology for water reuse application. *Desalination,* **23**, 495–514.

CHEN, J. C. and LIST, E. J. (1977). Spreading of buoyant discharges. In D. B. Spalding and N. Afghan (Eds), *Heat Transfer and Turbulent Buoyant Convection.* Hemisphere Washington. pp. 171–182.

CHENEY, W. O. and RICHARDS, G. V. (1966). Ocean temperature measurements for power plant design. In *Coastal Engineering Santa Barbara Speciality Conference, 1965.* American Society of Civil Engineering, New York. pp. 955–989.

CHESHER, R. H. (1975). Biological impact of a large-scale desalination plant at Key West, Florida. In E. J. Ferguson Wood and R. E. Johannes (Eds), *Tropical Marine Pollution,* Elsevier Oceanography Series, 12. Elsevier, Amsterdam. pp. 99–153.

CHIBA, K. (1981a). Bio-technical considerations of aquatic animal culture by using heat effluent and recirculating systems, especially with regard to stocking density. In K. Tiews (Ed.), *Aquaculture in Heated Effluents and Recirculation Systems,* Vol. II, Proceedings of a World Symposium, Stavanger, 1980. Heenemann, Berlin. pp. 41–51.

CHIBA, K. (1981b). Present status of flow-through and recirculation systems and their limitations in Japan. In K. Tiews (Ed.), *Aquaculture in Heated Effluents and Recirculation Systems,* Vol. II, Proceedings of a World Symposium, Stavanger, 1980. Heeneman, Berlin. pp. 343–355.

CLARK, J. (1977). *Coastal Ecosystem Management. A Technical Manual for the Conservation of Coastal Zone Resources,* Wiley, New York.

CLARK, J. and BROWNELL, W. (1973). Electric power plants in the coastal zone: environmental issues. *Spec. Publs Am. littoral Soc.* **7**, 1–149.

CLARK, L. J. (1975). The role of gas and the International Gas Union in world energy. In R. R. Ferber and R. A. Roxas (Eds), *Transactions 9th World Energy Conference, Detroit, 1974,* Vol. II. US National Committee of the World Energy Conference, New York. pp. 181–202.

CLARKE, W. D., JOY, J. W. and ROSENTHAL, R. J. (1970). Biological effects of effluent from a desalination plant at Key West, Florida. *Wat. Pollut. Control Res. Ser.,* Project No. 18050 DA1 02/70, pp. 1–94.

CLAUDE, G. (1930). Power from the tropical sea. *Mech. Engng,* **52** (12), 1039.

CLAUS, C., HOLDERBEKE, L. VAN, MAECKELBERGHE, H. and PERSOONE, G. (1981). Nursery culturing of bivalve spat in heated seawater. In K. Tiews (Ed.), *Aquaculture in Heated Effluents and Recirculation Systems,* Vol. II, Proceedings of a World Symposium, Stavanger, 1980. Heenemann, Berlin. pp. 465–480.

CLOKIE, J. J. P. and BONEY, A. D. (1980). The assessment of changes in intertidal ecosystems following major reclamation work: framework for interpretation of algal-dominated biota and the use and misuse of data. In J. H. Price, D. E. G. Irvine and W. F. Farnham (Eds), *The Shore*

Environment, Vol. 2, Ecosystems (Systematics Association, *Spec. Vol.*, **17(b)**). Academic Press, London. pp. 609–675.

CLYMER, E. A. (1979). A practical look at the recovery of rejected heat by combined heat and power techniques. In S. S. Lee and S. Sengupta (Eds), *Waste Heat Management and Utilization*, Vol. 3, Proceedings of a Conference, Miami, 1977. Hemisphere, Washington. pp. 2485–2495.

COLEBROOK, J. M. (1976). Trends in the climate of the North Atlantic Ocean over the last century. *Nature, Lond.*, **263**, 576–577.

COLEBROOK, J. M. and TAYLOR, A. H. (1978). Year-to-year changes in sea-surface temperature, North Atlantic and North Sea, 1948 to 1974. *Deep Sea Res.*, **26** (A), 825–850.

COLLINS, R. A. and DELMENDO, M. N. (1979). Comparative economics of aquaculture in cages, raceways and enclosures. In T. V. R. Pillay and W. A. Dill (Eds), *Advances in Aquaculture*, FAO Technical Conference on Aquaculture, Kyoto, 1976. Fishing News (Books) Ltd., Surrey. pp. 472–477.

COMMITTEE ON ENTRAINMENT (1978a). Introduction. In J. R. Schubel and B. C. Marcy (Eds), *Power Plant Entrainment: A Biological Assessment*. Academic Press, New York. pp. 1–18.

COMMITTEE ON ENTRAINMENT (1978b). On selecting the excess temperature to minimize the entrainment mortality rate. In J. R. Schubel and B. C. Marcy, Jr. (Eds), *Power Plant Entrainment: A Biological Assessment*. Academic Press, New York. pp. 211–277.

COMMITTEE ON ENTRAINMENT (1978c). Conclusions and recommendations. In J. R. Schubel and B. C. Marcy (Eds), *Power Plant Entrainment: A Biological Assessment*. Academic Press, New York. pp. 229–242.

CONSERVATION COMMISSION OF THE WORLD ENERGY CONFERENCE (1978). *World Energy: Looking Ahead to 2020*, IPC Science and Technology Press, Guildford.

CORY, R. L. and NAUMAN, J. W. (1969). Epifauna and thermal additions in the upper Patuxent river estuary. *Chesapeake Sci.*, **10**, 210–217.

COUGHLAN, J. (1970). Power stations and aquatic life. In E. B. Cowell (Ed.), *The Effects of Industry on the Environment*, Proceedings of a Symposium, Orielton Field Centre, South Wales. Field Studies Council, London. pp. 2–10.

COUGHLAN, J. and WHITEHOUSE J. W. (1977). Aspects of chlorine utilization in the United Kingdom. *Chesapeake Sci.*, **18**, 102–111.

COUTANT, C. C. (1968). Thermal pollution: biological effects. A review of the literature of 1967 on wastewater and water pollution control. *J. Wat. Pollut. Control. Fed.*, **40**, 1047–1052.

COUTANT, C. C. (1969a). Thermal pollution: biological effects. *J. Wat. Pollut. Control. Fed.*, **41**, 1036–1053.

COUTANT, C. C. (1969b). Temperature, reproduction and behavior. *Chesapeake Sci.*, **10**, 261–274.

COUTANT, C. C. (1970a). Thermal pollution: biological effects. *J. Wat. Pollut. Control. Fed.*, **42**, 1025–1057.

COUTANT, C. C. (1970b). Biological aspects of thermal pollution. 1. Entrainment and discharge canal effects. *CRC crit. Rev. environ. Control*, **1**, 341–381.

COUTANT, C. C. (1971a). Thermal pollution: biological effects. *J. Wat. Pollut. Control Fed.*, **43**, 1292–1334.

COUTANT, C. C. (1971b). Effects on organisms of entrainment in cooling water: steps toward predictability. *Nucl.Saf.*, **12**, 600–607.

COUTANT, C. C (1971c). Some biological considerations for the use of power plant cooling water in aquaculture. *Resour. Publs. U.S. Bur. Sport Fish. Wildl.* **102**, 103–114.

COUTANT, C. C. (1972). Biological aspects of thermal pollution. II. Scientific basis for water temperature standards at power plants. *CRC crit. Rev. environ. Control*, **3**, 1–24.

COUTANT, C. C. (1973). Effect of thermal shock on vulnerability of juvenile salmonids to predation. *J. Fish Res. Bd Can.*, **30**, 965–973.

COUTANT, C. C. (1977). Cold shock to aquatic organisms: guidance for power-plant siting, design and operation. *Nucl. Saf.*, **18**, 329–342.

COUTANT, C. C. and DEAN, J. M. (1972). Relationships between equilibrium loss and death as responses of juvenile chinook salmon and rainbow trout to acute thermal shock. US Atomic Energy Commission, Battelle-Northwest, Richland, Washington. *Res. Dev. Rep.*, BNWL–1520.

COUTANT, C. C. and GOODYEAR, C. P. (1972). Thermal effects. *J. Wat. Pollut. Control Fed.*, **44**, 1250–1294.

COUTANT, C. C. and KEDL, R. J. (1975). Survival of larval striped bass exposed to fluid-induced and thermal stresses in a simulated condenser tube (Oak Ridge National Laboratory, Oak Ridge, Tennessee, USA). *Publ. environ. Sci. Div.*, No. 637 (ORNL–TM–4695).

COUTANT, C. C. and PFUDERER, H. A. (1973). Thermal effects. *J. Wat. Pollut. Control Fed.*, **45**, 1331–1369.

COUTANT, C. C. and PFUDERER, H. A. (1974). Thermal effects. *J. Wat. Pollut. Control Fed.*, **46**, 1476–1540.

COUTANT, C. C. and SUFFERN, J. S. (1979). Temperature influences on growth of aquatic organisms. In S. S. Lee and S. Sengupta (Eds), *Waste Heate Management and Utilization*, Vol. 1, Proceedings of a Conference, Miami, 1977. Hemisphere, Washington. pp. 113–124.

COUTANT, C. C. and TALMAGE, S. S. (1975). Thermal effects. *J. Wat. Pollut. Control Fed.*, **47**, 1656–1711.

COUTANT, C. C. and TALMAGE, S. S. (1976). Thermal effects. *J. Wat. Pollut. Control Fed.*, **48**, 1487–1544.

COUTANT, C. C., DUCHARME, H. M., JR, and FISHER, J. R. (1974). Effects of cold shock on vulnerability of juvenile channel catfish (*Ictalurus punctatus*) and largemouth bass (*Micropterus salmoides*) to predation. *J. Fish Res. Bd Can.*, **31**, 351–354.

CRACKNELL, A. P. (Ed.) (1981). *Remote Sensing in Meteorology, Oceanography and Hydrology*, Ellis Horwood, Chichester.

CRISP, D. J. and MOLESWORTH, A. H. N. (1951). Habitat of *Balanus amphitrite* var. *denticulata* in Britain. *Nature, Lond.*, **167**, 489–490.

CRISP, D. J. and SOUTHWARD, A. J. (1964). The effects of the severe winter of 1962–1963 on marine life in Britain. South and Southwest coast. *J. Anim. Ecol.*, **33**, 179–183.

CUSHING, D. H. (1978). Biological effects of climate change. *Rapp. P.-v. Réun. Cons. int. Explor. Mer*, **173**, 107–116.

CUSHING, D. H. and DICKSON, R. R. (1976). The biological response in the sea to climatic changes. *Adv. mar. Biol.*, **14**, 1–122.

CUSHING, J. E. (1971). Immunology of fish. In W. S. Hoar and D. J. Randal (Eds), *Fish Physiology*. Academic Press, New York. pp. 465–500.

CUZON, G., COGNIE, D. and POULLAQUEC, G. (1981). Point sur la croissance de la crevette peneide (*P. japonicus*) en eaux rechaufees. In K. Tiews (Ed.), *Aquaculture in Heated Effluents and Recirculation Systems*, Vol. II, Proceedings of a World Symposium, Stavanger, 1980. Heenemann, Berlin, pp. 289–296.

DAUBERT, A. (1975). Echauffement des eaux par des centrales nucleaires en estuaire et bord de mer en France. In *Environmental Effects of Cooling Systems at Nuclear Power Plants*, Proceedings of a Symposium, Oslo, 1974. International Atomic Energy Agency, Vienna. pp. 151–162.

DAUGARD, S. J. and SUNDARAM, T. R. (1979). Engineering trade-offs governing waste heat management and utilization. In S. S. Lee and S. Sengupta (Eds), *Waste Heat Management and Utilization*, Vol. 1, Proceedings of a Conference, Miami, 1977. Hemisphere, Washington. pp. 553–564.

DAVIES, R. M. and JENSEN, L. D. (1974). Entrainment of zooplankton at three mid-Atlantic power plants. In L. D. Jensen (Ed.), *Entrainment and Intake Screening* (*EPRI Publ.*, No. 74–0049–00–5). EPRI, Palo Alto, California. pp. 157–162.

DAVIES, R. M. and JENSEN, L. D. (1975). Zooplankton entrainment at three mid-Atlantic power plants. *J. Wat. Pollut. Control. Fed.*, **47**, 2130–2142.

DAVIS, J. C. (1975). Minimal dissolved oxygen requirements of aquatic life with emphasis on Canadian species: a review. *J. Fish. Res. Bd Can.*, **32**, 2295–2332.

DAWSON, J. K. (1979). Copepods (Arthropoda: Crustacea: Copepoda). In C. W. Hart., Jr., and S. L. H. Fuller (Eds), *Pollution Ecology of Estuarine Invertebrates*. Academic Press, New York, 145–170.

DEEPAK, A. (Ed.) (1980). *Remote Sensing of Atmosphere and Oceans* (Proceedings of the Interactive Workshop on Interpretation of Remotely Sensed Data, Williamsburg, Virginia, 1979), Academic Press, New York.

DEPARTMENT OF THE ENVIRONMENT (1977). Possible ecological effects of desalination wastes on coastal and estuarine waters. *Tech. Note Cent. Wat. Plan. Unit* (Reading, U.K.), **18**, 1–19.

DESCAMPS, B., GROGNET, B. and FOULQUIER, L. (1981). Étude experimentale du grossissement des anguilles par l'utilisation des eaux rechauffees. In K. Tiews (Ed.), *Aquaculture in Heated Effluents and Recirculation Systems*, Vol. II, Proceedings of a World Symposium, Stavanger, 1980. Heenemann, Berlin. pp. 213–227.

DEVIK, O. (Ed.) (1975). *Harvesting Polluted Waters: Waste Heat and Nutrient-Loaded Effluents in the Aquaculture*, Plenum Press, New York.

DEWALLE, D. R. (1979). Utilization and dissipation of waste heat by soil warming. In S. S. Lee and S. Sengupta (Eds), *Waste Heat Management and Utilization*, Vol. 3, Proceedings of a Conference, Miami, 1977. Hemisphere, Washington. pp. 2261–2275.

DIAZ, L. F. and TREZEK, G. J. (1979). Energy recovery through utilization of thermal wastes in an energy-urban-agro-waste complex. In S. S. Lee and S. Sengupta (Eds), *Waste Heat Management and Utilization*, Vol. 1, Proceedings of a Conference, Miami, 1977. Hemisphere, Washington. pp. 565–583.

DICKSON, R. R., GARBUTT, P. A. and NARAYANA PILLAI, V. (1980). Satellite evidence of enhanced upwelling along the European continental slope. *J. phys. Oceanogr.* **10**, 813–819.

DIETRICH, J. R. (1975), The introduction of new power generation systems; implications of LWR experience. In R. R. Ferber and R. A. Roxas (Eds), *Transactions 9th World Energy Conference, Detroit, 1974*, Vol. V. US National Committee of the World Energy Conference, New York. pp. 254–277.

DINSMORE, A. F. (1979). Inland Florida Cooling Systems. In S. S. Lee and S. Sengupta (Eds), *Waste Heat Management and Utilization*, Vol. 2, Proceedings of a Conference, Miami, 1977. Hemisphere, Washington. pp. 1043–1071.

DITMARS, J. D., PADDOCK. R. A. and FRIGO, A. A. (1979). Observations of thermal plumes from submerged discharges in the Great Lakes and their implications for modelling and monitoring. In S. S. Lee and S. Sengupta (Eds), *Waste Heat Management and Utilization*, Vol. 2, Proceedings of a Conference, Miami, 1977. Hemisphere, Washington. pp. 1307–1328.

DOOLEY, H. D. (1974). A comparison of drogue and current meter measurements in shallow waters. *Rapp. P.-v. Réun. Cons. int. Explor. Mer*, **167**, 225–230.

DOUGLAS, P. A (1968). Heated discharges and aquatic life. *Bull. Sport Fish. Inst.* **198**, 1–5.

DOW, R. L. (1962). A method of predicting fluctuations in the sea scallop populations of Maine. *Comml Fish. Rev.*, **24**, 1–4.

DOW, R. L. (1964). A comparison among selected marine species of an association between sea water temperature and relative abundance. *J. Cons. int. Explor. Mer*, **28**, 425–431.

DOW, R. L. (1972). Fluctuations in Gulf of Maine sea temperature and specific molluscan abundance. *J. Cons. int. Explor. Mer*, **34**, 532–534.

DOW, R. L. (1977a). Relationships of sea surface temperature to American and European lobster landings. *J. Cons. int. Explor. Mer*, **37**, 186–191.

DOW, R. L. (1977b). Effects of climatic cycles on the relative abundance and availability of commercial marine and estuarine species. *J. Cons. int. Explor. Mer*, **37**, 274–280.

DOW, R. L. (1978). Effects of sea-surface temperature cycles on landings of American, European and Norway lobsters. *J. Cons. int. Explor. Mer*, **38**, 271–272.

DUNCAN, C. P., ATWOOD, D. K. and STALCUP, M. C. (1978). Energy from the sea: ocean thermal energy conversion possibilities in the Caribbean. In *CICAR II*, Symposium on Progress in Marine Research in the Caribbean and Adjacent Regions, Caracas, 1976 (*FAO Fisheries Rep.* No. 200, Suppl.). pp. 531–546.

DUNN, W. E., POLICASTRO, A. J. and PADDOCK, R. A. (1975). *Surface Thermal Plumes: Evaluations of Mathematical Models for the Near and Complete Field*, Waters Resources Research Programme, Energy and Environmental Systems Division Argonne National Laboratory, Argonne, Illinois (Parts 1 and 2; Rep. No. ANL/WR-75-3).

DZUNG, L. S. and MÜHLHAÜSER, H. (1975). Economic and environmental factors influencing the choice of steam turbines and heat exchangers for the combined power generation and district heating. In R. R. Ferber and R. A. Roxas (Eds), *Transactions 9th World Energy Conference, New York*, Vol. V. US National Committee of the World Energy Conference, New York, 786–805.

EDINGER, J. E. and POLK, E. M. (1971). Intermediate mixing of thermal discharges into a uniform current. *Wat. Air Soil Pollut.*, **1**, 7–31.

EDINGER, J. E., BRADY, D. K. and GEYER, J. C. (1974). *Heat Exchange and Transport in the Environment,* Electric Power Research Institute Cooling Water Discharges Project, Palo Alto, California (*EPRI Rep.,* No. 49 quoted by STAUFFER and EDINGER, 1980).

EDWARDSON, W. (1976). Energy demands of aquaculture: a worldwide survey. *Fish Farm. Int.,* **3** (4), 10–13.

EDWARDSON, W. (1977). Energy analysis and the fishing industry. In P. D. Wilmot and A. Slingerland (Eds), *Technology Assessment and the Oceans.* IPC Science and Technology Press, Guildford. pp. 60–66.

EFFERTZ, P. H. and FICHTE, W. (1976). The protection of condenser tubes in the cooling water side by ferrous sulphate dosing. *Maschinenschaden,* **49** (4), 163–172.

EGUSA, S. (1979). Notes on the culture of the European eel (*Anguilla anguilla* L.) in Japanese eel-farming ponds. *Rapp. P.-v. Réun. Cons. int. Explor. Mer,* **174,** 51–58.

EGUSA, S. (1981). Fish diseases and their control in intensive culture utilizing heated effluents or recirculating systems in Japan. In K. Tiews (Ed.), *Aquaculture in Heated Effluents and Recirculation Systems,* Vol. II, Proceedings of a World Symposium, Stavanger, 1980. Heenemann, Berlin. pp. 33–39.

E.I.F.A.C. (1973). Water quality criteria for European freshwater fish. Report on dissolved oxygen and inland fisheries. *Tech. Pap. E.I.F.A.C.,* **19,** 1–10.

ELLETT, D. J. and MARTIN, J. H. A. (1973). The physical and chemical oceanography of the Rockall Channel. *Deep Sea Res.,* **20,** 585–625.

ELLISON, T. H. and TURNER, J. S. (1959). Turbulent entrainment in stratified flow. *J. Fluid Mech.,* **6** (3), 423–448.

ELSER, H. J. (1965). Effects of warm-water discharge on angling in the Potomac River, Maryland, 1961–1962. *Prog. Fish Cult.,* **27,** 79–86.

ELTRINGHAM, S. K. and HOCKLEY, A. R. (1961). Migration and reproduction of the wood-boring isopod, *Limnoria,* in Southampton Water. *Limnol. Oceanogr.,* **6,** 467–482.

ENGLESSON, G. A., HU, M. C. and SAVAGE, W. F. (1979). Wet/dry cooling for water conservation. In S. S. Lee and S. Sengupta (Eds), *Waste Heat Management and Utilization,* Vol. 2, Proceedings of a Conference, Miami, 1977. Hemisphere, Washington. pp. 845–874.

ENRIGHT, J. T. (1977). Power plants and plankton. *Mar. Pollut. Bull. N.S.,* **8,** 158–161.

ESCH, G. W. and McFARLANE, R. W. (Eds) (1976). *Thermal Ecology, II,* Proceedings of a Symposium, Augusta, Georgia, 1975 (CONF-750425). Technical Information Center, US Energy Research and Development Administration, Oak Ridge, Tennessee.

ESTES, N. C. (1938). Chlorination of cooling waters. *Refiner and Natural Gasoline Manufacturer,* **17,** 191–197 (quoted by NASH and PAULSEN, 1981).

EWART, T. W. and BENDINER, W. P. (1974). Techniques for estuarine and open ocean dye dispersal measurement. *Rapp. P.-v. Rénun. Cons. int. Explor. Mer,* **167,** 201–212.

FAIRBANKS, R. B., COLLINGS, W. S. and SIDES, W. T. (1971). *An Assessment of the Effects of Electrical Power Generation on Marine Resources in the Cape Cod Canal,* Massachusetts Department of Nature Resources, Division of Marine Fisheries, Boston, Massachusetts.

FALKE, J. and SMITH, M. H. (1974). Effects of thermal effluent on the fat content of the mosquito fish. In J. W. Gibbons and R. R. Sharitz (Eds), *Thermal Ecology,* Proceedings of a Symposium, Augusta, Georgia, 1973 (CONF-730505). Technical Information Center, US Atomic Energy Commission, Oak Ridge, Tennessee. pp. 100–108.

FARMANFARMAIAN, A. and MOORE, R. (1978) Diseasonal thermal aquaculture. 1. Effect of temperature and dissolved oxygen on survival and growth of *Macrobrachium rosenbergii.* In J. W. Avault (Ed.), *Proceedings of the 9th Annual Meeting, World Mariculture Society.* Lousiana State University, Baton Rouge. pp. 55–66.

FELIX, J. R. (1978). *The Effects of Temperature Change on Growth and Survival of American Lobsters (Homarus americanus),* M.S. Thesis, San Diego State University (quoted by FORD and co-authors, 1979).

FISCHENKO, P. A., SERYAPOV, A. I. and POTAPENKO, I. A. (1973). The prevention of deposits in turbine condensers by ultrasonics. *Soviet Pwr Engrg.,* **4,** 300–301.

FISCHER, H. B., LIST, E. J., KOH, R. C. Y., IMBERGER, J. and BROOKS, N. H. (1979). *Mixing in Inland and Coastal Waters,* Academic Press, New York.

FLATOW, R. E. (1981). High dosage ultraviolet water purification: an indespensable tool for recycling, fish hatcheries and heated effluent aquaculture. In K. Tiews (Ed.), *Aquaculture in Heated Effluents and Recirculation Systems*, Vol. I, Proceedings of a World Symposium, Stavanger, 1980. Heenemann, Berlin. pp. 455–466.

FLETCHER, J. O. (1969). Controlling the plant's climate. *Impact Sci. Soc.*, **19** (2), 151–168.

FLOHN, H. (1963). *Klimaschwankungen und grossräumige Klimabeeinflussung* (Arbeitsgemeinschaft für Forschung des Landes Nordrhein-Westfalen, Heft 115), Köln and Opladen (Westdeutscher Verlag; quoted by LAMB, 1977).

FLÜGEL, H. (1972). Pressure: animals. In O. Kinne (Ed.), *Marine Ecology, Vol. I, Environmental Factors*, Part 3. Wiley, London. pp. 1407–1450.

FORD, R. F., FELIX, J. R., JOHNSON, R. L., CARLBERG, J. M. and OLST, J. C. VAN HALL (1979). Effects of fluctuating and constant temperatures and chemicals in thermal effluent on growth and survival of the American lobster (*Homarus americanus*). In J. W. Avault (Ed.), *Proceedings of the 10th Meeting, World Mariculture Society*. Louisiana State University, Baton Rouge. pp. 139–158.

FORREST, D. M. (1976). *Eel Capture, Culture, Processing and Marketing*, Fishing News (Books) Ltd., Surrey.

FORSTER, J. R. M., HARMAN, J. P. and SMART, G. R. (1977). Water economy: its effect on trout production. *Fish Farm. Int.*, **4**, 10–13.

FOULQUIER, L. and DESCAMPS, B. (1978). Donnees sur l'utilisation des eaux de refrigeration industrielle. Studes sur le grossissement d'*Anguilla anguilla* (L.) en eau rechauffee. In R. Lesel (Ed.), *Communications du 23-ème Congres National de l'Association Francaise de Limnologie organise par le Centre de Recherches Hydrobiologique, Biarritz (France) 22–25 Mai 1973. Bull. Cent. Etud. Rech. Scient. Biarritz*, **12** (3), 409–432.

FOURCY, A., DUMONT, M. and FREYCHET, A. (1979). Heating of greenhouses with tepid water. In S. S. Lee and S. Sengupta (Eds), *Waste Heat Management and Utilization*, Vol. 3, Proceedings of a Conference, Miami, 1977. Hemisphere, Washington. pp. 2085–2094.

FOX, J. L. and MOYER, M. S. (1973). Some effects of a power plant on marine microbiota. *Chesapeake Sci.*, **14**, 1–10.

FRAAS, A. P. (1975). Topping and bottoming cycles. In R. R. Ferber and R. A. Roxas (Eds), *Transactions 9th World Energy Conference, Detroit, 1974*, Vol. V. US National Committee of the World Energy Conference, New York. pp. 192–211.

FRAZER, W., BARR, D. I. H. and SMITH, A. A. (1968). A hydraulic model study of heat dissipation at Longannet Power Station. *Proc. Instn civ. Engrs.*, **39**, 23–44.

FREEMYERS, M. C. and INCROPERA, F. P. (1979). A simulation of waste heat utilization for greenhouse climate control. In S. S. Lee and S. Sengupta (Eds), *Waste Heat Management and Utilization*, Vol. 3, Proceedings of a Conference, Miami, 1977. Hemisphere, Washington. pp. 2321–2359.

FRY, F. E. J. (1947). Effects of the environment on animal activity. *Univ. Toronto Stud. biol. Ser.*, **55**, 1–62.

FRY, F. E. J. (1957a). The aquatic respiration of fish. In M. E. Brown (Ed.), *The Physiology of Fishes, Vol. I, Metabolism*. Academic Press, New York. pp. 1–63.

FRY, F. E. J. (1957b). The lethal temperature as a tool in taxonomy. *Année biol.*, **33**, 205–210. (Also in *Colloques Un. int. Sci. biol.*, **24**, 1958).

FRY, F. E. J., BRETT, J. R. and CLAWSON, G. H. (1941). Lethal limits of temperature for young goldfish. *Rev. can. Biol.*, **1**, 50–56.

FULLER, R. D. (1978). Ocean thermal energy conversion. *Ocean Mgmt*, **4**, 241–258.

FUSHIMI, K., (1975). MHD power generation development in Japan. In R. R. Ferber and R. A. Roxas (Eds), *Transactions 9th World Energy Conference*, Vol. V. US National Committee of the World Energy Conference, New York. pp. 78–97.

GABOR, D., COLOMBO, U., KING, A. and GALLI, R. (1978). *Beyond the Age of Waste. A Report to the Club of Rome*, Pergamon Press, Oxford.

GALLAWAY, B. J. and STRAWN, K. (1974). Seasonal abundance and distribution of marine fishes at a hot-water discharge in Galveston Bay, Texas. *Contr. mar. Sci.*, **18**, 71–137.

GALLAWAY, B. J. and STRAWN, K. (1975). Seasonal and areal comparisions of fish diversity indices at a hot-water discharge in Galveston Bay, Texas. *Contr. mar. Sci.*, **19**, 79–89.

GARSIDE, E. T. (1970). Temperature: fishes: structural responses. In O. Kinne (Ed.), *Marine Ecology, Vol. I, Environmental Factors*, Part 1. Wiley, London. pp. 561–573.

GAUCHER, T. A. (1968). Thermal enrichment and marine aquiculture. In W. J. MacNeil (Ed.), *Marine Aquicultrue*. Oregon State University Marine Science Center, Oregon State University Press, Corvallis, Oregon. pp. 141–152.

GAUDY, R., GUERIN, J. P. and MORAITOU-APOSTOLOPOULOU, M. (1982). Effect of temperature and salinity on the population dynamics of *Tisbe holothuriae*, Humes (Copepoda; Harpacticoida) fed two different diets. *J. exp. mar. Biol. Ecol.*, **57**, 257–271.

GEORGE, D. W. (1975). Australian contributions to research in energy conversion technology. In R. R. Ferber and R A. Roxas (Eds), *Transactions 9th World Energy Conference, Detroit, 1974*, Vol. V. US National Committee of the World Energy Conference, New York. pp. 98–111.

GERCHAKOV, S. M., ROOTH, C. G. H., SEGAR, D. A. and STEARNS, R. D. (1973). Rapid delineation of the mean plume intensity pattern from the sediment temperatures underlying a thermal discharge. *Bull. mar. Sci.*, **23**, 496–509.

GERSDORF, B. VON and SOMMER, W. (1975). Power plants and environmental interference in congested areas. In R. R. Ferber and R. A. Roxas (Eds), *Transactions 9th World Energy Conference, Detroit, 1974*. Vol. V. US National Committee of the World Energy Conference, New York. pp. 475–490.

GESSNER, F. (1970). Temperature: plants. In O. Kinne (Ed.), *Marine Ecology, Vol. 1, Environmental Factors*, Part 1. Wiley, London. pp. 363–406.

GESSNER, F. and SCHRAMM, W. (1971). Salinity: plants. In O. Kinne (Ed.), *Marine Ecology, Vol. 1, Environmental Factors*, Part 2. Wiley, London. pp. 705–820.

GIBBONS, J. W. and SHARITZ, R. R. (Eds) (1974). *Thermal Ecology*. Proceedings of a Symposium, Augusta, Georgia, 1973 (CONF–730505). Technical Information Center, U.S. Atomic Energy Commission, Oak Ridge, Tennessee.

GIBBONS, J. W., HOOK, J. T. and FORNEY, D. L. (1972). Winter responses of largemouth bass to a heated effluent from a nuclear reactor. *Prog. Fish Cult.*, **34**, 88–90.

GIORGETTI, G., CESCHIA, G. and BOVO, G (1981). Utilization of warm artesian fresh water for eel breeding: comparisons between two groups of eels with different origins. In K. Tiews (Ed.), *Aquaculture in Heated Effluents and Recirculation Systems*, Vol. II, Proceedings of a World Symposium, Stavanger, 1980. Heenemann, Berlin, pp. 229–237.

GLENDENNING, I. (1978). Wave power: a real alternative? *Ocean Mgmt*, **4**, 207–240.

GLUECKSTERN, P. and KANTOR, Y. (1977). Economic evaluation of using small sized nuclear reactors for seawater desalination. *Desalination*, **22**, 101–110.

GODFRIAUX, B. L., GUERRA, C. R. and RESH, R. E., (1977). Venture analyses for intensive waste heat aquaculture. In J. W. Avault (Ed.), *Proceedings of the 8th Annual Meeting, World Mariculture Society*. Louisiana State University, Baton Rouge. pp. 707–722.

GODSHALL, F. A., CORY, R. L. and PHINNEY, D. A. (1974). Measurement in a marine environment using low cost sensors of temperature and dissolved oxygen. *Chesapeake Sci.*, **15**, 178–181.

GONZÁLEZ, J. G. and YEVICH, P. P. (1977). Seasonal variation in the responses of estuarine populations to heated water in the vicinity of a steam generating plant. In M. Marois (Ed.), *Proceedings of the World Conference Towards a Plan of Actions for Mankind, Vol. 3, Biological Balance and Thermal Modifications* (Institut de la Vie, Paris, 1974). Pergamon Press, Oxford. pp. 115–127.

GORDON, L. (1975). The energy crisis and the world economy. In R. R. Ferber and R. A. Roxas (Eds), *Transactions 9th World Energy Conference, Detroit, 1974*, Vol. II. US National Committee of the World Energy Conference New York. pp. 952–975.

GOSS, W. P., McCOWAN, J. G. and CLOUTIER, P. D. (1977). Application of technology assessment methodology to ocean thermal energy conversion systems. In P. D. Wilmot and A. Slingerland (Eds), *Technology Assessment and the Oceans*. IPC Science and Technology Press, Guildford. pp. 122–129.

GOTO, T. and HAKUTA, T. (1977). Pollution problems in a distillation process. *Desalination*, **23**, 245–253.

GRAUBY, A. (1977). Quelques exemples d'utilisation des rejets thermiques sans le domaine de l'agriculture en France. In M. Marois (Ed.), *Proceedings of the World Conference Towards a Plan of Actions for Mankind, Vol. 3, Biological Balance and Thermal Modifications* (Institut de la Vie, Paris, 1974). Pergamon Press, Oxford. pp. 285–290.

GREEN, R. H. (1979). *Sampling Design and Statistical Methods for Environmental Biologists*, Wiley, New York.

GRIBBIN, J. (1981). Sun and weather: the stratospheric link. *New Scient.*, **91**, 669–671.

GRIMÅS, U. (1970). Warm water effluents in Sweden. *Mar. Pollut. Bull., N.S.*, **1**, 151–152.

GRIMES, C. B. (1971). Thermal addition studies of the Crystal River Steam Electric Station. *Prof. Pap. Ser. mar. Lab. Fla.*, **11**, 1–53.

GRIMES, C. B. and MOUNTAIN, J. A. (1971). Effects of thermal effluent upon marine fishes near the Crystal River Steam Electric Station. *Prof. Pap. Ser. mar. Lab. Fla.*, **17**, 1–64.

GROBERG, W. J., McCOY, R. H., PILCHER, K. S. and FRYER, J. L. (1978). Relation of water temperature to infections of coho salmon (*Oncorhynchus kisutch*), chinook salmon (*O. tshawytscha*), and steelhead trout (*Salmo gairdneri*) with *Aeromonas salmonicida* and *A. hydrophila*. *J. Fish. Res. Bd Can.*, **35**, 1–7.

GRONOW, G. (1974). Über die Anwendung des an Säugetieren erarbeiteten Begriffes 'Stress' auf Knochenfische. *Zool. Anz.*, **192**, 316–331.

GUERRA, C. R. (1975). *Power Plant Waste Heat Utilization in Aquaculture. Workshop I*, Public Service Electric and Gas Company, Newark, N.J. pp. 191–197.

GUERRA, C. R., GODFRIAUX, B. L., EBLE, A. F. and STOLPE, N. E. (1976). Aquaculture in thermal effluents from power plants. In G. Persoone and E. Jaspers (Eds), *Proceedings of the 10th European Symposium on Marine Biology, Ostend, 1975, Vol. 1, Mariculture.* Universa Press, Wetteren, Belgium. pp. 189–205.

GUERRA, C. R., GODFRIAUX, B. L. and SHEAHAN, C. J. (1979a). Utilization of waste heat from power plants by sequential culture of warm and cold weather species. In S. S. Lee and S. Sengupta (Eds), *Waste Heat Management and Utilization*, Vol. 3, Proceedings of a Conference, Miami, 1977. Hemisphere, Washington. pp. 2121–2140.

GUERRA, C. R., RESH, R. E., GODFRIAUX, B. L. and STEPHENS, C. A. (1979b). Venture analyses for a proposed commercial waste heat aquaculture facility. In J. W. Avault (Ed.), *Proceedings of the 10th Annual Meeting World Mariculture Society.* Lousiana State University, Baton Rouge. pp. 28–38.

GUIZERIX, J., MAGRITA, R. and MOLINARI, J. (1976). Etudes par traceurs des modalites de transfert dans un resau hydrographique. *Houille Blanche*, **3/4**, 277–284.

GULLAND, J. (1983). Fisheries resources. In O. Kinne (Ed.), *Marine Ecology, Vol. V, Ocean Management*, Part 2. Wiley, London. pp. 839–1061.

GUNDERSON, K. and BIENFANG, P. (1972). Thermal pollution: use of deep, cold, nutrient-rich sea water for power plant cooling and subsequent aquaculture in Hawaii. In M. Ruivo (Ed.), *Marine Pollution and Sea Life.* Fishing News (Books) Ltd., Surrey. pp. 513–516.

HADLEY, D. and STRAUGHAN, D. (1974). Tolerance of *Littorina planaxis* and *L. scutulata* to temperature changes. In D. F. Soule and M. Oguri (Eds), *Marine Studies of San Pedro Bay, California, Part III, Thermal Tolerance and Sediment Toxicity Studies.* Allan Hancock Foundation and The Office of Sea Grant Programs, Los Angeles. pp. 78–96.

HAIR, J. R. (1971). Upper lethal temperature and thermal shock tolerances of the opossum shrimp, *Meomysis awatschensis*, from the Sacramento-San Joaquin Estuary, California. *Calif. Fish Game*, **57**, 17–27.

HAMILTON, D. H., FLEMER, D. A., KEEFE, C. W. and MIHURSKY, J. A. (1970). Power plants: effects of chlorination on estuarine primary production. *Science, N.Y.*, **169**, 197–198.

HANNA, S. R. and GIFFORD, F. A. (1975). Meteorological effects of energy dissipation at large power parks. *Bull. Am. met. Soc.*, **56**, 1069–1076.

HANSEN, J. E., WANG, W.-C. and LACIS, A. A. (1978). Mount Agung eruption provides test of a global climatic perturbation. *Science, N.Y.*, **199**, 1065–1068.

HANSEN, J., JOHNSON, D., LACIS, A., LEBEDEFF, S., LEE, P., RIND, D. and RUSSELL, G. (1981). Climate impact of increasing atmospheric carbon dioxide. *Science, N.Y.*, **213**, 957–966.

HANSON, J. A. (1974). Energy for and with open sea mariculture. In J. A. Hanson (Ed.), *Open Sea Mariculture, Perspectives, Problems and Prospects.* Dowden, Hutchinson and Ross, Inc., Stroudsburg, Pennsylvania. pp. 334–355.

HARASHINA, H. (1977). 5 000 000 imp. gal./day seawater desalination plant for the Minsitry of Electricity and Water, Government of Kuwait. *Desalination*, **22**, 425–434.

HARDY, B. L. S. (1977). *The Effects of a Warm Water Effluent on the Biology of the Sand-dwelling Harpacticoid Copepod Asellopsis intermedia* (T. Scott), Ph.D. Thesis, University of Glasgow, Scotland.

HART, F. C. and DELANEY, B. T. (1978). The impact of RCRA (PL 94-580) on utility solid wastes. *Electric Power Research Institute Rep,*, EPR1 FP-878 and TPS-78-779 (quoted by NASH and PAULSEN, 1981).

HARTLEY, G. S. and MACLAUCHLAN, J. W. G. (1969). A simple integrating thermometer for field use. *J. Ecol.,* **57**, 151–154.

HARVEY, G. (1979). Wanted: fish farmers for the power stations. *Fish Farmer.,* **2**, (3), 8–9.

HATLE, Z. and LAMPAR, M. (1975). Beneficial exploitation of waste heat. In *Environmental Effects of Cooling Systems at Nuclear Power Plants*, Proceedings of a Symposium, Oslo, 1974. International Atomic Energy Agency, Vienna. pp. 731–740.

HAWES, F. B., COUGHLAN, J. and SPENCER, J. F. (1975). Environmental effects of the heated discharges from Bradwell Nuclear Power Station and of the cooling systems of other stations. In *Environmental Effects of Cooling Systems at Nuclear Power Plants*, Proceedings of a Symposium, Oslo, 1974. International Atomic Energy Agency, Vienna. pp. 423–448.

HEATH, A. G. (1977). Toxicity of intermittent chlorination to freshwater fish: influence of temperature and chlorine form. *Hydrobiologia,* **56**, 39–47.

HECHTEL, G. J., ERNST, E. J. and KALIN, R. J. (1970). *Biological Effects of Thermal Pollution*, Marine Science Research Centre, State University of New York, Northport, New York *(Tech. Rep.,* 3).

HEDGPETH, J. W. (1957). Estuaries and lagoons. II. Biological aspects. In J. W. Hedgpeth (Ed.), *Treatise on Marine Ecology and Palaeoecology, Vol. 1, Ecology (Mem. geol. Soc. Am.,* **67**, 693–729).

HEDGPETH, J. W. (1977). The limits of calefaction. In M. Marois (Ed.), *Proceedings of the World Conference Towards a Plan of Actions for Mankind, Vol. 3, Biological Balance and Thermal Modifications* (Institut de la Vie, Paris, 1974). Pergamon Press, Oxford. pp. 61–76.

HEDGPETH, J. W. and GONOR, J. J. (1969). Aspects of the potential effect of thermal alteration on marine and estuarine benthos. In P. A. Krenkel and P. L. Parker (Eds), *Biological Aspects of Thermal Pollution*, Proceedings of the National Symposium on Thermal Pollution, Portland, Oregon, USA, 1968. Vanderbilt University Press, Nashville, Tennessee. pp. 80–118.

HEFFNER, R. L., HOWELLS, G. P., LAUER, G. J. and HIRSCHFIELD, H. I. (1973). Effects of power plant operation on Hudson river estuary microbiota. In D. J. Nelson (Ed.), *Radionuclides in Ecosystems*, Vol. 1, Proceedings of the 3rd National Symposium on Radioecology, Oak Ridge, Tennessee, 1971. US Atomic Energy Commission, Oak Ridge, Tennessee. pp. 619–629.

HEIN, M. K. and KOPPEN, J. D. (1979). Effects of thermally elevated discharges on the structure and composition of estuarine periphyton diatom assemblages. *Estuar. coast. mar. Sci.,* **9**, 385–401.

HEINLE, D. R. (1969). Temperature and zooplankton. *Chesapeake Sci.,* **10**, 186–209.

HEINLE, D. R. (1976). Effects of passage through power plant cooling systems on estuarine copepods. *Environ. Pollut.,* **11**, 39–58.

HEINLE, D. (1977). In discussion to HEDGPETH (1977) p. 77.

HEITMAN, H., KELLY, D. F. and BOND, T. E. (1958). Ambient air temperature and weight gain in swine. *J. Anim. Sci.,* **17**, 62–67.

HENDERSON, C. D. (1979). The Turkey Point Cooling Canal System. In S. S Lee and S. Sengupta (Eds), *Waste Heat Management and Utilization*, Vol. 2, Proceedings of a Conference, Miami, 1977. Hemisphere, Washington. pp. 1219–1229.

HENRY, C. D. and FAZZOLARE, R. (1979). Thermodynamic analysis of Rankine cycle energy systems utilizing waste heat. In S. S. Lee and S. Sengupta (Eds), *Waste Heat Management and Utilization*, Vol. 3 Proceedings of a Conference, Miami, 1977. Hemisphere, Washington. pp. 2465–2472.

HESS, C. T., SMITH, C. W. and PRICE, A. H. (1975). Mode for the accumulation of radionuclides in oysters and sediments. *Nature, Lond.,* **258**, 225–226.

HESS, C. T., SMITH, C. W. and PRICE, A. H. (1977). A mathematical model of the accumulation of radionuclides by oysters (*C. virginica*) aquacultured in the effluent of a nuclear power reactor to include major biological parameters. *Hlth Phys.* **33**, 121–130.

HESS, C. T., SMITH, C. W., PRICE, A. H. and DARLING, I. C. (1979). Using heated effluent from a 835 MWe nuclear power reactor for shellfish aquaculture. In S. S. Lee and S. Sengupta (Eds), *Waste Heat Management and Utilization*, Vol. 3, Proceedings of a Conference, Miami, 1977. Hemisphere, Washington. pp. 2229–2241.

HETHERINGTON, J. A. (1976). Radioactivity in surface and coastal water of the British Isles, 1974. *Min. Agric. Fish Food, Radiobiol. Lab., Lowestoft, U.K.,* FRL 11, 1–34.

HETTLER, W. F. and CLEMENTS, L. C. (1978). Effects of acute thermal stress on marine fish embryos and larvae. In L. D. Jensen (Ed.), *4th National Workshop on Entrainment and Impingement, Chicago, Illinois, 1977.* Ecological Analysts Inc., New York. pp. 171–190.

HICKLING, C. F. and BROWN, P. L. (1973). *The Seas and Oceans,* Blandford, London.

HIDU, H., ROOSENBURG, W. H., DROBECK, K. G., McERLEAN, A. J. and MIHURSKY, J. A. (1974). Thermal tolerance of oyster larvae *Crassostrea virginica* Gmelin, as related to power plant operation. *Proc. natn. Shellfish, Ass.,* **64**, 102–110.

HILGE, V., DELVENTHAL, H. and KLINGER, H. (1981). Influences of heated, recirculated water and heated well water on several physiological and hematological parameters of channel catfish, *Ictalurus punctatus* Raf., reared at two stocking densities. In K. Tiews (Ed.), *Aquaculture in Heated Effluents and Recirculation Systems,* Vol. II, Proceedings of a World Symposium, Stavanger, 1980. Heenemann, Berlin. pp. 185–190.

HILL, K. M. (1977). Technology assessment and seawater resource engineering. In P. Wilmot and A. Slingerland (Eds), *Technology Assessment and the Oceans.* IPC Science and Technology Press, Guildford. pp. 93–101.

HILLMAN, R. E., DAVIS, N. W. and WENNEMER, J. (1977). Abundance, diversity, and stability in shore-zone fish communities in an area of Long-Island Sound affected by the thermal discharge of a nuclear power station. *Estuar. coast. mar. Sci.,* **5**, 355–381.

HIRAYAMA, K. and HIRANO, R. (1970a). Influences of high temperature and residual chlorine on the marine planktonic larvae. *Bull. Fac. Fish. Nagasaki Univ.,* **29**, 83–89 (Japanese: summary in English).

HIRAYAMA, K. and HIRANO, R. (1970b). Influences of high temperature and residual chlorine on marine phytoplankton. *Mar. Biol.,* **7**, 205–213.

HIRST, E. (1973). Environmental control in animal shelters using power plant thermal effluents. *J. environ. Qual.,* **2**, 166–171.

HOAK, R. D. (1961a). The thermal-pollution problem. *J. Wat. Pollut. Control Fed.,* **33**, 1267–1276.

HOAK, R. D. (1961b). Defining thermal pollution. *Pwr Engng, Chicago,* **65**, 39–42.

HOCKLEY, A. R. (1963). Some effect of warm water effluents in Southampton Water. *Rep. Challenger Soc.,* **3**(15), 37–38.

HOCUTT, C. H. (1980). Introduction. In C. H. Hocutt, J. R. Stauffer, J. E. Edinger, L. W. Hall and R. P. Morgan (Eds), *Power Plants; Effects on Fish and Shellfish Behavior,* Academic Press, New York. pp. 1–8.

HOCUTT, C. H., STAUFFER, J. R., EDINGER, J. E., HALL, L. W. and MORGAN, R. P. (Eds) (1980). *Power Plants, Effects on Fish and Shellfish Behaviour,* Academic Press, New York.

HOFFMAN, D. (1977). Second generation low temperature vapor compression plants. *Desalination,* **23**, 449–454.

HOLLIDAY, F. G. T. (1971). Salinity: fishes. In O. Kinne (Ed.), *Marine Ecology, Vol. I, Environmental Factors,* Part 2. Wiley, London. pp. 997–1083.

HOLMES, N. (1970a). Marine fouling in power stations. *Mar. Pollut. Bull., N.S.,* **1**, 105–106.

HOLMES, N. (1970b). Mussel fouling in chlorinated cooling systems. *Chemy Ind.,* **24**, 1244–1247.

HOLT, R. and STRAWN, K. (1977). Cage culture of seven fish species in a power plant effluent characterized by wide salinity fluctuations. In J. W. Avault (Ed.), *Proceedings of the 8th Annual Meeting, World Mariculture Society.* Lousiana State University, Baton Rouge. pp. 73–90.

HOOPER, D. J. (1979). Hydrographic surveying. In K. R. Dyer (Ed.), *Estuarine Hydrography and Sedimentation.* Cambridge University Press, Cambridge, pp. 41–56.

HOPKINS, S. R. and DEAN, J. M. (1975). The response of developmental stages of *Fundulus* to acute thermal shock. In F. J. Vernberg (Ed.), *Physiological Ecology of Estuarine Organisms,* Belle Baruch Library in Marine Science, **3**. South Carolina Press, Columbia. pp. 301–313.

HORST, J. M. A. VAN DER (1972). Waste heat use in greenhouses. *J. Wat. Pollut. Control Fed.,* **44**, 494–496.

HORST, T. J. (1977). Effects of power station mortality on fish population stability in relationship to life history strategy. In W. Van Winkle (Ed.), *Proceedings of the Conference on Assessing the Effects of Power-Plant-Induced Mortality on Fish Populations, Gatlinburg, Tennessee, 1977.* Pergamon Press, New York. pp. 297–310.

HOSS, D. E. and BLAXTER, J. H. S. (1979). The effect of rapid changes of hydrostatic pressure on the Atlantic herring *Clupea harengus* L. I. Larval survival and behaviour. *J. exp. mar. Biol. Ecol.,* **41**, 75–85.

Hoss, D. E., Coston, L. C. and Hettler, W. F. (1972). Effects of increased temperature on postlarval and juvenile estuarine fish. *Proc. a. Conf. S. East. Ass. Game Fish Commn*, **25**, 635–642.

Hoss, D. E., Hettler, W. F. and Coston, L. C. (1974). Effects of thermal shock on larval estuarine fish-ecological implications with respect to entrainment in power plant cooling systems. In J. H. S. Blaxter (Ed.), *The Early Life History of Fish*. Springer-Verlag, Berlin. pp. 357–371.

Hoss, D. E., Coston, L. C., Baptist, J. P. and Engel, D. W. (1975). Effects of temperature, copper and chlorine on fish during simulated entrainment in power plant condenser cooling systems. In *Environmental Effects of Cooling Systems at Nuclear Power Plants*, Proceedings of a Symposium, Oslo, 1974. International Atomic Energy Agency, Vienna. pp. 519–527.

Hoult, D. P. (1972). Oil spreading on the sea. *A. Rev. Fluid Mech.*, **4**, 341–368.

Hudec, P. P. and Sonnenfeld, P. (1974). Energy-power from brine. *Science, N.Y.*, **185**, 440.

Huguenin, J. E. (1975). *Indirect Use of a Power Plant Thermal Effluent in Marine Aquaculture*, Conference on Utilization of Thermal Effluents in Aquaculture, Sponsored by MIT/U Mass, Sea Grant Program, Univ. Massachusetts, October, 1975 (quoted by Aston, R. J., Brown, D. J. A. and Milner, A. G. P., 1976).

Huguenin, J. E. and Rothwell, G. N. (1979). The problems, economic potentials and system design of large future tropical marine fish cage systems. In J. W. Avault (Ed.), *Proceedings of the 10th Annual Meeting*, World Mariculture Society. Louisiana State University, Baton Rouge. pp. 162–181.

Huguenin. J. E. and Ryther, J. H. (1974). The use of power plant waste heat in marine aquaculture. *Proceedings 10th Annual Conference Marine Technology Society*, Washington, DC. pp. 431–445.

Huisman, J., Hoopen, H. J. G. Ten and Fuchs, A. (1980). The effect of temperature upon the toxicity of mercuric chloride to *Scenedesmus acutus*. *Environ. Pollut. (Series A)*, **22**, 133–148.

Hunt, G. J. (1979). Radioactivity in surface and coastal waters of the British Isles, 1977. *Min. Agric. Fish Food, Lowestoft, Aquatic Env. Monitoring Rep.*, **3**, 1–36.

Icanberry, J. W. and Adams, J. R. (1974). Zooplankton survival in cooling water systems of four thermal power plants on the California coast. In L. D. Jensen (Ed.), *Entrainment and Intake Screening* (EPRI Publ. 74-049-00-5), EPRI, Palo Alto, California. pp. 13–22.

Iles, R. B. (1963). Cultivating fish for food and sport in power station water. *New Scient.*, **17**, 227–229.

Ilus, E. and Keskitalo, J. (1980). First experiences of the environmental effects of cooling water from the nuclear power plant of Loviisa (South coast of Finland). Proceedings of the 6th Symposium of the Baltic Marine Biologists. *Ophelia*, **Suppl. 1**, 117–122.

Ingebrigtsen, O. and Torrissen, O. (1981). The use of effluent water from Matre power plant for raising salmonid fingerlings at Matre aquaculture station. In K. Tiews (Ed.), *Aquaculture in Heated Effluents and Recirculation Systems*, Vol. II, Proceedings of a World Symposium, Stavanger, 1980. Heenemann, Berlin. pp. 515–524.

Ingham, A. E. (ed.) (1975). *Sea Surveying*, Vol. 1 (Text), Vol. 2 (Illustrations), Wiley, London.

Institution of Engineers, Australia. (1972). *Thermal discharge Engineering and Ecology* (Papers, Thermo-fluids Conference, Sydney, 1972), Institution of Engineers, Sydney.

International Atomic Energy Agency (1971). *Environmental Aspects of Nuclear Power Stations*, Proceedings of a Symposium, New York, 1971, I.A.E.A.-SM-146. International Atomic Energy Agency, Vienna. 1–970.

International Atomic Energy Agency (1974). Thermal discharges at Nuclear Power Stations: their management and environmental impacts. *I.A.E.A. (Vienna) Tech. Rep. Series*, No. 155, 1–155.

International Atomic Energy Agency. (1975). *Environmental Effects of Cooling Systems at Nuclear Power Plants*, Proceedings of a Symposium, Oslo, 1974, I.A.E.A.-SM-187. International Atomic Energy Agency, Vienna. pp. 1–831.

Isaac, R. A. (1979). Power generation : effects on the aquatic environment in Massachusetts. In S. S. Lee and S. Sengupta (Eds), *Waste Heat Management and Utilization*, Vol. 1, Proceedings of a Conference, Miami, 1977. Hemisphere, Washington. pp. 243–259.

Isaacson, M. S., Koh, R. C. Y. and List, E. J. (1979). Hydraulic investigations of thermal diffusion during heat treatment cycles San Onofre Nuclear Generating Station Units 2 and 3.

In S. S. Lee and S. Sengupta (Eds), *Waste Heat Management and Utilization*, Vol. 2, Proceedings of a Conference, Miami, 1977. Hemisphere, Washington. pp. 1535–1558.

ISO, S. (1977). Problems on seawater intake and discharge facilities in coastal region. *Desalination*, **22**, 159–168.

IVERSON, E. S. (1968). *Farming the Edge of the Sea*, Fishing News (Books) Ltd., Surrey.

JAMES, W. G. (1967). Mussel fouling and the use of exomotive chlorination. *Chemy Ind.*, **24**, 994–996.

JARMAN, R. T. and TURVILLE, C. M. DE (1974). Dispersion of heat in Southampton Water. *Proc. Instn civ. Engrs.*, **57**, 129–142.

JEFFERS, F. J. (1972). A method of minimizing effects of waste heat discharges. *Int. J. environ. Stud.*, **3**, 321–327.

JENSEN, L. D. (Ed.) (1974). *Entrainment and Intake Screening*, Electric Power Research Institute (*Publ.* 74-049-00-5), Palo Alto, California.

JENSEN, L. D. (Ed.) (1976). *3rd National Workshop on Entrainment and Impingement* (Section 316(b)—Research and Compliance Considerations, February, 1976, New York), Ecological Analysts Inc., New York.

JENSEN, L. D. (Ed.) (1978). *4th National Workshop on Entrainment and Impingement*, *December 1977, Chicago, USA*, Ecological Analysts Inc., New York.

JENSEN, M. H. (1972). The use of waste heat in agriculture. In M. M. Yarosh (Ed.), *Waste Heat Utilization*, Proceedings of the National Conference, Gatlinburg, Tennessee, 1971 (CONF-711031) US National Technical Information Service, Springfield, Virginia. pp. 21–38.

JIJI, L. M. and HOCH, J. (1977). Flow establishment for buoyant and turbulent jets in cross-flow. In D. B. Spalding and N. Afgan (Eds), Heat Transfer and Turbulent Buoyant Convection. Hemisphere, Washington. pp. 263–274.

JOHNS, R., FOLWELL, R., DAILEY, R. and WIRTH, M. (1971). *Agricultural Alternatives for Utilizing Off-peak Electrical Energy and Cooling Water*, Dept. of Agric. Econ., Washington State University, Pullman, US (quoted by DEWALLE, 1979).

JOHNSON, J. D. (1977). Analytical problems in chlorination of saline water. *Chesapeake Sci.*, **18**, 116–118.

JOHNSON, W. C. (1979). Culture of freshwater prawns (*Macrobrachium rosenbergii*) using geothermal waste water. In J. W. Avault (Ed.), *Proceedings of the 10th Annual Meeting, World Mariculture Society*. Louisiana State University, Baton Rouge. pp. 385–391.

JOHNSON, W. J. (1979). Thermal impact reduction by dilution, Big Bend Station, Tampa, Florida. In S. S. Lee and S. Sengupta (Eds), *Waste Heat Management and Utilization*, Vol. 2, Proceedings of a Conference, Miami, 1977. Hemisphere, Washington. pp. 827–844.

JOLLEY, R. L. (1976). *Environmental Impact of Water Chlorination*, Proceedings of a Conference, Oak Ridge Tennessee, 1975 (CONF-751096). US National Technical Information Service, Springfield, Virginia.

JOLLEY, R. L., GORCHEV, H. and HAMILTON, D. H. (1978). *Water Chlorination, Environmental Impact and Health Effects*, Vol. 2, Proceedings of the 2nd Conference on the Environmental Impact of Water Chlorination, Gatlinburg, Tennessee, 1977. Science Publishers, Ann Arbor, Michigan, (Vol. 1 of this publication is a re-issue of JOLLEY, 1976).

JONES, A., BROWN, J. A. G., DOUGLAS, M. T., THOMPSON, S. J. and WHITFIELD, R. J. (1981). Progress towards developing methods for the intensive farming of turbot (*Scophthalmus maximus* L.) in cooling water from a nuclear power station. In K. Tiews (Ed.), *Aquaculture in Heated Effluents and Recirculation Systems*, Vol. II, Proceedings of a World Symposium, Stavanger, 1980. Heenemann, Berlin. pp. 481–496.

JONES, C. (1976). Cooling water farm set to go commercial. *Fish Farm. Int.*, **3** (4), 14–15.

JONES, C. (1980). Eels from Somerset. *Fish Farm. Int.*, **7** (1), 24–26.

JONES, L. T. (1963). The geographical and vertical distribution of British *Limnoria* [Crustacea : Isopoda]. *J. mar. biol. Ass. U.K.*, **43**, 589–603.

KALLE, K. (1971). Salinity : general introduction. In O. Kinne (Ed.), *Marine Ecology*, Vol. 1, *Environmental Factors*, Part 2. Wiley, London. pp. 683–688.

KALLE, K. (1972). Dissolved gases : general introduction. In O. Kinne (Ed.), *Marine Ecology*, Vol. 1, *Environmental Factors*, Part 3. Wiley, London. pp. 1451–1457.

KANTHA, L. H., PHILLIPS, O. M. and AZAD, R. S. (1977). On turbulent entrainment at a stable density interface. *J. Fluid Mech.*, **79**, 753–768.

KARABASHEV, G. S. and OZMIDOV, R. V. (1974). Investigation of admixture diffusion in the sea by means of luminescent tracers and towed fluorometer. *Rapp. P.-v. Réun. Cons. int. Explor. Mer*, **167**, 231–235.

KARAM, R. A. and MORGAN, K. Z. (1976). *Environmental Impact of Nuclear Power Plants*, Proceedings of a Conference, Atlanta, Georgia, 1974. Pergamon Press, New York.

KARKHECK, J. and POWELL, J. (1979). Prospects for the utilization of waste heat in large scale district heating systems. In S. S. Lee and S. Sengupta (Eds), *Waste Heat Management and Utilization*, Vol. 3, Proceedings of a Conference, Miami, 1977. Hemisphere, Washington. pp. 2191–2213.

KATO, J., NOMA, T. and UEKITA, Y. (1979). Design of floating breakwaters. In T. V. R. Pillay and W. A. Dill (Eds), *Advances in Aquaculture*, FAO Technical Conference on Aquaculture, Kyoto, 1976. Fishing News (Books) Ltd., Surrey. pp. 458–466.

KENDRICK, P. J. (1976). Remote sensing and water quality. *J. Wat. Pollut. Control Fed.*, **48**, 2243–2246.

KENNEDY, V. S. and MIHURSKY, J. A. (1967). *A Bibliography of the Effects of Temperature in the Aquatic Environment*, Natural Resources Institute, University of Maryland, USA (Contribution No. 326).

KENNEDY, V. S. and MIHURSKY, J. A. (1971). Upper temperature tolerance of some estuarine bivalves. *Chesapeake Sci.*, **12**, 193–204.

KENNEDY, V. S. and MIHURSKY, J. A. (1972). Effects of temperature on the respiratory metabolism of three Chesapeake Bay bivalves. *Chesapeake Sci.*, **13**, 1–22.

KENNEDY, V. S., ROOSENBURG, W. H., CASTAGNA, M. and MIHURSKY, J. A. (1974b). *Mercenaria mercenaria* (Mollusca : Bivalvia) : temperature–time relationships for survival of embryos and larvae. *Fish. Bull. U.S.*, **72**, 1160–1166.

KENNEDY, V. S., ROOSENBURG, W. H., ZION, H. H. and CASTAGNA, M. (1974a). Temperature–time relationships for survival of embryos and larvae of *Mulinia lateralis* (Mollusca : Bivalvia). *Mar. Biol.*, **24**, 137–145.

KENNISH, M. J. and OLSSON, R. K. (1975). Effects of thermal discharges on the microstructural growth of *Mercenaria mercenaria*. *Environ. Geol.*, **1**, 41–64.

KERR, J. E. (1953). Studies on fish preservation at the Contra Costa Steam Plant of Pacific Gas and Electric Company. *Calif. Fish Game Fish Bull.*, **92**, 66.

KERR, N. M. (1976a). Farming marine flatfish using waste heat from sea-water cooling. *Energy World*, **October**, 2–10.

KERR, N. M. (1976b). Recent technical advances in marine flatfish farming. *Proc. R. Soc. Edinb. (Section B)*, **75**, 263–270.

KERR, N. M. and KINGWELL, S. J. (1977). *Progress in Farming Marine Fish 1977*, White Fish Authority, Edinburgh.

KERR, N. M. and THAIN, B. P. (1977). *The Possible Use of Distillers By-products and Waste Heat in Fish Farming With Particular Reference to Marine Species* (Preprint of paper to be presented to the Pentlands Scotch Whisky Research Ltd., Distillers By-Products Symposium, Pitlochry, Scotland, 1977), White Fish Authority, Edinburgh. (Mimeo).

KEUP, L. E. (1981). Waste water aquaculture in the United States: potentials and constraints. In K. Tiews (Ed.), *Aquaculture in Heated Effluents and Recirculation Systems*, Vol. I, Proceedings of a World Symposium, Stavanger, 1980. Heenemann, Berlin. pp. 481–491.

KHALANSKI, M. (1975). Etudes réalisées en France sur les conséquences écologiques de la réfrigération des centrales thermiques en circuit ouvert. In *Environmental Effects of Cooling Systems at Nuclear Power Plants*, Proceedings of a Symposium, Oslo, 1974. International Atomic Energy Agency (SM-187), Vienna. pp. 461–476.

KINGWELL, S. J., DUGGAN, M. C. and DYE, J. E. (1977). The large scale handling of the larvae of the marine flatfish turbot, *Scophthalmus maximus* L., and Dover sole, *Solea solea* L., with a view to their subsequent fattening under farm conditions. 3rd Meeting of the I.C.E.S. Working Group on Mariculture, Brest, France. 1977. *Actes de Colloques du C.N.E.X.O.*, **4**, 27–34.

KINNE, O. (1960). Growth, food intake, and food conversion in a euryplastic fish exposed to different temperatures and salinities. *Physiol. Zool.*, **33**, 288–317.

KINNE, O. (1963). The effects of temperature and salinity on marine and brackish water animals. 1. Temperature. *Oceanogr. mar. Biol. A. Rev.*, **1**, 301–340.

KINNE, O. (1964). The effects of temperature and salinity on marine and brackish water animals. II. Salinity and temperature salinity combinations. *Oceanogr. mar. Biol. A. Rev.*, **2**, 281–339.

KINNE, O. (1970a). Temperature : general introduction. In O. Kinne (Ed.), *Marine Ecology, Vol. I, Environmental Factors*, Part 1. Wiley, London. pp. 321–346.

KINNE, O. (1970b). Temperature : invertebrates. In O. Kinne (Ed.), *Marine Ecology, Vol. I, Environmental Factors*, Part 1, Wiley, London. pp. 407–517.

KINNE, O. (1971). Salinity : invertebrates. In O. Kinne (Ed.), *Marine Ecology, Vol. I, Environmental Factors*, Part 2. Wiley, London. pp. 821–995.

KINNE, O. (1972). Pressure : general introduction. In O. Kinne (Ed.), *Marine Ecology, Vol. I, Environmental Factors*, Part 3. Wiley, London. pp. 1323–1360.

KINNE, O. (1976). Cultivation of marine organisms: water quality management and technology. In O. Kinne (Ed.), *Marine Ecology, Vol. III. Cultivation*, Part 1. Wiley, London. pp. 19–300.

KINNE, O. (1977). Cultivation of animals: research cultivation. In O. Kinne (Ed.), *Marine Ecology, Vol. III, Cultivation*, Part 2. Wiley, Chichester. pp. 579–1293.

KINNE, O. (1980). Diseases of marine animals: general aspects. In O. Kinne (Ed.), *Diseases of Marine Animals, Vol. I, General Aspects: Protozoa to Gastropoda*. Wiley, Chichester. pp. 13–73.

KINNE, O. (1982). Aquakultur: Ausweg aus der Ernährungskrise? *Spektrum der Wissenschaft*, **Dec. 12**, 46–57.

KINNE, O. (in press). *Realism in Aquaculture—the View of an Ecologist* (Keynote Address given at the World Conference on Aquaculture, Venice, September 1981; Symposium Proceedings).

KINNE, O. and ROSENTHAL, H. (1977). Commercial cultivation (aquaculture). In O. Kinne (Ed.), *Marine Ecology, Vol. III, Cultivation*, Part 3. Wiley, Chichester. pp. 1321–1398.

KITTELSEN, A. and GJEDREM, T. (1981). Cooling water from a hydro-power plant used in smolt production. In K. Tiews (Ed.), *Aquaculture in Heated Effluents and Recirculation Systems*, Vol. II, Proceedings of a World Symposium, Stavanger, 1980. Heenemann, Berlin. pp. 509–514.

KJERVE, B. (1979). Measurement and analysis of water current, temperature, salinity and density. In K. R. Dyer (Ed.), *Estuarine Hydrography and Sedimentation*. Cambridge University Press, Cambridge. pp. 186–226.

KJØLSETH, G. (1981). Heat exchange from seawater and its use in a smolt production plant in the polar region of Norway. In K. Tiews (Ed.), *Aquaculture in Heated Effluents and Recirculation Systems*, Vol. I, Proceedings of a World Symposium, Stavanger, 1980. Heenemann, Berlin. pp. 327–346.

KOMAGATA, S., IGA, H., NAKAMURA, H. and MINOHARA, Y. (1970). The status of geothermal utilization in Japan. *Geothermics*, **2**, 185–196 (quoted by PETERSON and SEO, 1977).

KONDRATYEV, K. YA. and NIKOLSKY, G. A. (1979). The stratospheric mechanism of solar and anthropogenic influences on climate. In B. M. McCormac and T. A. Seliga (Eds), *Solar–Terrestrial Influences on Weather and Climate*. Reidel, Dordrecht, Holland. pp. 317–322.

KOOPS, H. and KUHLMANN, H. (1981a). Annual variation of feeding and growth rate of eels farmed in thermal effluents of a conventional power station. In K. Tiews (Ed.), *Aquaculture in Heated Effluents and Recirculation Systems*, Vol. II, Proceedings of a World Symposium, Stavanger, 1980. Heenemann, Berlin. pp. 205–211.

KOOPS, H. and KUHLMANN, H. (1981b). Eel farming in the thermal effluents of a conventional power station in the harbour of Emden. In K. Tiews (Ed.), *Aquaculture in Heated Effluents and Recirculation Systems*, Vol. II, Proceedings of a World Symposium, Stavanger, 1980. Heenemann, Berlin. pp. 575–586.

KRENKEL, P. A. and PARKER, P. L. (Eds) (1969a). *Biological Aspects of Thermal Pollution*, Proceedings of the National Symposium on Thermal Pollution, Portland, Oregon, 1968. Vanderbilt University Press, Nashville, Tennessee.

KRENKEL, P. A. and PARKER, P. L. (Eds) (1969b). *Engineering Aspects of Thermal Pollution*, Proceedings of the National Symposium on Thermal Pollution, Nashville, 1968. Vanderbilt University Press, Nashville, Tennessee.

KUHLMANN, H. (1979a). Preliminary fish farming experiments in brackishwater thermal effluents. In T. V. R. Pillay and W. A. Dill (Eds), *Advances in Aquaculture*, FAO Technical Conference on Aquaculture, Kyoto, 1976. Fishing News (Books) Ltd., Surrey. pp. 502–505.

KUHLMANN, H. (1979b). The influence of temperature, food, initial size, and origin on the growth of elvers (Anguilla anguilla L.). Rapp. P.-v. Réun. Cons. int. Explor. Mer, **174**, 59–63.

KUHLMANN, H. and KOOPS, H. (1981). New technology for rearing elvers in heated waters. In K. Tiews (Ed.), Aquaculture in Heated Effluents and Recirculation Systems, Vol. I, Proceedings of a World Symposium, Stavanger, 1980. Heenemann, Berlin. pp. 301–308.

KULLENBERG, G. (1968). Measurements of horizontal and vertical diffusion in coastal waters. Rep. Inst. Fysisk Oceanogr. København Univ., **3**, 1–65.

KULLENBERG, G. (1974a). Investigations on dispersion in stratified vertical shear flow. Rapp. P.-v. Réun. Cons. int. Explor. Mer., **167**, 86–92.

KULLENBERG, G. (1974b). An experimental and theorteical investigation of the turbulent diffusion in the upper layer of the sea. Rep. Inst. Fysisk Oceanogr. København Univ., **25**, 1–289.

KULLENBERG, G. and TALBOT, J. W. (Eds) (1974). Physical processes responsible for dispersal of pollutants in the sea. Rapp. P.-v. Réun. Cons. int. Explor. Mer, **167**, 1–259.

KÜTÜKÇÜOĞLU, A. (1975). Environmental and engineering constraints influencing the site selection of nuclear power plants to be built in Turkey. In R. R. Ferber and R. A. Roxas (Eds), Transactions 9th World Energy Conference, Detroit, 1974, Vol. III. US National Committee of the World Energy Conference, New York. pp. 856–869.

LAMB, H. H. (1972). Climate: Present, Past and Future, Vol. 1, Fundamentals and Climate Now, Methuen, London.

LAMB, H. H. (1977). Climate: Present, Past and Future, Vol. 2, Climatic History and the Future, Methuen, London.

LAMB, H. H. (1982). Climate, History and the Modern World, Methuen, London.

LANDRY, A. M. and STRAWN, K. (1973). Annual cycle of sportfishing activity at a warm water discharge into Galveston Bay, Texas. Trans. Am. Fish. Soc., **102**, 573–577.

LANGFORD, T. E. (1972). A comparative assessment of thermal effects in some British and North American rivers. In R. T. Oglesby, C. A. Carlson and J. McCann (Eds), River Ecology and Man, Academic Press, New York. pp. 319–351.

LANGFORD, T. E. (1975). Ecology and cooling water from power stations. A review of recent biological research in Britain. In R. R. Ferber and R. A. Roxas (Eds), Transactions 9th World Energy Conference Detroit, 1974, Vol. III. US National Committee of the World Energy Conference, New York. pp. 283–310.

LANGFORD, T. E. (1977). Biological problems with the use of seawater for cooling. Chemy Ind., **16**, 612–616.

LANZA, G. R. and CAIRNS, J. (1972). Physio-morphological effects of abrupt thermal stress on diatoms. Trans. Am. microsc. Soc., **91**, 276–298.

LASKOWSKI, S. M. and WOODARD, K. (1979). Comparison of environmental effects due to operation of brackish and/or salt water natural and mechanical draft cooling towers. In S. S. Lee and S. Sengupta (Eds), Waste Heat Management and Utilization, Vol. 1, Proceedings of a Conference, Miami, 1977. Hemisphere, Washington, pp. 125–151.

LAUCKNER, G. (1980). Diseases of Protozoa to Mollusca: Gastropoda. In O. Kinne (Ed.), Diseases of Marine Animals, Vol. I, General Aspects: Protozoa to Gastropoda. Wiley, Chichester. pp. 75–424.

LAUER, G. J., WALLER, W. T., BATH, D. W. MEEKS, W., HEFFNER, D., GINN, T., ZUBARIK, L., BIBKO, P. and STORM, P. C. (1974). Entrainment studies on Hudson river organisms. In L. D. Jensen (Ed.), Entrainment and Intake Screening (EPRI Publ. 74-049-00-5). EPRI, Palo Alto, California. pp. 37–82.

LEACH, G. (1975). Energy costs of food production. In F. Steele and A. Bourne (Eds), The Man–Food Equation, Proceedings of a Symposium, London, 1973. Academic Press, London. pp. 139–163.

LEE, C. C. (1979). Potential research programmes in waste energy utilization. In S. S. Lee and S. Sengupta (Eds), Waste Heat Management and Utilization, Vol. 3, Proceedings of a Conference, Miami, 1977. Hemisphere, Washington. pp. 2277–2298.

LEE, S. S. and SENGUPTA, S. (Eds) (1979a). Waste Heat Management and Utilization, Vols 1, 2 and 3, Proceedings of a Conference, Miami, 1977. Hemisphere, Washington.

LEE, S. S. and SENGUPTA, A. (1979b). Preface to Waste Heat Management and Utilization, Vol. 1, Proceedings of a Conference, Miami, 1977. Hemisphere, Washington, p. xiii.

LEE, S., SENGUPTA, S., TSAI, C. and MILLER, H. (1979). Three-dimensional free surface model for thermal discharge. In S. S. Lee and S. Sengupta (Eds), *Waste Heat Management and Utilization*, Vol. 3. Proceedings of a Conference, Miami, 1977. Hemisphere, Washington. pp. 1615–1634.

LEIGHTON, D., NUSBAUM, I. and MULFORD, S. (1967). Effects of waste discharge from Point Loma Saline Water Conversion Plant on intertidal marine life. *J. Wat. Pollut. Control Fed.*, **29**, 1190–1202.

LEWIS, J. (1964). *The Ecology of Rocky Shores*, English Universities Press, London.

LEWIS, R. M. and HETTLER, W. F., JR. (1968). Effect of temperature and salinity on the survival of young Atlantic menhaden, *Brevoortia tyrannus*. *Trans. Am. Fish. Soc.*, **97**, 344–349.

LIAO, P. B. (1981). Treatment units in recirculation systems for intensive aquaculture. In K. Tiews (Ed.), *Aquaculture in Heated Effluents and Recirculation Systems*, Vol. I, Proceedings of a World Symposium, Stavanger, 1980. Heenemann, Berlin. pp. 183–197.

LIAO, P. B. and MAYO, R. D. (1972). Salmonid hatchery water reuse system. *Aquaculture*, **1**, 317–335.

LIEBMANN, H., OFFHAUS, K. and RIEDMÜLLER, S. (1960). Elektrophoretische Blutunter-suchungen bei normalen und bauchwassersucht-kranken Karpfen. *Schweiz. z. Hydrol.*, **21**, 507–517 (quoted from SNIESZKO, 1974).

LIGTERINGEN, H. (1979). Combined use of hydraulic and mathematical models in the design of a once-through cooling circuit along an estuary. In S. S. Lee and S. Sengupta (Eds), *Waste Heat Management and Utilization*, Vol. 3, Proceedings of a Conference, Miami, 1977. Hemisphere, Washington. pp. 1595–1613.

LOI, T.-n. and WILSON, B. J. (1979). Macroinfaunal structure and effects of thermal discharges in a mesohaline habitat of Chesapeake Bay, near a nuclear power plant. *Mar. Biol.*, **55**, 3–16.

LONG, R. R. (1978). A theory of mixing in a stably stratified fluid. *J. Fluid Mech.*, **84**, 113–124.

LOON, L. S. VAN, FRIGO, A. A. and PADDOCK, R. A. (1977). Thermal-plume measurement systems for Great Lakes coastal waters. In *Oceans '77*, Vol. 2, Proceedings of the 3rd Annual Combined Conference, Marine Technology Society and I.E.E.E. Council of Oceanic Engineering, New York. pp. 33C1–33C6.

LOUGH, R. G. (1975). A re-evaluation of the combined effects of temperature and salinity on survival and growth of bivalve larvae using response surface techniques. *Fish. Bull. U.S.*, **73**, 86–94.

LUCKOW, H. and REINKEN, G. (1979). The Agrotherm research project. In S. S. Lee and S. Sengupta (Eds), *Waste Heat Management and Utilization*, Vol. 3, Proceedings of a Conference, Miami, 1977. Hemisphere, Washington. pp. 2395–2408.

LUMLEY, J. L. ZEMAN, O. and SIESS, J. (1978). The influence of buoyancy on turbulent transport. *J. Fluid Mech.*, **84**, 581–597.

LYBERG, B., HAMBRAEUS, G. and GRADIN, R. (1975). The need or demand for energy—and analysis of the energy consumption. In R. R. Ferber and R. A. Roxas (Eds), *Transactions 9th World Energy Conference, Detroit, 1974*, Vol. II. US National Committee of the World Energy Conference, New York. pp. 109–123.

LYONS, W. G., COBB, S. P., CAMP, D. K., MOUNTAIN, J. A., SAVAGE, T., LYONS, L. and JOYCE, E. A., JR. (1971). Preliminary inventory of marine invertebrates collected near the electrical generating plant, Crystal River, Florida in 1969. *Prof. Pap. Ser. mar. Lab. Fla.*, No. 14, 1–45.

McCOY, J. W. (1974). *The Chemical Treatment of Cooling Water*, Chemical Publishing, New York.

MACCRIMMON, H. R., STEWART, J. E. and BRETT, J. R. (1974). Aquaculture in Canada. The practice and the promise. *Bull. Fish. Res. Bd Can.*, **188**, 1–84.

McERLEAN, A. J., O'CONNOR, S. G., MIHURSKY, J. A. and GIBSON, C. I. (1973). Abundance, diversity and seasonal patterns of estuarine fish populations. *Estuar. coast. mar. Sci.*, **1**, 19–36.

MACFADYEN, A. and WEBB, N. R. C. (1968). An improved temperature integrator for use in ecology. *Oikos*, **19**, 19–27.

MACLEOD, R. A. (1971). Salinity: bacteria, fungi and blue-green algae. In O. Kinne (Ed.), *Marine Ecology, Vol. 1, Environmental Factors*, Part 2. Wiley, London. pp. 689–703.

McGOWN, L. B. and BOCKRIS, J. O'M. (1980). *How to Obtain Abundant Clean Energy*, Plenum, New York.

McMILLAN, W. (1973). *Cooling from Open Water Surfaces: Lake Trawsfynydd Cooling Investigations*, Central Electricity Generating Board, UK (North-West Region Report NW/SSD/RR 1204/73).

McNEIL, W. J. (1970). Heated water from generators presents fish-culture possibilities. *Am. Fish Farm.*, **1**, 18–20.

MACQUEEN, J. F. (1978). Background water temperatures and power station discharges. *Adv. Wat. Resour.*, **1**, 195–203.

MACQUEEN, J. F. (1979). Turbulence and cooling water discharges from power stations. In C. J. Harris (Ed.), *Mathematical Modelling of Turbulent Diffusion in the Environment*. Academic Press, London. pp. 379–437.

McVICAR, A. H. (1975). Infection of plaice (*Pleuronectes platessa* L.) with *Glugea (Nosema) stephani* Hagenmuller, 1899 (Protozoa : Microsporidia) in a fish farm and under experimental conditions. *J. Fish Biol.*, **7**, 611–620.

MADDING, R. P., TOKAR, J. V. and MARMER, G. J. (1975). A comparison of aerial infrared and *in situ* thermal plume measurement techniques. In *Environmental Effects of Cooling Systems at Nuclear Power Plants*, Proceedings of a symposium, Oslo, 1974. International Atomic Energy Agency (SM-187-39), Vienna. pp. 163–185.

MADDOCK, L. and SWANN, C. L. (1977). A statistical analysis of some trends in sea temperature and climate in the Plymouth area in the last 70 years. *J. mar. biol. Ass. U.K.*, **57**, 317–338.

MANABE, S. and WETHERALD, R. T. (1967). Thermal equilibrium of the atmosphere with a given relative humidity. *J. Atmos. Sci.*, **24**, 241–259 (quoted by LAMB, 1977).

MANN, R. and RYTHER, J. H. (1977). Growth of six species of bivalve molluscs in a waste-recycling-aquaculture system. *Aquaculture*, **11**, 231–245.

MARCY, B. C., JR. (1971). Survival of young fish in the discharge canal of a nuclear power plant. *J. Fish. Res. Bd Can.*, **28**, 1057–1060.

MARCY, B. C., JR. (1973). Vulnerability and survival of young Connecticut River fish entrained at a nuclear power plant. *J. Fish. Res. Bd Can.*, **30**, 1195–1203.

MARCY, B. C., JR. (1975). Entrainment of organisms at power plants with emphasis on fish—an overview. In S. B. Saila (Ed.), *Fisheries and Energy Production: A Symposium*. Lexington Books, Lexington, Massachusetts. pp. 89–106.

MARCY, B. C. and GALVIN, R. C. (1973). Winter–Spring sport fishery in the heated discharge of a nuclear power plant. *J. Fish Biol.* **5**, 541–547.

MARCY, B. C., JR, BECK, A. D. and ULANOWICZ, R. E. (1978). Effects and impacts of physical stress on entrained organisms. In J. R. Schubel and B. C. Marcy, Jr. (Eds), *Power Plant Entrainment: A Biological Assessment*. Academic Press, New York. pp. 135–188.

MARCY, B. C., JR, JACOBSON, P. M. and NANKEE, R. L. (1972). Observations on the reactions of young American shad to a heated effluent. *Trans. Am. Fish. Soc.*, **101**, 740–743.

MARCELLO, R. A. and STRAWN, K. (1973). Cage culture of some marine fishes in the intake and discharge canals of a steam-electric generating station, Galveston Bay, Texas. In *Proceedings of the 4th Annual Workshop, World Mariculture Society*, Louisiana State University, Baton Rouge. pp. 97–114.

MARGRAF, F. J. (1977). The growth rate of oysters held in the intake and discharge canals of an electric generating station and in natural waters. In J. W. Avault (Ed.), *Proceedings of the 8th Annual Meeting, World Mariculture Society*. Louisiana State University, Baton Rouge. pp. 915–926.

MARKOWSKI, S. (1959). The cooling water of power stations: a new factor in the environment of marine and freshwater invertebrates. *J. Anim. Ecol.*, **28**, 243–258.

MARKOWSKI, S. (1960). Observations on the response of some benthonic organisms to power station cooling water. *J. Anim. Ecol.*, **29**, 349–357.

MARKOWSKI, S. (1962). Faunistic and ecological investigations in Cavendish Dock, Barrow-in-Furness. *J. Anim. Ecol.*, **31**, 43–52.

MARKOWSKI, S. (1966). The diet and infection of fishes in Cavendish Dock, Barrow-in-Furness. *J. Zool., Lond.*, **150**, 183–197.

MAROIS, M. (Ed.) (1977). *Proceedings of the World Conference Towards a Plan of Actions for Mankind, Vol. 3, Biological Balance and Thermal Modifications* (Institute de la Vie, Paris, 1974). Pergamon Press, Oxford.

MARSHALL, J. S. and TILLY, L. J. (1973). Temperature effects on phytoplankton productivity in a reactor cooling pond. In D. J. Nelson (Ed.), *Radionuclides in Ecosystems*, Vol. 1, Proceedings of the 3rd National Symposium on Radioecology, Oak Ridge, Tennessee, 1971. US Atomic Energy Commission, Oak Ridge, Tennessee. pp. 645–651.

MARSHALL, W. (1979). *Combined Heat and Electrical Power Generation in the United Kingdom*, Department of Energy (Energy Pap. 35), H.M. Stationery Office, London.

MARSHALL, W. L. (1971). Thermal discharges : characteristics and chemical treatment of natural waters used in power plants. *Oak Ridge Nat. Lab., Rep.* No. 4652. Oak Ridge National Laboratory, Oak Ridge, Tennessee.

MARTIN, J. H. A. (1972). Marine climatic changes in the north-east Atlantic, 1900–1966. *Rapp. P.-v. Réun. Cons. Int. Explor. Mer*, **162**, 213–219.

MARTIN, R. B. (1937). Chlorination of condenser cooling water. *Trans. Am. Soc. mech. Engrs.* **60** (Pap. FSP-60–16), 475–483 (quoted by NASH and PAULSEN, 1981).

MARTINO, P. A. and MARCHELLO, J. M. (1968). Using waste heat for fish farming. *Ocean Ind.*, **3** (4), 36–39.

MASNIK, M. T. (1979). The effects of thermal effluents on the populations of shipworms (Teredinidae:Mollusca) in the vicinity of a nuclear power station. In S. S. Lee and S. Sengupta (Eds), *Waste Heat Management and Utilization*, Vol. 1, Proceedings of a Conference, Miami, 1977. Hemisphere, Washington. pp. 301–321.

MATSON, J. V. (1977). Treatment of cooling tower blowdown. *Proc. Am. Soc. civ. Engrs, J. Envir. Engng*, **103** (EEI), 87–98.

MATTHEWS, M. A. (1959). The Earth's carbon cycle. *New Scient.*, **6**, 644–646.

MATTICE, J. S. and ZITTEL, H. E. (1976). Site specific evaluation of power plant chlorination. *J. Wat. Pollut. Control Fed.*, **48**, 2284–2308.

MAUCHLINE, J. (1980). Artificial radioisotopes in the marginal seas of north-western Europe. In F. T. Banner, M. B. Collins and K. S. Massie (Eds), *The North-West European Shelf Seas: The Sea Bed and the Sea in Motion, Vol. II, Physical and Chemical Oceanography, and Physical Resources (Elsevier Oceanogr. Series*, **24B**). Elsevier, Amsterdam. pp. 517–542.

MAYER, A. G. (1914). The effect of temperature upon tropical marine animals. *Pap. Tortugas Lab.*, **6**, 1–24.

MEADE, T. L. (1974). The technology of closed system culture of salmonids. *Mar. Tech. Rep.* (Sea Grant Publs Univ. Rhode Island, Narragansett), **30**, 1–30.

MEANEY, R. A. (1973). Development prospects for fish farming in Ireland. *Fish Farm. Int.*, **1**, 52–54.

MÉLARD, Ch. and PHILIPPART, J. C. (1981). Pisciculture intensive du tilapia *Sarotherodon niloticus* dans les effluents thermiques d'une centrale nucléaire en Belgique. In K. Tiews (Ed.), *Aquaculture in Heated Effluents and Recirculation Systems*, Vol. II, Proceedings of a World Symposium, Stavanger, 1980. Heenemann, Berlin. pp. 637–658.

MERCER, J. H. (1978). West Antarctic ice sheet and CO_2 greenhouse effect: a threat of disaster. *Nature, Lond.*, **271**, 321–325.

MERRIMAN, D. (1970). The calefaction of a river. *Scient. Am.* **222**, 42–52.

MERRIMAN, D. (1971). Does industrial calefaction jeopardize the ecosystem of a long tidal river? In *Environmental Aspects of Nuclear Power Stations*, Proceedings of a Symposium, New York, 1970. International Atomic Energy Agency (SM-131), Vienna. pp. 507–533.

MERRIMAN, D. (Ed.) (1972). *The Connecticut River Investigation, 1965–1972*, (Semi-annual Progress Reports to Connecticut Yankee Atomic Power Co., Haddam, Connecticut.

MESAROVIĆ, M. M. (1975). Waste-heat disposal from steam-electric plants with reference to the stochastic nature of some environmental conditions and to thermal pollution control regulations. In *Environmental Effects of Cooling Systems at Nuclear Power Plants*, Proceedings of a Symposium, Oslo, 1974. International Atomic Energy Agency (SM-187/25), Vienna. pp. 311–329.

MEYER, F. P. (1970). Seasonal fluctuations in the incidence of disease on fish farms. *Spec. Publs Am. Fish. Soc.*, **5**, 21–29.

MIHURSKY, J. A. (1967). Patuxent thermal studies. *Univ. Maryland Natur. Resour. Inst. Prog. Rep.*, (Ref. 67-13), 1–28.

MIHURSKY, J. A. (1969). Patuxent thermal studies: summary and recommendation. *Univ. Maryland Natur. Resour. Inst. Spec. Publ.*, **1** (Ref. 69-2), 1–20.

MIHURSKY, J. A. and CRONIN, L. E. (1967). Progress and problems in thermal pollution in Maryland. *Univ. Maryland Natur. Resour. Inst.*, (Ref. 67-112), 1–15.

MIHURSKY, J. A. and KENNEDY, V. S. (1967). Water temperature criteria to protect aquatic life. *Spec. Publs Am. Fish Soc.*, **4**, 20–32.

MIHURSKY, J. A. and PEARCE, J. B. (1969). Introduction. Proceedings of the 2nd Thermal Workshop of the U.S. International Biological Program. *Chesapeake Sci.*, **10**, 125–127.

MIHURSKY, J. A., MCERLEAN, A. J. and KENNEDY, V. S. (1970). Thermal pollution, aquaculture and pathobiology in aquatic systems. *J. Wildl. Dis.*, **6**, 347–355.

MIKKOLA, I. (1975). Heat discharges into the sea at the Olkiluoto site : laboratory model test results and reasons for selected arrangements. In *Environmental Effects of Cooling Systems at Nuclear Power Plants*, Proceedings of a Symposium, Oslo, 1974, International Atomic Energy Agency (SM-187/5), Vienna. pp. 141–149.

MIKOLA, J., TIAINEN, O. J. A., TOIVIAINEN, E., HAAVISTO, H., NEVANLINNA, L. and SEPPA, M. (1975). A study concerning different energy supply alternatives of the Helsinki metropolitan area in Finland. In R. R. Ferber and R. A. Roxas (Eds), *Transactions 9th World Energy Conference, Detroit, U.S.A., 1974*, Vol. V. US National Committee World Energy Conference, New York. pp. 152–172.

MILLER, E. R. (1966). *Trout Production Facilities in Pennsylvania: Analysis, Planning, and Design of Long-range Program*, Preprint: Northeast Fish and Wildl. Conf. Boston, Mass., 1966 (quoted by GAUCHER, 1968).

MILLER, D. C. and BECK, A. D. (1975). Development and application of criteria for marine cooling waters. In *Environmental Effects of Cooling Systems at Nuclear Power Plants*, Proceedings of a symposium, Oslo, 1974. International Atomic Energy Agency (SM-187/10), Vienna. pp. 639–657.

MILNE, P. H. (1972). *Fish and Shellfish Farming in Coastal Waters*, Fishing News (Books) Ltd., Surrey.

MILNE, P. H. (1975). Iceland opportunities for large-scale development. *Fish Farm. Int.*, **2**, 4–7.

MILNE, P. H. (1976). Engineering and the economics of aquaculture. *J. Fish. Res. Bd Can.*, **33**, 888–898.

MINER, R. M. and WARRICK, J. W. (1975). Environmental effects of cooling system alternatives at inland and coastal sites. *Nucl. Technol.*, **25**, 640–650.

MITCHELL, C. T. and NORTH, W. J. (1971). *Temperature–Time Effects on Marine Plankton Passing Through the Cooling System at San Onofre Generating Station*, Report to the Southern California Edison Co., Costa Mesa, Calif., Marine Biological Consultants Inc.

MITCHELL, N. T. (1967). Radioactivity in surface and coastal waters of the British Isles, *Minist. Agric. Fish. Food Radiobiol. Lab., Lowestoft, U.K., Tech. Rep.*, FRL 1, 1–45.

MITCHELL, N. T. (1968). Radioactivity in surface and coastal waters of the British Isles, 1967. *Minist. Agric. Fish. Food Radiobiol. Lab., Lowestoft, U.K., Tech. Rep.*, FRL 2, 1–41.

MITCHELL, N. T. (1969). Radioactivity in surface and coastal waters of the British Isles, 1968. *Minist. Agric. Fish. Food Radiobiol. Lab., Lowestoft, U.K., Tech. Rep.*, FRL 5, 1–38.

MITCHELL, N. T. (1971a). Radioactivity in surface and coastal waters of the British Isles, 1969. *Minist. Agric. Fish. Food Radiobiol. Lab., Lowestoft, U.K., Tech. Rep.*, FRL 7, 1–33.

MITCHELL, N. T. (1971b). Radioactivity in surface and coastal waters of the British Isles, 1970. *Minist. Agric. Fish. Food Radiobiol. Lab., Lowestoft, U.K., Tech. Rep.*, FRL 8, 1–34.

MITCHELL, N. T. (1973). Radioactivity in surface and coastal waters of the British Isles, 1971. *Minist. Agric. Fish Food Radiobiol. Lab., Lowestoft, U.K., Tech. Rep.*, FRL 9, 1–34.

MITCHELL, N. T. (1975). Radioactivity in surface and coastal waters of the British Isles, 1972–1973. *Minist. Agric. Fish. Food Radiobiol. Lab., Lowestoft, U.K., Tech. Rep..*, FRL 10, 1–40.

MITCHELL, N. T. (1977a). Radioactivity in surface and coastal waters of the British Isles, 1975. *Minist. Agric. Fish. Food Radiobiol. Lab., Lowestoft, U.K., Tech. Rep.*, FRL 12, 1–32.

MITCHELL, N. T. (1977b). Radioactivity in surface and coastal waters of the British Isles, 1976. 1. The Irish Sea and its environs. *Minist. Agric. Fish. Food Radiobiol. Lab. Lowestoft, U.K., Tech. Rep.*, FRL 13, 1–15.

MITCHELL, N. T. (1978). Radioactivity in surface and coastal waters of the British Isles, 1976. 2. Areas other than the Irish Sea and its environs. *Minist. Agric. Fish. Food Radiobiol. Lab.*, *Lowestoft, U.K., Tech. Rep.*, FRL 14, 1–20.

MIYAZAKI, T. and MASUDA, Y. (1978). Development of wave power generators. *Ocean Mgmt*, **4**, 259–271.

MOBERG, O. (1977). Development of desalination technology using warm discharge water. *Desalination*, **22**, 141–150.

MØLLER, B. and DAHL-MADSEN, K. I. (1979). The use of biological/chemical investigations for managing thermal effluents. In S. S. Lee and S. Sengupta (Eds), *Waste Heat Management and Utilization*, Vol. 1, Proceedings of a Conference, Miami, 1977. Hemisphere, Washington, pp. 223–241.

MØLLER, D. and BJERK, Ø. (1981). Smolt production in a recirculation system in northern Norway. In K. Tiews (Ed.), *Aquaculture in Heated Effluents and Recirculation Systems*, Vol. II, Proceedings of a World Symposium, Stavanger, 1980. Heenemann, Berlin. pp. 416–429.

MÖLLER, H. (1978a). Effects of power plant cooling on aquatic biota—and indexed bibliography. *Ber. Inst. Meeresk. Univ. Kiel*, **58**, 1–31.

MÖLLER, H. (1978b). Ecological effects of cooling water of a power plant at Kiel Fjord. *Meeresforschung*, **26**, 117–130.

MONAHAN, E. C. and PYBUS, M. J. (1978). Colour, ultraviolet absorbance and salinity of the surface waters off the west coast of Ireland. *Nature, Lond.*, **274**, 782–784.

MOORE, F. K. (1979). Problems of dry cooling. In S. S. Lee and S. Sengupta (Eds), *Waste Heat Management and Utilization*, Vol. II, Proceedings of a Conference, Miami, 1977. Hemisphere, Washington. pp. 659–686.

MOORE, D. J. and JAMES, K. W. (1973). Water temperature surveys in the vicinity of power stations with special reference to infrared techniques. *Wat. Res.*, **7**, 807–820.

MOORE, C. J., STEVENS, G. A., McERLEAN, A. J. and ZION, H. H. (1973). A sport fishing survey in the vicinity of a steam electric station on the Patuxent Estuary, Maryland. *Chesapeake Sci.*, **14**, 160–170.

MORGAN, R. P. (1969). Steam electric station effects on primary productivity in the Patuxent river estuary. *Univ. Maryland Natur. Resour. Inst.*, (Ref. 69-27), 1–24.

MORGAN, R. P. and CARPENTER, E. J. (1978). Biocides. In J. R. Schubel and B. C. Marcy, Jr., (Eds), *Power Plant Entrainment: A Biological Assessment*. Academic Press, New York. pp. 95–134.

MORGAN, R. P. and STROSS, R. G. (1969). Destruction of phytoplankton in the cooling water supply of a steam electric station. *Chesapeake Sci.*, **10**, 165–171.

MORTON, B. R. TAYLOR, G. I. and TURNER, J. S. (1956). Turbulent gravitational convection from maintained and instantaneous sources. *Proc. R. Soc. (Series A)*, **234**, 1–23.

MOUNTAIN, J. A. (1972). Further thermal addition studies at Crystal river, Florida with an annotated checklist of marine fishes collected 1969–1971. *Prof. Pap. Ser. mar. Lab. Fla.*, **20**, 1–103.

MUENCH, K. A. (1976). The role of the electric utilities industry in developing the use of thermal effluent in aquaculture. In J. W. Avault (Ed.), *Proceedings of the 7th Annual Meeting, World Mariculture Society*. Louisiana State University, Baton Rouge. pp. 535–541.

MUNRO, A. L. S. and FIJAN, N. (1981). Disease prevention and control. In K. Tiews (Ed.), *Aquaculture in Heated Effluents and Recirculation Systems*, Vol. II, Proceedings of a World Symposium, Stavanger, 1980. Heenemann, Berlin. pp. 19–32.

MURTHY, K. K. (1975). Population, power and pollution. In R. R. Ferber and R. A. Roxas (Eds), *Transactions 9th World Energy Conference, Detroit, 1974*, Vol. V. US National Committee of the World Energy Conference, New York. pp. 87–108.

MUSSELIUS, V. A. (1963). Diphyllobothriasis disease of pond fish. *Ref. Zh., Biol.*, No. 5K43.

NASH, C. E. (1968). Power stations as sea farms. *New Scient.*, **40**, 367–369.

NASH, C. E. (1969a). Thermal addition: planning for the future. *Chesapeake Sci.*, **10**, 279–296.

NASH, C. E. (1969b). Thermal aquaculture. *Sea Front.*, **15**, 268–276.

NASH, C. E. (1970a). Marine fish farming. *Mar. Pollut. Bull. N.S.*, **1**, 5–6.

NASH, C. E. (1970b). Marine fish farming (Part 2). *Mar. Pollut. Bull. N.S.*, **1**, 28–30.

NASH, C. E. (1974a). Residual chlorine retention and power plant fish farms. *Prog. Fish Cult.*, **36**, 92–95.

NASH, C. E. (1974b). Crop selection issues. In J. A. Hanson (Ed.). *Open Sea Mariculture: Perspectives, Problems and Prospects*. Dowden, Hutchinson and Ross, Inc., Stroudsburg, Pennsylvania. pp. 183–210.

NASH, C. E. and PAULSEN, C. L. (1981). Water quality changes relevant to heated effluents and intensive aquaculture. In K. Tiews (Ed.), *Aquaculture in Heated Effluents and Recirculation Systems*, Vol. I, Proceedings of a World Symposium, Stavanger, 1980. Heenemann, Berlin. pp. 3–15.

NATIONAL ACADEMY OF SCIENCES (1973). *Water Quality Criteria, 1972* EPA-RIII-73-033. US Environmental Protection Agency, Washington, DC. pp. 1–594.

NATIONAL ACADEMY OF SCIENCES (1975). *Understanding Climatic Change: A Program for Action* (prepared by the Panel on Climatic Variation: Co-Chairmen W. L. Gates and Y. Mintz of the US Committee for the Global Atmospheric Research Program), National Research Council, Washington, DC.

NATIONAL ACADEMY OF SCIENCES (1979). *Carbon Dioxide and Climate: a Scientific Assessment* (by an *ad hoc* study group of the Climate Research Board: Chairman, J. G. Charney of the National Research Council), National Academy of Sciences, Washington, DC. (Available from National Technical Information Service, Springfield, Virginia).

NAUMAN, J. W. and CORY, R. L. (1969). Thermal additions and epifaunal organisms at Chalk Point, Maryland. *Chesapeake Sci.*, **10**, 218–226.

NAYLOR, E. (1965a). Biological effects of a heated effluent in docks at Swansea, S. Wales. *Proc. zool. Soc.*, **144**, 253–268.

NAYLOR, E. (1965b). Effects of heated effluents upon marine and estuarine organisms. *Adv. mar. Biol.*, **3**, 63–103.

NEBEKER, A. V., STEVENS, D. G. and BRETT, J. R. (1976). Effects of gas supersaturated water on freshwater aquatic invertebrates. In D. H. Fickeisen and M. J. Schneider (Eds), *Gas Bubble Disease*, Proceedings of a Workshop, Richland, Washington, 1974 (CONF-741033). Technical Information Center, US Energy Research and Development Administration, Oak Ridge, Tennessee. pp. 51–65.

NEILL, W. H. (1969). *Ecological Responses of Lake Monona (Dode County, Wisconsin) Fishes to Heated Influent Water*, Annual Progress Report to Wisconsin Utilities Association, Wisconsin.

NEUDECKER, S. (1976). Effects of thermal effluent on the coral reef community at Tanguisson. *Tech. Rep. mar. Lab. Univ. Guam*, **30**, 1–55.

NEUDECKER, S. (1977). Development and environmental quality of coral reef communities near the Tanguisson Power Plant. *Tech. Rep. mar. Lab. Univ. Guam*, **41**, 1–68.

NIHOUL, J. C. J. and SMITZ, J. (1976). Dispersion of thermal pollution in the marine environment. *Bull. Cent. belge Étud. Docum. Eaux*, **29**, 326–328.

NORSE, E. A. (1974). Effects of subnormal temperatures on some common Los Angeles harbor animals. In D. F. Soule and M. Oguri (Eds), *Marine Studies of San Pedro Bay, California, Part III, Thermal Tolerance and Sediment Toxicity Studies*. Allan Hancock Foundation and The Office of Sea Grant Programs, Los Angeles. pp. 44–62.

NORTH, W. J. (1969). Biological effects of a heated water discharge at Morro Bay, California. *Proc. Int. Seaweed Symp.*, **6**, 275–286.

NORTH, W. J. (1977). Experimental introduction of a warm-tolerant strain of *Macrocystis pyrifera* (giant kelp) to the vicinity of a thermal discharge. In M. Marios (Ed.), *Proceedings of the World Conference Towards a Plan of Actions for Mankind, Vol. 3, Biological Balance and Thermal Modifications* (Institut de la Vie, Paris, 1974). Pergamon Press, Oxford. pp. 89–99.

NORTH, W. J. and ADAMS, J. R. (1969). The status of thermal discharges on the Pacific Coast. *Chesapeake Sci.*, **10**, 139–144.

OAKES, D., COOLEY, P., EDWARDS, L. L., HIRSCH, R. W. and MILLER, V. G. (1979). Ozone disinfection of fish hatchery waters: Pilot plant results, prototype design and control considerations. In J. W. Avault (Ed.), *Proceedings of the 10th Annual Meeting, World Mariculture Society*. Louisiana State University, Baton Rouge. pp. 854–870.

OKUBA, A. (1974). Some speculations of oceanic diffusion diagrams. *Rapp. P.-v. Réun. Cons. int. Explor. Mer*, **167**, 77–85.

OLSZEWSKI, M., SUFFERN, J. S., COUTANT, C. C. and COX, D. K. (1979). An overview of waste heat utilization research at Oak Ridge National Laboratory. In S. S. Lee and S. Sengupta (Eds), *Waste Heat Management and Utilization*, Vol. 3, Proceedings of a Conference, Miami, 1977. Hemisphere, Washington. pp. 2299–2320.

OPPENHEIMER, C. H. (1970). Temperature: bacteria, fungi and blue-green algae. In O. Kinne (Ed.), *Marine Ecology, Vol. 1, Environmental Factors*, Part 1. Wiley, London. pp. 347–361.

OPPENHEIMER, C. H. and BROGDEN, W. B. (1976). The ecology of a Texas bay. In O. Devik (Ed.), *Harvesting Polluted Waters, Waste Heat and Nutrient-Loaded Effluents in the Aquaculture*. Plenum Press, New York. pp. 237–274.

OSTRACH, S., PRAHL, J. and TONG, T. (1979). The discharge of a submerged buoyant jet into a stratified environment. In S. S. Lee and S. Sengupta (Eds), *Waste Heat Management and Utilization*, Vol. 2, Proceedings of a Conference, Miami, 1977. Hemisphere, Washington. pp. 1513–1533.

PAGE-JONES, R. M. (1971). The automatic detection of low levels of dissolved free chlorine in fish farming experiments using seawater effluents. *Prog. Fish Cult.*, **33**, 99–102.

PALFREMAN, D. A. (1973). The economics of marine fish farming. *Fish Farm. Int.*, **1**, 47–52.

PALMEGIANO, G. and SAROGLIA, M. G. (1981). Winter shrimp culture in thermal effluents. In K. Tiews (Ed.), *Aquaculture in Heated Effluents and Recirculation Systems*, Vol. II, Proceedings of a World Symposium, Stavanger, 1980. Heenemann, Berlin. pp. 297–302.

PANNELL, J. P. M., JOHNSON, A. E. and RAYMONT, J. E. G. (1962). An investigation into the effects of warmed water from Marchwood Power Station into Southampton Water. *Proc. Instn. civ. Engrs*, **23**, 35–62.

PARK, J. E., VANCE, J. M., CROSS, K. E. and WIE, N. H. VAN (1979). A computerized engineering model for evaporative water cooling towers. In S. S. Lee and S. Sengupta (Eds), *Waste Heat Management and Utilization*, Vol. 2, Proceedings of a Conference, Miami, 1977. Hemisphere, Washington. pp. 1007–1024.

PARKER, N. C. and STRAWN, K. (1976). *Aufwuchs* and sediment fouling rates in flow-through aquaria receiving a heated effluent from Upper Galveston Bay, Texas. In J. W. Avault (Ed.), *Proceedings of the 7th Annual Meeting, World Mariculture Society*. Louisiana State University, Baton Rouge. pp. 543–559.

PARKER, R. F. and UHL, A. E. (1975). Liquified natural gas and the environment. In R. R. Ferber and R. A. Roxas (Eds), *Transactions 9th World Energy Conference, Detroit, 1974*, Vol. III. US National Committee of the World Energy Conference, New York. pp. 882–901.

PARSONS, T. R., JANSSON, B.-O., LONGHURST, A. R. and SÆTERSDAL, G. (Eds) (1978). Marine ecosystems and fisheries oceanography. *Rapp. P.-v. Réun. Cons. int. Explor. Mer*, **173**, 1–240.

PEARCE, J. B. (1969). Thermal addition and the benthos, Cape Cod Canal. *Chesapeake Sci.*, **10**, 227–233.

PEARCY, W. G. and KEENE, D. F. (1974). Remote sensing of water color and sea surface temperatures off the Oregon coast. *Limnol. Oceanogr.*, **19**, 573–583.

PEARCY, W. G. and MUELLER, J. L. (1970). Upwelling, Columbia river plume and albacore tuna. *Proceedings of the 6th International Symposium on Remote Sensing of the Environment, Ann Arbor, Michigan*, pp. 1101–1113.

PEASE, T. E. and SKELLY, M. J. (1975). Statistical design basis for submerged diffuser. *J. Pwr Div. Am. Soc. civ. Engrs*, **101** (PO1), 11–22.

PERES, J.-M. (1982a). Structure and dynamics of assemblages in the pelagial. In O. Kinne (Ed.), *Marine Ecology, Vol. V, Ocean Management*, Part 1. Wiley, Chichester. pp. 67–117.

PERES, J.-M. (1982b). Structure and dynamics of assemblages in the benthal. In O. Kinne (Ed.), *Marine Ecology, Vol. V, Ocean Management*, Part 1. Wiley, Chichester. pp. 119–185.

PERES, J.-M. (1982c). Specific pelagic assemblages. In O. Kinne (Ed.), *Marine Ecology, Vol. V, Ocean Management*, Part 1. Wiley, Chichester. pp. 313–372.

PERSON-LE-RUYET, J., ALEXANDRE, J.-C. and ROUX, A. LE (1981). Methode de production de juveniles de sole (*Solea solea* L.) sur un aliment composé sec et en eau de mer chaufee et recyclée. In K. Tiews (Ed.), *Aquaculture in Heated Effluents and Recirculation Systems*, Vol. II, Proceedings of a World Symposium, Stavanger, 1980. Heenemann, Berlin. pp. 159–175.

PETERS, G., DELVENTHAL, H. and KLINGER, H. (1981). Stress diagnosis for fish in intensive culture systems. In K. Tiews (Ed.), *Aquaculture in Heated Effluents and Recirculation Sytems*, Vol. II, Proceedings of a World Symposium, Stavanger, 1980. Heenemann, Berlin. pp. 239–248.

PETERSON, R. E. and SEO, K. K. (1977). Thermal aquaculture. In J. W. Avault (Ed.), *Proceedings of the 8th Annual Meeting, World Mariculture Society*. Louisiana State University, Baton Rouge. pp. 491–503.

PICKFORD, G. E., SRIVASTAVA, A. K., SLICHER, A. N. and PANG, P. K. T. (1971). The stress response in abundance of circulating leucocytes in the killifish *Fundulus heteroclitus*. *J. exp. Zool.*, **177**, 89–96.

PILATI, D. A. (1976). Cold shock: biological implications and a method for approximating transient environmental temperatures in the near-field region of a thermal discharge. *Sci. Total Environ.*, **6**, 227–237.

PILLAY, T. V. R. (1979). The state of aquaculture 1976. In T. V. R. Pillay and W. A. Dill (Eds), *Advances in Aquaculture*, FAO Technical Conference on Aquaculture, Kyoto, Japan, 1976. Fishing News (Books) Ltd., Surrey. pp. 1–10.

PILLAY, T. V. R. and DILL, W. A. (Eds) (1979). *Advances in Aquaculture*, FAO Technical Conference on Aquaculture, Kyoto, Japan, 1976. Fishing News (Books) Ltd., Surrey.

PINCINE, A. B. and LIST, E. J. (1973). Disposal of brine into an estuary. *J. Wat. Pollut. Control Fed.*, **20**, 2335–2344.

PINGREE, R. D. and MADDOCK, L. (1977). Tidal eddies and coastal discharge. *J. mar. biol. Ass. U.K.*, **57**, 869–875.

PLASS, G. N. (1956). The carbon dioxide theory of climatic change. *Tellus*, **8**, 140–154.

POLLARD, D. A. (1976). Estuaries must be protected. *Aust. Fish.*, **35** (6), 6–10.

PORTER, R. W. and CHATURVEDI, S. K. (1979). Atmospheric spray-canal cooling systems for large electric power plants. In S. S. Lee and S. Sengupta (Eds), *Waste Heat Management and Utilization*, Vol. 2, Proceedings of a Conference, Miami, 1977. Hemisphere, Washington. pp. 949–989.

POSMENTIER, E. S. (1977). The generation of salinity fine-structures by vertical diffusion. *J. phys. Oceanogr.*, **7**, 298–300.

PRICE, A. H., II, HESS, C. T. and SMITH, C. W. (1976). Observations of *Crassostrea virginica* cultured in the heated effluent and discharged radionuclides of a nuclear power reactor. *Proc. natn. Shellfish. Ass.*, **66**, 54–68.

PRICE, B. L. (1972). Thermal water demonstration project. In M. M. Yarosh (Ed.), *Waste Heat Utilization*, Proceedings of the National Conference, Gatlinburg, Tennessee, 1971 (CONF-711031). US National Technical Information Service, Springfield, Virginia. pp. 166–185.

PRITCHARD, D. W. and CARPENTER, J. H. (1960). Measurements of turbulent diffusion in estuarine and inshore waters. *Bull. int. Ass. scient. Hydrol.*, **20**, 37–50.

QUICK, J. A., JR. (Ed.) (1971a). A preliminary investigation: the effect of elevated temperature on the American oyster *Crassostrea virginica* (Gmelin). A Symposium. *Prof. Pap. Ser. mar. Lab. Fla.*, **15**, 1–190.

QUICK, J. A., JR. (1971b). Pathological and parasitological effects of elevated temperatures on oysters with emphasis on *Labyrinthomyxa marina*. In J. A. Quick, Jr. (Ed.), *A Preliminary Investigation: The Effects of Elevated Temperature on the American Oyster Crassostrea virginica* (Gmelin). A Symposium. *Prof. Pap. Ser. mar. Lab. Fla.*, **15**, 105–171.

RANEY, E. C. and MENZEL, B. W. (1967). A bibliography: heated effluents and effects on aquatic life with emphasis on fishes. *Bull. Cornell Univ. Ichthyol. Ass.*, **1**, 1–89.

RANEY, E. C. and MENZEL, B. W. (1969). A bibliography: heated effluents and effects on aquatic life with emphasis on fishes. *Bull. Cornell Univ. Ichthyol. Ass.*, **2**, 1–470.

RANEY, E. C., MENZEL, B. W. and WELLER, E. C. (1973). A bibliography: heated effluents and effects on aquatic life with emphasis on fishes. *Bull. Cornell Univ. Ichthyol. Ass.*, **9**, 1–651.

RAYMOND, L. P., BIENFANG, P. K. and HANSON, J. A. (1974). Nutritional considerations of open sea mariculture. In J. A. Hanson (Ed.), *Open Sea Mariculture: Perspectives, Problems and Prospects*. Dowden, Hutchinson and Ross, Inc., Stroudsburg, Pennsylvania. pp. 129–182.

RAYMONT, J. E. G. (1972). Some aspects of pollution in Southampton Water. *Proc. R. Soc. (Series B)*, **180**, 451–468.

RAYMONT, J. E. G. (1976). The introduction of new species in habitats of heated effluents. In O. Devik (Ed.), *Harvesting Polluted Waters, Waste Heat and Nutrient-Loaded Effluents in the Aquaculture*. Plenum Press, New York. pp. 185–196.

RAYMONT, J. E. G. and CARRIE, B. G. A. (1964). The production of zooplankton in Southampton Water. *Int. Rev. ges. Hydrobiol.*, **49**, 185–232.

REAVES, R. S., HOUSTON, A. H. and MADDEN, J. A. (1968). Environmental temperature and the body fluid system of the freshwater Teleost Ionic regulation in rainbow trout, *Salmo gairdneri*, following abrupt thermal shock. *Comp. Biochem. Physiol.*, **25**, 849–860.

REEVE, M. R. (1970). Seasonal changes in the zooplankton of south Biscayne Bay and some problems of assessing the effects on the zooplankton of natural and artificial thermal and other fluctuations. *Bull. mar. Sci.*, **20**, 894–921.

REEVE, M. R. and COSPER, E. (1972). The acute effects of heated effluents on the copepod *Acartia tonsa* from a subtropical bay and some problems of assessment. In M. Ruivo (Ed.), *Marine Pollution and Sea Life*. Fishing News (Books) Ltd., Surrey. pp. 250–252.

REICHENBACH-KLINKE, H. H. (1981). The influence of temperature and temperature changes on the outbreak and intensity of fish diseases. In K. Tiews (Ed.), *Aquaculture in Heated Effluents and Recirculation Systems*, Vol. II, Proceedings of a World Symposium, Stavanger, 1980. Heenemann, Berlin. pp. 103–107.

REID, D. A. (1981). Development of Canadian thermal effluent aquaculture systems. In K. Tiews (Ed.), *Aquaculture in Heated Effluents and Recirculation Systems*, Vol. II, Proceedings of a World Symposium, Stavanger, 1980. Heenemann, Berlin. pp. 557–573.

REISH, D. J. (1977). In discussion to HEDGPETH (1977), p. 77.

REYNOLDS, J. Z. (1980). Power plant cooling systems: policy alternatives. *Science, N.Y.*, **207**, 367–372.

RHEINHEIMER, G. (1972). Dissolved gases: bacteria, fungi and blue-green algae. In O. Kinne (Ed.), *Marine Ecology Vol. I, Environmental Factors*, Part 3, Wiley, London. pp. 1459–1469.

ROBERTS, R. J. (1975). The effect of temperature on diseases and their histopathologic manifestations in fish. In W. E. Ribelin and G. Migaki (Eds), *The Pathology of Fishes*. University of Wisconsin Press, Madison, Wisconsin. pp. 477–496.

ROBERTS, R. J. (1976). Bacterial diseases of farmed fishes. In F. A. Skinner and J. G. Carr (Eds), *Microbiology in Agriculture*. Academic Press, London. pp. 55–62.

ROBERTS, R. J. (Ed.), (1978). *Fish Pathology*, Baillière Tindall, London.

ROBERTS, M. H., DIAZ, R. J., BENDER, M. E. and HUGGETT, R. J. (1975). Acute toxicity of chlorine to selected estuarine species. *J. Fish. Res. Bd Can.*, **32**, 2525–2528.

RODENHUIS, G. S. (1979). Numerical models in cooling water circulation studies: techniques, principle errors, practical applications. In S. S. Lee and S. Sengupta (Eds), *Waste Heat Management and Utilization*, Vol. 3, Proceedings of a Conference, Miami, 1977. Hemisphere, Washington, pp. 1577–1593.

ROELS, O. A., HAINES, K. C. and SUNDERLIN, J. B. (1976). The potential yield of artificial upwelling mariculture. In G. Persoone and E. Jaspers (Eds), *Proceedings of the 10th European Symposium on Marine Biology, Ostend, 1975, Vol. 1, Mariculture*. Universa Press, Wetteren, Belgium. pp. 381–390.

ROELS, O. A., SHARFSTEIN, B. A. and HARRIS, V. M. (1978). An evaluation of the feasibility of a temperate climate effluent-aquaculture-tertiary treatment system in New York City. In M. L. Wiley (Ed.), *Estuarine Interactions*. Academic Press, New York. pp. 146–156.

ROENSCH. L. F., GRIER, J. C. and KLEN, E. F. (1978). The achievement of slime control in utility surface condensers without impairing discharge water quality. *Combustion*, **50**, 16–20.

ROESSLER, M. A. (1971). Environmental changes associated with a Florida power plant. *Mar. Pollut. Bull., N.S.*, **2**, 87–90.

ROESSLER, M. (1977). Thermal additions in a tropical marine lagoon. In M. Marios (Ed.), *Proceedings of the World Conference Towards a Plan of Actions for Mankind, Vol. 3, Biological Balance and Thermal Modifications* (Institut de la Vie, Paris, 1974). Pergamon Press, Oxford. pp. 79–87.

ROGERS, A. and CANE, A. (1981). The operation of an 18-tonne rainbow trout rearing unit in power station cooling water. In K. Tiews (Ed.), *Aquaculture in Heated Effluents and Recirculation Systems*, Vol. II, Proceedings of a World Symposium, Stavanger, 1980. Heenemann, Berlin. pp. 545–555.

ROMERIL, M. G. (1977). Heavy metal accumulation in the vicinity of desalination plant. *Mar. Pollut. Bull.; N.S.*, **8**, 84–87.

ROMERIL, M. G. and DAVIS, M. H. (1976). Trace metals in eels grown in power station cooling water. *Aquaculture*, **8**, 139–150.

ROOSENBURG, W. H. (1969). Greening and copper accumulation in the American oyster, *Crassostrea virginica*, in the vicinity of a steam electric generating station. *Chesapeake Sci.*, **10**, 241–252.

ROSENTHAL, H. (1981). Ozonation and sterilization. In K. Tiews (Ed.), *Aquaculture in Heated Effluents and Recirculation Systems*, Vol. I, Proceedings of a World Symposium, Stavanger, 1980. Heenemann, Berlin. pp. 219–274.

ROSS, F. F. and WHITEHOUSE, J. W. (1973). Cooling towers and water quality. *Wat. Res.*, **7**, 623–631.

RUCKER, R. R. (1976). Gas bubble disease of salmonids: variation in oxygen-nitrogen ratio with constant total gas pressure. In D. H. Fickeisen and M. J. Schneider (Eds), *Gas Bubble Disease*, Proceedings of a Workshop, Richland, Washington, 1974 (CONF-741033). Technical Information Center, US Energy Research and Development Administration, Oak Ridge, Tennessee. pp. 85–88.

RUSSELL, F. S., SOUTHWARD, A. J., BOALCH, G. T. and BUTLER, E. I. (1971). Changes in biological conditions in the English Channel off Plymouth during the last half century. *Nature, Lond.*, **234**, 468–470.

RUTHERFORD, D. (1975). California shellfish hatchery supplies world market. *Fish Farm. Int.*, **2** (2), 26–29.

RYKBOST, K. A. and BOERSMA, L. (1973). Soil and air temperature changes induced by subsurface line heat sources. *Spec. Rep. Agric, exp. Stn*, **402**, Oregon State University, Corvallis, Oregon (quoted by DEWALLE, 1979).

RYKBOST, K. A., BOERSMA, L., MACK, H. J. and SCHMISSEUR, W. E. (1974). Crop response to warming soils above their natural temperatures. *Spec. Rep. Agric. exp. Stn*, **385**, Oregon State University, Corvallis, Oregon (quoted by DEWALLE, 1979).

RYLAND, J. S. (1960). The British species of *Bugula* (Polyzoa). *Proc. zool. Soc. Lond.*, **134**, 65–105.

RYTHER, J. H. and TENORE, K. R. (1976). Integrated systems of mollusk culture. In O. Devik (Ed.), *Harvesting Polluted Waters, Waste Heat and Nutrient-Loaded Effluents in the Aquaculture*. Plenum Press, New York. pp. 153–161.

SAILA, S. A. (Ed), 1975. *Fisheries and Energy Production. A Symposium*, Lexington Books, Lexington, Massachusetts.

SAKS, N. M., LEE, J. J., MULLER, W. A. and TIETJEN, J. H. (1974). Growth of salt-marsh microcosms subjected to thermal stress. In J. W. Gibbons and R. R. Sharitz (Eds), *Thermal Ecology*, Proceedings of a Symposium, Augusta, Georgia, 1973 (CONF-730505). Technical Information Center, US Atomic Energy Commission, Oak Ridge, Tennessee. pp. 391–398.

SALO, E. O. (1969). The concentration of Zn-65 by oysters maintained in the discharge canal of a nuclear power plant. In D. J. Nelson and F. C. Evans (Eds), *Proceedings of the 2nd National Symposium on Radioecology, Ann Arbor, Michigan, 1967* (CONF-670503). US Atomic Energy Commission, Oak Ridge, Tennessee. pp. 363–371.

SANDERS, J. E., PILCHER, K. S. and FRYER, J. L. (1978). Relation of water temperature to bacterial kidney disease in coho salmon (*Oncorhynchus kisutch*), sockeye salmon (*O. nerka*) and steelhead trout (*Salmo gairdneri*). *J. Fish Res. Bd Can.*, **35**, 8–11.

SARDAR, Z. (1981). Red Sea states unite against pollution. *New Scient.* **89**, 472.

SAROGLIA, M. G. and SCARANO, G. (1979). Influence of molting on the sensitivity to toxics of the crustacean *Penaeus kerathurus*. *Ecotoxicol. environ. Saf.*, **3**, 310–320.

SAROGLIA, M. G., QUEIRAZZA, G. and SCARANO, G. (1981). Water quality criteria for aquaculture in thermal effluents, heavy metals and residual antifouling products. In K. Tiews (Ed.), *Aquaculture in Heated Effluents and Recirculation Systems*, Vol. I, Proceedings of a World Symposium, Stavanger, 1980. Heenemann, Berlin. pp. 99–112.

SAUNDER, E. and ROSENTHAL, H. (1975). The application of ozone in water treatment for home aquaria, public aquaria and for agricultural purposes. In W. J. Blogoslawski and R. G. Rice (Eds), *Aquatic Application of Ozone*. International Ozone Institute, Syracuse, New York. pp. 103–114.

SAUNDERS, P. M. (1967). Aerial measurements of sea surface temperatures in the infrared. *J. geophys. Res.*, **72**, 4109–4117.

SAUNDERS, R. L. (1976). Heated effluent for the rearing of fry: for farming and release. In O. Devik (Ed.), *Harvesting Polluted Waters, Waste Heat and Nutrient-Loaded Effluents in the Aquaculyure*. Plenum Press, New York. pp. 213–232.

SAVAGE, P. D. (1969). Some effects of heated effluents on marine phytoplankton. In *Symposium on Marine Biology, Leatherhead, U.K., 1969* (Report RD/L/M 269). Central Electricity Research Laboratories, Leatherhead, UK. pp. 27–32.

SAWYER, J. S. (1965). Notes on the possible physical causes of long-term weather anomalies. *WMO Tech. Note*, **66**, 227–248.

SCHEFFLER, G. and LAMMERS, J. (1977). Integration of MSF desalination plants into large steel works complexes in coastal regions. *Desalination*, **23**, 433–447.

SCHLOTFELDT, H.-J. (1981). Some clinical findings of a several years survey of intensive aquaculture systems in northern Germany, with special emphasis on gill pathology and nephrocalcinosis. In K. Tiews (Ed.), *Aquaculture in Heated Effluents and Recirculation Systems*, Vol. II, Proceedings of a World Symposium, Stavanger, 1980. Heenemann, Berlin. pp. 109–119.

SCHNEIDER, S. H. (1974). A new world climate norm? Implications for future world needs. *Bull. Am. Acad. Arts Sci.*, **28** (3), 20–35 (quoted by LAMB, 1977).

SCHROEDER, P. B. (1975). *Thermal Stress in Thalassia testudinum*, Ph.D. Dissertation, University of Miami, Coral Gables, Florida.

SCHUBEL, J. R. (1974). Effects of exposure to time-excess temperature histories typically experienced at power plants on the hatching success of fish eggs. *Estuar. coast. mar. Sci.*, **2**, 105–116.

SCHUBEL, J. R. and MARCY, B. C. (Eds) (1978). *Power Plant Entrainment: A Biological Assessment*, Academic Press, New York.

SCHUBEL, J. R., COUTANT, C. C. and WOODHEAD, P. M. J. (1978). Thermal effects of entrainment. In J. R. Schubel and B. C. Marcy, Jr. (Eds), *Power Plant Entrainment: A Biological Assessment*. Academic Press, New York. pp. 19–93.

SCOTTISH MARINE BIOLOGICAL ASSOCIATION (1974). *Report for the Year 1973–1974*, Dunstaffnage Marine Research Laboratory, Oban, Argyll, p. 31.

SCOTTISH MARINE BIOLOGICAL ASSOCIATION (1979). *Report for the Year 1978–1979*, Dunstaffnage Marine Research Laboratory, Oban, Argyll, p. 45.

SCOTTISH MARINE BIOLOGICAL ASSOCIATION (1980). *Report for the Year 1979–1980*, Dunstaffnage Marine Research Laboratory, Oban, Argyll, p. 39.

SEARS, F. W. (1969). *Thermodynamics*, Addison-Wesley, Cambridge, Massachusetts.

SEDGWICK, S. D. (1970). *Thermodynamics*, Addison-Wesley, Cambridge, Massachusetts.

SEDGWICK, S. D. (1970). Rainbow trout farming in Scotland. Farming trout in salt water. *Scott. Agric.*, **1970**, 180–185.

SELYE, H. (1950). *Stress*, Acta Endocrinologica, Montreal.

SELYE, H. (1956). *General Physiology and Pathology of Stress* (5th report on stress), M. D. Publications, New York.

SELYE, H. (1973). The evolution of the stress concept. *Am. Scient.*, **61**, 692–699.

SENGUPTA, S., LEE, S., VENKATA, J. and CARTER, C. (1979). A three-dimensional rigid-lid model for thermal predictions. In S. S. Lee and S. Sengupta (Eds), *Waste Heat Management and Utilization*, Vol. 3, Proceedings of a Conference, Miami, 1977. Hemisphere, Washington. pp. 1731–1759.

SETCHELL, W. A. (1924). *Ruppia* and its environmental factors. *Proc. natn. Acad. Sci. U.S.A.*, **10**, 286–288.

SETCHELL, W. A. (1964). The genus *Ruppia*. *Proc. Calif. Acad. Sci.*, **25** (18), quoted by ANDERSON (1969).

SHABI, F. A. and HIBBERD, R. L. (1977). *Brit. Pat.*, 1476883 (quoted by SOWERBUTTS and FORSTER, 1981).

SHAPOT, R. M. (1978). An evaluation of power plant effects on initial patterns of fish distribution in a small Florida estuary. In *Proceedings of the 3rd Annual Tropical and Subtropical Fisheries Technological Conference of the Americas, New Orleans, La., USA, 1978 (Sea Grant Publ.*, TAMU-SG-79-101, Sea Grant Program). Texas A and M University, College Station, Texas. pp. 331–353.

SHEINDLIN, A. E. and JACKSON, W. D. (1975). MHD electrical power generation: an international status report. In R. R. Ferber and R. A. Roxas (Eds), *Transactions of the 9th World Energy Conference, Detroit, 1974*, Vol. V. US National Committee of the World Energy Conference, New York. pp. 212–233.

SIEBURTH, J. McN. (1967). Seasonal selection of estuarine bacteria by water temperature. *J. exp. mar. Biol. Ecol.*, **1**, 98–121.

SILL, B. L. (1979). Strategies for waste heat management of once-through cooling systems. In S. S. Lee and S. Sengupta (Eds), *Waste Heat Management and Utilization*, Vol. 2, Proceedings of a Conference, Miami, 1977. Hemisphere, Washington. pp. 1025–1042.

SILVERMAN, M. J. (1972). Tragedy at Northport. *Underwat. Natur.*, **7**, 15–18.

SIMMONS, G. M., JR., ARMITAGE, B. J. and WHITE, J. C. (1974). An ecological evaluation of heated water discharge on phytoplankton blooms in the Potomac river. *Hydrobiologia*, **45**, 441–465.

SIMPSON, J. H. (1981). Sea surface fronts and temperatures. In A. P. Cracknell (Ed.), *Remote Sensing in Meteorology, Oceanography and Hydrology*. Ellis Horwood, Chichester. pp. 295–311.

SMIDT, E. L. B. (1951). Animal production in the Danish Waddensea. *Meddr Kommn Danm. Fisk.-og Havunders.*, **11**, 1–151.

SMITH, R. C. and TEAS, H. J. (1979). Biological effects of thermal effluent from the Cutler Power Plant in Biscayne Bay, Florida. In S. S. Lee and S. Sengupta (Eds), *Waste Heat Management and Utilization*, Vol. 1, Proceedings of a Conference, Miami, 1977. Hemisphere, Washington. pp. 175–189.

SNIESZKO, S. F. (1974). Effects of environmental stress on infectious diseases of fishes. *J. Fish Biol.*, **6**, 197–208.

SNYDER, G. R. and BLAHM, T. H. (1971). Effects of increased temperature on cold-water organisms. *J. Wat. Pollut. Control Fed.*, **43**, 890–899.

SORGE, E. V. (1969). The status of thermal discharges east of the Mississippi river. *Chesapeake Sci.*, **10**, 131–138.

SOULE, D. F. (1974). Thermal effects and San Pedro Bay. In D. F. Soule and M. Oguri (Eds), *Marine Studies of San Pedro Bay, California, Part III, Thermal Tolerance and Sediment Toxicity Studies*. Allan Hancock Foundation and The Office of Sea Grant Programs, Los Angeles. pp. 1–20.

SOULE, D. F. and OGURI, M. (Eds), (1974). *Marine Studies of San Pedro Bay, California, Part III, Thermal Tolerance and Sediment Toxicity Studies*. Allan Hancock Foundation and The Office of Sea Grant Programs, Los Angeles. pp. 1–110.

SOUTH OF SCOTLAND ELECTRICITY BOARD (1977). *Annual Report and Accounts for the Year Ended 31 March, 1977*. South of Scotland Electricity Board, Glasgow.

SOUTH OF SCOTLAND ELECTRICITY BOARD (1978). *Annual Report and Accounts for the Year Ended 31 March, 1978*. South of Scotland Electricity Board, Glasgow.

SOUTH OF SCOTLAND ELECTRICITY BOARD (1980). *Annual Report and Accounts for the Year Ended 31 March, 1980*. South of Scotland Electricity Board, Glasgow.

SOUTHWARD, A. J. (1980). The Western English Channel—an inconsistent ecosystem? *Nature, Lond.*, **285**, 361–366.

SOUTHWARD, A. J., BUTLER, E. I. and PENNYCUICK, L. (1975). Recent cyclical changes in climate and in abundance of marine life. *Nature, Lond.*, **253**, 714–717.

SOWERBUTTS, B. S. and FORSTER, J. R. M. (1981). Gases exchange and reoxygenation. In K. Tiews (Ed.), *Aquaculture in Heated Effluents and Recirculation Systems*, Vol. I, Proceedings of a World Symposium, Stavanger, 1980. Heenemann, Berlin. pp. 199–217.

SPANHAAK, G., FINAN, M. A. and HARRIS, A. (1977). The application of a high temperature scale control additive to a European MSF plant. *Desalination*, **23**, 455–464.

SPEAKMAN, J. N. and KRENKEL, P. A. (1971). *Quantification of the Effects of Rate of Temperature Change on Aquatic Biota* (Report 16), Department of Environmental and Water Resources Engineering, Vanderbilt University, Nashville, Tennessee (quoted by FORD and co-authors, 1979).

SPURR, G. and SCRIVEN, R. A. (1975). United Kingdom experience of the physical behaviour of heated effluents in the atmosphere and in various types of aquatic systems. In *Environmental Effects of Cooling Systems at Nuclear Power Plants*, Proceedings of a Symposium, Oslo, 1974. International Atomic Energy Agency (SM-187/2), Vienna. pp. 227–246.

SQUIRE, J. L. (1967). Surface temperature gradients observed in marine areas receiving warm water discharges. *Tech. Pap. Bur. Sport Fish. Wildl. (U.S.)*, Washington, DC, No. 11, 8 pp.

STAUFFER, J. R. and EDINGER, J. E. (1980). Power plant design and fish aggregation phenomena. In C. H. Hocutt, J. R. Stauffer, J. E. Edinger, L. W. Hall, and R. P. Morgan (Eds), *Power Plants, Effects on Fish and Shellfish Behavior*. Academic Press, New York. pp. 9–28.

STEVENS, D. E., and FINLAYSON, B. J. (1978). Mortality of young striped bass entrained at two power plants in the Sacramento–San Joaquin delta, California. In L. D. Jensen (Ed.), *4th National Workshop on Entrainment and Impingement, Chicago, Illinois, 1977*. Ecological Analysts Inc., New York. pp. 57–69.

STEWART, V. N. (1973). Observation on the potential use of thermal effluent in mariculture. In *Proceedings of the 3rd Annual Workshop, World Mariculture Society*. St. Petersburg, Florida. pp. 173–178.

STICKNEY, R. R. (1979). *Principles of Warmwater Aquaculture*, Wiley, Chichester.

STOBER, Q. J. and HANSON, C. H. (1974). Toxicity of chlorine and heat to pink (*Oncorhynchus gorbuscha*) and chinook salmon (*O. tshwaytscha*). *Trans. Am. Fish. Soc.*, **103**, 569–576.

SUCHANEK, T. H., JR., and GROSSMAN, C. (1971). Viability of zooplankton. In G. C. Williams, J. B. Milton, T. H. Schanek, Jr., N. Gebelein, C. Grossman, J. Pearce, J. Young, C. E. Taylor, R. Mulstay and C. D. Hardy (Eds). *Studies on the Effects of a Steam-Electric Generating Plant in the Marine Environment at Northport* (Technical Report No. 9), Marine Science Research Center, State University of New York, New York. pp. 25–37.

SUNDARAM, T. R. and WU, J. (1973). Wind effects on thermal plumes in water bodies. In R. E. Arndt, R. Hickling and G. B. Wallace (Eds), *Flow-studies in Air and Water Pollution*. American Society of Mechanical Engineers, New York. pp. 25–37.

SUNDARAM, T. R., SAMBUCO, E., KAPUR, S. K. and SINNARWALLA, A. M. (1979). Laboratory investigations on some fundamental aspects of thermal plume behavior. In S. S. Lee and S. Sengupta (Eds), *Waste Heat Management and Utilization*, Vol. 2, Proceedings of a Conference, Miami, 1977. Hemisphere, Washington. pp. 1559–1574.

SUTTERLIN, A. M. (1981). Diversion methods and water quality problems associated with the utilization of hydro–electric waste heat in salmonid culture. In K. Tiews (Ed.), *Aquaculture in Heated Effluents and Recirculation Systems*, Vol. 1, Proceedings of a World Symposium, Stavanger, 1980. Heenemann, Berlin. pp. 63–76.

SVERDRUP, H. U., JOHNSON, M. W. and FLEMING, R. H. (1963). *The Oceans. Their Physics, Chemistry and General Biology*, Prentice-Hall, New Jersey.

SYLVA, D. P. DE (1963). Systematics and life history of the great barracuda, *Sphyraena barracuda* (Walbaum). *Stud. Trop. Oceanogr.*, **1**, 1–179.

SYLVA, D. P. DE (1969a). The unseen problems of thermal pollution. *Oceans Mag.*, **1**, 37–41.

SYLVA, D. P. DE (1969b). Theoretical consideration of the effects of heated effluents on marine fishes. In P. A. Krenkel and P. L. Parker (Eds), *Biological Aspects of Thermal Pollution*, Proceedings of the National Symposium on Thermal Pollution, Portland, Oregon, U.S.A., 1968. Vanderbilt University Press, Nashville, Tennessee. pp. 229–293.

SYLVESTER, J. R. (1975). Biological considerations on the use of thermal effluents for finfish aquaculture. *Aquaculture*, **6**, 1–10.

SZEKIELDA, K.-H. (1976). Fast temperature changes in the upwelling area along the N.W. coast of Africa. *J. Cons. int. Explor. Mer*, **36**, 199–204.

TAKANO, K. (1981). Ocean energy. *Tech. Ser. Intergovt. Oceanogr. Commn*, **21**, 15–17.

TALBOT, J. W. (1976). Diffusion data. *Fisheries Research Technical Report*, 28 (MAFF Fisheries Laboratory, Lowestoft).

TALBOT, J. W. and TALBOT, G. A. (1974). Diffusion in shallow seas and in English coastal and estuarine waters. *Rapp. P.-v. Réun. Cons. int. Explor. Mer*, **167**, 93–110.

TALMAGE, S. S. and COUTANT, C. C. (1978). Thermal effects. *J. Wat. Pollut. Control Fed.*, **50**, 1514–1553.

TANAKA, J. (1979). Utilization of heated discharge water from electric power plants in aquaculture. In T. V. R. Pillay and W. A. Dill (Eds), *Advances in Aquaculture*, FAO Technical Conference on Aquaculture, Kyoto, Japan, 1976. Fishing News (Books) Ltd., Surrey. pp. 499–502.

TARZWELL, C. M. (1970). Thermal requirements to protect aquatic life. *J. Wat. Pollut. Control Fed.*, **42**, 824–828.

TERJUNG, W. H. (1974). Climatic modification. In I. R. Manners and M. W. Midesell (Eds), *Perspectives on Environment* (Publ. 13). Association of American Geographers, Washington, D.C. pp. 105–151 (quoted by HEDGPETH, 1977).

THOMSON, D. A., MEAD, A. R. and SCHREIBER, J. R., JR. (1969). *Environmental Impact of Brine Effluents on Gulf of California*, *Research and Development Progress Rep.*, **387**, Office of Saline Water, US Department of the Interior, Washington, D.C.

THORHAUG, A. (1974). Effects of thermal effluents on the marine biology of southeastern Florida. In J. W. Gibbons and R. R. Sharitz (Eds), *Thermal Ecology*, Proceedings of a Symposium, Augusta, Georgia, 1973 (CONF-730505). Technical Information Center, US Atomic Energy Commission, Oak Ridge, Tennessee. pp. 518–531.

THORHAUG, A. and SCHROEDER, P. B. (1979). A comparison of the biological effects of heated effluents from two fossil fuel plants: Biscayne Bay, Florida, in the subtropics; Guayanilla Bay, Puerto Rico, in the tropics. In S. S. Lee and S. Sengupta (Eds), *Waste Heat Management and Utilization*, Vol. 1, Proceedings of a Conference, Miami, 1977. Hemisphere, Washington. pp. 191–221.

THORHAUG, A., SEGAR, D. and ROESSLER, M. A. (1973). Impact of a power plant on a subtropical estuarine environment. *Mar. Pollut. Bull.*, N.S., **4**, 166–169.

THORHAUG, A., BLAKE, N. and SCHROEDER, P. B. (1978). The effect of heated effluents from power plants on seagrass (*Thalassia*) communities quantitatively comparing estuaries in the subtropics to the tropics. *Mar. Pollut. Bull.*, N.S.., **9**, 181–187.

THORPE, S. A. (1973). Experiments on instability and turbulence in a stratified shear flow. *J. Fluid Mech.*, **61**, 731–751.

TIEWS, K. (Ed.) (1981). *Aquaculture in Heated Effluents and Recirculation Systems*, Vols. I and II, Proceedings of a World Symposium, Stavanger, 1980. Heenemann, Berlin.

TIMMERMAN, R. W. (1979). Utilization of power plant waste heat for heating. In S. S. Lee and S. Sengupta (Eds), *Waste Heat Management and Utilization*, Vol. 3, Proceedings of a Conference, Miami, 1977. Hemisphere, Washington. pp. 2065–2083.

TINSMAN, J. C. and MAURER, D. I. (1974). Effects of a thermal effluent on the American oyster. In J. W. Gibbons and R. R. Sharitz (Eds), *Thermal Ecology*, Proceedings of a Symposium, Augusta, Georgia, 1973 (CONF-730505.) Technical Information Center, US Atomic Energy Commission, Oak Ridge, Tennessee. pp. 223–236.

TOEVER, W. VAN and MACKAY, K. T. (1981). A modular recirculation hatchery and rearing system for salmonids utilizing ecological design principles. In K. Tiews (Ed.), *Aquaculture in Heated Effluents and Recirculation Systems*, Vol. II, Proceedings of a World Symposium, Stavanger, 1980. Heenemann, Berlin. pp. 403–413.

TOWNSEND, A. A. (1956). *The Structure of Turbulent Shear Flow*, Cambridge University Press, Cambridge.

TREMBLEY, F. J. (1960). *Research Project on Effects of Condenser Discharge Water on Aquatic Life. Progress Report 1950–1959*, Institute of Research, Lehigh University, Bethlehem, Pa.

TREMBLEY, F. J. (1965). Effects of cooling water from steam-electric power plants on stream biota. In C. M. Tarzwell (Ed.), *Biological Problems in Water Pollution* (*U.S. Public Health Serv. Publ.*, 999-WP-25). US Department of Health, Education and Welfare, US Printing Office, Washington, DC. pp. 334–345.

TURNER, R. D. (1974). In the path of a warm, saline effluent. Proceedings of the 39th Annual Meeting of the American Malacological Union, 1973. *Bull. Am. malac. Un.*, **39**, 36–41.

TUUNAINEN, P., WESTMAN, K., SUMARI, O. and VIRTANEN, E. (1981). Comparative rearing experiments with Baltic salmon (*Salmo salar*) fingerlings in heated brackish-water effluents and fresh water. In K. Tiews (Ed.), *Aquaculture in Heated Effluents and Recirculation Systems*, Vol. II, Proceedings of a World Symposium, Stavanger, 1980. Heenemann, Berlin. pp. 133–144.

UKELES, R. (1976). Cultivation of plants : unicellular plants. In O. Kinne (Ed.), *Marine Ecology*, Vol. III, Cultivation, Part 1. Wiley, London. pp. 367–466.

UNITED NATIONS SECRETARIAT (DEPARTMENT OF ECONOMIC AND SOCIAL AFFAIRS) (1974). Future energy requirements in the developing countries: problems and challenges. In R. R. Ferber and R. A. Roxas (Eds), *Transactions 9th World Energy Conference, Detroit, 1974*, Vol. II. US National Committee of the World Energy Conference, New York. p. 86 (summary only).

U.S. ENVIRONMENTAL PROTECTION AGENCY (1974). *Development Document for Proposed Effluents Limitations Guidelines and New Source Performance Standards for the Steam Electric Power Generating Point Source Category*, Effluent Guidelines Division, Office of Air and Water Programs, Washington, DC.

U.S. ENVIRONMENTAL PROTECTION AGENCY (1976). *Development Document for Best Technology Available for the Location, Design, Construction and Capacity of Cooling Water Intake Structures for Minimizing Adverse Environmental Impact* (440/1-76/015-a). Environmental Protection Agency, Washington, DC.

U.S. OFFICE OF SALINE WATER (1968). *A Study of the Disposal of the Effluent from a Large Desalination Plant (Res. Dev. Prog. Rep.*, **316**), US Department of the Interior, Washington, DC.

VAILLANCOURT, G. and COUTURE, R. (1975). Influence de l'apport thermique originaire de la centrale nucleaire Gentilly sur la temperature de l'eau et sur les gasteropodes. In *Environmental Effects of Cooling Systems at Nuclear Power Plants*, Proceedings of a Symposium, Oslo, 1974. International Atomic Energy Agency, Vienna (SM-187/51), 449–459.

VEERARAGHAVACHARY, K. (1975). Recent trends in siting of thermal power stations including environmental considerations. In R. R. Ferber and R. A. Roxas (Eds), *Transactions 9th World Energy Conference, Detroit, 1974*, Vol. V. US National Committee of the World Energy Conference, New York. pp. 566–587.

VERNBERG, F. J. (1972). Dissolved gases : animals. In O. Kinne (Ed.), *Marine Ecology, Vol. I, Environmental Factors*, Part 3. Wiley, London. pp. 1491–1515.

VIDAVER, W. (1972). Dissolved gases: plants. In O. Kinne (Ed.), *Marine Ecology, Vol. I, Environmental Factors*, Part 3. Wiley, London. pp. 1471–1489.

VIRTANEN, E., WESTMAN, K., SOIVIO, A. and TUUNAINEN, P. (1981). Physiological condition and smoltification of one-year-old Baltic salmon (*Salmo salar*) reared in heated brackish-water effluents and fresh water. In K. Tiews (Ed.), *Aquaculture in Heated Effluents and Recirculation Systems*, Vol. II, Proceedings of a World Symposium, Stavanger, 1980. Heenemann, Berlin. pp. 121–131.

VUGTS, H. F. and ZIMMERMAN, J. F. T. (1975). Interaction between the daily heat balance and the tidal cycle. *Nature, Lond.*, **255**, 113–117.

WADA, A., KATANO, N. and GOTO, T. (1977). Prediction of the diffusion of discharged brine by a simulation analytical method. *Desalination*, **22**, 91–100.

WANDSVIK, A. and WALLACE, J. (1981). An attempt to utilize the sea as a heat source for smolt production in northern Norway. In K. Tiews (Ed.), *Aquaculture in Heated Effluents and Recirculation Systems*, Vol. II, Proceedings of a World Symposium, Stavanger, 1980. Heenemann, Berlin. pp. 497–508.

WARINNER, J. E. and BREHMER, M. L. (1966). The effects of thermal effluents on marine organisms. *Int. J. Air Wat. Pollut.*, **10**, 277–289.

WAUGH, G. D. (1964). Observations on the effects of chlorine on the larvae of Oysters (*Ostrea edulis*) and Barnacles (*Elminius modestus*). *Ann. appl. Biol.*, **54**, 423–440.

WEATHERLY, A. H. (1976). Factors affecting maximization of fish growth. *J. Fish. Res. Bd Can.*, **33**, 1046–1058.

WEDEMEYER, G. A. (1969). Stress-induced ascorbic acid depletion and cortisol production in two salmonid fishes. *Comp. Biochem. Physiol.*, **29**, 1247–1251.

WEDEMEYER, G. A. (1970). The role of stress in disease resistance of fishes. In S. F. Snieszko (Ed.), *A Symposium on Diseases of Fishes and Shellfishes*. American Fisheries Society, Washington. pp. 30–35.

WEDEMEYER, G. A. (1973). Some physiological aspects of sublethal heat stress in juvenile steelhead trout, *Salmo gairdneri*, and coho salmon, *Oncorhynchus kisutch. J. Fish. Res. Bd Can.*, **30**, 831–834.

WEDEMEYER, G. A. (1981). The physiological response of fishes to the stress of intensive aquaculture in recirculation systems. In K. Tiews (Ed.), *Aquaculture in Heated Effluents and Recirculation Systems*, Vol. II, Proceedings of a World Symposium, Stavanger, 1980. Heenemann, Berlin. pp. 3–18.

WEINBERG, A. M. and HAMMOND, R. P. (1970). Limits to the use of energy. *Am. Scient.* **58**, 412–418.

WEINSTEIN, H., PORTER, R. W., CHATURVEDI, S., KULIK, R. A. and PAGANESSI, J. E. (1979). Dispersion of heat and humidity from atmospheric spray-cooling systems. In S. S. Lee and S. Sengupta (Eds), *Waste Heat Management and Utilization*, Vol. 2, Proceedings of a Conference, Miami, 1977. Hemisphere, Washington. pp. 1107–1147.

WHEATON, E. W. (1977). *Aquaculture Engineering*, Wiley–Interscience, New York.

WHITEHOUSE, J. W. (1975). *Chlorination of Cooling Water. A Review of Literature on the Effects of Chlorine on Aquatic Organisms*, Central Electricity Research Laboratories, Leatherhead, Surrey Rep. RD/L/M 496), 1–12.

WICKINS, J. F. (1981). Water quality requirements for intensive aquaculture: a review. In K. Tiews (Ed.), *Aquaculture in Heated Effluents and Recirculation Systems*, Vol. I, Proceedings of a World Symposium, Stavanger, 1980. Heenemann, Berlin. pp. 17–37.

WIDMER, M., MASCARELLO, J. M., VENTRE, E. and ROBERT, J. M. (1975). Nuclear power stations of very high capacity of tomorrow and their constraint on the environment : study of a solution. In R. R. Ferber and R. A. Roxas (Eds), *Transactions 9th World Energy Conference, Detroit, 1974.* Vol. V. US National Committee of the World Energy Conference, New York. pp. 745–765.

WIENBECK, H. (1981). On the oxygen balance of an experimental eel farm operated in thermal effluents of a conventional power station. In K. Tiews (Ed.), *Aquaculture in Heated Effluents and Recirculation Systems*, Vol. I, Proceedings of a World Symposium, Stavanger, 1980. Heenemann, Berlin. pp. 91–98.

WILCOX, H. A. (1977). The ocean food and energy farm project. In P. D. Wilmot and A. Slingerland (Eds), *Technology Assessment and the Oceans*. IPC Science and Technology Press, Guildford. pp. 67–76.

WILLIAMS, G. C., MITTON, J. B., SUCHANEK, T. H., JR, GEBELEIN, N., GROSSMAN, C., PEARCE, J., YOUNG, J., TAYLOR, C. E., MULSTAY, R. and HARDY, C. D. (Eds) (1971). *Studies on the Effects of a Steam-Electric Generating Plant on the Marine Environment at Northport* (Technical Report No. 9), Marine Science Research Center, State University of New York.

WILLIAMS, G. E. (1981). Sunspot periods in the late Precambrian glacial climate and solar-planetary relations. *Nature, Lond.*, **291**, 624–628.

WILLIAMS, J. M. (1974). A numerical model of the temperature distribution in the vicinity of Sizewell Power Station outfall. *Rapp. P.-v. Réun. Cons. int. Explor. Mer*, **167**, 171–176.

WINKLE, W. VAN (Ed.) (1977). *Proceedings of the Conference on Assessing the Effects of Power-Plant-Induced Mortality on Fish Populations, Gatlinburg, Tennessee, 1977*, Pergamon Press, New York.

WIUFF, R. (1977). *Experiments on the Surface Buoyant Jet*, Institute of Hydrodynamics and Hydraulic Engineering, Technical University of Denmark (Ser. Pap. 16), 1–168.

WOOD, E. J. F. (Ed.) (1973). Effects of thermal additions in temperature and cold waters. *Sci. Total Environ.*, **2**, 61–80.

WOOD, E. J. F. and ZIEMAN, J. C. (1969). The effects of temperature on estuarine plant communities. *Chesapeake Sci.*, **10**, 172–174.

WORLD METEOROLOGICAL ORGANIZATION (1972). *Climatic Fluctuation and the Problems of Foresight*, Unpublished Final Report of a Working Group of the Commission for Atmospheric Sciences, W.M.O., Geneva (quoted by LAMB, 1977).

WYNESS, A. W. (1960). Dispersal of warm effluent in coastal waters. Sea temperature studies. *Engineer, Lond.*, **209** (5431), 349–350.

YAMAZAKI, M. (1965). Problems in using sea water for condenser cooling in a thermal power station. *Adv. Wat. Pollut. Res.*, **3**, 117–132.

YAROSH, M. (Ed.) (1972). *Waste Heat Utilization. Proceedings of the National Conference, Gatlinburg, Tennessee, 1971* (CONF-711031) US National Technical Information Service, Springfield, Virginia.

YAROSH, M. M. (1977). Waste energy utilization—needs and effects. In M. Marois (Ed.), *Proceedings of the World Conference Towards a Plan of Actions for Mankind, Vol. 3, Biological Balance and Thermal Modifications* (Institut de la Vie, Paris, 1974). Pergamon Press, Oxford. pp. 267–282.

YAROSH, M. M., HIRST, E. A., MICHEL, J. W., NICHOLS, B. L. and LEE, W. C. (1972). *Productive Use of Waste Heat from Steam Generating Electric Power Plants* (CF 72-1-35), Oak Ridge National Laboratory, Oak Ridge, Tennessee, (quoted by YAROSH, 1977).

YEE, W. C. (1971). Thermal aquaculture design. In *Proceedings of the 2nd Annual Workshop*, World Mariculture Society, Galveston, Texas. pp. 55–65.

YEE, W. C. (1972). Thermal aquaculture : engineering and economics. *Environ. Sci. Technol.*, **6**, 232–237.

YIH, C.-S. (1978). Buoyant plumes in a transverse wind. In *Proceedings of the 12th Symposium on Naval Hydrodynamics, Session VIII*. National Academy of Sciences, Washington, D.C. pp. 607–614.

YOUNG, J. S. (1974). Menhaden and power plants—a growing concern. *Mar. Fish. Rev.*, **36**, 19–23.

YOUNG, J. S. and GIBSON, C. I. (1973). Effects of thermal effluent on migrating menhaden. *Mar. Pollut. Bull., N.S.*, **4**, 94–96.

YOUNGBLUTH, M. J. (1976). Zooplankton populations in a polluted embayment. *Estuar. coast. mar. Sci.*, **4**, 481–496.

ZECHELLA, A. P. and HAMMOND, R. P. (1975). Floating nuclear power plants—present and future. In R. R. Ferber and R. A. Roxas (Eds), *Transactions 9th World Energy Conference, Detroit, 1974*, Vol. III. US National Committee of the World Energy Conference, New York. pp. 557–568.

ZEITOUN, I. H. and REYNOLDS, J. Z. (1978). Power plant chlorination. *Environ. Sci. Technol.*, **12**, 780–783.

ZEITOUN, M. A., MANDELLI, E. F. and MCILHENNY, W. F. (1969a): *Disposal of the Effluents from Desalination Plants into Estuarine Waters* (*Res. Dev. Prog. Rep.*, 415), Office of Saline Water, US Department of the Interior, Washington, DC.

ZEITOUN, M. A., MANDELLI, E. F., MCILHENNY, W. F. and REID, R. O. (1969b). *Disposal of the Effluent from Desalination Plants. The Effects of Copper Content, Heat and Salinity* (*Res. Dev. Prog. Rep.*, 437), Office of Saline Water, US Department of the Interior, Washington, DC.

ZIEMAN, J. C. and WOOD, E. J. F. (1975). Effects of thermal pollution on tropical-type estuaries, with emphasis on Biscayne Bay, Florida. In E. J. Ferguson Wood and R. E. Johannes (Eds), *Tropical Marine Pollution* (*Elsevier Oceanography Series*, **12**). Elsevier, Amsterdam. pp. 99–153.

AUTHOR INDEX

Numbers in italics refer to those pages on which the author's work is stated in full.

ABBOTT, O. J., 1751, *1764*
ABELL, P. R., 1802, 1810, 1813, 1862, *1931*
ABRAHAM, G., 1790, *1926*
ACKER, L., 1655, 1674, 1705, *1707*
ADAMS, E. E., 1855, 1856, *1926*
ADAMS, J. R., 1802, 1805, 1816, 1817, 1820, 1825, 1827, 1836, 1839, 1926, *1942, 1952*
ADAMS, W. J., 1683, *1700*
ADAMSON, R. H., 1652, *1695*
ADDISON, R. F., 1652, 1654, 1668, 1693, *1695, 1709*
ADEMA, D. M. M., 1683, *1695*
AHNHOFF, M., 1629, *1695*
AHOKAS, J. T., 1652, *1695*
AINSWORTH, K. A., 1688, 1689, *1709*
AKITAKE, H., 1658, *1703*
AKIYAMA, K., 1664, 1668, 1705, *1709*
ALABASTER, J. S., 1811, 1812, 1885, 1886, 1893, 1903, *1926*
ALDEN, R. W., 1802, 1803, 1809, *1926*
ALDERDICE, D. F., 1690, 1692, 1695, 1810, 1830, 1831, *1927*
ALDERSON, R., 1834, 1835, 1889, 1890, *1927*
ALEXANDER, H. C., 1644, *1696*
ALEXANDER, M., 1656, *1702*
ALEXANDER, S. V., 1688, *1709*
ALEXANDER, W. B., 1711, 1740, 1741, 1742, *1759*
ALEXANDRE, J. C., 1882, *1953*
ALLEMAN, R. T., 1853, *1927*
ALLEN, G. H., 1810, 1903, *1927*
ALLEN, M. J., 1726, *1759*
ALLRED, E. R., 1905, *1927*
ALMEIDA, S. S., 1746, *1761*
ALPERS, C. E., 1751, *1766*
ALZIEU, C., 1668, *1695*
AMERICAN NUCLEAR SOCIETY, 1892, *1927*
AMERICAN PUBLIC HEALTH ASSOCIATION, 1715, 1719, 1721, 1725, 1726, *1759*
AMICO, V., 1668, *1695*
AMIEL, A. J., 1734, *1759*
ANDERSEN, N. R., 1915, *1927*
ANDERSEN, R. B., 1665, *1696*
ANDERSEN, R. J., 1639, *1699*
ANDERSON, J., 1687, *1704*
ANDERSON, J. W., 1642, 1683, *1706, 1707*
ANDERSON, L., 1686, *1698*

ANDERSON, R. R., 1819, *1927*
ANDERSON, S. J., 1797, 1799, 1800, 1802, 1803, 1804, 1808, 1809, 1834, 1837, 1838, *1931*
ANDERSON, V. T., Jr., 1751, *1765*
ANDERSSON, K., 1628, *1705*
ANDERSSON, Ö., 1668, *1696*
ANDRE, H., 1922, *1927*
ANDREWS, J. H., 1817, *1927*
ANDREWS, J. W., 1878, 1893, *1927*
ANONYMOUS, 1746, 1759, 1874, 1877, 1923, 1924, *1927*
ANRAKU, M., 1797, 1799, *1927*
ANSELL, A. D., 1820, 1822, 1825, 1826, 1829, 1835, 1863, 1864, 1882, 1886, 1889, 1902, *1927*
ANSUINI, F. J., 1899, *1928*
ANWAR, H. O., 1790, *1928*
ANZION, C. J. M., 1638, *1707*
APPAN, S. G., 1719, 1759, *1765*
ARAB LEAGUE EDUCATIONAL, SCIENTIFIC AND CULTURAL ORGANIZATION, 1860
ARGYLE, R. L., 1642, *1696*
ARMITAGE, B. J., 1808, *1958*
ASH, G. C., 1813, *1928*
ASHLEY, G. C., 1907, 1908, *1928*
ASTON, R. J., 1797, 1885, 1886, 1887, 1888, 1889, 1890, 1891, 1893, 1894, 1895, *1928*
ATLAS, E. L., 1661, 1662, *1700*
ATWOOD, D. K., 1922, 1923, *1935*
AUBERT, J., 1749, 1751, *1759*
AUBERT, M., 1749, 1751, 1759
AUDUNSON, T., 1788, 1789, 1795, *1928*
AURICH, H., 1736, *1762*
AYLES, G. B., 1874, 1902, *1928*
AZAD, R. S., 1791, *1944*

BACHE, C. A., 1665, *1696*
BACK, P. A. A., 1921, *1928*
BACKUS, R. H., 1633, 1668, 1700, *1701*
BADER, R. G., 1816, 1817, 1818, 1819, 1838, 1839, 1910, *1928*
BAGGE, P., 1741, 1742, *1759*
BAHNER, L. H., 1684, 1685, *1705*
BAIRD, R. D., 1851, *1928*
BAKER, A. L., 1799, 1834, *1930*
BAKER, R. W. R., 1638, *1696*
BALABAN, M., 1781, *1928*

BALCH, N., 1733, 1745, *1759*
BALL, L. M., 1653, *1696*
BALLARD, A. J., 1745, *1763*
BALLIGAND, P., 1879, *1928*
BALLSCHMITER, K., 1638, 1655, 1664, 1668, 1696, 1705, *1709*
BAPTIST, J. P., 1806, 1833, 1834, 1835, 1892, 1928, *1942*
BARATZ, D., 1890, *1930*
BARBER, R. T., 1722, 1739, *1766*
BARICA, J., 1874, 1902. *1928*
BARKER, S. L., 1802, *1928*
BARKLEY, S. W., 1903, *1928*
BARNARD, J. L., 1715, 1719, 1746, *1759*
BARNES, H., 1717, *1759*
BARNETT, P. R. O., 1770, 1791, 1794, 1805, 1814, 1822, 1823, 1824, 1826, 1859, 1864, *1927, 1929, 1930*
BARR, D. I. H., 1788, *1929, 1937*
BASSINDALE, R., 1711, 1740, 1741, 1742, *1759*
BASSON, P. W., 1831, 1911, *1929*
BATH, D, W., 1797, 1798, 1799, 1800, 1801, 1802, 1803, 1804, 1805, 1806, 1834, 1837, 1863, *1946*
BAUGHMAN, G. J., 1651, *1709*
BAUGHMAN, J. L., 1711, *1759*
BAUMANN OFSTAD, E. 1638, 1672, 1696, *1704*
BAYNE, B. L., 1755, *1759*
BEALL, S. E., 1909, *1929*
BEARDSLEY, G. L., 1754, *1761*
BEAUCHAMP, R. S. A., 1797, 1888, 1889, *1929*
BECK, A. D., 1837, 1855, *1948, 1950*
BECKER, C. D., 1777, 1887, 1880, 1889, 1891, *1929*
BEER, L. O., 1801, *1929*
BEERS, J. R., 1734, 1736, *1759*
BELLAN, G., 1729, 1730 1732, 1745, 1747, 1751, 1757, *1759*
BELLAN-SANTINI, D., 1743, 1747, *1759*
BELTER, W. G., 1773, 1775, 1852, 1875, 1909, 1910, *1929*
BELTZ, J. R., 1862, *1929*
BEND, J. R., 1652, 1653, 1655, 1696, 1701, *1706*
BEND, S. G., 1652, *1696*
BENDER, M. E., 1719, 1723, 1759, 1765, 1888, *1955*
BENDINER, W. P., 1786, *1935*
BENEZET, H. J., 1655, 1659, *1696*
BENGTSSON, B.-E., 1683, 1691, *1696, 1703*
BENJAMIN, T. B., 1790, *1929*
BERG, O., 1668, *1697*
BERGAN, P., 1807, *1929*
BERGE, G., 1660, *1702*
BERGSTEDT, B. V., 1645, *1708*
BERRISFORD, C. D., 1745, *1763*
BERNER, M., 1665, *1704*
BEVENUE, A., 1634, 1661, 1696, *1708*

BEWTRA, J. K., 1893, *1929*
BIBKO, P., 1797, 1798, 1799, 1800, 1801, 1802, 1803, 1804, 1805, 1806, 1834, 1837, 1863, *1946*
BIDLEMAN, T. F., 1661, 1663, *1696*
BIENFANG, P., 1799, 1800, 1808, 1837, 1859, 1883, 1891, 1929, *1939, 1954*
BIGGS, D. C., 1686, 1687, *1696*
BIGGS, R. B., *1929*
BIKSEY, 1757
BILLI, B., 1810, *1929*
BIRKELAND, C., 1826, *1929*
BISSET, K. A., 1896, *1929*
BJERK, ?, 1874, *1951*
BLACK, E. C., 1864, *1929*
BLACKMAN, R. R., 1645, *1705*
BLAHM, T. H., 1886, *1958*
BLAKE, N., 1803, 1818, *1959*
BLAND, R. A., 1785, *1929*
BLAU, G. E., 1644, 1645, *1696, 1705*
BLAXTER, J. H. S., 1837, 1929, *1941*
BLEGVAD, H., 1711, 1745, *1760*
BLOCK, R. M., 1835, 1888, *1930*
BLOMKVIST, G., 1634, 1664, 1668, *1696, 1702*
BOALCH, G. T., 1921, *1956*
BOARD, P. A., 1821, *1930*
BOCKRIS, J. O'M., 1921, *1947*
BODMAN, R. H., 1629, *1696*
BODOY, A., 1826, 1864, *1927*
BOERSMA, L., 1905, 1906, 1907, *1930, 1956*
BOESCH, D. F., 1728, *1760*
BOGAN, J., 1664, 1674, 1675, *1701*
BOIES, D. B., 1890, *1930*
BOLUS, R. L., 1784, 1785, *1930*
BOND, T. E., 1909, *1940*
BONEY, A. D., 1817, *1932*
BORISOV, P. M., 1914, *1930*
BORTHWICK, P. W., 1647, *1708*
BOTSFORD, L. W., 1878, 1898, *1930*
BOUSH, G. M., 1654, 1659, *1706*
BOVO, G., 1874, *1938*
BOWDEN, N., 1828, 1829, *1931*
BOWEN, V. T., 1629, 1633, 1661, 1668, *1696, 1700, 1701*
BOWES, G. W., 1654, 1668, 1687, *1696, 1697*
BOWMAN, M. J., 1734, *1760*
BOYDSTUN, L. B., 1810, 1903, *1927*
BRAAMS, C. M., 1843, *1930*
BRADY, D. K., 1854, 1936
BRAESTRUP, L., 1668, *1697*
BRANCH, M. R., 1881, 1901, *1930*
BRANSON, D. R., 1644, 1645, 1693, *1696, 1705*
BREHMER, M. L., 1799, 1827, 1838, *1961*
BRETT, J. R., 1676, 1696, 1769, 1771, 1861, 1862, 1864, 1878, 1894, 1902, 1930, *1937, 1947, 1951*
BREWER, G., 1831, *1930*

BRIAND, F. J.-P., 1797, 1799, 1800, 1801, 1802, 1808, 1837, *1930*
BRINKMANN, U. A. TH., 1638, *1707*
BRODFIELD, B., 1854, 1855, 1856, *1930*
BRODIE, B. B., 1652, *1697*
BROGDEN, W. B., 1878, *1953*
BROOK, A. J., 1799, 1834, 1930
BROOKS, C. J. W., 1635, *1697*
BROOKS, N. H., 1731, 1733, 1758, 1760, 1761, 1788, 1855, *1936*
BROWN, D. J. A., 1779, 1878, 1880, 1885, 1887, 1889, 1890, 1891, 1893, 1894, 1895, 1896, *1928*
BROWN, E. W., 1749, 1754, *1760*
BROWN, J. A. G., 1878, 1880, 1885, 1889, 1890, 1893, 1894, 1896, *1943*
BROWN, P. L., 1883, *1941*
BROWN, R. S., 1749, 1754, *1760*
BROWNELL, W., 1779, 1778, 1782, 1795, 1797, 1798, 1800, 1801, 1802, 1803, 1804, 1805, 1806, 1809, 1810, 1811, 1812, 1813, 1815, 1816, 1818, 1819, 1822, 1834, 1835, 1836, 1837, 1838, 1839, 1848, 1849, 1850, 1851, 1852, 1853, 1857, 1858, 1913, *1932*
BROWNIE, A. C., 1652, *1699*
BRUNGS, W. A., 1797, 1810, 1834, 1835, 1889, *1930*
BUCK, J. D., 1798, 1801, *1930*
BUDD, J., 1642, *1705*
BUDYKO, M. I., 1920, *1930*
BULLIMORE, B., 1828, 1829, *1931*
BULNHEIM, H.-P., 1627, *1703*, 1736, *1762*
BURCHARD, J. E., 1831, 1911, *1929*
BURNETT, J. M., 1791, 1792, *1931*
BURNS, E. R., 1907, 1908, *1931*
BURRIDGE, L., 1683, *1704*
BURTON, B. W., 1786, 1810, *1931*
BURTON, D. T., 1802, 1810, 1813, 1862, *1931*
BURWITZ, P., 1785, *1931*
BUTLER, I., 1917, *1931*
BUTLER, E. I., 1921, *1956, 1958*
BUTLER, P. A., 1646, 1647, 1691, *1697*

CAMP, D., 1827, *1947*
CAIRNS, J., 1797, 1798, 1800, 1801, 1810, 1813, 1816, 1835, 1836, 1838, *1931, 1946*
CALDWELL, R. S., 1668, *1697*
CALIFORNIA STATE RESOURCES CONTROL BOARD, *1760*
CANE, A., 1880, 1900, *1955*
CAPELLI, R., 1668, *1697*
CAPERON, J., 1734, *1760*
CAPIZZI, T. P., 1810, 1813, *1931*
CAPUZZO, J. M., 1834, 1835, 1888, *1931*
CAREY, A. E., 1659, *1697*
CARLBERG, J. M., 1885, *1937*
CARLUCCI, A. F., 1721, 1736, *1760*

CARPENTER, E. J., 1686, 1699, 1797, 1799, 1800, 1802, 1803, 1804, 1808, 1809, 1834, 1835, 1837, 1838, 1888, *1931, 1951*
CARPENTER, J. H., 1786, *1954*
CARRIE, B. G. A., 1809, 1820, 1954
CARROLL, B. B., 1883, 1884, 1898, 1900, 1901, 1902, *1931*
CARSTENS, T., 1795, *1931*
CARTER, C., 1769, 1787, *1957*
CARTER, L., 1757, 1760, *1932*
CASPERS, H., 1798, *1932*
CASTAGNA, M., 1804, *1944*
CASTILLO, J. G., 1745, *1761*
CATTELL, S. A., 1734, *1760*
CECH, J. J., 1898, *1932*
CESCHIA, G., 1874, *1938*
CHADWICK, H. K., 1838, *1932*
CHAHINE, M. T., 1785, *1932*
CHAMBERLAIN, G., 1875, 1895, 1898, 1899, *1932*
CHAMPION, E. R., 1852, *1932*
CHAN, H. S., 1638, 1661, 1662, 1668, *1700*
CHANNABASSAPPA, K. C., 1779, *1932*
CHATURVEDI, S. K., 1851, *1936, 1961*
CHEN, J. C., 1790, *1932*
CHEN, K. Y., 1737, 1738, *1765*
CHENEY, W. O., 1782, 1785, *1932*
CHESHER, R. H., 1832, 1833, *1932*
CHEW, K. K., 1745, *1766*
CHIA, S. N., 1784, 1785, *1930*
CHIANTORE, G., 1810, *1929*
CHIBA, K., 1771, 1874, 1875, 1879, 1880, 1881, 1882, 1885, 1899, 1900, 1901, *1932*
CHILDRESS, R., 1691, *1697*
CHIOU, C. T., 1645, *1697*
CHOI, P. M. K., 1668, *1709*
CHYMKO, N. R., 1813, *1928*
CIANTORE, G., 1810, *1929*
CIMBERG, R., 1743, *1760, 1763*
CLAEYS, R. R., 1659, 1668, *1697, 1703*
CLARK, J., 1777, 1778, 1782, 1795, 1797, 1798, 1800, 1801, 1802, 1803, 1804, 1805, 1806, 1809, 1810, 1811, 1812, 1813, 1815, 1816, 1818, 1819, 1822, 1834, 1835, 1836, 1837, 1838, 1839, 1848, 1849, 1850, 1851, 1852, 1853, 1857, 1858, 1913, *1932*
CLARK, L. J., 1841, 1843, *1932*
CLARKE, W. D., 1832, 1833, *1932*
CLAUDE, G., 1883, 1922, *1932*
CLAUS, C., 1883, 1902, *1932*
CLAUSEN, J., 1668, *1697*
CLAWSON, G. H., 1864, *1937*
CLAYES, R. C., *1697*
CLEMENTS, L. C., 1806, 1813, *1940*
CLIFTON, R. J., 1633, *1701*
CLOKIE, J. J. P., 1817, *1932*
CLOUTIER, P. D., 1922, *1938*
CLYMER, E. A., 1845, *1933*

COBB, S. P., 1827, *1947*
COGNETTI, G., 1745, *1760*
COGNIE, D., 1884, *1934*
COHEN, D. L., 1862, *1929*
COLEBROOK, J. M., 1917, *1933*
COLLINGS, W. S., 1811, 1835, 1903, *1936*
COLLINS, R. A., 1899, *1933*
COLOMBO, U., 1914, 1921, *1937*
COLWELL, R. R., 1656, *1707*
COMMITTEE ON ENTRAINMENT, 1849, 1859, 1867, 1868, *1933*
CONROY, D. A., 1749, *1760*
CONSERVATION COMMISSION OF THE WORLD ENERGY CONFERENCE, 1843, 1924, 1925, *1933*
CONTARDI, V., 1667, *1697*
COOK, G. H., 1657, 1683, *1697*, 1706
COOKE, M., 1674, *1697*
COOLEY, P., 1897, *1952*
COPELAND, T. L., 1683, *1697*
COPIUS PEEREBOOMS, J. W., 1638, *1707*
COPPAGE, D. L., 1657, *1697*
CORY, R. L., 1782, 1824, 1826, 1828, 1835, *1933*, *1938*, *1952*
COSPER, E., 1802, 1803, *1955*
COSTON, L. C., 1806, 1807, 1815, 1833, 1834, 1835, 1838, *1941*, *1942*
COUCH, J. A., 1683, 1690, 1691, *1697*, *1706*, 1751, *1760*
COUGHLAN, J., 1793, 1801, 1820, 1825, 1826, 1827, 1828, 1829, 1833, 1835, 1837, 1872, 1888, 1890, *1927*, *1933*, *1940*
COURTNEY, W. A. M., 1642, *1697*
COUTANT, C. C., 1778, 1779, 1782, 1802, 1805, 1806, 1807, 1810, 1811, 1813, 1815, 1837, 1838, 1839, 1849, 1860, 1861, 1862, 1863, 1864, 1865, 1866, 1867, 1868, 1869, 1870, 1871, 1872, 1896, 1902, 1904, 1906, 1907, 1909, 1913, 1914, *1933*, *1934*, *1952*, *1957*, *1959*
COUTURE, R., 1785, *1961*
COX, D. K., 1906, 1907, 1909, *1952*
COX, J. L., 1650, *1697*
CRACKNELL, A. P., 1784, *1934*
CRANDALL, M., 1745, *1764*
CRISP, D. J., 1770, 1828, 1829, *1934*
CRONIN, L. E., 1811, 1826, 1833, *1960*
CROSBY, D. G., 1656, *1699*
CROSS, K. E., 1852, *1953*
CURTIS, M. W., 1683, *1697*
CUSHING, D. H., 1896, 1915, 1916, 1917, 1921, *1934*
CUTSHALL, N. H., 1668, *1697*
CUZON, G., 1884, *1934*

DAHL-MADSEN, K. I., 1825, 1836, *1951*
DAILEY, R., 1906, *1942*
DALLAIRE, E. E., 1758, *1760*
DAMICO, J. N., 1638, *1697*

DANSETTE, P. M., 1652, *1696*
DARLING, I. C., 1824, 1825, 1875, 1882, 1893, *1940*
DAUBERT, A., 1788, 1789, *1934*
DAUGARD, S. J., 1910, *1934*
DAVIDOV, *1914*
DAVIDSON, J. A., 1834, 1835, 1888, *1931*
DAVIES, R. M., 1802, 1803, *1934*
DAVIS, J. C., 1893, *1934*
DAVIS, M. H., 1892, *1955*
DAVIS, N. W., 1810, *1941*
DAVIS, W. P., 1835, 1888, *1930*
DAWE, C. J., 1749, *1762*
DAWSON, J. K., 1802, *1934*
DAWSON, E. Y., 1724, 1743, 1749, *1760*
DAWSON, R., 1634, 1638, 1661, *1697*
DEAN, J. M., 1807, 1863, *1932*, *1941*
DEBERNARDIS, J. F., 1639, 1662, *1708*
DEFOE, D. L., 1645, *1708*
DEEPAK, A., 1784, *1934*
DELANEY, B. T., 1777, 1891, *1939*
DELAPPE, B. W., 1629, 1638, 1668, *1697*, *1706*
DELEON, J. R., 1638, *1697*
DELMENDO, M. N., 1899, *1933*
DELVENTHAL, H., 1897, *1941*, *1953*
DEMETER, J., 1638, *1697*
DEPARTMENT OF THE ENVIRONMENT, 1779, 1780, 1831, 1832, 1833, 1834, 1857, 1859, *1934*
DERENBACH, J., 1662, *1698*
DESCAMPS, B., 1875, 1879, *1928*, *1934*, *1937*
DETHLEFSEN, V., 1691, *1709*
DEVEL, D. G., 1751, *1763*
DEVIK, O., 1782, 1873, *1935*
DEWALLE, D. R., 1905, 1906, 1907, *1935*
DEXTER, R. N., 1674, 1689, *1697*
DIAZ, L. F., 1872, *1935*
DIAZ, R. J., 1888, *1955*
DICKSON, A. G., 1638, 1642, 1665, *1698*
DICKSON, R. R., 1915, 1916, 1917, 1921, *1934*, *1935*
DIEK, J. J. VAN, 1683, *1698*
DIETRICH, J. R., 1843, *1935*
DILL, W. A., 1899, *1954*
DINSMORE, A. F., 1852, *1935*
DITMARS, J. D., 1784, 1791, *1935*
DODSON, A. N., 1722, *1766*
DOMINIK, J., 1674, *1705*
DOOLEY, H. D., 1786, *1935*
DORSEY, J. H., 1743, *1760*
DOUGHERTY, R. C., 1638, *1698*
DOUGLAS, M. T., 1878, 1880, 1885, 1887, 1889, 1893, 1894, 1896, *1943*
DOUGLAS, P. A., 1903, *1935*
DOW, R. L., 1921, *1935*
DOYLE, M. L. JR., 1836, *1926*
DRAXLER, A. F. J., 1734, *1760*
DRESCHER, H. E., 1668, *1698*

DROBECK, K. G., 1803, 1804, 1850, *1941*
DUBOUL-RAZAVET, C., 1633, 1674, *1704*
DUCE, R. A., 1663, *1698*
DUCHARME, H. M., 1815, *1934*
DUEDALL, I. W., 1734, *1760*
DUGGAN, M. C., 1902, *1944*
DUMONT, M., 1907, 1908, *1937*
DUNCAN, C. P., 1922, 1923, *1935*
DUNN, W. E., 1788, *1935*
DUPREE, H. K., 1642, *1696*
DYBERN, B. I., 1665, *1698*
DYE, J. E., 1902, *1944*
DYRYNDA, P. E. J., 1828, 1829, *1931*
DZUNG, L. S., 1843, 1935

EBLE, A. F., 1877, 1880, 1882, 1901, *1939*
EDDY, 1711, *1763*
EDER, G., 1633, 1634, 1638, 1642, 1643, 1644, 1649, 1655, 1665, 1668, 1674, *1698*, *1699*, *1700*, *1707*
EDINGER, J. E., 1782, 1792, 1848, 1854, 1855, *1935*, *1941*, *1958*
EDWARDS, L. L., 1897, *1952*
EDWARDSON, W., 1875, 1876, *1936*
EFFERTZ, P. H., 1887, *1936*
EGUSA, S., 1895, 1896, *1936*
EHRHARDT, M., 1629, 1632, 1633, 1662, *1698*
E.I.F.A.C., 1899, *1936*
EISLER, R., 1675, 1676, 1683, *1698*
ELDER, D. L., 1645, 1661, 1674, *1698*, *1699*
ELDER, J. H., 1686, *1698*
ELKINS, D., 1645, *1703*
ELLETT, D. J., 1917, *1936*
ELLIS, D., 1733, 1745, *1759*
ELLISON, T. H., 1790, *1936*
ELMAMLOUK, T. H., 1652, 1653, *1696*, *1698*, *1699*
ELSER, H. J., 1810, 1902, 1903, *1936*
ELTRINGHAM, S. K., 1821, *1836*
ENGEL, D. W., 1806, 1833, 1834, 1835, *1942*
ENGLESSON, G. A., 1853, *1936*
ENRIGHT, J. T., 1806, 1838, 1913, *1936*
ENVIRONMENTAL PROTECTION AGENCY, 1628, 1699, 1772, 1777, 1802, *1960*
EPPLEY, R. W., 1721, 1736, *1760*
ERNST, E. J., 1816, *1940*
ERNST, W., 1633, 1634, 1638, 1642, 1643, 1644, 1645, 1646, 1647, 1648, 1649, 1650, 1653, 1654, 1656, 1657, 1658, 1665, 1668, 1691, 1694, *1695*, *1698*, *1699*, *1700*, *1707*, *1709*
ESCH, G. W., 1781, *1936*
ESTES, N. C., 1888, *1936*
EVERHART, W. H., 1665, *1696*
EWART, T. W., 1786, *1936*

FAAS, L., 1648, 1683, 1688, 1689, *1707*, *1709*
FAIR, G. M., *1760*
FAIRBANKS, R. B., 1811, 1835, 1903, *1936*

FALKE, J., 1810, *1936*
FANG, C. S., 1784, 1785, *1930*
FARLEY, C. A., 1754, *1760*
FARMANFARMAIAN, A., 1875, 1880, 1882, 1884, *1936*
FAULKNER, D. J., 1639, *1699*
FAZZOLARE, R., 1843, *1940*
FELIX, J. R., 1885, 1886, *1936*, *1937*
FENDERSON, O. C., 1665, *1696*
FENICAL, W., 1662, *1704*
FICHTE, W., 1887, *1936*
FIJAN, N., 1895, 1896, 1897, *1951*
FILICE, F. P., 1746, *1760*
FINAN, M. A., 1780, *1958*
FINLAYSON, B. J., 1797, *1958*
FISCHENKO, P. A., 1891, *1936*
FISCHER, H. B., 1788, 1855, *1936*
FISHBEIN, L., 1638, *1699*
FISHER, J. R., 1815, *1934*
FISHER, N. S., 1686, 1687, 1690, *1699*, *1705*, *1706*
FITCH, J. E., 1749, *1760*
FLATOW, R. E., 1890, 1895, 1897, *1936*
FLEMER, D. A., 1834, 1838, *1939*
FLEMING, R. H., 1894, 1914, 1915, *1959*
FLETCHER, J. O., 1914, *1937*
FLOHN, H., 1914, *1937*
FLÜGEL, H., 1837, *1937*
FOLGER, D. W., 1737, *1761*
FOLWELL, R., 1906, *1943*
FORBES, S. A., 1711, 1740, *1761*
FORD, R. F., 1885, *1937*
FORESTER, J., 1647, 1648, 1683, 1690, 1692, *1701*, *1703*, *1706*, *1707*, *1708*
FORESTER, G. E., 1687, *1702*
FORNEY, D. L., 1902, 1903, *1938*
FORREST, D. M., 1901, *1937*
FORSTER, J. R. M., 1901, 1937, *1958*
FORTIER, J., 1745, *1764*
FOSTER, M., 1743, *1761*
FOULQUIER, L., 1875, 1879, *1928*, *1935*, *1937*
FOURCY, A., 1907, 1908, *1937*
FOUTS, J. R., 1652, *1696*, *1706*
FOWLER, S. W., 1645, 1674, *1698*, *1699*
FOX, J. L., 1798, 1799, 1800, 1837, *1937*
FRAAS, A. P., 1842, *1937*
FRAZER, W., 1788, *1937*
FREELAND, G. L., 1737, *1761*
FREED, V. H., 1645, *1697*
FREEMYERS, M. C., 1908, *1937*
FREYCHET, A., 1907, 1908, *1937*
FRIGO, A. A., 1783, 1784, 1791, *1935*, *1947*
FRY, F. E. J., 1862, 1864, *1937*
FRYER, J. L., 1896, *1938*, *1956*
FUCHS, A., 1834, *1942*
FULLER, R. D., 1922, 1923, 1924, *1937*
FULLER, S. L. H., 1740, 1746, *1761*
FUSHIMI, K., 1842, *1937*

GABOR, D., 1914, 1921, *1937*
GALLI, R., 1914, 1921, *1937*
GALLARDO, V. A., 1745, *1761*
GALLAWAY, B. J., 1810, 1811, 1814, 1836, 1957, 1903, *1937*
GALLOWAY, J. M., 1738, *1761*
GALLUP, D. N., 1813, *1928*
GALVIN, R. C., 1903, *1948*
GARBUTT, P. A., 1784, *1935*
GARCIA, F. G., 1810, 1903, *1927*
GARD, K. L., 1664, 1665, 1672, 1673, *1708*
GARNAS, R. L. 1656, *1699*
GARRETT, W. D., 1633, *1700*
GARSIDE, E. T., 1676, 1700, 1769, 1771, *1937*
GASPARINI, R., 1810, *1929*
GAUCHER, T. A., 1885, 1901, *1937*
GAUDY, R., 1831, *1938*
GAUFIN, A. R., 1740, *1761*
GEBELEIN, N., 1799, 1800, *1962*
GEESEY, G. G., 1755, 1757, *1764*
GELDIAY, R., 1754, *1762*
GEORGE, D. W., 1843, *1938*
GERCHAKOV, S. M., 1792, *1938*
GERLACH, S. A., 1627, *1700*
GERSDORF, B. VON, 1845, *1938*
GESSER, H., 1629, 1652, *1708*
GESSNER, F., 1676, 1700, 1769, 1771, 1801, 1871, *1938*
GESSNER, T., 1652, *1698*, *1699*
GETHER, J., 1639, 1661, *1704*
GEYER, J. C., 1854, *1936*
GHIRADELLI, E., 1745, *1761*
GIAM, G. S., 1633, 1638, 1657, 1661, 1662, 1668, 1674, 1676, 1683, *1700*, *1701*, *1703*, *1705*, *1707*, *1709*
GIBBONS, J. W., 1781, 1902, 1903, *1938*
GIBBONS, G., 1893, *1927*
GIBSON, C. I., 1810, 1812, 1813, 1839, *1947*, *1962*
GIFFORD, F. A., 1853, 1860, *1939*
GILCHRIST, J. R. S., 1751, *1764*
GINN, T., 1797, 1798, 1799, 1800, 1801, 1802, 1803, 1804, 1805, 1806, 1834, 1837, 1863, *1946*
GIORGETTI, G., 1874, *1938*
GJEDREM, T., 1874, 1894, 1895, *1945*
GLEDHILL, W. E., 1683, *1700*
GLENDENNING, I., 1922, *1938*
GLICKMAN, A. H., 1658, *1700*
GLOOSCHENKO, W. A., 1686, *1704*
GLUECKSTERN, P., 1781, *1938*
GODFRIAUX, B. L., 1875, 1877, 1880, 1882, 1884, 1890, 1891, 1901, 1902, *1938*, *1939*
GODSHALL, F. A., 1782, *1938*
GOERKE, H., 1634, 1638, 1642, 1643, 1644, 1647, 1649, 1653, 1654, 1656, 1657, 1665, 1668, *1698*, *1699*, *1700*, *1707*
GUERLITZ, D. F., 1634, *1700*

GOLDBERG, E. D., 1638, 1693, *1700*
GOLDMAN, J. C., 1834, 1835, 1888, *1931*
GONOR, J. J., 1781, 1804, 1904, *1940*
GONZÁLEZ, J. G., 1797, 1802, 1803, 1804, 1805, 1809, 1816, 1819, 1821, 1825, 1826, 1827, 1834, *1938*
GOOCHEY, T. K., 1747, 1748, *1763*
GOODMAN, C. H., 1852, *1932*
GOODMAN, L. R., 1647, 1685, 1690, *1697*, *1701*
GOODYEAR, C. P., 1782, 1862, *1933*
GORCHEV, H., 1888, *1943*
GORDON, J. A., 1651, *1709*
GORDON, L., 1841, 1842, *1938*
GORMLY, H. J., 1836, *1926*
GOSS, W. P., 1922, *1938*
GOSSARD, T. W., 1878, 1898, *1930*
GOTO, T., 1857, 1859, *1938*, *1961*
GRADIN, R., 1842, *1947*
GRAHAM, L. B., 1686, *1699*
GRAUBY, A., 1879, 1905, 1907, *1928*, *1938*
GRASSHOFF, K., 1633, *1700*
GRASSLE, J. F., 1747, *1761*
GRASSLE, J. P., 1747, *1761*
GRAY, J. S., 1724, 1742, 1761
GREEN, R. H., 1781, *1938*
GREENBERG, A. E., 1711, *1761*
GREENE, C. S., 1727, 1737, 1761, *1765*
GRIBBIN, J., 1916, *1939*
GRICE, G. D., 1633, 1668, *1700*, *1701*
GRIER, J. C., 1888, *1955*
GRIMÅS, U., 1810, 1811, *1939*
GRIMES, C. B., 1810, 1811, 1814, 1826, 1833, *1939*
GRIMM, 1826, *1919*
GROBERG, W. J., 1896, *1939*
GROGNET, B., 1879, *1935*
GRONOW, G., 1897, *1939*
GROSSMAN, C., 1799, 1800, 1802, 1803, 1804, *1959*, *1962*
GRUGER, E. H., 1642, *1700*
GRZENDA, A. R., 1642, *1701*
GUARINO, A. M., 1652, 1655, 1683, *1696*, *1701*, *1703*, *1706*
GUARINO, C. F., 1746, 1761
GUERIN, J. P., 1831, *1938*
GUERRA, C. R., 1875, 1877, 1880, 1882, 1884, 1890, 1891, 1892, 1901, 1902, *1938*, *1939*
GUIZERIX, J., 1786, *1939*
GULLAND, J., 1913, *1939*
GUNDERSON, K., 1859, 1883, *1939*
GUTEMANN, W. H., 1665, *1709*

HAAVISTO, H., 1841, *1950*
HADLEY, D., 1830, *1939*
HAGER, L. P., 1662, *1709*
HAINES, K. C., 1859, 1883, *1955*
HAIR, J. R., 1862, *1939*

HAKUTA, T., 1857, *1938*

HALCROW, W., 1664, 1674, 1675, 1737, 1745, *1761*

HALL, C. E. VAN, 1638, *1697, 1701*

HALL, L. W., 1782, 1848, *1941*

HAMBRAEUS, G., 1842, *1947*

HAMILTON, D. H., 1800, 1834, 1838, 1888, *1939, 1942*

HAMILTON, E. I., 1633, *1701*

HAMMOND, R. P., 1860, 1920, *1961, 1963*

HANKS, A. R., 1642, *1848*

HANNA, S. R., 1853, 1860, *1939*

HANSCH, C., 1645, *1703*

HANSEN, D. J., 1646, 1647, 1648, 1683, 1684, 1685, 1690, 1691, *1701, 1706, 1707*

HANSEN, J. E., 1915, 1916, 1917, 1918, 1919, 1920, *1939*

HANSEN, P. D., 1691, *1709*

HANSON, C. H., 1835, *1959*

HANSON, J. A., 1860, 1883, 1922, *1939, 1954*

HARASHINA, H., 1779, 1780, 1859, *1939*

HARDING, L. W., 1650, 1689, *1701*

HARDY, B. L. S., 1799, 1800, 1805, 1814, 1822, 1823, 1824, 1911, *1929, 1939*

HARDY, C. D., 1799, 1800, *1962*

HARDY, J. T., 1734, 1735, *1761*, 1831, *1929*

HARDY, S. A., 1734, 1735, *1761*

HARMAN, J. P., 1894, *1937*

HARMS, U., 1668, 1691, 1697, *1709*

HARRIS, A., 1780, *1958*

HARRIS, V. M., 1884, *1955*

HARRIS, R. C., 1686, 1687, 1689, 1704, *1705*

HARSHBARGER, J. C., 1749, *1762*

HART, C. W., JR., 1740, 1746, *1761*

HART, F. C., 1777, 1891, *1939*

HARTLEY, G. S., 1782, *1940*

HARTMAN, O., 1746, *1759*

HARTNOLL, R. G., 1742, *1761*

HARTMAN, O., 1723, *1761*

HARVEY, G., 1885, *1940*

HARVEY, G. R., 1629, 1633, 1659, 1661, 1668, 1697, 1700, *1701*

HASLER, A. D., 1735, *1761*

HATCHER, R. F., 1633, *1701*

HATLE, Z., 1908, *1940*

HAWES, F. B., 1793, 1801, 1825, 1826, 1828, 1833, 1837, 1872, *1940*

HAWKINS, S. J., 1742, *1761*

HAYASHIDA, F., 1751, *1763*

HEATH, A. G., 1836, 1889, 1931, *1940*

HECHTEL, G. J., 1816, *1940*

HEDGPETH, J. W., 1781, 1797, 1804, 1818, 1819, 1825, 1826, 1904, 1914, 1920, *1940*

HEESEN, T. C., 1650, 1709, 1738, 1739, 1755, *1767*

HEFFNER, D., 1797, 1798, 1799, 1800, 1801, 1802, 1803, 1804, 1805, 1806, 1834, 1837, 1863, *1946*

HEFFNER, R. L., 1799, 1800, 1801, 1837, *1940*

HEIN, M. K., 1816, *1940*

HEINLE, D. R., 1799, 1802, 1803, 1804, 1805, 1834, 1858, 1911, 1940

HEITMAN, H., 1909, *1940*

HELZ, G. R., 1650, 1663, 1701, 1761, 1835, 1888, *1930*

HENDERSON, C. D., 1851, *1940*

HENRY, C. D., 1843, *1940*

HERBST, E., 1653, *1701*

HERSHELMAN, G. P., 1737, 1738, 1739, *1761*

HESS, C. T., 1824, 1825, 1875, 1882, 1892, 1893, 1940, *1954*

HESS, E., 1754, *1761*

HETHERINGTON, J. A., 1892, *1940*

HETTLER, W. F., 1806, 1807, 1812, 1813, 1815, 1838, *1940, 1941, 1942, 1947*

HEYNDRICX, A., 1638, *1697*

HIBBERD, R. L., 1894, *1957*

HICKLING, C. F., 1779, 1883, *1941*

HICKS, O., 1683, *1700*

HIDU, H., 1803, 1804, 1850, *1941*

HIETALA, J. S., 1907, 1908, *1941*

HILGE, V., 1897, *1941*

HILL, J. M., 1734, *1761*

HILL, K. M., 1774, 1773, 1775, *1941*

HILLMAN, R. E., 1810, *1941*

HIRAIZUMI, Y., 1674, *1701*

HIRANO, R., 1801, 1808, 1809, 1835, *1941*

HIRAYAMA, K., 1801, 1808, 1809, 1835, *1941*

HIRSCH, R. W., 1897, *1952*

HIRSCHFELD, H. I., 1799, 1800, 1801, 1808, 1837, *1940*

HIRST, E., 1874, 1909, *1941, 1962*

HISER, H. W., 1785, *1929*

HOAK, R. D., 1811, 1814, 1839, *1941*

HOCH, J., 1790, *1943*

HOCKLEY, A. R., 1808, 1821, *1936, 1941*

HOCUTT, C. H., 1771, 1782, 1848, *1941*

HOFFMAN, D., 1780, *1941*

HOFFMANN, M. J., 1642, *1702*

HOGAN, J. W., 1634, *1708*

HOLDERBEKE, L. VAN, 1883, 1902, *1932*

HOLDEN, A. V., 1634, 1661, 1668, 1675, 1676, 1683, *1702, 1709*

HOLLAND, R. E., 1721, *1761*

HOLLIDAY, F. G. T., 1676, 1701, 1810, 1871, *1941*

HOLLISTER, T. A., 1687, *1702*

HOLME, N. A., 1717, 1719, *1761*

HOLM-HANSEN, O., 1721, 1736, *1760*

HOLMES, N., 1778, 1788, 1789, *1941*

HOLT, R., 1850, 1875, 1886, 1898, 1900, *1941*

HOLTON, R., 1668, *1697*

HOOFTMAN, R. N., 1691, *1702*

HOOK, J. T., 1902, 1903, *1938*

HOOPEN, H. J. G., TEN, 1834, *1942*

HOOPER, D. J., 1783, *1941*

HOPKINS, C. L., 1691, *1702*
HOPKINS, S. R., 1807, *1941*
HORII, S., 1664, 1668, 1705
HORST, J. M. VAN DER, 1839, 1902, 1905, 1906, 1908, 1909, *1941*
HORST, T. J., 1839, *1941*
HOSS, D. E., 1806, 1807, 1815, 1833, 1834, 1835, 1837, 1838, 1892, 1928, *1941*, *1942*
HOULT, D. P., 1790, *1942*
HOUSTON, A. H., 1807, *1954*
HOWELLS, G. P., 1799, 1800, 1801, 1808, 1837, *1940*
HOWSE, H. D., 1751, *1764*
HRUBY, T., 1642, 1700
HSU, R. Y., 1663, *1701*
HU, M. C., 1853, *1936*
HUBERT, W. A., 1883, 1884, 1898, 1900, 1901, *1931*
HUDEC, P. P., 1911, *1942*
HUGGETT, K. J., 1734, *1761*
HUGGETT, R. J., 1888, *1955*
HUGUENIN, J. E., 1886, 1894, 1899, 1902, *1928*, *1942*
HUISMAN, J., 1834, *1942*
HULINGS, N. C., 1724, *1761*
HUNT, G. J., 1892, *1942*
HURNI, W., 1633, *1706*
HUSCHENBETH, E., 1668, 1697
HUTZINGER, O., 1634, 1635, 1638, 1645, *1702*, *1707*, *1708*, *1709*
HYLIN, J. W., 1634, 1661, *1696*, 1708

ICANBERRY, J. W., 1802, *1942*
IGA, H., 1874, *1945*
ILES, R. B., 1810, 1814, *1942*
ILUS, E., 1808, 1816, 1828, *1942*
IMBERGER, J., 1788, 1855, *1936*
IMPELLIZERI, G., 1668, *1695*
INCROPERA, F. P., 1908, *1937*
INGERBRIGTSEN, O., 1883, 1894, 1895, *1942*
INGHAM, A. E., 1783, *1942*
INGRAM, W. III, 1802, 1803, *1926*
INSTITUTION OF ENGINEERS, AUSTRALIA, 1781, *1942*
INTERNATIONAL ATOMIC ENERGY AGENCY, 1781, 1798, 1801, 1811, 1814, 1815, 1816, 1834, 1873, 1907, *1942*
INTERNATIONAL COUNCIL FOR THE EXPLORATION OF THE SEA, CHARLOTTENLUND, 1668, *1702*
IRION, G., 1674, *1705*
ISAAC, R. A., 1799, 1800, 1805, 1808, 1810, 1834, 1835, 1836, 1838, *1942*
ISAACSON, M. S., 1788, 1790, 1855, *1942*
ISO, S., 1857, *1943*
IVERSEN, E. J., 1754, *1761*
IVERSON, E. S., 1898, *1943*

JACKSON, G. A., 1642, 1702, 1731, 1758, *1761*
JACKSON, W. D., 1842, *1957*
JACOBSON, P. M., 1811, *1948*
JALEES, K., 1693, *1702*
JAMES, K. W., 1789, *1951*
JAMES, M. O., 1652, 1653, 1655, *1696*, *1701*
JAMES, W. G., 1785, 1797, *1943*
JAN, T. K., 1737, 1738, 1739, *1761*, *1767*
JANICKI, R. H., 1683, *1703*
JANSSON, B., 1634, 1652, 1653, 1654, 1664, *1702*, *1708*
JANSSON, B.-O., 1921, *1953*
JARMAN, R. T., 1783, 1792, 1809, *1943*
JARVINEN, A. W., 1642, *1702*
JEFFERS, F. J., 1858, *1943*
JENSEN, L. D., 1781, 1802, 1803, 1848, 1934, *1943*
JENSEN, M. H., 1907, 1908, 1909, *1943*
JENSEN, S., 1627, 1634, 1638, 1652, 1653, 1654, 1655, 1660, 1664, 1665, 1668, *1698*, *1702*, *1708*
JERNELOV, A., 1660, *1702*
JIJI, L. M., 1790, *1943*
JOH, H., 1674, *1701*
JOHNELS, A. G., 1634, 1638, 1668, *1702*
JOHNS, R., 1906, *1943*
JOHNSON, A. E., 1795, 1808, 1809, 1820, 1821, 1857, *1953*
JOHNSON, B. M., 1853, *1927*
JOHNSON, D., 1915, 1916, 1917, 1918, 1919, 1920, *1939*
JOHNSON, J. D., 1888, *1943*
JOHNSON, J. E., 1862, *1929*
JOHNSON, J. L., 1634, *1708*
JOHNSON, M. W., 1894, 1914, 1915, *1959*
JOHNSON, R. D., 1639, *1706*
JOHNSON, R. L., *1937*
JOHNSON, W., 1799, 1800, 1837, 1891, *1929*
JOHNSON, W. C., 1874, *1943*
JOHNSON, W. J., 1850, *1943*
JOHNSTON, R., 1627, *1702*
JOLLEY, R. L., 1888, *1943*
JONES, A., 1878, 1880
JONES, C., 1779, *1943*
JONES, G. F., 1715, 1717, 1718, 1719, *1759*, *1762*
JONES, L. T., 1829, *1943*
JONES, R., 1642, *1702*
JONKEL, C. J., 1668, *1696*
JOSEFSSON, B., 1629, 1639, 1661, *1695*, *1704*
JOY, J. W., 1832, 1833, *1932*
JOYCE, E. A. JR., 1827, *1947*
JUENGST, F. W., 1656, *1702*

KABATA, Z., 1749, *1762*
KALEY, R. G., 1683, *1700*
KALIN, R. J., 1816, *1940*
KALLE, K., 1780, *1943*
KANEKO, S., 1664, 1668, *1705*

KANTHA, L. H., 1791, *1943*
KANTOR, Y., 1781, *1938*
KAPUR, S. K., 1788, 1789, 1790, *1959*
KARABASHEV, G. S., 1786, *1944*
KARAM, R. A., 1781, *1944*
KARINEN, J. F., 1655, 1659, *1702*
KARKHECK, J., 1774, 1844, 1845, 1849, *1944*
KARRICK, N. L., 1642, *1700*
KATANO, N., 1857, 1859, *1961*
KATO, J., 1899, *1944*
KATOR, H., 1633, *1705*
KAUWLING, T. J., 1755, 1757, *1764*
KAWANO, Y., 1661, *1696*
KAYSER, M., 1734, *1762*
KEDL, R. J., 1806, 1807, 1837, *1934*
KEEFE, C. W., 1834, 1838, *1939*
KEENE, D. F., 1784, *1953*
KEIL, J. E., 1654, 1686, *1703*
KELLERY, T. W., 1661, *1696*
KELLY, D. F., 1909, *1940*
KENAGA, E. E., 1665, *1703*
KENDRICK, P. J., 1784, 1785, *1944*
KENNEDY, V. S., 1802, 1803, 1804, 1811, 1812, 1815, 1823, 1825, 1826, 1862, *1944*, *1950*
KENNISH, M. J., 1823, 1825, 1850, *1944*
KERR, J. E., 1806, *1944*
KERR, N. M., 1849, 1974, 1875, 1877, 1878, 1879, 1880, 1885, 1886, 1887, 1889, 1890, 1900, 1901, *1944*
KESKITALO, J., 1808, 1816, 1828, *1942*
KEUP, L. E., 1884, 1886, 1897, 1910, *1944*
KHALANSKI, M., 1785, *1944*
KHAN, M. A. Q., 1652, *1703*
KHORRAM, S., 1683, *1703*
KIEFER, D., 1721, 1736, *1760*
KINDIG, A. C., 1749, *1762*
KING, A., 1914, 1921, *1937*
KINGWELL, S. J., 1875, 1877, 1902, *1944*
KINNE, O., 1627, 1676, 1692, 1694, *1703*, 1736, 1742, 1746, 1747, 1749, 1754, *1762*, 1769, 1770, 1771, 1794, 1801, 1803, 1804, 1810, 1814, 1819, 1825, 1826, 1830, 1831, 1834, 1835, 1837, 1840, 1861, 1862, 1871, 1872, 1873, 1874,.1875, 1877, 1878, 1879, 1882, 1886, 1887, 1894, 1895, 1896, 1897, 1899, 1901, 1916, *1944*, *1945*
KINNER, P., 1746, *1766*
KINTER, W. B., 1683, *1703*
KISH, T., 1756, 1758, *1762*
KITAMORI, R., 1745, *1762*
KITTELSEN, A., 1874, 1894, 1895, *1945*
KJERVE, B., 1782, 1783, *1945*
KJOLSETH, G., 1874, 1902, *1945*
KLEIN, W., 1653, *1701*
KLEN, E. F., 1888, *1955*
KLINGER, H., 1897, *1941*, *1953*
KNIGHT, A. W., 1683, *1703*

KNIGHT, L. H., 1878, *1927*
KOBAYASHI, K., 1658, *1703*
KOCATAS, A., 1745, *1762*
KOH, R. C. J., 1731, 1758, 1761, 1788, 1790, 1855, *1936*, *1942*
KOHNERT, R. L., 1645, *1697*
KOLKWITZ, R., 1711, 1740, 1746, *1762*
KOLPACK, R. L., 1733, *1762*
KOMAGATA, S., 1874, *1945*
KONDRATYEV, K. YA., 1916, 1917, *1945*
KOOPS, H., 1875, 1879, *1945*, *1946*
KOPPEN, J. D., 1816, *1940*
KORN, S., 1642, *1704*
KORRINGA, P., 1736, *1762*
KORTE, F., 1628, 1653, 1701, *1703*
KOSAKA, M., 1667, *1679*
KOZASA, E., 1797, 1799, *1927*
KRASNICK, G., 1734, *1760*
KRAYBILL, H. F., 1749, *1762*
KREJCAREK, G. E., 1639, *1706*
KRENKEL, P. A., 1781, 1886, *1945*, *1958*
KUBOTA, T., 1751, *1763*
KUDINOV, J. D., 1745, *1764*
KUDOH, S., 1751, *1763*
KUHLMANN, H., 1875, 1879, *1945*, *1946*
KULIK, R. A., 1851, *1961*
KULLENBERG, G., 1786, *1946*
KÜTÜKCÜOGLU, A., 1774, *1946*
KVALVAG, J., 1665, *1708*

LACIS, A. A., 1915, 1916, 1917, 1918, 1919, 1920, *1939*
LAMB, H. H., 1914, 1915, 1916, 1917, 1919, 1920, 1921, 1926, *1946*
LAMBERTON, J. G., 1655, 1659, *1702*, *1703*
LAMMERS, J., 1771, 1779, *1957*
LAMPAR, M., 1908, *1940*
LAND, J., 1788, 1789, 1795, *1928*
LANDER, K. F., 1820, 1822, 1825, 1826, 1929, *1927*
LANDRY, A. M., 1810, 1903, *1946*
LANGE, R., 1660, *1702*
LANGFORD, T. E., 1772, 1802, 1805, 1810, 1811, 1812, 1814, 1821, 1826, 1827, 1835, 1889, 1902, *1946*
LANGSTON, W. J., 1645, *1697*
LANE, C. E., 1684, *1703*
LANZA, G. R., 1798, 1800, 1931, *1946*
LARK, J. G. I., 1874, 1902, *1928*
LARSSON, K., 1633, 1660, *1703*
LASETER, J. L., 1633, 1638, *1697*, *1705*
LASKOWSKI, S. M., 1852, *1946*
LAUCKNER, G., 1749, 1754, 1762, 1826, 1840, *1946*
LAUER, G. J., 1797, 1798, 1799, 1800, 1801, 1802, 1803, 1804, 1805, 1806, 1808, 1834, 1837, 1865, *1940*, *1946*

LAUGHLIN, R. B., 1676, *1703*
LAW, L. M., 1634, *1700*
LAWRENCE, S. A., 1834, 1888, *1931*
LEACH, G., 1875, *1946*
LEATHEM, W., 1746, *1766*
LEBEDEFF, S., 1915, 1916, 1917, 1918, 1919, 1920, *1939*
LEBLANC, G. A., 1683, *1700*
LECH, J. J., 1652, 1656, *1704*, *1708*
LECH, J. L., 1658, *1700*
LEDET, E. J., 1633, *1705*
LEE, C. C., 1846, *1946*
LEE, J. J., *1956*
LEE, P., 1915, 1916, 1917, 1918, 1920, *1939*
LEE, R. F., 1652, 1690, 1703, *1707*
LEE, S., 1787, 1788, *1957*
LEE, S. S., 1775, 1781, 1782, 1785, 1788, 1840, 1873, *1929*, *1946*
LEE, W. C., 1874, *1962*
LEEPER, G., 1633, *1706*
LEIGHTON, D. 1831, 1832, 1857, *1947*
LEMBI, C. A., 1686, *1698*
LENZ, J., 1721, *1762*
LEO, A. J., 1645, *1703*
LEONI, V., 1661, 1664, *1706*
LEPPÄKOSKI, E., 1740, 1742, 1745, *1762*
LETTERMANN, E. F., 1629, *1697*
LEVIN, J. E., 1890, *1930*
LEWIS, C. W., 1770, 1892, *1928*
LEWIS, J., 1793, *1947*
LEWIS, R. M., 1812, *1947*
LIAO, P. B., 1890, 1893, 1894, 1895, 1902, *1947*
LIEBMANN, H., 1897, *1947*
LIGTERINGEN, H., 1787, *1947*
LINDEN, E., 1683, *1703*
LINKO, R. R., 1668, *1703*
LINTON, T. G., 1719, *1765*
LISK, D. J., 1665, *1696*, *1709*
LIST, E. J., 1788, 1790, 1855, *1857*, *1932*, *1942*, *1954*
LITTLEPAGE, J. M., 1733, 1745, *1759*
LITTLER, M. M., 1725, 1743, 1749, *1762*, *1763*
LIVINGSTON, R. J., 1684, *1703*
LIZARRAGA-PARTIDA, M. L., 1745, *1763*
LOI, T.-N., 1827, *1947*
LOM, J., 1749, *1763*
LONG, B., 1633, 1674, *1704*
LONG, R. R., 1791, *1947*
LONGHURST, A. R., 1921, *1953*
LOON, L. S. VAN, 1783, *1947*
LOOSMORE, F. A., 1820, 1825, 1826, 1829, *1927*
LOS ANGELES COUNTY SANITATION DISTRICT, *1762*
LOUGH, R. G., 1882, *1947*
LOVELOCK, J. E., 1652, 1660, 1662, *1703*
LOWE, J. I., 1648, 1691, 1692, *1701*, *1704*

LU, P.-Y., 1644, *1704*
LUCKAS, B., 1665, *1704*
LUCKOW, J., 1904, 1905, *1947*
LUDWIG, H. F., 1733, 1736, *1763*
LUMLEY, J. L., *1947*
LUNDE, G., 1638, 1639, 1661, 1672, *1696*, *1704*
LYBERG, B., 1842, *1947*
LYONS, L., 1827, *1947*
LYONS, W. G., 1827, *1947*

MABBETT, A. N., 1735, *1762*
McCAIN, B. B., 1751, *1766*
McCARTHY, J. J., 1721, 1736, *1760*
McCLURE, V. E., 1661, *1707*
McCONNELL, G., 1628, 1638, 1648, 1660, 1662, 1663, 1665, 1672, 1689, *1706*
McCONNELL, O. J., 1662, *1704*
McCOWAN, J. G., 1922, *1938*
McCOY, J. W., 1888, *1947*
McCOY, R. H., 1896, *1939*
MacCRIMMON, H. R., 1902, *1947*
McDERMOTT, D. J., 1650, *1709*, 1716, *1763*
MACEK, K. J., 1642, 1676, *1704*
McERLEAN, A. J., 1803, 1804, 1810, 1813, 1815, 1850, 1903, *1941*, *1947*, *1950*, *1951*
McFADEN, W. H., 1635, *1704*
MacFADYEN, A., 1782, *1947*
MacFARLANE, R. B., 1686, *1704*
McFARLANE, R. W., 1781, *1936*
McGOWN, L. B., 1921, *1947*
MacGREGOR, J. S., 1633, 1645, 1665, *1704*
McILHENNY, W. F., 1831, 1833, 1834, 1857, *1963*
MACK, H. J., 1905, 1906, *1956*
MacKAY, D. W., 1737, 1745, *1761*
MacKAY, K. T., 1874, 1876, *1960*
MacKAY, O. W., 1664, 1674, 1675, *1701*
McKIM, J. M., 1684, *1704*
MacLAUCHLAN, J. W. G., 1782, *1940*
McLEESE, D. W., 1648, 1683, *1704*
MacLeod, R. A., 1676, 1704, 1871, *1947*
McLUSKY, D. S., 1742, *1763*
McMILLAN, J. A., 1639, *1706*
McMILLAN, W., 1791, 1792, 1931, *1947*
McNEIL, W. J., 1886, *1948*
McNULTY, J. K., 1736, *1763*
MacQUEEN, J. F., 1788, 1789, 1790, 1791, 1792, 1931, 1948
McVICAR, A. H., 1896, *1948*
McWILLIAM, D. C., 1745, *1763*
MADDEN, J. A., 1807, *1954*
MADDING, R. P., 1785, 1786, *1948*
MADDOCK, L., 1855, 1921, *1948*, *1954*
MADEWELL, C. E., 1907, 1908, *1931*
MAECKELBERGHE, H., 1853, 1902, *1932*
MAGGS, R. J., 1662, *1703*

MAGRITA, R., 1786, *1939*
MAHONEY, J. B., 1751, *1763*
MAICKEL, R. P., 1652, *1697*
MAITRA, N., 1652, *1703*
MALAHOFF, A., 1915, *1927*
MALISCH, R., 1674, *1705*
MANABE, S., 1915, *1948*
MANABE, T., 1674, *1701*
MANDELLI, E. F., 1831, 1833, 1834, 1857, *1963*
MANN, R., 1882, *1948*
MANN, S., 1743, *1760*
MANFREDI, E., 1750, *1763*
MARCELLO, R. A., *1948*
MARCHELLO, J. M., 1883, 1894, 1899, *1949*
MARCY, B. C., JR., 1782, 1797, 1802, 1806, 1897, 1811, 1837, 1903, *1948, 1957*
MARE, M. F., 1722, *1763*
MARGRAF, F. J., 1891, 1892, *1948*
MARGREY, S. L., 1802, 1862, *1931*
MARKOWSKI, S., 1781, 1802, 1804, 1806, 1809, 1814, 1815, 1817, 1819, 1824, 1830, 1834, 1837, *1948*
MARLES, E., 1733, 1745, *1759*
MARMER, G. J., 1785, 1786, *1948*
MAROIS, M., 1781, 1873, *1948*
MARSDEN, K., 1634, *1701*
MARSHALL, J. S., 1773, 1799, *1949*
MARSHALL, W., 1843, 1844, 1845, 1846, 1909, *1949*
MARSHALL, W. L., 1889, *1949*
MARSSON, M., 1711, 1740, 1746, *1762*
MARTIN, A., 1748, 1749, *1763*
MARTIN, J. H. A., 1888, 1917, *1936, 1949*
MARTIN, R. B., *1949*
MARTINO, P. A., 1883, *1949*
MARTINSEN, K., 1638, 1672, *1696*
MASCARELLO, J. M., 1853, *1962*
MASNIK, M. T., 1820, 1821, 1827, 1828, 1829, *1949*
MASON, J. W., 1646, 1647, *1704*
MASSE, H., 1826, 1864, *1927*
MASUDA, Y., 1922, *1951*
MATSON, J. V., 1777, 1888, 1890, *1949*
MATSUMURA, F., 1654, 1655, 1659, *1696*
MATTHEWS, M. A., 1915, *1949*
MATTICE, J. S., 1888, 1889, *1949*
MATURO, F. J. S. JR., 1802, 1803, *1926*
MAUCHLINE, J., 1892, 1949
MAUGH, T. H., 1627, *1704*
MAUER, D., 1746, *1766*
MAURER, D. I., 1882, 1892, *1960*
MAWDESLEY-THOMAS, L. E., 1749, *1763*
MAYER, A. G., 1819, 1912, *1949*
MAYER, F. L., 1642, *1704*
MAYO, R. D., 1894, *1947*
MEAD, A. R., 1831, 1857, *1959*
MEADE, T. L., 1894, *1949*

MEANEY, R. A., 1874, *1949*
MEARNS, A. J., 1726, 1744, 1747, 1748, 1750, 1751, 1752, 1755, 1757, *1763, 1764, 1765, 1766*
MEEKS, W., 1797, 1798, 1799, 1800, 1801, 1802, 1803, 1804, 1805, 1806, 1834, 1837, 1863, *1946*
MEHRLE, P. M., 1642, *1704*
MELANCON, M. J., 1652, *1704*
MÉLARD, CH., 1892, 1893, *1949*
MENDOLA, J. T., 1629, *1697*
MENZEL, B. W., 1782, 1862, *1954*
MENZEL, D. W., 1687, *1704*
MERCER, J. H., 1918, *1949*
MERKENS, L. S., 1683, *1703*
MERRIMAN, D., 1770, 1815, *1949*
MESAROVIĆ, M. M., 1853, *1949*
MESTRES, R., 1633, 1674, *1705*
METCALF, 1711, *1763*
METCALF, R. L., 1644, 1652, 1658, 1683, *1703, 1704, 1705*
METCALFE, C. D., 1683, *1704*
MEYER, F. P., 1896, 1897, *1949*
MICHAEL, P. R., 1683, *1700*
MICHAELIS, R., 1687, 1690, *1706*
MICHEL, J. W., 1874, *1962*
MIDDLEDITCH, B. S., 1635, *1697*
MIDLIGE, F. H., 1751, *1763*
MIGET, R., 1633, *1705*
MIHURSKY, J. A., 1781, 1797, 1802, 1803, 1804, 1810, 1811, 1812, 1813, 1815, 1823, 1825, 1826, 1833, 1834, 1838, 1850, 1862, 1864, 1903, 1912, *1939, 1944, 1947, 1949, 1950*
MIKKOLA, I., 1788, *1950*
MIKLAS, H. P., 1661, 1668, *1701*
MIKOLA, J., 1841, *1950*
MILLER, B. S., 1751, *1766*
MILLER, D. C., 1855, *1950*
MILLER, E. R., 1901, *1950*
MILLER, H., 1787, 1788, *1947*
MILLER, L., 1838, *1932*
MILLER, V. G., 1897, *1951*
MILLS, P. A., 1634, *1705*
MILNE, P. H., 1874, 1899, 1901, *1950*
MILNER, A. G. P., 1885, 1887, 1891, 1895, *1928*
MINAGAWA, K., 1638, *1708*
MINER, R. M., 1776, 1853, *1950*
MINOHARA, Y., 1874, *1944*
MITCHELL, A. I., 1654, *1705*
MITCHELL, C. T., 1802, *1950*
MITCHELL, N. T., 1892, 1893, *1949*
MITTON, J. B., 1799, 1800, *1962*
MIYAZAKI, T., 1664, 1668, 1705, 1922, *1951*
MIZUSHIMA, T., 1751, *1763*
MOBERG, O., 1847, 1860, 1910, *1951*
MOLESWORTH, 1828, 1829. *1934*
MOLINARI, J., 1786, *1939*
MØLLER, B., 1825, 1836, *1951*

Møller, D., 1874, *1951*

Möller, H., 1782, 1815, 1836, *1951*

Monahan, E., 1784, *1951*

Monod, J. L., 1629, *1697*

Moore, C. J., 1875, 1880, 1882, 1884, 1903, *1951*

Moore, D. J., 1657, 1686, 1698, 1785, 1791, 1792, *1931, 1951*

Moore, F. K., 1853, *1951*

Moore, J. C., *1697*

Moore, R., *1936*

Moore, R. E., 1662, *1705*

Moore, R. P., *1951*

Moore, S. A., 1687, 1689, *1705*

Moraitou-Apostolopoulou, M., 1831, 1938

Morgan, J. J., 1731, 1758, *1761*

Morgan, K. Z., 1781, *1944*

Morgan, R. P., 1782, 1799, 1804, 1834, 1835, 1848, 1888, *1941, 1951*

Morita, M., 1668, *1709*

Moriseta, M., 1729, *1763*

Morton, B. R., 1790, *1951*

Moyer, M. S., 1798, 1799, 1800, 1837, *1937*

Mosser, J. L., 1686, 1690, *1705*

Mount, D. I., 1684, *1705*

Mountain, J. A., 1810, 1811, 1827, 1833, 1939, *1947, 1951*

Mühlhaüser, H., 1843, *1935*

Mueller, J. L., 1785, *1953*

Muench, K. A., 1875, 1878, 1882, 1883, 1886, 1887, 1893, *1951*

Müller, G., 1674, *1705*

Mulford, S., 1831, 1832, 1857, *1947*

Muller, W. A., *1956*

Mulstay, R., 1799, 1800, *1962*

Munro, A. L. S., 1895, 1896, 1897, *1951*

Murai, T., 1878, 1893, *1927*

Murakami, H., 1668, *1709*

Murchelano, R. A., 1751, *1763, 1767*

Murray, A. J., 1638, 1660, 1663, *1705*

Murray, H. E., 1662, 1668, 1674, *1705*

Murray, S. N., 1725, 1743, *1762, 1763*

Murthy, K. K., 1847, *1951*

Musselius, V. A., 1815, *1951*

Musty, P. R., 1629, *1705*

Myers, D. M., 1851, *1928*

Nagai, A., 1751, *1763*

Nakai, A., 1751, *1763*

Nakamura, H., 1874, *1945*

Nakazima, M., 1736, *1763*

Nankee, R. L., 1811, *1948*

Narayana Pillay, V., 1784, *1935*

Nash, C. E., 1814, 1835, 1872, 1875, 1878, 1885, 1887, 1888, 1889, 1890, 1891, 1892, 1897, 1898, *1951, 1953*

National Academy of Sciences, 1864, 1865, 1915, 1921, *1952*

National Research Council, 1628, *1705*

Nauman, J. W., 1824, 1826, 1828, 1835, *1933, 1952*

Navrot, J., 1734, *1759*

Naylor, E., 1781, 1782, 1808, 1809, 1814, 1819, 1820, 1823, 1825, 1827, 1828, 1829, *1903, 1953*

Nazansky, B., 1674, *1708*

Nebeker, A. V., 1894, *1952*

Neely, W. B., 1644, 1645, *1696, 1705*

Neff, G. S., 1638, 1657, 1661, 1662, 1668, 1676, 1683, *1700, 1707, 1709*

Neff, J. M., 1657, *1703, 1709*

Neill, W. H., 1805, 1811, *1952*

Nelson, M. D., 1746, *1761*

Nestel, H., 1642, *1705*

Neu, H. J., 1638, 1655, *1696, 1705*

Neudecker, A., 1826, *1952*

Neudecker, S., 1826, *1952*

Neushul, M., 1743, *1761*

Nevanlinna, L., 1841, *1896*

Nichols, B. L., 1874, *1962*

Nicholson, N., 1743, *1763*

Nickless, G., 1629, 1674, *1697, 1705*

Nihoul, J. C. J., 1788, *1952*

Nikolsky, G. A., 1952, 1953, *1945*

Nilsson, C.-A., 1628, 1705

Nimmo, D. R., 1645, 1684, 1685, *1707*, 1751, *1760*

Nishida, K., 1674, *1675*

Nishimura, H., 1674, *1675*

Noma, T., 1899, *1944*

Nordtröm, A., 1628, *1705*

Norse, E. A., 1830, *1952*

North, W. J., 1802, 1817, 1827, 1839, *1950, 1952*

Nusbaum, I., 1831, 1832, 1857, *1947*

Oakes, D., 1897, *1952*

O'Brodovich, H., 1629, *1708*

O'Connor, S. G., 1810, 1813, *1947*

O'Connors, H. B., Jr., 1686, 1687, *1696*, 1734, *1760*

Odham, G., 1633, 1660, *1703*

Offhaus, K., 1897, *1947*

Officer, C. B., 1758, *1763*

Ogura, M., 1751, *1763*

Oguri, M., 1813, 1830, *1958*

Olst, J. C., van, 1885, *1937*

Onley, C. E., 1663, *1698*

Okuba, A., 1786, *1952*

Oliff, W. D., 1745, *1764*

Olney, C. E., 1661, 1663, *1696*

Olsson, M., 1634, 1638, 1664, 1668, *1702*

Olsson, R. K., 1823, 1825, 1850, *1944*

OLSZEWSKI, M., 1906, 1907, 1909, *1952*
OPPENHEIMER, C. H., 1633, 1676, *1704*, *1706*, 1769, 1771, 1798, 1878, *1952*, *1953*
ORCEL, L., 1749, 1751, *1759*
ORIENTE, G., 1638, *1695*
OSHIDA, P. S., 1747, 1748, 1755, 1757, *1764*
OSTERROHT, C., 1629, 1645, 1661, 1665, *1706*, *1707*
OSTRACH, S., 1788, *1953*
OTTERLIND, G., 1634, 1638, 1668, *1702*
OVERSTREET, R. M., 1751, *1764*
OZMIDOV, R. V., 1786, *1944*

PADDOCK, R. A., 1783, 1784, 1788, 1791, *1935*, *1947*
PAGANESSI, J. E., 1851, *1961*
PAGE-JONES, R. M., 1890, *1953*
PAGNON, M., 1674, *1704*
PALFREMAN, D. A., 1876, *1953*
PALMEGIANO, G., 1882, *1953*
PALMER, C. M., 1721, *1764*
PALMER, H. D., 1737, *1760*
PALMORK, K. H., 1660, *1702*
PANG, P. K. T., 1807, *1954*
PANNELL, J. P. M., 1795, 1808, 1809, 1820, 1821, 1857, *1953*
PARIS, D. F., 1642, *1700*
PARK, J. E., 1852, *1953*
PARKER, B. C., 1633, 1701, *1706*
PARKER, B. D., 1771, 1779, 1781, 1836, *1931*
PARKER, J. G., 1747, *1764*
PARKER, N. C., 1900, *1953*
PARKER, P. L., 1781, *1945*
PARKER, R. F., 1858, *1953*
PARRISH, P. R., 1647, 1648, 1683, 1684, 1685, 1691, 1692, *1701*, *1703*, *1706*
PARSI, P., 1645, 1674, *1699*
PARSONS, T. R., 1921, *1953*
PATIL, K. C., 1654, 1659, *1706*
PATRICK, J. M., 1647, 1648, 1683, 1690, 1692, *1703*, *1706*, *1707*
PAUL, I. C., 1639, *1706*
PAULSEN, C. L., 1885, 1887, 1888, 1889, 1891, 1892, *1952*
PAVLOU, S. P., 1674, *1698*
PAYNE, J. R., 1629, *1697*
PEAKALL, D. B., 1628, *1706*
PEARCE, J., 1781, 1799, 1800, 1812, 1826, 1835, *1962*
PEARCE, J. B., 1746, 1749, 1754, 1767, 1912, *1950*, *1953*
PEARCY, W. G., 1784, 1785, *1953*
PEARSON, C. R., 1628, 1638, 1648, 1660, 1662, 1663, 1665, 1672, 1683, 1689, *1706*
PEARSON, E. A., 1733, *1764*
PEARSON, T. H., 1719, 1728, 1740, *1764*
PEASE, T. E., 1855, *1953*

PECK, B. B., 1797, 1799, 1800, 1802, 1803, 1804, 1808, 1809, 1834, 1837, 1838, *1931*
PELLACANI, T., 1668, *1697*
PENNYCUICK, L., 1921, *1958*
PEPPLE, S. K., 1656, *1708*
PERES, J. M., 1745, 1759, 1917, *1953*
PERKINS, E. J., 1751, *1764*
PERRIN, C., 1903, *1943*
PERSON-LE-RUYET, J., 1882, *1953*
PERSOONE, G., 1883, 1902, *1932*
PERTTILÄ, M., 1668, *1709*
PETERS, G., 1897, *1953*
PETERS, N., 1749, *1764*
PETERSEN, C. G. J., 1723, *1764*
PETERSON, R. E., 1851, 1874, 1875, 1880, 1882, 1883, 1884, 1892, 1900, 1904, *1953*
PETROCELLI, S. R., 1642, 1683, 1706, *1707*
PFUDERER, H. A., 1782, 1862, *1934*
PHILIPPART, J. C., 1892, 1893, *1949*
PHILLIPS, J. H., 1650, *1701*
PHILLIPS, O. M., 1791, *1944*
PHILPOT, R. M., 1653, *1696*
PHINNEY, D. A., 1782, *1938*
PHIZACKLEA, P., 1742, *1763*
PIATELLI, M., 1638, *1695*
PICARD, A., 1732
PICER, M., 1674, *1708*
PICER, N., 1674, *1708*
PICKFORD, G. E., 1807, *1954*
PIELOU, E. C., 1727, *1764*
PIGNATTI, S., 1745, *1761*
PILATI, D. A., 1856, *1954*
PILCHER, K. S., 1896, 1939, *1956*
PILE, R. S., 1907, 1908, *1931*
PILLAY, T. V. R., 1899, *1954*
PINCINE, A. B., 1857, *1954*
PINGREE, R. D., 1859, *1954*
PINTO, J. S., 1736, *1764*
PIOTROWICZ, S. R., 1663, *1698*
PIPES, W. O., 1801, *1929*
PLACK, P. A., 1654, *1705*
PLASS, G. N., 1915, *1954*
PLUMMER, *1906*
POHL, R. J., 1652, *1706*
POLIKARPOW, G. G., 1645, 1674, *1698*, *1699*
POLICASTRO, A. J., 1788, *1935*
POLK, E. M., 1792, *1935*
POLLARD, D. A., 1858, *1954*
POORE, G. C. B., 1745, *1764*
PORTER, R. W., 1851, *1954*, *1961*
PORTMANN, J. E., 1683, *1706*
POSMENTIER, E. S., 1791, *1954*
POTAPENKO, I. A., 1891, *1936*
POULLAQUEC, G., 1884, *1934*
POVEY, A., 1674, *1697*
POWELL, J., 1774, 1844, 1845, 1849, *1944*
POWERS, C. D., 1686, 1687, 1690, *1706*

PRAHL, J., 1788, *1953*
PRATT, F. B., 1862, *1929*
PRICE, A. H., 1824, 1825, 1875, 1882, 1892, 1893, *1940, 1954*
PRICE, A. R. G., 1831, 1911, *1929*
PRICE, B. L., 1905, *1954*
PRIESTER, L. E., 1654, 1686, *1702*
PRITCHARD, D. W., 1786, *1954*
PSCHEIDL, H., 1665, *1704*
PUCCETTI, G., 1661, 1664, *1706*
PYBUS, M. J., 1784, *1951*
PYM, R., 1733, 1745, *1759*
PYYSALO, H., 1668, *1709*

QUEIRAZZA, G., 1887, 1888, 1889, 1892, *1956*
QUICK, J. A., 1823, 1824, 1831, 1840, *1954*
QUINN, J. G., 1663, *1698, 1707*

RACHFORD, *1906*
RAINIO, K., 1668, *1703*
RALL, D. P., 1652, *1696*
RANDTKE, A., 1687, *1704*
RANEY, E. C., 1782, 1862, *1954*
RANDALL, R. H., 1828, *1929*
RANKIN, J. S., 1798, *1930*
RANTAMKI, P., 1668, *1703*
RAPPE, C., 1628, *1709*
RAY, B. J., 1663, *1698*
RAY, L. E., 1662, 1668, 1674, *1705*
RAYMOND, L. P., 1883, *1954*
RAYMONT, J. E. G., 1795, 1801, 1808, 1809, 1820, 1821, 1822, 1825, 1828, 1829, 1857, *1953, 1954*
REAVES, R. S., 1807, *1954*
REED, F. M. H., 1734, 1736, *1759*
REEVE, M. R., 1802, 1803, *1954, 1955*
REICHENBACH-KLINKE, H. H., 1896, 1897, *1955*
REID, D. A., 1775, 1900, *1955*
REID, R. O., 1831, 1833, 1834, 1857, *1963*
REINERT, R. E., 1665, *1706*
REINKE, J., 1629, *1708*
REINKEN, G., 1904, 1905, *1947*
REISH, D. J., 1719, 1723, 1724, 1727, 1737, 1740, 1742, 1744, 1745, 1746, 1747, 1755, 1757, 1759, 1764, 1765, 1858, 1911, *1955*
REMSEN, C. C., 1686, *1699*
RENBERG, L., 1660, *1702*
RESH, R. E., 1875, 1880, 1882, 1884, 1890, 1901, *1938, 1939*
RESIG, J. M., 1747, *1764*
REUTHER, R., 1674, *1705*
REYNOLDS, J. Z., 1850, 1853, 1854, 1888, 1889, 1891, *1955, 1963*
REYNOLDS, L. M., 1634, *1706*
RHEINHEIMER, G., 1893, *1955*
RICHARDS, G. V., 1782, 1785, *1932*
RICHARDSON, L. B., 1802, 1862, *1931*

RICHARDSON, R. E., 1711, 1740, *1761*
RICHTER-BERNBERG, 1917
RICE, C. P., 1650, *1706*
RICE, T. R., 1892, *1928*
RIEDMÜLLER, S., 1897, *1947*
RIGBY, R. A., 1684, 1685, *1705*
RILEY, J. P., 1629, 1638, 1660, 1661, 1663, 1665, *1697, 1698, 1705, 1706*
RINEHART, K. L., 1639, *1706*
RIND, D., 1915, 1916, 1917, 1918, 1919, 1920, *1939*
RISEBROUGH, R. W., 1629, 1638, 1668, *1697, 1707*
RISTICH, S. S., 1745, *1765*
RITCHIE, A. R., 1691, *1702*
ROBERT, J. M., 1853, *1962*
ROBERTS, D. J., 1674, *1697*
ROBERTS, M. H., 1888, *1955*
ROBERTS, R. J., 1896, 1897, *1955*
ROBERTSON, K. J., 1663, *1709*
RODENHUIS, G. S., 1787, 1788, *1955*
RODGERS, C. R., 1642, *1704*
ROELS, O. A., 1859, 1883, 1884, *1955*
ROENSCH, L. F., 1888, *1955*
ROESIJADI, G., 1683, *1707*
ROESSLER, M., 1828, 1850, 1851, *1955*
ROESSLER, M. A., 1816, 1817, 1818, 1819, 1838, 1839, 1840, 1910, *1928, 1955, 1960*
ROGERS, A., 1880, 1900, *1955*
ROGERS, T. C., 1734, 1735, *1764*
ROHATGI, N. K., 1737, 1738, *1765*
ROHDE, K., 1749, *1765*
ROLLWAGEN, J., 1674, *1708*
ROMERIL, M. G., 1779, 1892, *1955*
ROOSENBURG, W. H., 1803, 1804, 1850, 1892, 1941, 1944, *1955*
ROOTH, G. C. H., 1792, *1938*
ROSEN, J. D., 1652, *1703*
ROSENBERG, R., 1719, 1728, 1740, 1746, 1747, 1757, *1763, 1765*
ROSENTHAL, H., 1691, 1709, 1873, 1874, 1875, 1877, 1878, 1879, 1882, 1886, 1887, 1895, 1897, 1899, *1945, 1955, 1956*
ROSENTHAL, R. J., 1832, 1838, *1932*
ROSS, F. F., 1893, *1956*
ROSSI, S. S., 1755, 1757, *1764*
ROTHWELL, G. N., 1899, *1942*
ROUX, A. LE, 1882, *1953*
ROWE, D. R., 1646, 1647, *1704*
ROWE, R. C., 1743, *1765*
ROWLAND, R. G., 1686, 1687, 1690, *1696, 1706*
RUCKER, R. R., 1894, *1956*
RUDLING, L., 1638, *1707*
RUIVO, M., 1627, *1707*
RUSSELL, F. S., 1921, *1956*
RUSSEL, G., 1915, 1916, 1917, 1918, 1919, 1920, *1939*

RUTHERFORD, D., 1882, 1883, *1956*
RYE, H., 1788, 1789, 1795, *1928*
RYTHER, J. H., 1758, 1763, 1882, 1883, 1886, 1887, 1894, *1942, 1948*
RYKBOST, K. A., 1905, 1906, 1907, *1956*
RYLAND, J. S., 1821, *1956*
RYTHER, J. H., 1880, 1883, 1887, *1956*

SAEGER, V. W., 1683, *1700*
SAETERSDAL, G., 1921, *1953*
SAFE, S. S., 1634, 1635, 1638, 1702, 1707
SAILA, S. A., 1782, *1956*
SAILA, S. B., 1749, 1754, *1760*
SAKS, N. M., *1956*
SALO, E. O., 1892, *1956*
SAMBUCO, E., 1788, 1789, 1790, *1959*
SANDERS, H. L., 1728, 1745, *1765*
SANDERS, H. O., 1642, *1704*
SANDERS, J. E., 1896, *1956*
SANDIFER, S. H., 1686, *1702*
SANDISON, A., *1793*
SARDAR, Z., 1859, 1860, *1956*
SAROGLIA, M. G., 1882, 1887, 1888, 1889, 1892, *1953, 1956*
SAUNDER, E., 1897, *1956*
SAUNDERS, P. M., 1785, 1880, 1883, *1956*
SAUNDERS, R. L., 1878, *1956*
SAVAGE, P. D., 1797, 1799, 1801, 1807, 1808, 1833, *1956*
SAVAGE, T., 1827, *1947*
SAVAGE, W. F., 1853, *1936*
SAWYER, J. S., *1956*
SAYLER, G. S., 1656, *1707*
SCHAEFER, R. G., 1634, 1638, 1642, 1643, 1644, 1653, 1665, 1668, *1698, 1699, 1707*
SCHAFER, C. T., 1716, 1745, 1747, *1765*
SCHAFER, H. A., 1712, 1714, 1737, 1738, 1739, 1745, *1761*
SCHEFFLER, G., 1771, 1779, *1957*
SCHEUER, P. J., 1639, *1707*
SCHIMMEL, S. C., 1647, 1648, 1656, 1683, 1690, *1706, 1707*
SCHLOTFELDT, H. J., 1754, 1765, 1895, 1896, *1957*
SCHMEDDING, D. W., 1645, *1697*
SCHMIDT, T. T., 1629, *1697*
SCHMIDT-BLEEK, F., 1693, *1707*
SCHMISSEUR, W. E., 1905, 1906, *1956*
SCHNEIDER, R., 1665, *1707*
SCHNEIDER, S. H., 1920, *1957*
SCHOMAKER, K., 1633, *1708*
SCHRAMM, W., 1676, 1700, 1871, *1938*
SCHREIBER, J. R., JR., 1831, 1857, *1959*
SCHROEDER, P. B., 1803, 1816, 1817, 1818, 1819, 1840, *1957, 1960*
SCHUBEL, J. R., 1782, 1797, 1806, 1807, 1838,

1849, 1862, 1863, 1864, 1865, 1866, 1867, 1868, 1869, 1872, 1913, 1914, *1957*
SCHULTE, E., 1655, *1705, 1707*
SCHULTZ, D. M., 1660, *1707*
SCHWARTZ, H. E., 1638, *1707*
SCARANO, G., 1887, 1888, 1889, 1892, *1956*
SCOTT, K. R., 1874, 1902, *1928*
SCOTTISH MARINE BIOLOGICAL ASSOCIATION, 1823, *1957*
SCIUTO, S., 1668, *1695*
SCRIVEN, R. A., 1792, *1958*
SCURA, E. D., 1661, *1707*
SEARS, F. W., *1957*
SEDGWICK, S. D., 1880, *1957*
SEGAR, D., 1816, 1817, 1818, 1839, 1840, *1960*
SEGAR, D. A., 1792, *1938*
SELYE, H., 1897, *1957*
SENGUPTA, S., 1775, 1781, 1782, 1785, 1787, 1788, 1840, 1873, *1929, 1946, 1947, 1957*
SEO, K. K., 1851, 1874, 1875, 1880, 1882, 1883, 1884, 1892, 1900, 1903, *1953*
SEPPA, M., 1841, 1950
SERGEANT, D. B., 1648, 1683, *1704*
SERUM, J. W., 1665, *1696*
SERYAPOV, A. I., 1891, *1936*
SETCHELL, W. A., 1819, *1957*
SHABI, F. A., 1894, *1957*
SHAPOT, R. M., 1810, 1811, 1850, *1957*
SHARFSTEIN, B. A., 1884, *1955*
SHARITZ, R. R., 1781, *1938*
SHARP, J. M., 1719, 1745, *1759*
SHAW, P. D., 1639, *1706*
SHEAHAN, C. J., 1875, 1880, 1882, 1891, 1901, 1902, *1939*
SHEINDLIN, A. E., 1842, *1957*
SHELBOURN, J. E., 1878, *1930*
SHELTON, R. G. J., 1746, *1765*
SHEPHERD, J. G., 1791, 1792, *1931*
SHEPPARD, J. M., 1684, 1685, *1705*
SHERWOOD, M. J., 1749, 1750, 1751, 1752, 1753, 1763, *1765*
SHOOP, C. T., 1878, *1930*
SHROPSHIRE, J. C., 1629, *1697*
SIEBERT, D. L. R., 1722, *1766*
SIEBURTH, J. McN., 1798, *1957*
SIESS, J., 1947
SIDES, W. T., 1811, 1835, 1903, *1936*
SIKKA, H. C., 1605, *1706*
SILL, B. L., 1790, *1957*
SILVA, E. S., 1736, *1764*
SILVERMAN, M. J., 1871, *1957*
SIMMONS, G. M., JR., 1808, *1958*
SIMPSON, J. H., 1785, *1958*
SINDERMANN, C. J., 1694, *1708*, 1749, 1751, *1765*
SINGER, S. C., 1690, *1708*
SINNARWALLA, A. M., 1788, 1789, 1790, *1959*
SIUDA, J. F., 1639, 1662, *1706, 1708*

SIVADAS, P., 1882, *1927*
SKELLY, M. J., 1855, *1953*
SLABAUGH, L. V., 1629, *1696*
SLATER, R. H., 1737, *1760*
SLAWSON, P. R., 1852, *1932*
SLICHER, A. N., 1807, *1954*
SMART, G. R., 1894, *1937*
SMETACEK, V., 1645, *1706*
SMIDT, E. L. B., 1792, *1958*
SMITH, A. A., 1788, *1937*
SMITH, C. W., 1824, 1825, 1875, 1882, 1892, 1893, 1940, *1953*
SMITH, G. C., 1853, *1927*
SMITH, M. H., 1810, *1936*
SMITH, R. C., 1816, 1817, 1818, 1819, *1958*
SMITH, R. W., 1727, *1765*
SMITH, T. G., 1668, *1695*
SMITZ, J., 1788, *1952*
SMOKLER, P. E., 1664, 1665, 1672, 1673, *1708*
SNIESZENKO, S. F., 1896, 1897, *1958*
SNYDER, G. R., 1886, *1958*
SØOERGREN, A., 1633, 1660, *1703*
SØRENSEN, O., 1634, *1708*
SOIVIO, A., 1883, *1961*
SOLLY, S. R. B., 1691, *1702*
SOMMER, W., 1845, *1938*
SONNENFELD, P., 1911, *1942*
SORGE, E. V., 1770, *1958*
SOULE, D. F., 1764, 1813, 1814, 1830, 1835, *1958*
SOULE, J. D., 1757, *1764*
SOUTH CALIFORNIA COASTAL WATER RESEARCH PROJECT, 1712, 1714, 1720, 1725, 1726, 1731, 1739, 1752, 1753, *1765, 1766*
SOUTHGATE, B. A., 1711, 1740, 1741, 1742, *1759*
SOUTH OF SCOTLAND ELECTRICITY BOARD, 1774, 1885, 1886, *1958*
SOUTHWARD, A. J., 1770, 1916, 1917, 1921, *1931, 1958*
SOUTHWARD, A. J., 1921, *1934, 1956, 1958*
SOWERBUTTS, B. S., 1894, *1958*
SPANHAAK, G., 1780, *1958*
SPEAKMAN, J. N., 1886, *1958*
SPENCER, J. F., 1793, 1801, 1825, 1826, 1828, 1833, 1837, 1872, *1940*
SPRINGER, A. M., 1629, *1697*
SPURR, G., 1792, *1958*
SQUIRE, J. L., 1785, *1858*
SRIVASTAVA, A. K., 1807, *1954*
STADLER, D., 1633, 1661, 1662, *1708*
STAINKEN, D., 1674, *1708*
STALCUP, M. C., 1922, 1923, *1935*
STALLING, D. L., 1634, 1642, *1704, 1708*
STANTON, R. H., 1652, *1703*
STATHAM, C. N., 1656, 1658, *1700, 1708*
STAUFFER, J. R., 1782, 1848, 1854, 1855, *1941, 1958*

STEARNS, R. D., 1792, *1938*
STEEDMAN, H. F., 1722, *1765*
STEINESS, E., 1639, *1704*
STEINHAUER, W. G., 1661, 1668, *1701*
STENERSEN, J., 1665, *1708*
STEPHAN, C. E., 1684, *1705*
STEPHENS, C. A., 1884, 1890, *1939*
STEVENS, D. E., 1797, 1838, 1932, *1958*
STEVENS, D. G., 1894, *1952*
STEVENS, G. A., 1903, *1951*
STEWARD, N. E., 1655, 1659, *1702*
STEWART, G. L., 1734, 1736, *1759*
STEWART, J. E., 1902, *1947*
STEWART, J. R., 1802, *1928*
STEWART, V. N., 1897, *1959*
STICKNEY, R. P., 1878, 1885, 1886, 1890, 1895, 1897, 1899, 1901, *1959*
STOBER, Q. J., 1835, *1959*
STOLPE, N. E., 1877, 1880, 1882, 1901, *1939*
STOLZENBACH, K. D., 1855, 1856, *1926*
STOPFORD, S., 1711, *1766*
STORM, P. C., 1797, 1798, 1799, 1800, 1801, 1802, 1803, 1804, 1805, 1806, 1834, 1837, 1863, *1946*
STORRS, P. N., 1733, 1736, *1762*
STOUT, V. F., 1665, *1708*
STRAUGHAN, D., 1743, 1760, 1830, *1939*
STRAWN, K., 1810, 1811, 1812, 1814, 1836, 1850, 1857, 1875, 1881, 1894, 1895, 1898, 1899, 1900, 1901, 1903, *1930, 1932, 1937, 1941, 1946, 1948, 1953*
STROSS, R. G., 1799, 1834, *1951*
SUCHANEK, T. H., JR., 1799, 1800, 1802, 1803, 1804, *1959, 1962*
SUFFERN, J. S., 1815, 1906, 1907, 1909, *1934, 1952*
SUMARI, O., 1883, 1894, 1895, *1960*
SUNDARAM, T. R., 1788, 1789, 1790, 1791, 1859, 1910, 1934, *1959*
SUNDERLIN, J. B., 1859, 1883, *1955*
SUNDSTRØM, G., 1652, 1655, 1683, *1702, 1703, 1708*
SUTHERLAND, D. J., 1652, *1703*
SUTTERLIN, A. M., 1874, *1959*
SVANBERG, O., 1683, *1703*
SVERDRUP, H. U., 1894, 1914, 1915, *1959*
SWANN, C. L., 1921, *1948*
SWIFT, D. P., 1737, *1761*
SYLVA, D. P. DE, 1814, 1815, 1903, *1959*
SYLVESTER, J. R., 1782, 1810, 1884, 1888, 1893, 1894, *1959*
SZEKIELDA, K.-H., 1784, *1959*

TAGATZ, M. E., 1647, *1708*
TAKABATAKE, E., 1638, *1708*
TAKAHASHI, M., 1668, 1674, *1701*
TAKAHASHI, R., 1668, *1709*

TAKANO, K., 1916, 1920, 1922, 1925, *1959*
TAKESHITA, R., 1638, *1708*
TAKIZAWE, Y., 1638, *1708*
TALBOT, G. A., 1786, 1791, *1959*
TALBOT, J. W., 1786, 1946, *1959*
TALMAGE, S. S., 1782, 1810, 1862, *1934, 1959*
TANAKA, J., 1875, 1877, 1880, 1881, 1884, 1900, 1901, *1959*
TARDIFF, R. G., 1749, *1762*
TARZWELL, C. M., 1740, 1761, 1770, 1803, 1814, 1815, 1826, *1959*
TAYLOR, A. H., 1917, *1933*
TAYLOR, C. E., 1799, 1800, *1962*
TAYLOR, D., 1629, *1706*
TAYLOR, G. I., 1790, *1951*
TAYLOR, W. J., 1642, *1700*
TEARE, M., 1742, *1763*
TEAS, H. J., 1816, 1817, 1818, 1819, *1958*
TEICHMAN, J., 1634, *1708*
TENG, T. C., 1690, *1705*
TENORE, K. R., 1880, 1883, 1887, *1956*
TERJUNG, W. H., 1925, *1959*
TERRIERE, L., 1655, 1659, *1702*
THAIN, B. P., 1874, *1944*
THATCHER, T. O., 1777, 1887, 1888, 1889, 1890, *1929*
THOM, R. M., 1743, 1745, *1766*
THOMAS, R., 1656, *1707*
THOMAS, W. H., 1722, 1736, 1766
THOMSON, D. A., 1831, 1857, *1959*
THOMSON, I. M., 1654, *1705*
THOMPSON, S. J., 1878, 1880, 1885, 1889, 1890, 1893, 1894, 1896, *1943*
THORHAUG, A., 1803, 1816, 1817, 1819, 1820, 1838, 1839, 1840, *1928, 1959, 1960*
THORNTON, I., 1737, 1745, *1761*
THORPE, S. A., 1791, *1960*
THORSLUND, T. W., 1642, *1702*
THORSON, G., 1727, *1766*
TIAINEN, O. J. A., 1841, *1950*
TIETJEN, J. H., *1956*
TIEWS, K., 1782, 1840, 1873, 1899, *1960*
TILLY, L. J., 1799, *1949*
TIMMERMAN, R. W., 1846, 1902, *1960*
TINDLE, R. C., 1634, *1708*
TINSMAN, J. C., 1882, 1892, *1960*
TOBIAS, W., 1785, *1931*
TOEVER, W. VAN, 1874, 1876, *1960*
TOIVIAINEN, E., 1841, *1950*
TOKAR, J. V., 1785, 1786, *1948*
TOMIYAMA, T., 1658, *1703*
TONG, T., 1788, *1953*
TOPPING, G., 1668, *1702*
TORRISSEN, O., 1883, 1894, 1895, *1940*
TOWNSEND, A. A., 1790, *1960*
TREMBLEY, F. J., 1814, 1903, *1960*
TREZEK, G. J., 1872, *1935*

TRINGALI, C., 1638, *1695*
TRONEL-PEYROZ, J., 1879, *1928*
TSAI, C., 1787, 1788, *1947*
TULP, M. TH. M., 1645, *1708*
TULKKI, P., 1741, 1745, *1766*
TURNER, J. S., 1790, 1936, *1951*
TURNER, R. D., 1821, *1960*
TURNER, W. D., 1745, *1763*
TURVILLE, C. M. DE, 1783, 1792, 1809, *1943*
TUUNAINEN, P., 1883, 1894, 1895, *1960*, 1961

UEKITA, Y., 1899, *1944*
UHL, A. E., 1771, 1779, 1858 *1953*
UKELES, R., 1801, *1960*
ULANOWICZ, R. E., 1837, *1948*
UNESCO, 1722, 1766, 1922
UNITED NATIONS SECRETARIAT, 1841, *1960*
UOTANI, I., 1751, *1763*
URPO, K., 1668, *1703*
US ENVIRONMENTAL PROTECTION AGENCY, *see* EPA
US OFFICE OF SALINE WATER, 1857, *1961*
UTHE, J. F., 1629, *1708, 1709*

VADUS, 1923
VAILLANCOURT, G., 1785, *1961*
VALENTINE, D. W., 1754, *1766*
VANCE, J. M., 1852, *1953*
VAN DER MER, C., 1683, *1698*
VAN PETGEHAM, C., 1638, *1697*
VAN VLIET, H. P. M., 1638, *1707*
VAZ, R., 1634, 1664, 1668, *1702*
VEERARAGHAVACHARY, K., 1843, *1961*
VEITH, G. D., 1645, *1709*
VEMURI, R., 1693, *1702*
VENKATA, J., 1787, *1957*
VENTRE, E., 1853, *1962*
VERNBERG, F. J., 1893, *1961*
VIDAVER, W., 1893, *1961*
VILICIC, D., 1674, *1709*
VILLENEUVE, J.-P., 1661, *1698, 1699*
VINK, G. J., 1683, 1691, *1695, 1702*
VINK, G. L., 1683, *1695*
VIRTANEN, E., 1883, 1894, 1895, *1960, 1961*
VREELAND, V., 1648, *1709*
VUGTS, H. F., 1793, 1794, *1961*

WADA, A., 1857, 1859, *1961*
WADE, B. A., 1745, 1747, *1766*
WADE, R. J., 1662, *1703*
WADE, T. L., 1663, *1698*
WAGENKNECHT, P., 1693, *1707*
WALKER, W., 1638, 1668, *1706*
WALLACE, J., 1902, *1961*
WALLER, W. T., 1797, 1798, 1799, 1800, 1801, 1802, 1803, 1804, 1805, 1806, 1834, 1837, 1863, *1946*

WALSH, G. E., 1687, 1688, 1689, *1702*, *1709*
WALTHER, W. G., 1690, *1705*
WANDSVIK, A., 1902, *1961*
WANG, W.-C., 1917, *1939*
WARE, D. M., 1668, *1709*
WARE, R. R., 1728, 1757, *1766*
WARD, C. H., 1683, 1697, 1719, 1723, 1759, *1765*
WARDEN, R. L., 1883, 1884, 1898, 1900, 1901, 1902, *1931*
WARINNER, J. E., 1799, 1827, 1838, *1961*
WARREN, V., 1638, *1697*
WARRICK, J. W., 1776, 1853, *1950*
WATLING, L., 1746, *1766*
WATSON, J., 1822, *1929*
WAUGH, G. D., *1961*
WEATHERLY, A. H., 1878, 1898, *1961*
WEBB, N. R. C., 1782, 1947
WEBER, K., 1633, 1638, 1642, 1643, 1644, 1791, 1793, 1795, 1800, 1804, 1810, 1816, *1840*, *1841*, *1842*, *1851*
WEDEMEYER, G. A., 1749, *1766*, 1824, 1897, *1961*
WEINBERG, A. M., 1920, *1961*
WEINSTEIN, H., 1851, *1961*
WEISGERBER, I., 1653, *1701*
WELLER, E. C., 1782, 1862, *1943*
WELLINGS, S. R., 1751, *1766*
WENNEMER, J., 1810, *1941*
WESTERNHAGEN, H. VON, 1691, *1709*
WESTMAN, K., 1883, 1894, 1895, *1960*, *1961*
WESTOLL, 1917
WETHE, C., 1746, *1766*
WETHERALD, R. T., 1915, *1948*
WHEATON, E. W., 1897, *1961*
WHITE, J. C., 1808, *1958*
WHITE, R. H., 1662, *1709*
WHITEHOUSE, J. W., 1835, 1888, 1890, 1893, *1933*, *1956*, *1961*
WHITFIELD, R. J., 1878, 1880, 1885, 1889, 1890, 1893, 1894, 1896, *1943*
WICKENS, J. F., 1894, 1895, *1961*
WICKSTRO, K., 1668, *1709*
WIDDOWSON, T. B., 1743, *1765*, *1766*
WIDMER, M., 1853, *1962*
WIE, N. H. VAN, 1852, *1953*
WIENBECK, H., 1893, 1894, *1962*
WIJNANS, M., 1683, *1698*
WILCOX, H. A., 1919, 1920, 1921, *1962*
WILHELMI, J., 1711, 1745, *1766*
WILKES, F. G., 1755, 1757, 1764
WILLIAMS, G. C., 1642, 1696, 1799, 1800, *1962*
WILLIAMS, G. E., 1916, *1962*
WILLIAMS, J. M., 1787, 1788, *1962*
WILLIAMS, P. M., 1663, *1709*, 1721, 1736, *1760*
WILLIAMS, R., 1668, 1709
WILLIAMS, S. J., 1737, 1766

WILLIS, D. E., 1652, 1654, 1695, 1709
WILSEY, C. D., 1657, 1709
WILSON, A. J., JR., 1645, 1646, 1647, 1648, 1683, 1684, 1685, 1690, 1691, *1697*, *1701*, *1705*, *1707*
WILSON, B. J., 1827, *1947*
WILSON, P. D., 1645, 1648, 1691, *1701*, *1705*
WILSON, K. W., 1683, *1706*
WINKLE, W. VAN, 1782, 1809, *1962*
WINSTEAD, J. T., 1690, *1697*
WIRTH, M., 1906, *1943*
WIUFF, R., 1774, 1775, 1849, *1962*
WOFFORD, H. W., 1657, *1709*
WOHLSCHLAG, D. E., 1898, *1932*
WOLF, K. E., 1749, *1766*
WOLFE, I. N. I., 1651, *1709*
WOLKE, R. E., 1749, 1754, *1760*
WOOD, E. J. F., 1815, 1816, 1817, 1818, 1819, 1821, 1822, *1962*, *1963*
WOODARD, K., 1852, *1946*
WOODHEAD, P. M. J., 1806, 1807, 1849, 1862, 1863, 1864, 1865, 1866, 1867, 1868, 1869, 1872, 1913, 1914, *1957*
WORD, J. Q., 1719, 1720, 1721, 1729, 1730, 1731, 1744, 1745, 1763, 1766
WORLD METEOROLOGICAL ORGANIZATION, 1915, *1962*
WYNESS, A. W., 1783, *1962*
WU, A., 1658, *1700*
WU, J., 1791, 1859, *1959*
WURSTER, C. F., 1686, 1687, 1690, *1699*, *1705*, *1706*, *1709*

YAMAGISHI, T., 1664, 1668, *1705*, *1709*
YAMAZAKI, M., 1772, *1962*
YANEZ, L. A., 1745, *1761*
YAROSH, M., 1782, 1847, 1874, 1904, 1905, 1907, 1909, *1962*
YEE, W. C., 1886, 1901, *1962*
YEVICH, P. P., 1797, 1803, 1804, 1805, 1809, 1816, 1819, 1821, 1825, 1826, 1827, 1834, *1938*
YIH, C.-S., 1859, *1962*
YOUNG, D. K., 1714, 1715, 1722, 1726, 1749, 1753, 1754, 1755, *1766*
YOUNG, D. L. K., 1739, *1767*
YOUNG, D. R., 1650, 1664, 1665, 1672, 1673, 1708, 1709, 1738, 1739, *1767*
YOUNG, J., 1799, 1800, *1962*
YOUNG, J. S., 1767, 1796, 1802, 1806, 1813, 1836, 1838, 1839, *1962*
YOUNG, R. A., 1737, *1761*
YOUNGS, W. D., 1665, *1696*, *1709*
YOUNGBLUTH, M. J., 1802, 1803, 1804, 1809, *1963*

ZANICCHI, G., 1668, *1697*

ZECHELLA, A. P., 1860, *1963*

ZEITOUN, I. H., 1831, 1833, 1834, 1857, *1963*

ZEITOUN, M. A., 1888, 1889, *1963*

ZELL, M., 1638, 1655, 1664, 1668, 1696, *1702, 1709*

ZEMAN, O., *1947*

ZEPP, R. G., 1651, *1709*

ZIEBARTH, U., 1661, 1663, *1708*

ZIEMAN, J. C., 1816, 1817, 1818, 1819, 1822, *1962, 1963*

ZIMMERMAN, J. F. T., 1793, 1794, *1961*

ZITTEL, H. E., 1888, 1889, *1949*

ZION, H. H., 1804, 1903, 1944, *1951*

ZIRGMARH, R., 1743, 1761

ZISKOWSKI, J., 1742, *1763, 1765, 1767*

ZITKO, V., 1634, 1638, 1648, 1668, 1683, *1702, 1704, 1709*

ZUBARIK, L., 1797, 1798, 1799, 1800, 1801, 1802, 1803, 1804, 1805, 1806, 1834, 1837, 1863, *1946*

TAXONOMIC INDEX

abalone, 1749, 1754, 1901
Acartia clausi, 1804, 1805, 1809
A. grani, 1809
A. tonsa, 1804, 1809
Achromobacter sp., 1799
Acrobicularia plana, 1741
actinian, 1670
Acanthogobius flavimanus, 1671
Aeromonas, 1751, 1799, 1896
Agonus cataphractus, 1677–1679, 1683
Alburnus alburnus, 1677, 1682, 1683, 1703
Alcaligenes sp., 1799
algae, 1645, 1650 1654, 1662, 1685, 1698, 1702,
    1705, 1709, 1722, 1743, 1747, 1749, 1760,
    1762, 1764, 1816
algae, blue-green, 1704, 1706, 1808, 1816, 1817,
    1832, 1838, 1839, 1927
algae, brown, 1743, 1816, 1817
algae, green, 1816, 1832, 1839
algae, macro, 1819
algae, red, 1704, 1816, 1822, 1832
Amphidinium carteri, 1687
Amphiodia sp., 1718
Amphipoda (amphipods), 1729, 1745, 1747,
    1759, 1821, 1822–1824, 1826, 1828
Amphiura chiajei, 1742
A. filiformis, 1742
Anchoa mitchilli, 1836
anchovy, 1757, 1813, 1836
anchovy, Northern, 1752
Anguilla anguilla, 1671, 1836, 1879, 1881, 1889,
    1890, 1892, 1893, 1896, 1897
A. japonica, 1879, 1881, 1896
A. rostrata, 1677–1680
annelids, 1700, 1764, 1765
Anoplopoma fimbria, 1752
anthomedusa, 1830
Aphrodite aculeata, 1643
Arenicola marina, 1823
Arius felis, 1810–1812, 1814
armed bullhead, 1677–1679, 1683
Ascidia sp., 1821, 1829
A. aspersa, 1821
A. nigra, 1832, 1833
Asellopsis intermedia, 1822–1824
Asparagopsis armate, 1662, 1704
Atherinopsis californiensis, 1903
Atlantic croaker, 1928
Atlantic menhaden, 1807

Atlantic silverside, 1677–1680

bacteria, 1656, 1704, 1706, 1725, 1751, 1754,
    1798, 1799, 1880
Balanus amphitrite, 1828, 1829
B. crenatus, 1829
B. improvisus, 1824
Bankia gouldi, 1820, 1828
barnacles, 1809, 1820, 1821, 1824, 1828, 1829,
    1832, 1889
bass, 1892
bass, kelp, 1664
bass, striped, 1664, 1811, 1838
Bellerochia sp., 1686
bigmouth sole, 1752
birds, 1757
Bivalvia (bivalves), 1642, 1650, 1658, 1760,
    1804, 1805, 1820, 1824, 1829, 1830, 1835,
    1863, 1883, 1889, 1902, 1927, 1928
black perch, 1664
bleak, 1677, 1682, 1683
bluegills, 1684
bluehead, 1677–1680
bocaccio, 1755
Bossiella orgigniana, 1749
Bothriocephalus gowkongenis, 1815
Botryllus schlosseri, 1821
Botrytis sp., 1908
Bowerbankia gracillima, 1829
Brachynotus sexdentatus, 1828
Brevoortia patronus, 1810, 1812
B. tyrannus, 1807, 1812, 1836
Bryozoa, 1829
Bugula neritina, 1821, 1829
B. stolonifera, 1821
Bursatella sp., 1832

Callinects sapidus, 1671, 1681, 1707, 1826,
    1829
Capitella sp., 1741, 1747, 1761
C. capitata, 1711, 1742, 1745, 1747, 1757
Caprella actifrons, 1747
Carassius auratus, 1700
Carcinus maenas, 1678, 1681, 1682, 1823, 1829
Cardita sp., 1718
Cardium corbis, 1805, 1825
C. edule, 1677, 1680–1683, 1764
C. lamarcki, 1742
carp, 1876

catfish, 1671, 1696, 1702, 1851, 1876, 1878, 1884, 1899, 1927
Cerastoderma edule, 1666, 1669, 1741
C. glaucum, 1829
Cerianthus sp., 1747
Cestoda (cestodes), 1725
Chilara taylori, 1752
Chironomus sp., 1741
C. plumosus, 1742, 1828
Chlamydomonas sp., 1829, 1801
Chlorococcum sp., 1829–1831
chlorophyta, 1724
Choleia sp., 1718
Chrysophrys major, 1781, 1782
Ciona intestinalis, 1821, 1829
Citharichthys sp., 1755
C. fragilis, 1752
C. sordidus, 1752
Cladonema sp., 1830
clams, 1642, 1646–1648, 1704, 1706, 1823, 1825, 1884, 1927
clams, coot, 1804
clams, hard-shell, 1804
clams, soft-shell, 1754, 1760
Clupea harengus, 1667, 1669–1671, 1837
C. sprattus, 1672
Coccolithus huxleyi, 1687
cockle, 1677, 1680–1683, 1805
cod, 1642, 1665, 1670–1672, 1702, 1705, 1707, 1708, 1836, 1876
    North Sea, 1642, 1643
combfish, longspine, 1752
combfish, shortspine, 1752
Copepoda (copepods), 1722, 1804, 1805, 1809, 1822, 1824, 1834, 1926
Corallina officinalis, 1749
Corophium sp., 1742, 1747
C. acherusicum, 1821
C. volutator, 1742
Corynebacterium sp., 1896
Coryphaena equiselis, 1668
C. hippurus, 1668
crab, 1645, 1754, 1767, 1826, 1828, 1829, 1832, 1836
crab, blue, 1671, 1681, 1690, 1706, 1707, 1835, 1878
crab, fiddler, 1706
crab, mud, 1676, 1703
crab, shore, 1678, 1681, 1682, 1823
crab, stone, 1690
Crangon crangon, 1669, 1677–1683, 1754, 1765
C. franciscorum, 1678, 1703
C. septemspinosa, 1682
Crassostrea angulata, 1826
C. gigas, 1882
C. virginica, 1648, 1677, 1678, 1681, 1703, 1804, 1823–1825, 1831, 1882, 1883, 1892

crayfish, 1884
Crepidula fornicata, 1827
croaker, Atlantic, 1928
croaker, white, 1752, 1757, 1766
Crustacea (crustaceans), 1666, 1669, 1676, 1698, 1722, 1747, 1761, 1802, 1881, 1882
curlfin sole, 1752
cusk-eel, spotted, 1752
cyanophyta, 1724
Cyclotella nana, 1686, 1687
Cymatogaster aggregata, 1752
Cynoscion nothus, 1671
Cyprinodon macularius, 1814, 1878
C. variegatus, 1647, 1678–1680, 1683–1685, 1690, 1701, 1707
Cystophora cristata, 1668
Cystoseira stricta, 1743

dab, 1642, 1682
dab, sand, 1755
Decapoda (decapods), 1681, 1749, 1755
desert pupfish, 1878
diatoms, 1702, 1704, 1705, 1800, 1816, 1838, 1839
Dicentrarchus labrax, 1889, 1892
dinoflagellates, 1764, 1808
Diplosoma listerianum, 1821
Dissostichus eleginoides, 1670, 1671
dogfish shark, 1655, 1701
dog whelk, 1793
Donax semistriatus, 1927
D. trunculus, 1927
D. vittatus, 1927
Dorvillea rudolphi, 1729
Dover sole, 1664, 1739, 1750–1752, 1927
Dunaliella sp., 1688
D. euchlora, 1801
D. tertiolecta, 1686–1689

Echinodermata (echinoderms), 1722, 1832
eels, 1671, 1836, 1874, 1891, 1892, 1893, 1896, 1927, 1928
eelpout, blackbelly, 1752
eels, American, 1677–1680
Eggerella advena, 1747
Elminius modestus, 1809, 1820, 1829, 1834
Elphidium clavatum, 1747
E. incertum, 1747
Embiotoca jacksoni, 1664
English sole, 1752, 1766
Engraulis mordax, 1752, 1757, 1813
Entamoeba histolytica, 1725
Enteromorpha sp., 1816, 1817, 1819, 1832
Eopsetta jordani, 1752
Erignatus barbatus, 1668, 1669
Euilyodrilus hammoniensis, 1742
Euphausiacea (euphausiids), 1645, 1666

Euterpina acutifrons, 1809
Exuviella baltica, 1687, 1690

Fantail sole, 1752
fishes, 1633, 1644–1648, 1650, 1652–1655, 1657, 1658, 1664, 1665, 1667, 1669–1671, 1675, 1676, 1694, 1695, 1698, 1700–1702, 1707, 1708, 1722, 1726, 1749, 1754, 1757, 1764, 1765, 1773, 1798, 1806, 1807, 1809, 1810, 1816, 1820, 1832–1844, 1836, 1848–1850, 1855, 1856, 1863, 1866, 1867, 1874, 1878, 1892, 1899, 1902, 1903, 1912, 1913, 1926–1928
flagfish, 1684, 1685
flatfish, 1642, 1643, 1789
flatworms, 1722, 1815, 1826, 1828
flounder, 1670, 1681, 1751, 1807, 1814
flounder, Baltic, 1691, 1709
flounder, starry, 1766
flounder, winter 1767, 1812, 1839
foraminiferans, 1747
Fragilania pinnata, 1686
Fundulus heteroclitus, 1675, 1677–1680
F. majalis, 1677–1680
F. similis, 1670, 1677–1681
fungi, 1704, 1706

Gadus morhua, 1643, 1667, 1669–1672, 1702, 1854
Gambusia affinis, 1658
gametophytes, 1817
gammarids, 1804
Gammarus sp., 1805
G. locusta, 1829
Gastropoda (gastropods), 1703, 1729, 1762, 1822, 1832
gastrotrichs, 1722
Genyonemus lineatus, 1752, 1757, 1766
Glugea stephani, 1896
Glyptocephalus cynoglossus, 1644
G. zachirus, 1751–1753
Gobius minutus, 1681
goby fish, 1671, 1681, 1705
goldfish, 1700
Gonyaulax sp., 1736
G. polyedra, 1802, 1808
gorgonians, 1832
guillemot, 1702, 1708
gulf menhaden, 1810

haddock, 1702
Halichoerus gryphus, 1671
Harpacticoids, 1677, 1681, 1683, 1703, 1830
herring, 1670, 1671, 1702, 1876
herring, Atlantic, 1837
Hippoglossina stomata, 1752
Homarus americanus, 1653, 1680, 1696, 1698, 1699, 1704, 1761

hydroids, 1826, 1828
Hydroides norvegica, 1821
hydrozoans, 1821
Hydrurga leptonyx, 1668
Hyperprosopon argenteum, 1903

Ictalurus punctatus, 1696, 1851, 1878, 1881, 1884, 1897
invertebrates, 1657, 1664, 1740, 1749, 1754, 1862, 1863
invertebrates, macro, 1866
Isochrysis galbana, 1801
Isopoda (isopods), 1741, 1821, 1828, 1829

jacksmelt, 1893
Jassa falcata, 1747
Jordanella floridae, 1684, 1685

kamptozons, 1828
kelp, 1662, 1666, 1722, 1839
kelp, giant, 1817
kelp bass, 1664
killifish, 1681
killifish, longnose, 1670, 1677–1679
killifish, striped, 1677–1680

Labyrinthomyxa marina, 1824
Lagodon rhomboides, 1646–1648, 1677, 1680, 1682, 1690
Laminaria digitata, 1662
Lanice conchilega, 1644, 1647, 1648, 1670
Leiostomus xanthurus, 1648, 1677–1681, 1690, 1806
Lepomis macrochirus, 1684, 1685
Limanda limanda, 1643, 1648, 1682
Limnoria quadripunctata, 1821, 1829
L. tripunctata, 1821, 1828, 1829
limpets, 1832
Listriolobus sp., 1718
L. pelodes. 1745, 1746
Lithothrix aspergillum, 1749
Littorina planaxis, 1830
L. scutulata, 1830
lobsters, 1653, 1680, 1696, 1704, 1767, 1883
lobsters, American, 1653, 1680, 1696, 1698, 1699, 1704, 1754, 1761
Loligo forbesi, 1666
Loxosomella kefersteinii, 1828
lugworm, 1823
Luidia clathrata, 1671
Lycodopsis pacifica, 1752
Lyopsetta exilis, 1752
Lyrodus pedicellatus, 1821

Macoma sp., 1747
M. balthica, 1740, 1742, 1823
Macrobrachium rosenbergii, 1882, 1884, 1891

Macrocystis pyrifera, 1817, 1839, 1921
Macrophyta (macrophytes), 1724, 1743
Macropipus puber, 1829
mammals, 1668, 1669, 1692
mammals, Arctic, 1697
Meganyctiphanes norvegica, 1645
Melanogrammus aeglefinus, 1702
Melosira sp., 1761
menhaden, 1811, 1812, 1836, 1838
menhaden, Atlantic, 1807, 1812
menhaden, gulf, 1810, 1812
menhaden, winter, 1813
Menidia menidia, 1677–1680
Menippe mercenaria, 1832
Mercenaria mercenaria, 1646, 1647, 1804, 1820,
    1822, 1823, 1825, 1828, 1829, 1835, 1839,
    1882
Mesidotea entomon, 1742
metazoans, 1762
microfauna, 1722, 1724
Microgadus tomcod, 1806
microplankton, 1759
Micropogon undulatus, 1646, 1881, 1928
Microstomus pacificus, 1664, 1750–1753
milkfish, 1876
minnow, 1642, 1656, 1691, 1742, 1748
minnow, fathead, 1684, 1685, 1702, 1747, 1749
minnow, sheepshead, 1678–1680, 1683, 1690,
    1701, 1709
molluscs, 1653, 1665, 1666, 1669, 1676, 1722,
    1747, 1760, 1825, 1835, 1881, 1889
Morone saxatilis, 1806
Mugil sp., 1810
M. cephalus, 1677–1680, 1810, 1811, 1836, 1881,
    1889, 1898
M. curema, 1677–1681
Mulinia lateralis, 1804
mullet, 1884
mullet, grey, 1810
mullet, striped, 1677–1680, 1810, 1811, 1836,
    1898
mullet, white, 1677–1681
mummichog, 1677–1680
mussels, 1642, 1644–1649, 1670, 1671, 1681,
    1682, 1683, 1698, 1736, 1743, 1820, 1821,
    1826, 1832, 1889
Mya sp., 1747
M. arenaria, 1646–1648, 1666, 1669, 1741, 1742,
    1754, 1760, 1823, 1826
Mylio macrocephalus, 1881, 1882
mysids, 1804, 1862
Mysidopsis bahia, 1683, 1684
Mytilus sp., 1747
M. edulis, 1641, 1642, 1644, 1646, 1647, 1648,
    1666, 1669–1671, 1681–1683, 1699, 1740,
    1824–1826, 1830, 1882
M. galloprovincialis, 1666, 1669

Nassarius obsoletus, 1824
N. reticulatus, 1822, 1823
Navicula seminulum, 1800
Neanthes sp., 1747
N. arenaceodentata, 1748, 1749, 1763
Nebalia pugettensis, 1745
Nematoda (nematodes), 1722, 1725
Neomysis americana, 1804, 1811
N. awatchensis, 1862
Neopanope texanasayi, 1828, 1829
Nephrops norvegicus, 1666, 1669
Nephtys incisa, 1742
Nereis diversicolor, 1698, 1699
N. virens, 1641, 1647, 1653, 1658, 1690, 1699
Nitocra spinipes, 1677, 1683, 1703
Nitzschia sp., 1688, 1689
N delicatissima, 1686, 1704
Nordotis discus hannai, 1881, 1882
Nucula nitidu, 1742

Oligochaeta (oligochaetes), 1711
Oncorhynchus sp., 1883
O. kisutch, 1880
O. nerka, 1878
O. tshawytscha, 1815
Ophiuroidea (Ophiuroides), 1642, 1718, 1741
Ophryotrocha diadema, 1681, 1682, 1691, 1702
Ostracoda (ostracods), 1729
Ostrea edulis, 1805, 1825–1827, 1834, 1882
O. lurida, 1805, 1820, 1825
oyster, 1646, 1647, 1708, 1709, 1736, 1804,
    1805, 1820, 1824, 1826, 1833, 1834, 1883,
    1891, 1892, 1893
oyster, American, 1677, 1678, 1703, 1823
oyster, Eastern, 1681

Pachygrapsus crassipes, 1830
Palaemon elegans, 1681
Palaemonetes pugio, 1677–1680, 1682
Pandalus borealis, 1666, 1669
P. jordani, 1666
P. montagni, 1681, 1682
Panulirus argus, 1883
Paralabrax clathratus, 1664
P. nebulifer, 1751, 1752, 1754
Paralichthys sp., 1807, 1836
Parophyrys vetulus, 1752, 1766
Pectinaria sp., 1718
Pelecypoda (pelecypods), 1718, 1729, 1740,
    1741, 1749
pelican, 1655
pelican, brown, 1755
Peloscolex sp., 1747
Penaeus aztecus, 1670, 1677, 1681
P. duorarum, 1647, 1678, 1679, 1681
P. japonicus, 1881, 1882, 1884

P. kerathurus, 1882, 1889, 1892
P. setiferus, 1682
perch, pile, 1902
perch, shiner, 1752
perch, walleye surf, 1902
Peridinium trochoideum, 1687
Petrale sole, 1752
Phaecodactylum tricornutum, 1689
Phaeophyta, 1724
Phanerodon furcatus, 1752
Phoca hispida, 1668, 1669
P. vitulina, 1668, 1669, 1698
Phoxinus phoxinus, 1691, 1696
phytoplankton, 1686, 1687, 1689, 1696, 1697,
    1699, 1701, 1704–1706, 1709, 1721, 1722,
    1736, 1760, 1766, 1784, 1798–1802, 1808,
    1837, 1838, 1846, 1882
Pilumnus hirtellus, 1828
Pimephales promelas, 1642, 1684, 1685, 1702,
    1747, 1748
pinfish, 1677, 1680, 1682
plaice, 1677, 1679–1683, 1814, 1835, 1878,
    1893, 1896, 1927
plainfin midshipman, 1752
plants, 1662
Platichthys flesus, 1682, 1691, 1699, 1709, 1814
P. stellatus, 1766
Platynereis dumerili, 1729
Plecoglossus alteivelis, 1882
Pleuronectes platessa, 1648, 1677, 1679–1683,
    1814, 1835, 1878, 1889, 1893, 1896, 1927
Pleuronichthys decurrens, 1752
P. verticalis, 1752
poacher, blacktip, 1752
Podocerus variegatus, 1747
Poecilia latipinna, 1694, 1703
Pogonais cromis, 1881
Polychaeta (polychaetes), 1642–1645, 1647,
    1648, 1649, 1653, 1654, 1670, 1690, 1691,
    1699, 1700, 1702, 1711, 1718, 1722, 1729,
    1741, 1743, 1747, 1749, 1760, 1761, 1828
Polydora sp., 1747
P. ciliata, 1742
P. ligni, 1824
P. websteri, 1824
Polynoidae, 1643
Polyphysia crassa, 1643
polyzoans, 1828, 1829
Pontoporeia affinis, 1742, 1828
Porichthys notatus, 1752
Porphyridium sp., 1687
Porphyridium cruentum, 1688
Potamogeton perfoliatus, 1819
Potamothrix hammoniensis, 1828
prawns, 1881
Prorocentrum sp., 1736
P. micans, 1764

protistans, 1762
Protozoa (protozoans), 1703, 1725, 1762
Pseudomonas sp., 1656, 1751
Pseudopleuronectes americanus, 1812
Pyranimonas sp., 1687

Rex sole, 1751, 1752
Rhabdosargus sarba, 1881
Rhacochilus vacca, 1903
Rhithropanopeus harrisii, 1703
Rhodophyta, 1724
rockfish, calico, 1752
rockfish, flag, 1752
rockfish, greenblotched, 1752
rockfish, greenstriped, 1751, 1752
rockfish, halfbanded, 1752
rockfish, shortbelly, 1752
rockfish, vermilion, 1752
Ruppia maritima, 1819

sablefish, 1752
sailfin molly, 1684, 1703
Salmo gairdnerio, 1644, 1695, 1705, 1708, 1815,
    1863, 1864, 1878, 1880, 1881, 1884, 1891,
    1900
Salmo salar, 1696, 1882, 1894
salmon, 1811, 1838, 1895, 1926
salmon, Atlantic, 1696, 1874
salmon, Baltic, 1894, 1895
salmon, chinook, 1815
salmon, coho, 1700
salmonids, 1893, 1896
Salvelinus fontinalis, 1684
sand bass, barred, 1751, 1752, 1754
sand dab, Gulf, 1752
sand dab, Pacific, 1752
Sargassum sp., 1666
Sargatia troglodytes, 1670
Sarotherodon niloticus, 1892, 1893
Scenedesmus acutus, 1834
scallops, 1755
Scolelepis sp., 1747
S. fuliginosa, 1742
Scophthalmus maximus, 1878–1880, 1893
Scorpaena guttato, 1755
scorpionfish, 1755
sea anemones, 1747, 1832
sea bream, 1882, 1892
sea catfish, 1810–1812, 1814
seaperch, pink, 1752
seaperch, white, 1752
sea slug, 1832
sea trout, 1811, 1926
sea urchins, 1748, 1763, 1832
seals, 1652, 1654, 1664, 1702
seal, Arctic ringed, 1695

seal grey, 1671
seal, harbour, 1698
seal, wild, 1708
sea star snail, 1657
seaweeds, 1662, 1883, 1884, 1921
Sebastes dalli, 1753
S. elongatus, 1751, 1752, 1753
S. jordani, 1752
S. miniatus, 1752
S. paucipinis, 1755
S. rosenblatti, 1752
S. rubrivinctus, 1752
S. semincinctus, 1752
Sebastolobus alascanus, 1752
Seriola quinqueradiata, 1880, 1881, 1899
shark, dogfish, 1655, 1701
sheepshead minnow, 1678–1680, 1683, 1685,
   1690, 1701, 1707, 1709
shellfish, 1694, 1703, 1707, 1709, 1711, 1725,
   1749, 1763, 1765, 1884
shipworms, 1820, 1821, 1828
shrimp, 1645, 1647, 1671, 1677, 1682, 1754,
   1760, 1884, 1891, 1892
shrimp, brown, 1670, 1677–1683, 1706, 1709
shrimp, grass, 1677–1681, 1703
shrimp, mysid, 1683
shrimp, opossum, 1684, 1685, 1811
shrimp, pink, 1678, 1679, 1681, 1682, 1706
Skeletonema costatum, 1687–1689, 1801
slender sole, 1752
sole, 1814, 1835, 1878, 1879, 1885, 1890
sole, bigmouth, 1752
sole, curlfin, 1752
sole, Dover, 1664, 1739, 1750, 1751, 1752,
   1927
sole, English, 1752, 1766
sole, Petrale, 1752
sole, Rex, 1751, 1752
sole, slender, 1752
Solea solea, 1699, 1814, 1835, 1878, 1879, 1882,
   1889, 1927
Spermatophyta, 1724
spot, 1677–1681
sprat, 1662, 1672
Squalus acanthias, 1655
Squilla empusa, 1666, 1671
starfish, 1671
striped bass, 1806
Strombus gigas, 1883
Strongylocentrotus purpuratus, 1748, 1763, 1832
Symphurus atricauda, 1752
Syndosmya alba, 1742
S. filiformis, 1742

Tapes philippinarum, 1703
teleosts, 1766

Tellina sp., 1718
T. fabula, 1927
T. tenuis, 1741, 1805, 1824, 1927
Teredo bartschi, 1828
T. furcifera, 1828
T. navalis, 1821, 1828
Thais lapillus, 1793
Thalassia testudinum, 1816–1819, 1836, 1839,
   1840
Thalassiosira fluviabilis, 1687, 1689
T. pseudonana, 1686, 1688, 1690
Thalassoma bifasciatum, 1677–1680
Theostoma oerstedi, 1729
thornyhead, shortspine, 1752
Thunnus alalunga, 1668
T. thynnus, 1667
Tilapia sp., 1874, 1875, 1892, 1893, 1899
Tisbe sp., 1830
T. holothuriae, 1831
tomcod, 1806
tonguefish, California, 1752
Trematoda (trematodes), 1725
tube worms, 1821, 1832
Tubifex sp., 1711
Trochammina pacifica, 1747
trout, 1702, 1709, 1876
trout, brook, 1684, 1685
trout, Cayuga lake, 1696
trout, rainbow, 1653, 1675, 1691, 1695, 1696,
   1700, 1704, 1705, 1708, 1815, 1863, 1876,
   1878, 1880, 1884, 1891, 1900
trout, sand, 1671
turbot, 1878, 1885, 1890, 1894
turbot, hornyhead, 1752
turtle grass, 1816–1819, 1828, 1836, 1839

Ulva sp., 1816
Urothöe brevicornis, 1822–1824

Venerupis pollustra, 1741
Venus mercenaria, 1927
Vibrio sp., 1656, 1751, 1799, 1896
viruses, 1725

white croaker, 1752, 1757, 1766
wildfowl, 1851
worms, 1676, 1681, 1682, 1691, 1698,
   1815
worms, bristle, 1764
worms, echiuroid, 1745
worms, flat, 1722, 1815, 1826, 1828
worms, lug, 1823
worms, sabellid, 1833
worms, ship, 1820, 1821, 1828
worms, tube, 1821, 1832

Xeneretmus latifrons, 1752
Xystreurys liolepis, 1752

yellowtail, 1880, 1899

Zalembius rosaceous, 1752

Zaniolepis frenata, 1752
Z. *latipinnis*, 1752
zooplankton. 1645, 1666, 1686, 1698, 1721, 1722, 1736, 1765, 1766, 1798, 1802, 1803, 1805, 1811, 1815, 1834, 1836–1838, 1866, 1867

SUBJECT INDEX

Acute toxicity, 1677
Aeration, 1894
AF—*see* Application factor
Aldrin, 1646, 1679
Aliphatic chlorinated hydrocarbons, 1663,
    1672, 1682
Alkyl phenols, 1682
Ametryne, 1688
Analysis,
    of biota, 1632
    of organic pollutants,
        methods, 1638
    of sediment, 1632
    of water, 1632
Animal shelters, 1909
Anti-fouling agents, 1797
Application factor, 1683, 1685
    no-effect concentration, 1685
Aquaculture,
    extensive, 1873, 1874, 1899
    heat utilization, 1873, 1902
    in greenhouses, 1901
    in ponds, 1900, 1901
    in raceways, 1900, 1901
    in tanks, 1900, 1901
    intensive, 1874, 1891
    use of heated effluents, 1874–1885
Aroclor, 1677, 1685, 1689
Atrazin, 1681, 1688

Bacteria, 1798, 1799
Bacterial kidney disease, 1896, 1897
BCFs, 1640, 1643, 1645, 1646, 1650,
    1651
Benthic samplers,
    effectiveness, 1720
Bioconcentration,
    organic pollutants, 1639, 1640
Bioconcentration factors—*see* BCFs
    organic pollutants, 1640, 1643
Biodegradability, 1653
Biological fouling, 1888
Biological surveys,
    of domestic wastes, 1719
Biomagnification, 1642
    organic pollutants, 1639, 1640
Biphenyl, 1674
Black lists, 1619

Blow-down, 1850
BOD, 1713, 1715, 1798

Calefaction, 1770
Carbaryl, 1685
Carbophenothion, 1688
Chlophen, 1677
Chloramines, 1888
Chlordane, 1647, 1680, 1687
Chlorinated hydrocarbons, 1672, 1716
Chlorinated naphthalenes, 1689
Chlorinated pesticides, 1661
Chlorination, 1778, 1797, 1834, 1835,
        1888–1890, 1897, 1898, 1900, 1902
Chlorine tolerances, 1889
Chloroform, 1672
Chlorophenols, 1662
Clean-up,
    organic pollutants, 1634
Cold effluents, 1771, 1813, 1830, 1872, 1911,
    1912
Cold shock, 1813, 1839, 1871, 1872
Combined Heat and Power Generation (CHP),
    1843–1847
Committee on Entrainment, 1868
Condensers, 1773
Cooling,
    closed circuit, 1776, 1777 1858, 1914
    in canals, 1850–1852, 1858
    in lakes, 1850–1852, 1899, 1913
    in ponds, 1850–1852, 1858, 1899, 1913
    in reservoirs, 1850–1852, 1913
    in towers, 1852–1854, 1858, 1902, 1907
    open circuit, 1776, 1777, 1858, 1898, 1902,
        1912, 1913
    open cycle, 1776
    screens, 1777
    with ammonia, 1853
Cooling systems, 1773, 1777
Cooling water, 1771–1775
Criteria for mitigating effects of temperature,
    1860–1872
Critical thermal maximum temperature, 1862

DDD, 1646, 1661, 1668
DDE, 1661, 1668, 1674, 1687, 1690
DDT, 1620, 1621, 1642, 1643, 1645, 1646,
        1650, 1653, 1654, 1656, 1658, 1661, 1666,

DDT—*continued*
    1668, 1672–1674, 1677, 1686, 1687, 1690,
    1693, 1715, 1738, 1755
DDVP, 1676
Decachlorophenyl, 1637
Degradation,
    organic pollutants, 1651, 1652
    in animals, 1652
    in DDT, 1652
    in micro-organisms, 1656–1659
    in PCBs, 1652, 1653
    pathways, 1651
    phthalate esters, 1657
Depuration,
    organic pollutants, 1641
Desalination, 1779, 1780, 1910, 1911
    effluents, 1831–1833
    plant, 1859, 1860, 1910
        effluents, 1779, 1780
    techniques, 1847, 1848
Diazinon, 1685
Dichlobenil, 1682
Dichlorethane, 1689
Dichloroaniline, 1691
Dichlorobiphenyl, 1636, 1689
Dichlorophenols, 1674
Dieldrin, 1629, 1631, 1669, 1674, 1677, 1687,
    1691
Diethyl phthalate, 1674
Dibutyl phthalate, 1674
Dimethoate, 1681
Diquat, 1681, 1688, 1689
Disease, 1896, 1897
Discharge effects
    fish, 1809–1811
    phytoplankton, 1808, 1809
    zooplankton, 1809
Diseases,
    domestic wastes, 1749–1754
    due to thermal effluents 1895–1898
    fin erosion, 1813
    fish, 1751
    gas embolism, 1813
Domestic wastes, 1619–1625, 1711–1767
    biological surveys, 1719
    chemical analysis, 1715, 1716
    diseases, 1749–1754
    dispersion, 1730–1733
    effects on organisms, 1747
    fate, 1730–1733
    identification, 1715
    interaction with sediments, 1736
    monitoring, 1719
    quantification, 1715
Dry cooling towers, 1852–1854, 1914

Ecosystem effects, 1837–1840

Effects on living systems, 1675
Effluents,
    cold, 1771, 1796, 1872
    desalination plant, 1779, 1796, 1857
    heated 1771–1773, 1795, 1874–1910
    sewage, 1857
Effluent temperatures, 1849
Electricity generation, 1773
Elimination,
    organic pollutants, 1640, 1641
Endosulfan, 1647, 1679, 1685
Endrin, 1646, 1647, 1679, 1685, 1687
Energy parks, 1775, 1860
Energy use, 1847
    improvement, 1842–1848
Entrainment, 1796, 1802–1808, 1848, 1849,
        1866, 1867, 1912
    effects, 1797, 1798
EPN, 1688
Ethyl parathion, 1680
Eutrophication, 1736
Extraction,
    organic pollutants, 1633

Factors affecting toxicity, 1675
Fast-breeder reactors, 1913
Fenitrothion, 1648
Fentin acetate, 1681
Fin erosion, 1739, 1750–1754
Fish, 1806, 1807, 1809
Fish farming,
    gross energy requirements, 1875–1877
Fish kills, 1812, 1839
Float, 1899
Food conversion ratios, 1878

Gas embolism, 1813
Gas liquefaction,
    natural, 1779
Gas supersaturation, 1893, 1894
General adaptation syndrome, 1897
Geological analysis,
    of domestic wastes, 1717
Geothermal heat utilization, 1874
Global cooling, 1916, 1917
Global warming, 1916–1921, 1925
Greenhouses, 1901, 1902, 1915, 1916
    heating, 1904, 1906–1909
Greening,
    in oysters, 1892
Grey list, 1619
Gross energy requirements (GER),
    in fish farming, 1875–1877
Growth enhancement, 1878–1884

Halogenated aliphatic hydrocarbons, 1631,
    1638

Halogenated aromatics, 1638
Halogenated ethers, 1630
Halogenated pesticides, 1638
Halogenated phenols, 1638
α-HCH, 1647, 1669
γ-HCH, 1647, 1661, 1669, 1678
Heat dissipation in soil, 1906
Heat shock, 1826, 1827
Heat sinks, 1921
Heated effluents, 1771–1773
Heavy metals, 1737, 1739
    in domestic wastes, 1737–1739
Heptachlor, 1647, 1674, 1679, 1685
Heptachlorepoxide, 1647
Heptachlorobiphenyls, 1636
Heptachlorostyrene, 1672
Hexachlorobenzene, 1670, 1672, 1674
Hexachlorobiphenyls, 1636
Hexachlorobutadiene, 1648, 1672
Hydrocarbons, 1631, 1714

Identification,
    organic pollutants, 1634, 1635
Impingement, 1848, 1849, 1912
Impingement effects, 1836
Indicator organisms, 1746, 1747
Infrared scanning techniques, 1782, 1784, 1785
Infrared surveys, 1785
Infrared thermometer, 1785
Instrumentation,
    measurement of temperature changes, 1782

Kelthane, 1678
Kepone, 1647, 1681, 1688

LC50, 1676
Lethal effects,
    cold shock, 1810, 1839
    heat shock, 1811–1813, 1826
Limits of detection,
    organic pollutants, 1635, 1638
Lindane, 1685
Long-term toxicity, 1676

Magnetohydrodynamic generation—see MHD
Malathion, 1680, 1685
Management,
    emphasis of policies, 1841, 1842
    industrial energy use, 1842–1848
    of waste heat impacts, 1840–1872
        combined heat and power generation,
            1843–1847
    of waste heat uses, 1872–1911
        in agriculture, 1904–1910
        in aquaculture, 1873–1902
        in sport fishing, 1902–1904
        other uses, 1910, 1911

of waste disposal, 1756–1758
power station efficiencies, 1842, 1843, 1888
Mariculture—see aquaculture
MATC—see Maximum acceptable toxicant
    concentration
Maximum acceptable toxicant concentration,
    1684, 1685
Methidathion, 1680
Methyl parathion, 1680
Metabolites of PCBs and DDT, 1652
Methods,
    analysis of organic pollutants, 1638
Methoxychlor, 1647, 1678
Methylparathion, 1676
MFO—see Mixed function oxidasis
MHD, 1842, 1843
Minamata disease, 1755
Mirex, 1647, 1681, 1687, 1688
Mixed function oxidasis (MFO), 1690
Modelling,
    effluent analysis, 1787
Monitoring,
    chemical dyes, 1786
    floats, 1786
    hydraulic models, 1787, 1788
    infrared techniques, 1784
    methods, 1782, 1786
    modelling, 1787
    of domestic wastes, 1719
    physical models, 1787
    radioisotopes, 1786
    rhodamine B, 1786
    tracers, 1786
Monochlorobiphenyls, 1636
Monocyclic aromatics, 1630
Multiport diffuser, 1855, 1857
Municipal waste,
    management of disposal, 1756–1758

Nitrosamines, 1630
Nonachlorobiphenyls, 1637

Ocean thermal energy conversion (OTEC),
    1922–1924
Octachlorobiphenyls, 1637
Octachlorostyrene, 1672
Organic chemicals,
    distribution,
        conclusions, 1665
        in organisms, 1664
        in sea water, 1660
        in sediments, 1665
        in surface slicks, 1663, 1664
    occurrence, 1659
Organic pollutants,
    analysis of, 1628, 1629
    clean-up, 1634

Organic pollutants—*continued*
  conclusions, 1639, 1692
  effects on living systems, 1675
    acute toxicity, 1675
  effects on algae, 1687, 1688
  effects on man, 1692
  extraction, 1633
  future prospects, 1692–1695
  identification, 1634
  limits of detection, 1635
  quantification, 1634
  sampling, 1629
  separation, 1634
Organisms, 1664, 1665
OTEC—*see* Ocean thermal energy conversion
Other constructive uses of thermal effluents,
    1910, 1911
Oxygenation, 1894
Ozonation, 1897

Paraquat, 1681
Parathion, 1680
Particulate matter,
  organic pollutants, 1645
PCBs, 1621, 1627, 1628, 1630, 1636, 1638,
    1642, 1643, 1645, 1648, 1650, 1656, 1661,
    1666, 1668, 1672, 1674, 1686, 1690, 1691,
    1693, 1715, 1738, 1755
PCP, 1648
Pentachlorobenzene, 1672
Pentachlorobiphenyls, 1636
Pentachlorophenol, 1670, 1674, 1681, 1691
Perchloroethylene, 1648
Pesticides, 1619–1625, 1627–1709
  degradation, 1651
Phosphamidon, 1680
Phthalate esters, 1631, 1638
  degradation, 1657
Phthalic acid esters, 1662, 1683
Phytobenthos, 1816–1820
Phytoplankton, 1799–1802
Plumes, 1784, 1785, 1854, 1857, 1858
  buoyant surfaces, 1788
  entrainment, 1790
  thermal, 1789
Pollutants,
  priority substances, 1630, 1631
Pollution, 1619–1963
  due to pesticides, 1627–1709
  due to domestic wastes, 1627–1709
  due to technical organic chemicals,
    1627–1709
  due to thermal deformations, 1627–1709
  index of, 1729
  organic,
    analysis of, 1628, 1629
    sampling, 1628, 1629

Polychlorinated biphenyls—*see* PCBs
Polycyclic aromatic hydrocarbons, 1631
Power generation, 1910
  improvements, 1913
Power stations, 1885, 1893
  cooling waters, 1887
  efficiencies, 1774, 1842, 1843, 1888
  effluents, 1773
  nuclear, 1913
  siting considerations, 1858–1860
  temperature, 1849
Pressure effects, 1837
Priority pollutants, 1630, 1631
Protection of the sea, 1619–1963
Paper pollutants, 1631
Pulp pollutants, 1631

Quantification,
  organic pollutants, 1634, 1635

Radiometers, 1785
Rarefaction technique, 1728
Red tide, 1802, 1808, 1838
Remote sensing devices, 1784
Russell cycle, 1916, 1921

Safety margins, 1695
Sampling,
  organic pollutants, 1629
Scanners,
  infrared, 1784
  multispectral, 1784
SCCWRP, 1712, 1725, 1726, 1765
Sea water, 1660
  density, 1780
Sediments, 1665
  classification, 1717
  distribution of, 1718
  organic pollutants, 1645
Separation,
  organic pollutants, 1634, 1635
Sewage,
  domestic, 1711–1767
  effluents, 1857
Sewage treatment, 1712
  primary, 1712, 1714
  secondary, 1712, 1714, 1758
  tertiary, 1712, 1714, 1715
Simazine, 1681
Soil heating, 1908
Soil warming, 1904, 1906, 1908
Southern California Coastal Waters Research
    Project—*see* SCCWRP
Sport fishing,
  effects due to heated effluents, 1902–1904
Steam electricity, 1773

Steam turbine, 1773
Stress, 1897
Surface foam, 1811
Surface slicks, 1660, 1663
Synergistic effects, 1831–1840

Technical organic chemicals, 1627–1709
   degradation, 1651
Temperature integrator, 1782
Temperature–mortality data, 1866
Temperature–time exposures, 1866
Temperature–time graph, 1865
Temperature–time mortality, 1865
Tempering, 1850
Tetrachlorethylene, 1672
Tetrachlorobenzene, 1672
Tetrachlorobiphenyls, 1636
Tetrachlorophenols, 1674
Thermal addition, 1770
Thermal aquaculture, 1873–1902, 1914
Thermal deformations, *see also* Thermal
      effluents, 1619–1626, 1627–1709,
      1769–1963
   biological effects, 1796–1840
     bacteria, 1798, 1799
     entrainment effects, 1797, 1798
     fish, 1806, 1807
     phytoplankton, 1799–1802
     zooplankton, 1802–1806
   cold effluents, 1771, 1911
   conclusions and the future, 1911–1926
   definition, 1624
   discharge area effects, 1807–1830
     fish, 1809–1816
     phytobenthos, 1816–1820
     phytoplankton, 1808
     zoobenthos, 1820–1830
     zooplankton, 1809
   ecosystem effects, 1837–1840
   effects on,
     bacteria, 1798, 1799
     fish, 1806, 1809
     phytobenthos, 1816–1820
     phytoplankton, 1799, 1808
     zoobenthos, 1820–1830
     zooplankton, 1799, 1809
   ecosystem effects, 1837–1840
   future aspects, 1911–1926
   heated effluents, 1771–1773
   impingement effects, 1836, 1837
   management of effects, 1840–1872
   pressure effects, 1837
   synergistic effects, 1831–1840
Thermal discharges,
   dilution of heated effluent, 1855
   long-term exposure, 1869–1872
   mitigation of effects, 1860

   reduction of impact, 1854–1858
   short-term exposure, 1862–1869
Thermal effluents—*see also* Thermal
   deformations,
   buoyant, 1788
   cold effluents, 1813, 1814
   cold shock, 1813
   diseases, 1895–1898
   parasites, 1895–1898
   stress, 1895–1898
   enhancement of growth,
     in fish, 1878–1884
     in molluscs, 1882–1884
   far field, 1789, 1791, 1792
   fish kills, 1812, 1839
   heat shock, 1811–1813
   in situ distribution, 1788
   lethal effects, 1811–1813
   mid-field, 1790, 1791
   near field, 1789
   negatively buoyant, 1795
   sublethal effects, 1814, 1815
Thermal enrichment, 1770
Thermal mixing zones, 1859
Thermal pollution, 1769–1963
Thermistors, 1782–1784
   probes, 1782
   strings, 1785
Thermographs, 1782–1784
Thermometers,
   maximum and minimum, 1782
   mercury, 1782
Tidal-plug, 1792
Toxaphene, 1681, 1685
Toxicants,
   combined effects, 1690
   miscellaneous effects, 1690–1692
   movement through food chain, 1754
Toxicity,
   effecting factors, 1675, 1676
   long-term, 1676
   maximum acceptable concentration
     (MATC), 1684
Trace metals, 1716, 1737, 1738
Transformation in micro-organisms, 1656
Transformation in animals, 1652
Trichlorethylene, 1672, 1689
Trichlorobenzene, 1672
Trichlorobiphenyls, 1637
Trichlorophenols, 1674
Turbo-generators, 1773, 1775

Ultimate incipent lethal temperature,
   1862
Ultraviolet sterilization, 1890
Use of heated effluent in aquaculture,
   1874–1885

Volatility,
  organic compounds, 1644
  pesticides, 1649
  in agriculture, 1904–1910
  in horticulture, 1904–1910

Waste-heat utilization, 1841–1910

  animal shelters, 1909, 1910
  other applications, 1910, 1911
Waste-heat regulation, 1841
Water, 1660
Water-temperature standards, 1860
Wet cooling towers, 1852

Zooplankton, 1802, 1809